Foundation
MATHEMATICS
for AQA GCSE

Tony Banks and David Alcorn

Pearson Education Limited
Edinburgh Gate
Harlow
Essex
CM20 2JE
England

ISBN-13: 978-1-4058-3139-0
ISBN-10: 1-4058-3139-1

Exam questions
Past exam questions, provided by the *Assessment and Qualifications Alliance*, are denoted by the letters AQA. The answers to all questions are entirely the responsibility of the authors/publisher and have neither been provided nor approved by AQA.

Every effort has been made to locate the copyright owners of material used in this book. Any omissions brought to the notice of the publisher are regretted and will be credited in subsequent printings.

Page design
Billy Johnson

Reader
Barbara Alcorn

Artwork
David Alcorn

Cover design
Raven Design

Typesetting by Billy Johnson, San Francisco, California, USA

Printed and bound by Scotprint, Haddington, Scotland

preface

Foundation Mathematics for AQA GCSE has been written to meet the requirements of the National Curriculum Key Stage 4 Programme of Study and provides full coverage of the new **AQA Specifications for the Foundation Tier of entry**.

The book is suitable for students preparing for assessment at the Mathematics Foundation Tier of entry on either a 1-year or 2-year course or as a revision text.

In preparing the text, full account has been made of the requirements for students to be able to use and apply mathematics in written examination papers and be able to solve problems both with and without a calculator. Some chapters include ideas for investigational, practical and statistical tasks and give the student the opportunity to improve and practice their skills of using and applying mathematics.

The planning of topics within chapters and sections has been designed to provide efficient coverage of the specifications. Depending on how the book is to be used you can best decide on the order in which chapters are studied.

Chapters   1 - 11   Number
Chapters 12 - 20   Algebra
Chapters 21 - 33   Shape, Space and Measures
Chapters 34 - 40   Handling Data

Each chapter consists of fully worked examples with explanatory notes and commentary; carefully graded questions, a summary of what you need to know and a review exercise.
The review exercises provide the opportunity to consolidate topics introduced within the chapter and consist of exam-style questions, which reflect how the work is assessed, plus lots of past examination questions (marked AQA).

Further opportunities to consolidate skills acquired over a number of chapters are provided with section reviews, which have been organised into two parts for non-calculator and calculator practice.

As final preparation for the exams a further compilation of exam and exam-style questions, organised for non-calculator paper and calculator paper practice, has been included.

The book has been designed so that it can be used in conjunction with the companion book
Foundation Mathematics for AQA GCSE - Student Support Book
Without Answers: ISBN 1-405834-90-0
With Answers: ISBN 1-405834-91-9

---

contents

CHAPTER 5 Fractions

CHAPTER 6 Working with Number

CHAPTER 7 Percentages

CHAPTER 8 — Time and Money

CHAPTER 9 — Personal Finance

CHAPTER 10 — Ratio and Proportion

CHAPTER 11 — Speed and Other Compound Measures

SR — Section Review - Number

Whole Numbers

The numbers 0, 1, 2, 3, 4, 5, ... can be used to count objects.

"I have 3 pound coins in my pocket."

"There are 8 tables in the room."

"There are 0 students absent today."

Such numbers are called **whole numbers**.

Even numbers and odd numbers

All whole numbers are either even numbers or odd numbers.

Numbers that end in 0, 2, 4, 6 or 8 are called **even numbers**.

12, 746 and 3514 are examples of even numbers.

Numbers that end in 1, 3, 5, 7 or 9 are called **odd numbers**.

17, 425 and 4527 are examples of odd numbers.

Place value

Our number system is made up of the digits 0, 1, 2, 3, 4, 5, 6, 7, 8 and 9.

The position a digit has in a number is called its **place value**.

In the number 5384 the digit 8 is worth 80, but in the number 4853 the digit 8 is worth 800.

$$17 = 10 + 7 \qquad = 1 \times 10 + 7$$
$$567 = 500 + 60 + 7 \quad = 5 \times 100 + 6 \times 10 + 7$$
$$2060 = 2000 + 60 \qquad = 2 \times 1000 + 6 \times 10$$

EXAMPLES

1 Which of these numbers are even numbers?

| 28 | 99 | 165 | 356 | 2001 |

The numbers 28 and 356 are even numbers.

2 What is the value of the digit 2 in the number 6234?

The 2 is worth 200.

3 Which of these numbers is the largest?

| 374 | 276 | 375 | 357 | 283 |

The largest number is 375.

4 Write the following numbers in descending order.
54, 49, 123, 98, 1001.

1001, 123, 98, 54, 49.

5 Using the digits 5, 6 and 8
(do not use the same digit more than once)
make as many three-digit numbers as you can.
Place your numbers in ascending order.

568, 586, 658, 685, 856, 865.

| Smallest number | **ascending order** | Largest number |
| Largest number | **descending order** | Smallest number |

1 Which of these numbers are even numbers?

| 6 | 9 | 12 | 21 | 77 | 110 |

2 Copy and complete the following.
(a) $923 = 900 + 20 + 3 = 9 \times 100 + 2 \times 10 + 3$
(b) $54 =$ =
(c) $456 =$ =
(d) $1872 =$ =

3 In the number 173: the digit 1 is worth 100, the digit 7 is worth 70, the digit 3 is worth 3.
Give the value of each digit in the following numbers.
(a) 53   (b) 341   (c) 673   (d) 1897   (e) 1052

4 In the number 3<u>8</u>4 the value of the underlined figure is 80.
Give the value of the underlined figure in the following.
(a) 62<u>3</u>4   (b) 12<u>3</u> 456 789   (c) 95 <u>6</u>70   (d) <u>2</u>003   (e) 9<u>4</u> 705   (f) 423<u>6</u>

5 For these pairs of numbers say which digit has the greater value.
(a) The 1 in 512 or the 8 in 648.
(b) The 7 in 745 or the 9 in 892.
(c) The 5 in 599 or the 6 in 769.

6 Look at these numbers. | 97 | 32 | 23 | 28 | 302 | 203 |
(a) Which are odd numbers?
(b) Which is the largest number?
(c) Which is the smallest number?

7 Write the following numbers in ascending order.
(a) 74, 168, 39, 421.
(b) 555, 545, 544, 554.
(c) 3842, 5814, 3874, 3801, 4765.

8 Write the following numbers in descending order.
(a) 399, 425, 103, 84, 429.
(b) 234, 239, 349, 324.
(c) 9434, 9646, 9951, 9653.

9 Is 1000 an odd number or an even number?
Give a reason for your answer.

10 Using the digits 2, 3 and 7 make as many three-digit numbers as you can.
Use each digit just once in each three-digit number, e.g. 237.
Put your three-digit numbers in descending order.
How many of your numbers are even numbers?

11 Using the digits 8, 5, 4 and 3 make as many four-digit numbers as you can.
Use each digit just once in each four-digit number, e.g. 8543.
Put your numbers in ascending order.

12 By using each of the digits 5, 4 and 7 make:
(a) the largest three-digit number,
(b) the smallest three-digit number.
In each case explain your method.

13 By using each of the digits 5, 1, 6 and 2 make:
(a) the largest four-digit **even** number,
(b) the smallest four-digit **odd** number.

$543 = 5 \times 100 + 4 \times 10 + 3 \times 1$

The number 543 is written or read as, "five hundred and forty-three."

For numbers bigger than one thousand:
- split the number into groups of three digits, starting from the units,
- combine the numbers of *millions* and *thousands* with the number less than 1000.

$18543 = 18\ 543$ and is written or read as, "eighteen thousand five hundred and forty-three."

EXAMPLES

1 Write these numbers in words.

32	thirty-two
919	nine hundred and nineteen
$45237 = 45\ 237$	forty-five *thousand* two hundred and thirty-seven
$1234567 = 1\ 234\ 567$	one *million* two hundred and thirty-four *thousand* five hundred and sixty-seven

2 Write these numbers in figures.

Four hundred and seventy   470

Two thousand five hundred and seventeen   2517

Two million eight hundred and fifty thousand four hundred and sixty-one   2 850 461

Exercise 1.2

1 Write these numbers in words.
(a) 17   (b) 88   (c) 187   (d) 2045   (e) 5612
(f) 7802   (g) 8888   (h) 92000   (i) 132045   (j) 1500000

2 Write these numbers in figures.
(a) (i) one   (ii) ten   (iii) one hundred   (iv) one thousand   (v) ten thousand
(b) What do you notice?
(c) What are the next two numbers in the sequence? Write them in words as well.

3 Write the following numbers in figures.
(a) five hundred and forty-six   (b) six hundred and seven
(c) one thousand and ten   (d) seventy thousand two hundred
(e) one million two hundred thousand and fifty-two

4 One million is 1 000 000.
Write these numbers in figures.
(a) two million   (b) ten million
(c) half a million   (d) one and a half million

5 In the following report numbers are written in words.
Rewrite the report showing the numbers as figures.

The attendance at the football match was *forty-eight thousand*.
The pitch measured *one hundred and nineteen* yards by *sixty-two* yards.
After *twenty-five* minutes the centre forward (who cost *fifteen million* pounds)
scored from *eighteen* yards.

6 Write answers to the following using figures.
(a) ten more than seven thousand and twenty
(b) one hundred less than five hundred and sixty-three
(c) one thousand more than ten thousand

Non-calculator methods for addition

Writing numbers in columns

Write the numbers in tidy columns according to place value.
Add together the numbers of units, tens, hundreds, etc.
If any of these answers are 10 or more, something is carried to the next column.

> **EXAMPLE**
>
> Work out   4567 + 835.
>
> $$\begin{array}{r} 4\,5\,6\,7 \\ +\quad 8\,3\,5 \\ \hline 5\,4\,0\,2 \\ \hline \scriptstyle 1\ 1\ 1 \end{array}$$
>
> $7 + 5 = 12$ which is 2 carry 1.
> $6 + 3 + \text{carried } 1 = 10$,   which is 0 carry 1.
> $5 + 8 + \text{carried } 1 = 14$,   which is 4 carry 1.
> $4 + \text{carried } 1 = 5$.

Using a number line

A number line shows a different method for adding numbers.
With practice you should not need to draw a number line.

> **EXAMPLE**
>
> Work out   26 + 37.
>
>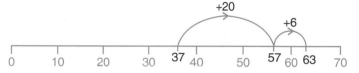
>
> 26 + 37   is the same as   37 + 26.
> $37 + 20 = 57$ (adding 20)
> $57 + 6 = 63$ (adding 6)
> So, $26 + 37 = 63$.

Working in context

1. Identify the calculation required by the question.
2. Do the calculation.
3. Give the answer to the question using a short sentence.

> **EXAMPLE**
>
> In Year 7 the class attendances were as follows:
>
>   28   30   27   30   25
>
> What was the total attendance?
>
> $$\begin{array}{r} 2\,8 \\ 3\,0 \\ 2\,7 \\ 3\,0 \\ +\ 2\,5 \\ \hline 1\,4\,0 \\ \hline \scriptstyle 1\ 2 \end{array}$$
>
> The total attendance was 140.

Exercise **1.3**   Do not use a calculator for this exercise.

1   Here is a number line for   18 + 15.

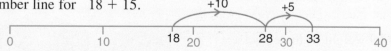

   (a)   What is   18 + 10?
   (b)   What is   18 + 15?

2   Draw a number line for each of the following sums and work out the answers.
   (a)   14 + 15   (b)   18 + 25   (c)   7 + 36   (d)   24 + 29

3   Work these out in your head.
   (a)   7 + 5   (b)   9 + 6   (c)   15 + 12   (d)   24 + 32
   (e)   19 + 16   (f)   26 + 48   (g)   29 + 41   (h)   13 + 99

4 What must be added to each of these numbers to make 100?
(a) 9   (b) 96   (c) 45   (d) 37   (e) 62   (f) 83   (g) 24   (h) 77

5 This signpost is between Poole and London.
How far is it from Poole to London?

POOLE 47 Miles   LONDON 65 Miles

6

Café Enfant
Price List
Drink.................. 54p
Doughnut............ 35p
Packet of Crisps.. 27p

How much does it cost to buy:
(a)   a drink and a doughnut,
(b)   a drink and a packet of crisps?

7 What is the total cost of the holiday?

Holiday Special   CYPRUS
B & B: 7 Nights   £269
Insurance   £37

8 The costs to build a fence are: labour £95,   materials £127.
What is the total cost of building the fence?

9 The class attendance in Year 7, Year 8 and Year 9 is shown.

Year 7	Year 8	Year 9
86	85	85

What is the total attendance?

10 Look at these number cards.   2   3   5   7   9

Use all five number cards to make the largest possible answer to this sum.

☐ ☐ + ☐ ☐ + ☐ = ...

Copy and complete the cards and give the answer.

11 Work these out by writing the numbers in columns.
(a)   765 + 23   (b)   27 + 56   (c)   76 + 98   (d)   324 + 628
(e)   1273 + 729   (f)   3495 + 8708   (g)   67 + 89 + 45   (h)   431 + 865 + 245

12 Marcus drove 1754 miles on his holiday. His milometer reading was 38 956 at the start.
What was his milometer reading at the end?

13 What is the total weight of these packages?

49 grams   257 grams   724 grams

14 The table shows the number of tickets sold each day last week at a cinema.

Mon	Tue	Wed	Thu	Fri	Sat	Sun
125	87	95	105	278	487	201

How many tickets were sold altogether?

15 The last four attendances at a football stadium were: 21 004, 19 750, 18 009, 22 267.
What is the total attendance?

16 Last month Mr Ahmed had the following household bills to pay.

| Mortgage | £429 | Insurance | £26 | Council Tax | £135 |
| Gas | £39 | Electricity | £18 | Telephone | £23 |

What is the total cost of these bills?

17 The chart shows the distances in kilometres between some towns.
Kathryn drives from Bath to Woking and then from Woking to York.

Bath

| 153 | Woking |
| 362 | 367 | York |

Calculate the total distance she drives.

Non-calculator methods for subtraction

Writing numbers in columns

Write the numbers in columns according to place value.
The order in which the numbers are written down is important.
Then, in turn, subtract the numbers of units, tens, hundreds, etc.
If the subtraction in a column cannot be done, because the number being subtracted is greater,
borrow 10 from the next column.

Work out $7238 - 642$.

$$
\begin{array}{r}
{}^{6}\;{}^{11}\,{}^{1}\\
7\,2\,3\,8\\
-\;\;\;6\,4\,2\\
\hline
6\,5\,9\,6
\end{array}
$$

Units: $8 - 2 = 6$
Tens: $3 - 4$ cannot be done, so borrow
10 from the 2 in the next column.
Now $10 + 3 - 4 = 9$.
Hundreds: $1 - 6$ cannot be done, so borrow
10 from the 7 in the next column.
Now $10 + 1 - 6 = 5$.
Thousands: $6 - 0 = 6$.

> You can use addition
> to check your subtraction.
> Does $6596 + 642 = 7238$?
>
> $$
> \begin{array}{r}
> 6\,5\,9\,6\\
> +\;\;\;6\,4\,2\\
> \hline
> 7\,2\,3\,8\\
> {}_{1}\;\;{}_{1}
> \end{array}
> $$

EXAMPLES

1
$$
\begin{array}{r}
{}^{0}\,{}^{16}\,{}^{15}\,{}^{1}\\
1\,7\,6\,2\\
-\;\;\;\;8\,7\,3\\
\hline
8\,8\,9
\end{array}
$$

2
$$
\begin{array}{r}
{}^{9}\;{}^{9}\\
{}^{2}\,3\,0\,0\,{}^{1}6\\
-\;\;1\,8\,4\,7\\
\hline
1\,1\,5\,9
\end{array}
$$

3
$$
\begin{array}{r}
{}^{9}\\
{}^{8}\,9\,0\,{}^{10}\,1\,2\\
-\;\;\;5\,6\,7\,8\\
\hline
3\,3\,3\,4
\end{array}
$$

Check the answers by addition.

> Addition is the opposite
> (inverse) operation to
> subtraction.
> If $a - b = c$,
> then $c + b = a$.

Subtracting numbers in context

EXAMPLE

A holiday costs £429.
A deposit of £95 is paid when the holiday is booked.
How much is left to pay?

$$
\begin{array}{r}
{}^{3}\,{}^{12}\\
4\,2\,9\\
-\;\;\;9\,5\\
\hline
3\,3\,4
\end{array}
$$
£334 is left to pay.

Do not use a calculator for this exercise.

1 Here is a number line for 43 − 16.
  (a) What is 43 − 10?
  (b) What is 43 − 16?

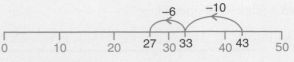

2 Draw a number line for each of the following questions and work out the answers.
  (a) 58 − 26   (b) 44 − 17   (c) 37 − 19   (d) 51 − 25

3 Write down the answers to the following by working them out in your head.
  (a) 100 − 95   (b) 100 − 8   (c) 100 − 57   (d) 100 − 32
  (e) 100 − 24   (f) 100 − 83   (g) 100 − 41   (h) 100 − 79

4 A batsman has scored 69 runs. How many more runs must he score to get a century?

5 Sylvia takes £100 on holiday. On the first day she spends £13.
How much money has she got left?

6 Work these out in your head.
  (a) 9 − 5   (b) 26 − 10   (c) 26 − 9   (d) 87 − 37   (e) 87 − 38
  (f) 200 − 110   (g) 204 − 99   (h) 500 − 350   (i) 500 − 199   (j) 1003 − 999

7 Ivan has saved £53. He spends £27. How much has he got left?

8 Mandy wins £1000. She spends £185 on a DVD player. How much has she got left?

9 Work these out by writing the numbers in columns. Use addition to check your answers.
  (a) 978 − 624   (b) 843 − 415   (c) 1754 − 470   (d) 407 − 249
  (e) 5070 − 2846   (f) 2345 − 1876   (g) 8045 − 1777   (h) 10 000 − 6723

10 A secretary has 67 letters to post. She has nineteen stamps.
How many more stamps does she need?

11 A car park has spaces for 345 cars. On Tuesday 256 spaces are used.
How many spaces are not used?

12 A school has 843 pupils. How many are boys if there are 459 girls?

13 (a) Ricky buys a bunch of spring onions.
He pays with 50p.
How much change is he given?
  (b) Liz buys a cucumber and a lettuce.
She pays with £1.
How much change is she given?

Salad Specials

Cucumber	39p each
Lettuce	55p each
Spring onions	33p a bunch

14 A shop records the number of tins of soup it has on its shelves.

	Tomato	Oxtail	Chicken
Start of the week	67	54	81
End of the week	28	36	26

  (a) How many tins of each type of soup did the shop sell?
  (b) How many tins of soup did the shop sell altogether?

15 The table shows the milometer readings for three cars at the start and end of a year.

	Car A	Car B	Car C
Start	2501	55667	48050
End	10980	67310	61909

Which car has done the most miles in the year?

Multiplication of whole numbers

It is very useful to know your Multiplication Tables up to 10×10.

×	1	2	3	4	5	6	7	8	9	10
1	1	2	3	4	5	6	7	8	9	10
2	2	4	6	8	10	12	14	16	18	20
3	3	6	9	12	15	18	21	24	27	30
4	4	8	12	16	20	24	28	32	36	40
5	5	10	15	20	25	30	35	40	45	50
6	6	12	18	24	30	36	42	48	54	60
7	7	14	21	28	35	42	49	56	63	70
8	8	16	24	32	40	48	56	64	72	80
9	9	18	27	36	45	54	63	72	81	90
10	10	20	30	40	50	60	70	80	90	100

Activity

How quickly can you answer the following questions?

$$9 \times 5$$

$$6 \times 6$$

$$7 \times 4$$

$$9 \times 6$$

$$8 \times 7$$

$$9 \times 7$$

$$8 \times 8$$

Working with a partner ask each other questions from the table.

Non-calculator method for short multiplication

Short multiplication is when the multiplying number is less than 10,   e.g. 165×7.
One method multiplies the units, tens, hundreds, etc. in turn.

```
  1 6 5
×     7
-------
1 1 5 5
  1 4 3
```

Units: $7 \times 5 = 35$,   which is 5 carry 3.
Tens: $7 \times 6 = 42 +$ carried $3 = 45$,   which is 5 carry 4.
Hundreds: $7 \times 1 = 7 +$ carried $4 = 11$,   which is 1 carry 1.
There are no more digits to be multiplied by 7, the carried 1 becomes 1 thousand.

EXAMPLES

1
```
  1 6 2
×     4
-------
  6 4 8
      2
```

2
```
  9 0 7 1
×       7
---------
6 3 4 9 7
        4
```

3
```
  4 8 3 5
×       8
---------
3 8 6 8 0
    6 2 4
```

Multiplying numbers in context

EXAMPLE

A minibus holds 16 people.
How many people will 6 minibuses hold?

```
  1 6
×   6
-----
  9 6
    3
```

6 minibuses hold 96 people.

Exercise 1.5

Do not use a calculator for this exercise.

1 I get 8 doughnuts for £1.
How many doughnuts will I get for £5?

2 In one packet there are 4 muffins.
How many muffins are there in 7 packets?

3 Marbles are packed in bags of 6.
How many marbles are there in 8 bags?

4 When John uses a store card he gets 4 points for every pound he spends. He spends £18.
How many points does he get?

5 A machine makes 24 jigsaws in an hour.
How many jigsaws will it make in 6 hours?

6 Linda is paid 3p for each leaflet she delivers. She delivers 184 leaflets.
How much is she paid?

7 Look at these prices.

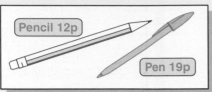

Pencil 12p

Pen 19p

(a) What is the cost of 5 pencils?
(b) What is the cost of 6 pens?
(c) What is the total cost of 4 pens and 7 pencils?

8 Work these out using a method you find easiest.
(a) 21 × 4 (b) 17 × 5 (c) 36 × 7 (d) 183 × 3
(e) 264 × 8 (f) 3179 × 5 (g) 4012 × 6 (h) 6012 × 7

9 The table shows the maximum number of people that can be carried on some buses.

(a) How many people can 4 minibuses carry?
(b) How many people can 5 coaches carry?
(c) How many people can 9 double-decker buses carry?
(d) A trip for 174 people is planned.
4 coaches are ordered.
How many empty seats will there be?

Minibus	17
Coach	55
Double-Decker Bus	74

10 A caretaker puts out 7 rows of chairs.
There are 13 chairs in each row.
How many more chairs are needed for 120 chairs to be put out?

Multiplying a whole number by 10, 100, 1000, …

When you multiply a whole number by:
10 The units become 10s, the 10s become 100s, the 100s become 1000s, and so on.
100 The units become 100s, the 10s become 1000s, the 100s become 10 000s, and so on.
1000 The units become 1000s, the 10s become 10 000s, the 100s become 100 000s, and so on.

EXAMPLES
753 × 10 = 7530
753 × 100 = 75 300
753 × 1000 = 753 000

We can show these multiplications in a table.

100 000s	10 000s	1000s	100s	10s	Units	
			7	5	3	
		7	5	3	0	←753 × 10
	7	5	3	0	0	←753 × 100
7	5	3	0	0	0	←753 × 1000

Explain any patterns you can see.

100 = 10 × 10
Multiplying a number by 100 is the same as multiplying the number by 10 and then by 10 again.

Multiplying a whole number by multiples of 10 (20, 30, 40, …)

Work out 753 × 20.

This can be written as:
753 × 20 = 753 × 10 × 2
= 7530 × 2
= 15 060

20 = 10 × 2

Exercise 1.6   Do not use a calculator for this exercise.

1 Write down the answers to the following questions. You do not have to show any working.
 (a) 132×10 (b) 123×100 (c) 47×1000 (d) 384×100

2 What number should be put in the box to make each of these statements correct?
 (a) $231 \times 10 = \boxed{}$ (b) $\boxed{} \times 1000 = 514\,000$ (c) $172 \times \boxed{} = 17\,200$

3 Here is a price list from an office catalogue.
 (a) What is the cost of 10 desks?
 (b) What is the cost of 100 chairs?
 (c) What is the total cost of 100 desks and 1000 chairs?

Desk	£120
Chair	£59

4 A packet of 24 custard cream biscuits costs 49 pence. I buy 10 packets.
 (a) How many biscuits do I buy?
 (b) How much will they cost altogether?

5 I get 15 euros for £10. How many euros will I get for £100?

6 A bag of compost costs £12. $1\,\text{m}^2$ of turf costs £7.
 (a) What is the cost of 10 bags of compost?
 (b) What is the cost of $100\,\text{m}^2$ of turf?
 (c) What is the total cost of 100 bags of compost and $1000\,\text{m}^2$ of turf?

7 There are 7 classrooms in a school. In each classroom there are 30 chairs and 20 tables. Find the total number of   (a) chairs,   (b) tables.

8 Calculate the number of seconds in   (a) 5 minutes,   (b) 9 minutes.

9 A gardener plants 15 rows of lettuces. In each row he plants 20 lettuces. How many lettuces does he plant altogether?

10 A day trip to France costs £30.
 What is the total cost for a party of 25 Scouts to go on the day trip?

11 What is the total weight of 20 coins if each coin weighs 28 grams?

12 A farmer has 32 bags of turnips. Each bag weighs 50 kg. What is the total weight?

13

PLAYHOUSE THEATRE
PRESENT
CABARET
Tickets: £19   Programmes: £3

On the first night of the show, 70 tickets and 40 programmes are sold. How much money is paid in total for the tickets and programmes?

14 Work out.
 (a) 357×20 (b) 632×30 (c) 537×40
 (d) 260×50 (e) 186×70 (f) 239×90

15 (a) Describe a method of multiplying by 200, 300, 400, and so on.
 (b) Describe a method of multiplying by 2000, 3000, 4000 and so on.
 (c) Work out (i) 67×200,   (ii) 35×3000,   (iii) 174×400,   (iv) 287×5000.

16 There are 400 metres in one lap of a running track. How many metres are there in 25 laps?

17 A secretary buys one dozen boxes of staples. There are 5000 staples in each box. How many staples does she buy?

Non-calculator method for short division

The process of dividing a number by a number less than 10 is called **short division**.
Short division relies on knowledge of the Multiplication Tables.
What is $32 \div 8$, $42 \div 7$, $72 \div 9$, $54 \div 6$?

Work out $882 \div 7$.

$7)8\,^18\,^42$
  $1\ 2\ 6$

Starting from the left:
$8 \div 7 = 1$ remainder 1,   which is 1 carry 1.
$18 \div 7 = 2$ remainder 4,   which is 2 carry 4.
$42 \div 7 = 6$, with no remainder.
So, $882 \div 7 = 126$.

You can check your division by multiplying.
Does $126 \times 7 = 882$?

You may set your
working out like
this:
$$\begin{array}{r} 1\ 2\ 6 \\ \hline 7)8\,^18\,^42 \end{array}$$

Multiplication is the
opposite (inverse) operation
to division.
If $a \div b = c$,
then $c \times b = a$.

> **EXAMPLES**
>
> **1** $\dfrac{2\ 4\ 5}{}$ ($6)1\ 4\,^27\,^30$)
>
> **2** $9)2\ 7\ 6\,^63$, $3\ 0\ 7$

Dividing numbers in context

> **EXAMPLES**
>
> **1**   8 children share 112 sweets.
> How many sweets does each child receive?
>
> $112 \div 8$ $8)1\ 1\,^32$
> $1\ 4$
>
> Each child receives 14 sweets.
>
> **2**   Pencils are boxed in packs of 5.
> I have 87 pencils.
> How many boxes can I fill?
>
> $87 \div 5$ $5)8\,^37$
> $1\ 7$ remainder 2
>
> I can fill 17 boxes.
> There will be 2 pencils left over.

Exercise **1.7**   Do not use a calculator for this exercise.

1   Two tins of cat food cost 98p. What is the cost of one tin?

2   Flower pots cost £5 each. I have £30. How many flower pots can I buy?

3   Pam buys 3 watches for £42. How much is each watch?

4   A packet of 8 crayons is 96p. What is the cost of one crayon?

5   Candles are sold in packets of 4. I need 60 candles.
How many packets do I need to buy?

6   How many chews costing 6p each can I buy for 90p?

7   Calculate these divisions. Show your working clearly.
State the remainder if there is one.
Use multiplication to check your answers.
(a) $85 \div 5$ (b) $471 \div 3$ (c) $816 \div 6$ (d) $455 \div 6$
(e) $3146 \div 8$ (f) $824 \div 4$ (g) $9882 \div 9$ (h) $80\ 560 \div 4$

8   A lollipop costs 7p. Maxine has 95p.
(a)   How many lollipops can she buy?
(b)   What amount of money will she have left?

Dividing a whole number by 10, 100, 1000, ...

When you divide a whole number by:

10   The 10s   become units,   the 100s   become 10s,   the 1000s   become 100s, and so on.
100   The 100s become units,   the 1000s   become 10s,   the 10 000s become 100s, and so on.
1000 The 1000s become units,   the 10 000s become 10s,   the 100 000s become 100s, and so on.

Examples

$7530 \div 10 = 753$
$12\ 400 \div 100 = 124$
$631\ 000 \div 1000 = 631$

10 000s	1000s	100s	10s	Units
	7	5	3	0
		7	5	3

←$7530 \div 10$

Draw tables to show the other division sums.
Explain any patterns you can see.

Dividing a whole number by multiples of 10 (20, 30, 40, ...)

Work out $7530 \div 30$.

$7530 \div 30$
$= (7530 \div 10) \div 3$
$= 753 \div 3$ (dividing by 10)
$= 251$ (dividing by 3)

> $30 = 10 \times 3$
> Dividing by 30 is the same as
> dividing by 10 and then dividing by 3.

Exercise 1.8

Do not use a calculator for this exercise.

1   Work out.
    (a) $4560 \div 10$ (b) $465\ 000 \div 1000$ (c) $64\ 000 \div 1000$ (d) $65\ 400 \div 100$

2   Raffle tickets cost 10p each. Jo has 60 pence.
    How many raffle tickets can she buy?

3   What number should be put in the box to make each of these statements correct?
    (a) $56\ 400 : \square = 564$ (b) $\square \div 1000 = 702$ (c) $35\ 000 \div \square = 3500$

4   How many £20 notes are needed to pay a bill of £240?

5   A coach can carry 50 people.
    How many coaches are needed to carry 350 people?

6   A school has 480 pupils.
    How many classes are there if there are 30 pupils in each class?

7   A box holds 40 matches.
    How many boxes are needed to hold 1000 matches?

8   Hannah completes a puzzle in 420 seconds.
    How many minutes is this?

9   The total cost of a New Year's Eve party for 80 people is £2800.
    What is the cost for each person?

10   Work out.
    (a) $7590 \div 30$ (b) $7110 \div 90$ (c) $21\ 480 \div 40$
    (d) $7560 \div 60$ (e) $900 \div 20$ (f) $30\ 650 \div 50$

11   (a)   Describe a method to divide by multiples of 100 (200, 300, 400, ...)
       (b)   Work out: (i) $13\ 000 \div 500$ (ii) $329\ 600 \div 800$

12   A computer can print 50 000 characters per minute.
    How many minutes would it take to print one million characters?

Whole Numbers . . . . Whole Numbers . . . .

Long multiplication

Long multiplication is used when the multiplying number is greater than 10, e.g. 24 × 17.

Work out 24 × 17.

```
    2 4
  ×  1 7
  ───────
  1 6 8   ←24 ×  7 = 168
+ 2 4 0   ←24 × 10 = 240
  ───────
  4 0 8
```

A standard non-calculator method for doing long multiplication multiplies the number by:

- the units figure, then
- the tens figure, then
- the hundreds figure, and so on.

All these answers are added together.

EXAMPLES

1
```
      1 4 5
    ×   6 2
  ─────────
      2 9 0   ←145 × 2
+   8 7 0 0   ←145 × 60
  ─────────
    8 9 9 0
```

2
```
        2 7 3
    ×   2 3 4
  ───────────
      1 0 9 2   ←273 ×   4
      8 1 9 0   ←273 ×  30
+   5 4 6 0 0   ←273 × 200
  ───────────
    6 3 8 8 2
```

Exercise 1.9

Do not use a calculator for this exercise.

1 Work out.
   (a) 17 × 12 (b) 23 × 15 (c) 42 × 32 (d) 76 × 32 (e) 143 × 34 (f) 718 × 54

2 A box holds 12 tins of soup.
   How many tins will 18 boxes hold?

3 A computer prints 17 symbols on a line.
   How many symbols will it print on 15 lines?

4 A gardener buys 36 flower pots at 25 pence each.
   What is the total cost?

5 A restaurant charges £23 for dinner.
   What is the total charge for 42 dinners?

6 A company has 47 shops. Each shop employs 28 people.
   How many people are employed altogether?

7 A coach trip costs £29 per person. 48 people go on the trip.
   How much money is paid altogether?

8 A company is organising a Christmas party for 128 workers.
   Each worker must pay £12 towards the cost.
   Work out the total amount paid by workers.

9 Here is part of a price list.

Office Supplies

Desk	**£126**
Cabinet	**£149**

What is the total cost of 14 desks and 23 cabinets?

10 The organisers of a concert sell 1624 tickets at £27 each.
   How much money is collected from the sale of tickets?

Long division

Long division works in exactly the same way as short division, except that all the working out is written down.

Consider   952 ÷ 7.

Using short division:

$7\overline{)9^2 5^4 2}$
   1 3 6

```
    1 3 6
7)9 5 2
  7
  ‾‾
  2 5
  2 1
  ‾‾
    4 2
    4 2
    ‾‾
     0
     ‾
```

9 ÷ 7 = 1 and a remainder.
What is the remainder?
1 × 7 = 7   (write below the 9).
9 − 7 = 2   (which is the remainder).
Bring down the next figure (5) to make 25.
Repeat the above process.
25 ÷ 7 = 3 and a remainder.
3 × 7 = 21,   25 − 21 = 4 (remainder).
Bring down the next figure (2) to make 42 and repeat the process.
42 ÷ 7 = 6, but there is no remainder.
6 × 7 = 42,   42 − 42 = 0 (remainder).
There are no more figures to be brought down and there is no remainder.
So,   952 ÷ 7 = 136.

Long division process:

→ ÷ (obtain biggest answer possible)
  × } calculates
  − } the remainder
— Bring down the next figure

Repeat process until there are no more figures to be brought down.

EXAMPLE

How many 27p stamps can I buy for £5?
What change am I given?

£5 is the same as 500p.

500 ÷ 27

```
      1 8
27)5 0 0
   2 7
   ‾‾‾
   2 3 0
   2 1 6
   ‾‾‾‾
     1 4
```

27 goes into 50 once with a remainder of 23.
Bring down the 0 to make 230.
27 × 8 = 216, so, 230 ÷ 27 = 8 remainder 14.

18 stamps with 14p change.

Exercise 1.10

Do not use a calculator for this exercise.

1 Work out.
  (a) 473 ÷ 11
  (b) 480 ÷ 15
  (c) 324 ÷ 12
  (d) 544 ÷ 17
  (e) 624 ÷ 13
  (f) 943 ÷ 23
  (g) 777 ÷ 37
  (h) 841 ÷ 29

2 Work these out and state the remainder each time.
  (a) 410 ÷ 25
  (b) 607 ÷ 24
  (c) 800 ÷ 45
  (d) 754 ÷ 57

3 A pint of milk costs 35 pence. How many pints of milk can be bought for £5?
How much change will there be?

4 Tins of beans are packed in boxes of 24. A supplier has 1000 tins of beans.
  (a) How many full boxes is this?
  (b) How many tins are left over?

5 I have £5 to spend.
  (a) How many packets of crisps can I buy?
  (b) How many biscuits can I buy?
  (c) How many cans of drink can I buy?
In each case state the change, if there is any.

price List
Packet of Crisps   25p
Biscuit   14p
Can of Drink   45p

6 John collects £17 each from his friends to buy tickets for a football match.
He collects a total of £391.
How many tickets does he buy?

Order of operations in a calculation

What is $4 + 3 \times 5?$ It is not sensible to have two possible answers.
It has been agreed that calculations are done obeying certain rules:

First	Brackets and Division line
Second	Divide and Multiply
Third	Addition and Subtraction

EXAMPLES

1 $4 + 3 \times 5 \qquad = \qquad 4 + 15 \qquad = \qquad 19$

2 $10 \div 2 + 3 \qquad = \qquad 5 + 3 \qquad = \qquad 8$

3 $10 \div (2 + 3) \qquad = \qquad 10 \div 5 \qquad = \qquad 2$

4 $(5 + 6) \times 3 + 4 \quad = \quad 11 \times 3 + 4 \quad = \quad 33 + 4 \quad = \quad 37$

5 $\dfrac{12}{11 - 8} - 3 \qquad = \qquad \dfrac{12}{3} - 3 \qquad = \qquad 4 - 3 \qquad = \qquad 1$

This is the same as $12 \div (11 - 8) - 3.$

Exercise 1.11   Do not use a calculator for this exercise.

1 Work out the value of $17 - 3 \times (6 - 1).$

2 Work these out.

(a) $7 + 6 \times 5$
(b) $7 - (6 - 2)$
(c) $24 \div 6 + 5$

(d) $7 \times 6 + 8 \times 2$
(e) $10 \div 5 + 8 \div 2$
(f) $(5 - 2) \times 7 + 9$

(g) $60 \div (5 + 7)$
(h) $60 \div 5 + 7$
(i) $4 \times 3 + 2$

(j) $4 \times (3 + 2)$
(k) $12 \times (20 - 2) \div 9$
(l) $36 \div (5 + 4)$

(m) $4 \times 12 \div 8 - 6$
(n) $\dfrac{15}{18 - 3} + 4$
(o) $\dfrac{22 - 4}{9} + 12 \div 3$

3 Put brackets into the expression $6 + 3 \times 5$ to give the value 45.

4 Use brackets and the signs $+$, $-$, $\times$ and $\div$ to complete these sums.

(a) $7 \quad 2 \times 3 = 15$
(b) $3 + 5 \quad 2 = 4$
(c) $(4 \quad 1) \times 7 \quad 2 = 25$

5 Choose from the four signs $+$, $-$, $\times$ and $\div$ to make these sums correct.

(a) $5 \quad 6 \quad 7 = 37$
(b) $5 \quad 6 \quad 7 = 47$
(c) $15 \quad 8 \quad 9 = 87$

(d) $15 \quad 8 \quad 9 = 129$
(e) $15 \quad 8 \quad 9 = 111$
(f) $15 \quad 5 \quad 3 = 6$

(g) $5 \quad 24 \quad 6 = 1$
(h) $19 \quad 19 \quad 7 = 8$
(i) $4 \quad 4 \quad 7 \quad 2 = 30$

6 Using all the numbers 6, 3, 2 and 1 in this order, brackets and the signs $+$, $-$, $\times$ and $\div$, make all the numbers from 1 to 10.

$$6 - 3 \times 2 + 1 = 1, \qquad 6 - 3 - 2 + 1 = 2, \qquad \text{and so on.}$$

Problems involving number

The number skills you have met in Exercise 1.11 can be applied to practical situations.

EXAMPLE

Harold loads 5 parcels each weighing 3 kg and 4 parcels each weighing 7 kg onto a trolley.
The unloaded trolley weighs 18 kg.
What is the total weight of the trolley and the parcels?

$$\text{Total weight} = (5 \times 3) + (4 \times 7) + 18 \text{ kg}$$
$$= 15 + 28 + 18 \text{ kg}$$
$$= 61 \text{ kg}$$

Set your working out clearly so that someone else can follow what you are doing.

The total weight of the trolley and parcels is 61 kg.

How could a calculator be used to answer this problem?

Exercise 1.12

You should be able to do this exercise without using your calculator.
Having completed the exercise use a calculator to check your working.

1. Claire is 16 cm taller than Rachel. Their heights add up to 312 cm.
How tall is Rachel?

2. Adrian is 6 kg lighter than Richard. Their weights add up to 152 kg.
How heavy is Richard?

3. A box containing 6 packets of tea weighs 750 g. Each packet of tea weighs 120 g.
What is the weight of the box?

4. Look at this price list.

Price List
Can of drink   55p
Packet of crisps 32p
Bar of chocolate 28p

   (a) What change does Harry get from £2, if he buys 2 bars of chocolate and a can of drink?
   (b) How much does Harry save if he buys a packet of crisps and a can of drink instead of 2 bars of chocolate and a can of drink?

5. The caretaker set out 17 rows of chairs. There are 15 chairs in each row.
How many more chairs are needed to provide seats for 280 people?

6. The total weight of a carton which contains 6 eggs is 520 g. The carton weighs 70 g.
What is the weight of each egg?

7. A cupboard is 90 cm wide. It is placed between two walls which are 160 cm apart.
The gap between the cupboard and each wall is the same.
What is the size of the gap?

8. A roll of wire is 550 cm long. From the roll, Hilary cuts 3 pieces which each measure 85 cm and 4 pieces which each measure 35 cm.
How much wire is left on the roll?

9. A box, which contains 48 matches, has a total weight of 207 g.
If each match weighs 4 g, what is the weight of the empty box?

10. The admission charges to a zoo are £7 for a child and £12 for an adult.
Zoe is organising a trip to the zoo for a group of people and worked out that the total cost would be £336. She collected £84 from the adults in the group.
   (a) How many children are in the group?
   (b) What is the total number of people in the group?

What you need to know

You should be able to:

- Read and write whole numbers expressed in figures and words.
- Order whole numbers.
- Recognise the place value of each digit in a number.
- Use mental methods to carry out addition and subtraction.
- Carry out accurately non-calculator methods for addition and subtraction.
- Know the Multiplication Tables up to 10×10.
- Carry out multiplication by a number less than 10 (short multiplication).
- Multiply whole numbers by 10, 100, 1000, … and by 20, 30, 40, …
- Carry out division by a number less than 10 (short division).
- Divide whole numbers by 10, 100, 1000, … and by 20, 30, 40, …
- Carry out long multiplication and long division.
- Know the order of operations in a calculation.

IDEAS FOR INVESTIGATION

Write down a three-digit number.
Write the number in words.
Count the letters.
Write this number in words.
Count the letters.
Write this number in words, and so on.

Start with the number 207.
What number do you end up with?

Investigate further.

Example:
569, five hundred and sixty-nine
23, twenty-three
11, eleven
6, six
3, three
5, five
4, four

Review Exercise 1

Do not use a calculator for this exercise.

1 A list of numbers is given:

| 4 | 5 | 7 | 12 | 15 | 19 |

State which of these numbers are: (a) even numbers, (b) odd numbers. AQA

2 (a) Write 870 302 in words.
(b) Write three million twenty-seven thousand four hundred and nine in figures.

3 (a) In 4690 the 9 represents 9 tens. What does the 6 represent?
(b) What does the 6 represent in the **answers** to the following calculations?
(i) 4690×10 (ii) $4690 \div 10$ AQA

4 Write these numbers in order, smallest first. 2295, 92 345, 480, 120 000, 408. AQA

5 (a) Here is a list of numbers: 15 150 1500 15 000 150 000 1 500 000
Write down the number from this list which is:
(i) Fifteen hundred. (ii) One hundred and fifty thousand.
(b) What are the missing numbers?
(i) $1500 \div \boxed{} = 150$ (ii) $15 \times \boxed{} = 150\ 000$ AQA

6 (a) What must be added to 33 to make 100?
(b) What numbers are needed to complete these sums?
(i) $100 - 17 = \boxed{}$ (ii) $55 + \boxed{} = 100$ (iii) $100 - \boxed{} = 29$
(c) (i) Work out 5×100. (ii) Work out $2500 \div 100$.

7 Work out   (a) 78×3,   (b) $78 \div 3$.

8 Jamie's grandad said,

> "When I was at school there were 12 pennies in 1 shilling and 20 shillings in £1."

How many pennies were there in £1 when Jamie's grandad was at school?   AQA

9 You can use the four cards to make different numbers.

5 **9** **4** and **1**

Each card can only be used **once**. All the cards must be used.
(a) Write down the **largest** number.
(b) Write down the **smallest** number.   AQA

10 Gaby stated: **"The sum of three consecutive numbers is always an odd number."**
By means of an example, show that this statement is not true.

11 In the game of darts the scores of the three darts are added together.
These are then taken away from the current total to calculate the new total.
In each case work out the score of the three darts and the new total.

	Current Total	1st dart	2nd dart	3rd dart	Score	New Total
(a)	501	60	18	19	[ ]	[ ]
(b)	420	19	57	38	[ ]	[ ]
(c)	301	50	25	17	[ ]	[ ]

12 Work out the following. Show your working.
(a) $465 + 12 + 1582$ (b) $2465 - 1878$

13 Write down the answers to these questions.
(a) 735×100 (b) 214×30 (c) $3\,020\,000 \div 1000$ (d) $18\,480 \div 40$

14 Using only $+10$ -10 $\times 10$ $\div 10$

complete the following to make a correct mathematical statement.

Here is an example: $39 \longrightarrow \boxed{-10} \longrightarrow \boxed{\times 10} \longrightarrow 290$

(a) $390 \longrightarrow \boxed{} \longrightarrow 39$

(b) $390 \longrightarrow \boxed{} \longrightarrow \boxed{} \longrightarrow 49$

(c) $38 \longrightarrow \boxed{} \longrightarrow \boxed{} \longrightarrow \boxed{} \longrightarrow 37$ AQA

15 Work out the following. Show your working clearly.
(a) 718×9 (b) $1446 \div 6$

16 The chart shows the distances in miles between some towns.

Liverpool			
109	Nottingham		
77	44	Sheffield	
102	87	61	York

Isaac drives from Liverpool to Nottingham, from Nottingham to York and then from York back to Liverpool. Calculate the total distance he drives.

17 One thousand chocolate biscuits are packed in boxes of 6.
  (a) How many full boxes will there be?
  (b) How many biscuits will be left over?

18 Calculate these. Remember to do the operations in the right order.
  (a) $2 + 6 \times 8$ (b) $3 \times 6 - 4$ (c) $72 \div 8 + 1$ (d) $(9 - 4) \times (3 + 7)$

19 For each of the following write down whether it is true or false **and** give a reason for your answer.
  (a) $5 \times 7 \times 6 = 7 \times 6 \times 5$
  (b) $20 \div 4 = 4 \div 20$
  (c) $5 \times 2 + 7 \times 2 = 12 \times 2$
                                                                                              AQA

20 A plank of wood is 396 cm in length. The plank is cut into two pieces.
  One piece is 28 cm longer than the other.
  How long is the shorter piece of wood?

21 A supermarket orders 1800 kg of potatoes.
  The potatoes are delivered in 15 kg bags.
  How many bags are delivered?

22 (a) Work out the values of $(1 + 3) \times 5 + 5$
  (b) Add brackets ( ) to make each statement correct.
      You may use more than one pair of brackets in each statement.
      (i) $1 + 3 \times 5 + 5 = 40$ (ii) $1 + 3 \times 5 + 5 = 31$

23 Alys is given this sum: $84 + 16 \times 5$
  She works out the answer to be 500.
  Is she right? Explain your answer.

24 Use the information that $45 \times 230 = 10\,350$ to find the value of
  (a) $10\,350 \div 23$,   (b) 44×230.

25 Lauren does a paper round.
  She delivers 47 newspapers, 6 days a week for 52 weeks a year.
  How many newspapers does she deliver altogether in one year?

26 A company is organising a holiday for 274 workers.
  Each worker pays £53 as part of the cost.
  Work out exactly the total amount paid by the workers.
  You **must** show all your working.   AQA

27 Angelique is taking 149 pupils to the theatre to see the Royal Ballet.
  The cost of the theatre ticket is £23 per person.
  Work out the total cost of the tickets.   AQA

28 Joe wants to save £884 to pay for his holiday.
  He saves the same amount each week for 52 weeks.
  How much does Joe save each week?   AQA

29 Sue has collected £544 from her friends at work for theatre tickets.
  The tickets cost £17 each.
  Work out the number of theatre tickets she can buy.   AQA

30 A survey counted the number of visitors to a website on the Internet.
  Altogether it was visited 30 million times.
  Each day it was visited 600 000 times.
  Based on this information, for how many days did the survey last?   AQA

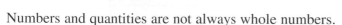

Numbers and quantities are not always whole numbers.

The number system you met in Chapter 1 can be extended to include **decimal numbers**, such as tenths, hundredths, thousandths, and smaller numbers.

A **decimal point** is used to separate the whole number part from the decimal part of the number.

73.26   This number is read as seventy-three point two six.

whole number
73

decimal part
2 tenths + 6 hundredths
(which is the same as 26 hundredths)

Place value

In the number 1.53   the digit 1 is worth 1 unit = 1
the digit 5 is worth 5 tenths = 0.5
the digit 3 is worth 3 hundredths = 0.03

$1.53 = 1 + 0.5 + 0.03$

1 unit = 10 tenths
1 tenth = 10 hundredths
1 hundredth = 10 thousandths
… and so on.

The first digit after the decimal point represents **tenths**.

1.50   1.51   1.52   1.53   1.54   1.55   1.56   1.57   1.58   1.59   1.60

The second digit after the decimal point represents **hundredths**.

The number 1.53 can be represented by a diagram.

1 unit

5 tenths

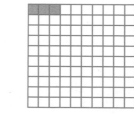
3 hundredths

Ordering decimals

Compare the numbers 52.359 and 52.36. Which number is the bigger?

You can use a grid to compare the numbers.

tens	units	•	tenths	hundredths	thousandths
5	2	•	3	5	9
5	2	•	3	6	

Start by comparing the digits with the greatest place value, the tens.
Both numbers have 5 tens, so move down to compare the units.
Both numbers have 2 units, so move down to compare the tenths.
Both numbers have 3 tenths, so move down to compare the hundredths.
52.359 has 5 hundredths but 52.36 has 6 hundredths.

So, 52.36 is bigger than 52.359.

A similar method can be used to place a list of decimal numbers in order.

EXAMPLES

1 What numbers are arrows *A*, *B* and *C* pointing to on this scale?

There are 10 marks between 2 and 3.
Each mark represents 0.1.
Arrow *A*: 2.2   Arrow *B*: 2.6   Arrow *C*: 2.9

The scale shows the main numbers. The distance between the main numbers is divided up using marks. To read a scale you must first work out what the distance between two marks represents.

2 What numbers are arrows *P*, *Q* and *R* pointing to on this scale?

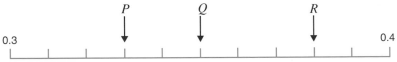

There are 10 marks between 0.3 and 0.4.
Each mark represents 0.01.
Arrow *P*: 0.33   Arrow *Q*: 0.35   Arrow *R*: 0.38

Exercise 2.1

1 2.564 = 2 + 0.5 + 0.06 + 0.004.
Write these numbers in the same way.
(a) 4.7   (b) 5.55   (c) 7.62   (d) 37.928   (e) 7.541

2 In the number 17.4$\underline{6}$2 the value of the underlined figure is 0.06.
Give the value of the underlined figures in the following.
(a) 2.$\underline{7}$ (b) 3.5$\underline{2}$ (c) 27.$\underline{4}$3   (d) 36.42$\underline{9}$ (e) 2$\underline{8}$5.03

3 Write down the numbers shown by these diagrams.
(a)

(b)

4 On these scales what numbers are shown by the arrows?

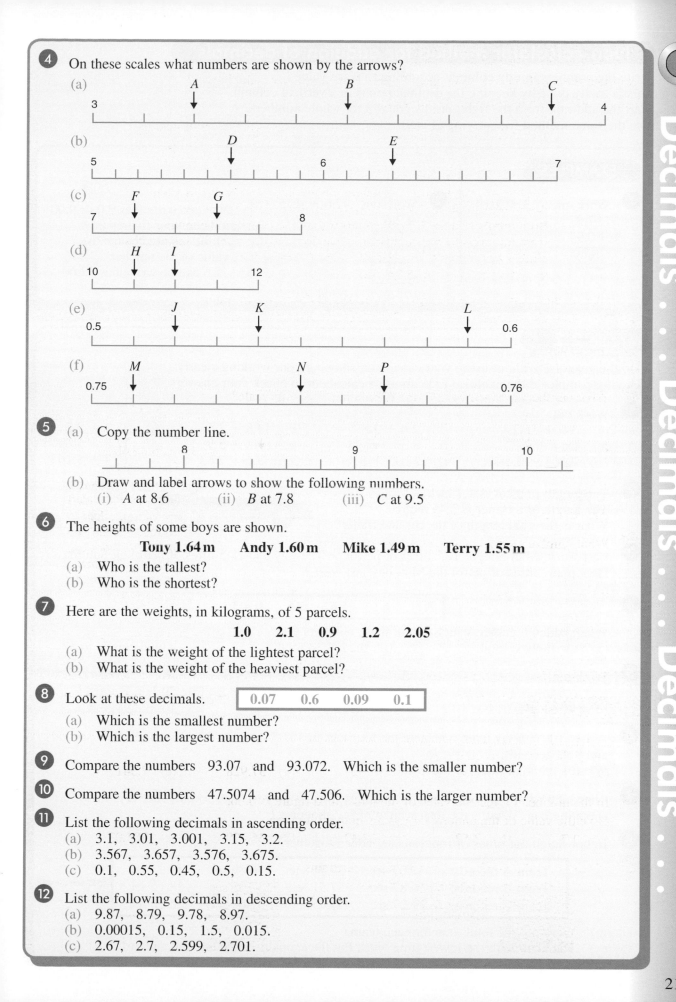

(a)

            A   *B*   *C*

3   4

(b)   *D*   *E*

5   6   7

(c) *F*   *G*

7   8

(d) *H* *I*

10   12

(e)   *J*   *K*   *L*

0.5   0.6

(f)   *M*   *N*   *P*

0.75   0.76

5 (a) Copy the number line.

8   9   10

   (b) Draw and label arrows to show the following numbers.
     (i) *A* at 8.6   (ii) *B* at 7.8   (iii) *C* at 9.5

6 The heights of some boys are shown.

       Tony 1.64 m   **Andy 1.60 m**   **Mike 1.49 m**   **Terry 1.55 m**

   (a) Who is the tallest?
   (b) Who is the shortest?

7 Here are the weights, in kilograms, of 5 parcels.

       1.0   2.1   0.9   1.2   2.05

   (a) What is the weight of the lightest parcel?
   (b) What is the weight of the heaviest parcel?

8 Look at these decimals. | 0.07   0.6   0.09   0.1 |

   (a) Which is the smallest number?
   (b) Which is the largest number?

9 Compare the numbers 93.07 and 93.072. Which is the smaller number?

10 Compare the numbers 47.5074 and 47.506. Which is the larger number?

11 List the following decimals in ascending order.
   (a) 3.1, 3.01, 3.001, 3.15, 3.2.
   (b) 3.567, 3.657, 3.576, 3.675.
   (c) 0.1, 0.55, 0.45, 0.5, 0.15.

12 List the following decimals in descending order.
   (a) 9.87, 8.79, 9.78, 8.97.
   (b) 0.00015, 0.15, 1.5, 0.015.
   (c) 2.67, 2.7, 2.599, 2.701.

Non-calculator method for addition of decimals

Write the numbers in tidy columns according to place value.
This is easily done by keeping the decimal points in a vertical column.
Start the addition from the right, just as you did for whole numbers.
Use the same method for carrying as well.

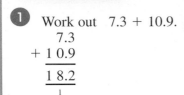

1 Work out 7.3 + 10.9.

```
    7.3
+ 1 0.9
  1 8.2
    1
```

2 Work out   42.6 + 0.75 + 9.

```
  4 2.6
    0.7 5
+   9.0
  5 2.3 5
    1 1
```

You can write 9 as 9.0 or 9.00 to keep your figures tidy. This does not change the value of the number. 42.6 can be written as 42.60

Exercise 2.2

Do this exercise without using your calculator, showing your working clearly.
Having completed the exercise you can use a calculator to check your answers.

1 Work out.
(a) 3.6 + 15.2   (b) 2.6 + 3.8   (c) 14.8 + 3.5   (d) 23.4 + 9.7
(e) 5.14 + 3.72   (f) 8.36 + 4.74   (g) 6.48 + 5.9   (h) 11.8 + 5.69
(i) 7.065 + 5.384 (j) 17.93 + 8.09 (k) 5.06 + 27.3   (l) 12.7 + 5.463

2 The length of a car is 4.7 metres.
The length of a trailer is 2.45 metres.
What is the total length of the car and trailer?

3 Last week, Matt bought 17.6 litres of petrol on Tuesday and 18.5 litres of petrol on Saturday.
How many litres of petrol did Matt buy last week?

4

3.2	4.1	1.6	2.5	0.8

When added together, which two of these numbers give:
(a)   the highest total,   (b)   the lowest total,   (c)   a total closest to 5?

5 Work out.
(a) 6.54 + 0.27 + 0.03   (b) 2.22 + 0.78 + 0.07   (c) 79.1 + 7 + 0.23
(d) 5.564 + 0.017 + 10.2   (e) 9.123 + 0.71 + 6.2   (f) 16 + 2.98 + 5.9

6 A 4 × 100 m relay team complete the four legs in 10.05 seconds, 10.13 seconds, 9.89 seconds and 9.92 seconds.
What is the total time for the team?

7 I have 2.5 kg of potatoes, 0.5 kg of butter, 0.75 kg of grapes and 0.6 kg of cheese in my shopping bag. What is the total weight of my shopping?

8 In bobsleigh the times of four runs are added together.

Team A records:	37.03 sec	37.76 sec	36.89 sec	37.25 sec
Team B records:	36.87 sec	37.51 sec	37.03 sec	38.12 sec
Team C records:	37.27 sec	37.45 sec	37.64 sec	36.72 sec

(a)   Work out the total time for each team.
(b)   The team with the lowest time wins. Put the teams in order 1st, 2nd and 3rd.

Non-calculator method for subtraction of decimals

Write the numbers in tidy columns according to place value.
This is easily done by keeping the decimal points in a vertical column.
Start the subtraction from the right, just as you did for whole numbers.
Use the same method for borrowing as well.

EXAMPLES

1 Work out 5.6 − 3.8.

$$\begin{array}{r} {\overset{4}{\cancel{5}}}.{\overset{1}{6}} \\ - \ 3.8 \\ \hline 1.8 \end{array}$$

You can use addition to check your subtraction.
Does 1.8 + 3.8 = 5.6?

2 Work out 17.1 − 8.72.

$$\begin{array}{r} 1\,{\overset{6}{\cancel{7}}}.{\overset{10}{\cancel{1}}}{\overset{1}{0}} \\ - \quad 8.7\,2 \\ \hline 8.3\,8 \end{array}$$

Check the answer by addition.

Useful tip:
Writing 17.1 as 17.10 can make the working easier.
This does not change the value of 17.1.

Money

1360p can be written as £13.60

£13.60

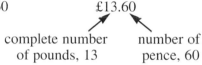

complete number
of pounds, 13

number of
pence, 60

£6 can be written as £6.00
There must be exactly **two** figures after the decimal point when a decimal point is used to record amounts of money.

EXAMPLE

I buy a newspaper for 45p, a set of batteries for £2.50 and a book of stamps for £2.
What is the total cost? How much change should I get from £5?

Working in pounds.

$$\begin{array}{r} 0.4\,5 \\ 2.5\,0 \\ + \ 2.0\,0 \\ \hline 4.9\,5 \end{array}$$

The total cost is £4.95.

$$\begin{array}{r} 5.0\,0 \\ - \ 4.9\,5 \\ \hline 0.0\,5 \end{array}$$

The change is £0.05 or 5p.

Other uses of decimal notation

Many measurements are recorded using decimals, including time, distance, weight, volume, etc.
The metric and imperial measures you need to know are given in Chapter 34.
The same rules for addition and subtraction can be applied if all the measurements involved are recorded using the same units.

Exercise 2.3

Do this exercise without using your calculator, showing your working clearly.
Having completed the exercise you can use a calculator to check your answers.

1 (a) Work out.
    (i) 6.7 − 2.3   (ii) 9.47 − 3.24   (iii) 7.4 − 2.8   (iv) 24.5 − 9.7
    (v) 12.48 − 7.52   (vi) 37.6 − 16.8   (vii) 14.15 − 3.07   (viii) 45.04 − 20.36
   (b) Show how addition can be used to check each of the answers to part (a).

Decimals . . . Decimals . . . Decimals . . .

2 Work out.
(a) $4.7 - 2.56$ (b) $10 - 4.78$ (c) $9.57 - 4.567$ (d) $9.13 - 7.89$
(e) $17.1 - 8.82$ (f) $9.123 - 2.85$ (g) $14.2 - 5.16$ (h) $3.1 - 1.204$

3 Add these amounts of money. Calculate the change from the given amount.
(a) (i)   45p, 63p, 79p, £1.43   (ii)   What is the change from £5?
(b) (i)   £2.47, £6, £1.50, £1.27   (ii)   What is the change from £15?
(c) (i)   31p, £0.25, 27p   (ii)   What is the change from £10?
(d) (i)   £12, £3.57, 67p   (ii)   What is the change from £50?

4 Fred cuts three pieces of wood of length 0.95 m, 1.67 m and 2.5 m from a plank 10 m long. How much wood is left?

5 Kevin is 0.15 m shorter than Sally. Sally is 1.7 m tall. How tall is Kevin?

6 Swimmer A finishes the 100 m freestyle in 51.371 seconds.
Swimmer B finishes in 52.090 seconds.
How long after Swimmer A does Swimmer B finish?

Working mentally

Addition and subtraction of decimals can be carried out mentally, in your head.
For example, using place value, we know that $2.5 = 2 + 0.5$.
So, adding 2.5 to a number is the same as adding 2 and then adding 0.5.

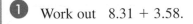

> **EXAMPLES**
>
> **1** Work out $8.31 + 3.58$.
>
> $3.58 = 3 + 0.5 + 0.08$
> $8.31 + 3.58$
> $= 8.31 + 3 + 0.5 + 0.08$
> $= 8.39 + 3 + 0.5$ (adding 8.31 and 0.08)
> $= 8.89 + 3$ (adding 8.39 and 0.5)
> $= 11.89$
>
>
>
> **2** Work out $25.4 - 8.7$.
>
> $25.4 - 8.7$
> $= 25.4 - 8 - 0.7$
> $= 24.7 - 8$ (subtracting 0.7 from 25.4)
> $= 16.7$

> $3.58 = 3 + 0.5 + 0.08$
> The adding of 3, 0.5 and 0.08 can be carried out in any order.
> Here, we have added the numbers in order of size, starting with the smallest. Choose a method you find easiest.

> $8.7 = 8 + 0.7$
> To subtract 8.7, first subtract 0.7 and then **subtract** 8.
> Alternatively, first subtract 8 and then **subtract** 0.7.

Exercise 2.4

Work these out in your head.
Having completed the exercise you can use a calculator to check your answers.

1 $0.7 + 0.6$ **2** $2.5 + 8.4$ **3** $0.7 + 0.95$

4 $0.36 + 0.54$ **5** $6.47 + 4.53$ **6** $12.06 + 5.72$

7 $2.7 - 1.5$ **8** $1.3 - 0.7$ **9** $2.6 - 0.9$

10 $0.48 - 0.16$ **11** $15.87 - 6.43$ **12** $4 - 0.8$

Multiplying and dividing decimals by powers of 10 (10, 100, 1000, . . .)

When you multiply a decimal by:

10   Each figure moves 1 place to the left.
100 Each figure moves 2 places to the left.
1000 Each figure moves 3 places to the left.
         … and so on.

When you divide a decimal by:

10   Each figure moves 1 place to the right.
100 Each figure moves 2 places to the right.
1000 Each figure moves 3 places to the right.
         … and so on.

EXAMPLES

276 has the same value as 276.0

Noughts can be used as place fillers to locate the decimal point, as in $2.76 \times 1000 = 2760$.
If the nought was omitted the value of all other figures would change.

Multiplication

		2 •	7	6	
		2	7 •	6	
		2	7	6 •	
	2	7	6	0 •	

←$2.76 \times 10 = 27.6$
←$2.76 \times 100 = 276$
←$2.76 \times 1000 = 2760$

Division

3 •	4	5			
0 •	3	4	5		
0 •	0	3	4	5	
0 •	0	0	3	4	5

←$3.45 \div 10 = 0.345$
←$3.45 \div 100 = 0.0345$
←$3.45 \div 1000 = 0.00345$

Exercise 2.5   Do this exercise without using a calculator.

1 Work out the following.
   (a) 25.06×10 (b) 25.06×100 (c) 25.06×1000
   (d) 0.93×10 (e) 0.93×100 (f) 0.93×1000
   (g) 0.0623×10 (h) 0.0623×100 (i) 0.0623×1000
   (j) 9.451×10 (k) 9.451×100 (l) 9.451×1000

2 Work out the following.
   (a) $37.7 \div 10$ (b) $37.7 \div 100$ (c) $37.7 \div 1000$
   (d) $0.27 \div 10$ (e) $0.27 \div 100$ (f) $0.27 \div 1000$
   (g) $189.02 \div 10$ (h) $189.02 \div 100$ (i) $189.02 \div 1000$
   (j) $9 \div 10$ (k) $9 \div 100$ (l) $9 \div 1000$

3 (a) Multiply 0.064 by   (i) 10   (ii) 100   (iii) 1000
   (b) Divide 6.4 by   (i) 10   (ii) 100   (iii) 1000

4 A biro costs 25 pence.
   (a) How much will 10 cost?   (b) How much will 100 cost?   (c) How much will 1000 cost?

5 One lap of a cycling track is 0.504 km.
   (a) How far is 10 laps?   (b) How far is 100 laps?   (c) How far is 1000 laps?

6 (a) 100 calculators cost £795. How much does one cost?
   (b) 1000 pencils cost £120. How much does one cost?
   (c) 10 litres of petrol cost £8.69. How would the cost of 1 litre be advertised?

7 Write down pairs of calculations which give the same answer.
   12.3×1000 $12.3 \div 100$ 12.3×0.1 $12.3 \div 0.01$ 12.3×10 $12.3 \div 0.1$
   12.3×100 12.3×0.001 $12.3 \div 0.001$ $12.3 \div 10$ $12.3 \div 1000$ 12.3×0.01

The result of multiplying two numbers is called the **product**.

Activity

Use a calculator to multiply these decimals.

$$5.924 \times 2.34 \qquad 5.2 \times 6.4 \qquad 6 \times 3.7 \qquad 5.1 \times 6.02 \qquad 2.16 \times 5.79$$

Count the total number of decimal places in the numbers to be multiplied together.
For example, 5.924 has three decimal places (there are three figures to the right of the decimal point) and 2.34 has two decimal places. The product of 5.924 and 2.34 has five decimal places.
Can you find a rule?

How many decimal places does your rule predict 0.5×0.5 should have?

Non-calculator method for multiplying decimals

To multiply decimals without using a calculator:

1. Ignore the decimal points and multiply the numbers using long multiplication.

2. Count the total number of decimal places in the numbers being multiplied together.

3. Place the decimal point so that the answer has the same total number of decimal places.

EXAMPLES

1. Work out 1.7×0.4.

$$
\begin{array}{r}
1.7 \\
\times\, 0.4 \\
\hline
0.6\,8 \\
\hline
\scriptstyle 2
\end{array}
$$

1.7 has 1 decimal place.
0.4 has 1 decimal place.
The answer has 2 decimal places.

$1.7 \times 0.4 = 0.68$

2. Work out 4.25×0.18.

$$
\begin{array}{r}
4.2\,5 \\
\times\, 0.1\,8 \\
\hline
3\,4\,0\,0 \\
4\,2\,5\,0 \\
\hline
0.7\,6\,5\,0
\end{array}
$$

$\leftarrow 425 \times 8$
$\leftarrow 425 \times 10$

The answer must have 4 decimal places because 4.25 has 2 and 0.18 has 2.

$4.25 \times 0.18 = 0.7650$ This can be written as 0.765 which has the same value as 0.7650.

3. Work out 0.2×0.4.

$$
\begin{array}{r}
0.2 \\
\times\, 0.4 \\
\hline
0.0\,8
\end{array}
$$

0.2 has 1 decimal place.
0.4 has 1 decimal place.
The answer has 2 decimal places.

○○○○○○○○○○○○○○○

$4 \times 2 = 8$
The answer must have 2 decimal places.
Noughts are used in the answer to locate the decimal point and to preserve place value.

$0.2 \times 0.4 = 0.08$

Do this exercise without using a calculator, showing your working clearly.
Having completed the exercise you can use a calculator to check your answers.

1 Work these out in your head.
  (a) 0.6×2 (b) 1.7×5 (c) 3.2×4 (d) 12×0.3 (e) 5×2.6
  (f) 8×2.2 (g) 6×1.8 (h) 4.3×7 (i) 3×9.6 (j) 87×0.4

2 A puzzle costs £1.90.
  How much will 4 puzzles cost?

3 A cup of coffee costs £1.15.
  (a) How much will I have to pay for 7 cups?
  (b) How much change will I get from £10?

4 I buy 5 kites which cost £2.99 each.
  (a) What is the total cost?
  (b) How much change will I get from £20?

5 Here is a price list:
  (a) What is the cost of 5 geometry sets?
  (b) What is the cost of 7 calculators?
  (c) I buy 3 geometry sets and 4 calculators with £50.
      What is my change?

Geometry Set   £2.45
Calculator   £4.99

6 Calculate these products.
  (a) 0.7×0.6 (b) 0.2×0.3 (c) 2.5×3.5 (d) 8.7×1.9
  (e) 54×0.36 (f) 4.1×0.25 (g) 0.9×4.32 (h) 13.4×0.7
  (i) 0.7×5.4 (j) 0.06×0.72 (k) 0.35×0.08 (l) 0.07×0.02

7 (a) Multiply each of these numbers by 0.6.
      (i) 5   (ii) 2.5   (iii) 0.4   (iv) 25
  (b) What do you notice about the original numbers and each of your answers?

8 Work out the cost of these vegetables.
  (a) 0.6 kg of carrots at 35p per kilogram.
  (b) 4.6 kg of potatoes at 40p per kilogram.
  (c) 1.2 kg of cabbage at 65p per kilogram.

9 What is the cost of each of these lengths of material?
  (a) 7 metres of sheeting at £2.99 a metre.
  (b) 4.5 metres of linen at £13.50 a metre.
  (c) 7.8 metres of satin at £26.90 a metre.
  (d) 3.2 metres of silk at £38.20 a metre.

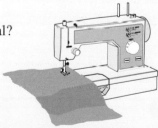

10

Select
CHEESES
Price per kilogram

Cheddar	£4.80
Cotherstone	£6.20
Sage Derby	£4.50
Stilton	£7.60

  (a) Work out the cost for each of these portions of cheese.
      (i) 0.7 kg of Stilton.
      (ii) 1.6 kg of Cheddar.
      (iii) 0.8 kg of Sage Derby.
      (iv) 0.45 kg of Cotherstone.
  (b) Corrine buys:
      0.25 kg of Stilton and 1.4 kg of Cheddar.
      She pays with a £10 note.
      How much change will she get?

Non-calculator method for dividing decimals

Work out $2.4 \div 0.4$.

$2.4 \div 0.4$ can be written as $\frac{2.4}{0.4}$

$\frac{2.4}{0.4} = \frac{2.4 \times 10}{0.4 \times 10} = \frac{24}{4} = 6$

It is easier to divide by a whole number than by a decimal.

To divide by a decimal:
1. Multiply the dividing number by a power of 10 (10, 100, 1000, …) so that it becomes a whole number.
2. Multiply the number to be divided by the same number.
3. If necessary the answer will have a decimal point in the same place.

EXAMPLES

1 Work out the following.
(a) $4 \div 0.8$
Multiply both numbers by 10.
$40 \div 8 = 5$
So, $40 \div 0.8 = 5$

(b) $9 \div 4$

$4 \overline{)9.0^{1}0^{2}}$
$2.2\,5$

So, $9 \div 4 = 2.25$

Noughts are added until the division is finished.

2 8 bandages cost £14.
How much does each bandage cost?

You must work out $14 \div 8$.

$8 \overline{)1\,4.^{6}0^{4}0}$
$1.7\,5$

One bandage costs £1.75.

Noughts can be added to the end of a decimal. Adding noughts does not change the value of the number. 14 has the same value as 14.00. Continue dividing until either there is no remainder or the required accuracy is obtained.

3 Work out $11.06 \div 0.7$.

$0.7 \times 10 = 7 \quad 11.06 \times 10 = 110.6$
$110.6 \div 7$ has the same value as $11.06 \div 0.7$.

$7 \overline{)1\,1^{4}0.^{5}6}$
$1\,5.8$

So, $11.06 \div 0.7 = 15.8$

Use the same method of working as you used for dividing whole numbers.
The decimal point moves vertically to the same position in the answer.

Exercise 2.7

Do this exercise without using a calculator, showing your working clearly.
Having completed the exercise you can use a calculator to check your answers.

1 Work these out in your head.
(a) $0.9 \div 3$ 　　(b) $7.5 \div 5$ 　　(c) $6.8 \div 4$ 　　(d) $22.4 \div 7$ 　　(e) $35.2 \div 8$

2 Work out.
(a) $7 \div 4$ 　　(b) $8 \div 5$ 　　(c) $1.2 \div 8$ 　　(d) $18.2 \div 7$ 　　(e) $10.5 \div 6$

3 What number should be put in the box to make each of these statements correct?
(a) $8 \div 0.5 = 80 \div \Box$ 　　(b) $1.2 \div 0.3 = \Box \div 3$ 　　(c) $3.5 \div 0.07 = \Box \div 7$

4 Work out.
 (a) $2 \div 0.5$ (b) $3 \div 0.2$ (c) $6 \div 0.4$
 (d) $10 \div 2.5$ (e) $60 \div 1.2$

5 Given that $221 \div 13 = 17$ write down the value of:
 (a) $22.1 \div 13$ (b) $22.1 \div 1.3$ (c) $2.21 \div 1.3$

6 Work out.
 (a) $2.46 \div 0.2$ (b) $0.146 \div 0.05$ (c) $2.42 \div 0.4$ (d) $100.1 \div 0.07$
 (e) $0.0025 \div 0.05$ (f) $0.05 \div 0.004$ (g) $4.578 \div 0.7$ (h) $0.3 \div 0.008$

7 Use long division to work out the following.
 (a) $81.4 \div 2.2$ (b) $15.12 \div 2.7$ (c) $7 \div 0.16$
 (d) $11.256 \div 0.24$ (e) $0.1593 \div 0.15$

8 (a) Divide each of these numbers by 0.6.
   (i) 6 (ii) 3.6 (iii) 0.18
 (b) What do you notice about the original numbers and each of your answers?

9 A 3-litre bottle of lemonade costs £1.41.
 What is the cost of 1 litre of lemonade?

10 A pack of 7 tubes of oil paint costs £9.45.
 How much does each tube of oil paint cost?

11 13 oranges cost £1.43.
 How much does each orange cost?

12 12 rolls cost £1.08.
 How much does each roll cost?

13 25 litres of petrol cost £22.70.
 What is the cost of 1 litre of petrol?

14 A bottle holds 0.25 litres.
 How many bottles can be filled from a tank holding 30 litres?

15 A steel bar is 12.73 metres long.
 How many pieces, each 0.19 metres long, can be cut from it?

16 A jug holds 1.035 litres. A small glass holds 0.023 litres.
 How many of the small glasses would be required to fill the jug?

Changing decimals to fractions

How to change a decimal to a fraction:

Change 0.12 to a fraction.   0.12

Write the decimal without the decimal point.   12
This will be the numerator (top number).

The denominator (bottom number) is a power of 10. $\frac{12}{100}$
The number of noughts is the same as the number of
decimal places in the original decimal.

Fractions are covered
in further detail in
Chapter 5.

Divide both the numerator and denominator by   Divide by 4
the largest possible number.
This gives the fraction in its simplest form. $\frac{3}{25}$

Write the following decimals as fractions in their simplest form.

(a) $0.3 = \frac{3}{10}$ (b) $0.6 = \frac{6}{10} = \frac{3}{5}$ (c) $0.45 = \frac{45}{100} = \frac{9}{20}$

(d) $1.5 = 1 + 0.5$

$= 1 + \frac{5}{10}$

$= 1 + \frac{1}{2}$

$= 1\frac{1}{2}$

$1\frac{1}{2}$ is called a **mixed number**. It is a mixture of whole numbers and fractions.

Recurring decimals

Some decimals have recurring digits.

For example, $\frac{1}{3} = 0.3333\ldots$.

The number $0.3333\ldots$ is called a **recurring decimal**.

Recurring decimals are covered in more detail in Chapter 5.

Exercise 2.8

1 Write the following decimals as fractions in their simplest form.
(a) 0.25   (b) 0.5   (c) 0.75   (d) 0.1

2 Write the following decimals as fractions in their simplest form.
(a) 0.7   (b) 0.4   (c) 0.01   (d) 0.2
(e) 0.05   (f) 0.15   (g) 0.52   (h) 0.07
(i) 0.125   (j) 0.65   (k) 0.6   (l) 0.95

3 Change these decimals into mixed numbers.
(a) 1.7   (b) 2.3   (c) 1.4   (d) 3.25
(e) 4.8   (f) 12.1   (g) 16.75   (h) 5.05

4 What fraction is equal to each of these recurring decimals?
(a) 0.6666…   (b) 0.1111…   (c) 0.5555…

What you need to know

You should be able to:
- Place decimals in order by considering place value.
- Add and subtract decimals.
- Use decimal notation for money and other measures.
- Multiply and divide decimals by powers of 10 (10, 100, 1000, …)
- Multiply decimals by other decimals.
- Divide decimals by other decimals.
- Change decimals to fractions.
- Carry out a variety of calculations involving decimals.
- Know that:
  when a number is **multiplied** by a number between 0 and 1 the result will be **smaller** than the original number,
  when a number is **divided** by a number between 0 and 1 the result will be **larger** than the original number.

1 List the following decimals in order, smallest first.

| 0.7 | 0.5 | 0.8 | 0.85 | 0.55 |

2 Write 0.45 as a fraction. Give your answer in its simplest form.

3 Work out.
(a) 2.94 + 9.47   (b) 10 − 5.67
(c) Check the subtraction in part (b) with an addition.

4 Work out.
(a) 3.6 × 4   (b) 14 × 0.3   (c) 7.8 ÷ 6   (d) 8 ÷ 0.4

5 (a) Which two of these numbers multiply together to give the **smallest** possible answer?
(b) Which two of these numbers, when multiplied together, give the answer which is closest to 1?

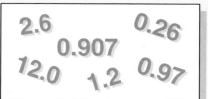

2.6   0.26   0.907   12.0   1.2   0.97

AQA

6 Four parcels weigh 1.6 kg, 0.8 kg, 0.55 kg and 1.25 kg.
What is the total weight of the parcels?

7 A 4 × 100 m relay team runs the four stages in: 10.01 s, 9.93 s, 10.15 s and 9.91 s.
What is the overall time for the team?

8 Two pieces of wood of length 0.97 m and 1.78 m are sawn from a plank 5.12 m long.
How much wood is left?

9 To change gallons into litres: **Multiply the number of gallons by 4.5**
Change 9 gallons into litres.

AQA

10 (a) Multiply 87.3 by 30.   (b) Divide 87.3 by 30.

11 (a) A calculator costs £4.95. How much will 50 cost?
(b) 20 textbooks cost £159.80. How much does one cost?

12 Given that 275 × 41 = 11 275 work out the value of:
(a) 275 × 0.0041   (b) 112.75 ÷ 4.1

13 Calculate these products.
(a) 7.4 × 6.3   (b) 3.76 × 2.7   (c) 176.5 × 0.6

14 Calculate these divisions.
(a) 16.56 ÷ 2.3   (b) 98.8 ÷ 0.08   (c) 5480 ÷ 0.4

15 Goldfish cost 85p each. How many goldfish can you buy with £10?

AQA

16 Korky cat food costs 54p a tin. Alex buys 18 tins of Korky cat food. He pays with a £10 note.
How much change should he receive?

AQA

17 Work out $\dfrac{24.076}{1.5 + 3.7}$

AQA

18 Tony uses his calculator to work out $\dfrac{4.2 \times 86}{3.2 \times 0.47}$
What answer should he get?

AQA

19 Work out $\dfrac{4.7 \times 20.1}{5.6 - 1.8}$
Write down your full calculator display.

AQA

Approximation and Estimation

Approximation

In real-life it is not always necessary to use exact numbers.
A number can be **rounded** to an **approximate** number.
Numbers are rounded according to how accurately we wish to give details.
For example, the distance to the Sun can be given as 93 million miles.

Can you think of other situations where approximations might be used?

Rounding to the nearest 10, 100, 1000

The number line shows the position of 87.

87 is between 80 and 90, but is closer to 90.

87 rounded to the nearest 10 is 90.

If there were 7487 people at a football match a newspaper report could say,

"7000 at the football match".

The number 7487 can be approximated as 7490, 7500 or 7000 depending on the degree of accuracy required.

7487 rounded to the nearest 10 is 7490.

7487 rounded to the nearest 100 is 7500.

7487 rounded to the nearest 1000 is 7000.

> It is a convention to round a number which is in the middle to the higher number.
>   75 to the nearest 10 is 80.
>   450 to the nearest 100 is 500.
>   8500 to the nearest 1000 is 9000.

EXAMPLES

1 Round the numbers: 7547, 973, 62 783 and 9125
to the nearest 10, to the nearest 100, to the nearest 1000.

	Number			
	7547	973	62 783	9125
Rounded to the nearest 10	7550	970	62 780	9130
Rounded to the nearest 100	7500	1000	62 800	9100
Rounded to the nearest 1000	8000	1000	63 000	9000

2 A lifeguard says, "There are 120 people in the pool today."
This figure is correct to the nearest 10.
What is the smallest and largest possible number of people in the pool?

The smallest whole number that rounds to 120 is 115.
The largest whole number that rounds to 120 is 124.
So, the smallest possible number of people is 115 and the largest possible number of people is 124.

1 Write each of these numbers to the nearest 10.   (a) 47   (b) 53   (c) 65

2
  (a) Which of these numbers: 4850, 4860, 4870, 4880, 4890 is closest to 4872?
  (b) Which of these numbers: 4600, 4700, 4800, 4900, 5000 is closest to 4872?
  (c) Which of these numbers: 3000, 4000, 5000, 6000   is closest to 4872?

3 Round the number 364   (a) to the nearest 10,   (b) to the nearest 100.

4 Round the number 7475
  (a) to the nearest 10,   (b) to the nearest 100,   (c) to the nearest 1000.

5 Copy and complete this table.

	Number								
	7613	977	61 115	9714	623	9949	5762	7509	7499
Round to the nearest 10									
Round to the nearest 100									
Round to the nearest 1000									

6
  (a) Rearrange the cards: 5 4 6 3 to make the smallest possible number.
  (b) Round your number to the nearest 100.

7 64 537 people signed a petition. A newspaper report stated: '65 000 people sign petition'
To what degree of accuracy is the number given in the newspaper report?

8 Write down these figures to appropriate degrees of accuracy.
  (a) There were 19 141 people at a football match.
  (b) There were 259 people on a plane.
  (c) The class raised £49.67 for charity.
  (d) The population of a town is 24 055.
  (e) The land area of a country is 309 123 km².
  (f) The distance to London is 189 km.

9 Write down a number each time which fits these roundings.
  (a) It is 750 to the nearest 10 but 700 to the nearest 100.
  (b) It is 750 to the nearest 10 but 800 to the nearest 100.
  (c) It is 8500 to the nearest 100 but 8000 to the nearest 1000.
  (d) It is 8500 to the nearest 100 but 9000 to the nearest 1000.

10 Pete says, "I had 40 birthday cards." The number is correct to the nearest 10.
  (a) What is the smallest possible number of cards Pete had?
  (b) What is the largest possible number of cards Pete had?

11 The number of people at a concert is 2000 to the nearest 100.
  (a) What is the smallest possible number of people at the concert?
  (b) What is the largest possible number of people at the concert?

12   **"43 000 spectators watch thrilling Test Match."**
The number reported in the newspaper was correct to the nearest thousand.
What is the smallest possible number of spectators?

13 Carl has 140 postcards in his collection. The number is given to the nearest ten.
What is the smallest and greatest number of postcards Carl could have in his collection?

14 "You require 2700 tiles to tile your swimming pool." This figure is correct to the nearest 100.
What is the greatest number of tiles needed?

Rounding in real-life problems

In a real-life problem a rounding must be used which gives a sensible answer.

Penny is arranging a BBQ. 50 people have been invited. She caters for everyone to have one burger. Burgers are sold in packs of 12. How many packs of burgers should she buy?

The answer is found by working out $50 \div 12$.
In 4 packs, there are $4 \times 12 = 48$ burgers.
In 5 packs, there are $5 \times 12 = 60$ burgers.
$50 \div 12 = 4$ remainder 2.
Penny must buy 5 packs in order that everybody has one burger.
(In fact she will have 10 left over for those who might want a second burger.)

EXAMPLES

1. A Year group in a school are going to Alton Towers.
   There are 242 students and teachers going.
   Each coach can carry 55 passengers. How many coaches should be ordered?

   $242 \div 55 = 4.4$ This should be rounded up to 5.

   4 coaches can only carry 220 passengers $(4 \times 55 = 220)$.

2. Filing cabinets are to be placed along a wall. The available space is 460 cm.
   Each cabinet is 80 cm wide. How many can be fitted in?

   $460 \div 80 = 5.75$ This should be rounded down to 5.

   Although the answer is nearly 6 the 6th cabinet would not fit in.

Exercise 3.2   Do not use a calculator for this exercise.

1. 49 students are waiting to go to the Sports Stadium.
   A minibus can take 15 passengers at a time. How many trips are required?

2. A classroom wall is 700 cm long.
   How many tables, each 120 cm long, could be fitted along the wall?

3. 76 people are waiting to go to the top of Canary Wharf. The lift can only take 8 at a time.
   How many times must the lift go up?

4. A group of 175 people are going to Margate. Coaches can take 39 passengers.
   How many coaches should be ordered?

5. There are 210 students in a year group. They each need an exercise book.
   The exercise books are sold in packs of 25. How many packs should be ordered?

6. Car parking spaces should be 2.5 m wide.
   How many can be fitted into a car park which is 61 m wide?

7. A sweet manufacturer puts 17 sweets in a bag.
   How many bags can be made up if there are 500 sweets?

8. How many 30p stamps can be bought for £5?

9. How many grapefruit, each costing 29p, can be bought for £1.50?

10. Kim needs 26 candles for a cake. The candles are sold in packs of 4.
    How many packs must she buy?

11. Lauren needs 50 doughnuts for a party. Doughnuts are sold in packs of 12.
    How many packs must she buy?

Rounding using decimal places

What is the cost of 1.75 metres of material costing £3.99 a metre?

$$1.75 \times 3.99 = 6.9825$$

The cost of the material is £6.9825 or 698.25p.

As you can only pay in pence, a sensible answer is £6.98, correct to two decimal places (nearest penny).

This means that there are only two decimal places after the decimal point.

> Often it is not necessary to use an exact answer. Sometimes it is impossible, or impractical, to use the exact answer.

To round a number to a given number of decimal places

When rounding a number to one, two or more decimal places:

1. Write the number using one more decimal place than asked for.
2. Look at the last decimal place and
   - if the figure is 5 or more round up,
   - if the figure is less than 5 round down.
3. When answering a problem remember to include any units and state the degree of approximation used.

EXAMPLES

1 Write 2.76435 to
   (a) 2 decimal places,
   (b) 1 decimal place.

   (a) Look at the third decimal place. **4** This is less than 5, so round down. Answer 2.76

   (b) Look at the second decimal place. **6** This is 5 or more, so round up. Answer 2.8

2 Write 7.104 to 2 decimal places.
   7.104 = 7.10 to 2 d.p.
   The zero is written down because it shows the accuracy used, 2 decimal places.

> **Notation:**
> Often decimal place is shortened to d.p.

3 5.98 = 6.0 to 1 d.p.
   Notice that the next tenth after 5.9 is 6.0.

Exercise 3.3

1 Write the number 3.9617 correct to
   (a) 3 decimal places,   (b) 2 decimal places,   (c) 1 decimal place.

2 Write the number 567.654 correct to
   (a) 2 decimal places,   (b) 1 decimal place,   (c) the nearest whole number.

3 The display on a calculator shows the result of 34 ÷ 7.

$$\boxed{4.857142857}$$

What is the result correct to two decimal places?

4 68.847 kg   The scales show Gary's weight.
   Write Gary's weight correct to one decimal place.

5 Copy and complete this table.

Number	2.367	0.964	0.965	15.2806	0.056	4.991	4.996
d.p.	1	2	2	3	2	2	2
Answer	2.4						

6 Carry out these calculations giving the answers correct to
   (a) 1 d.p.   (b) 2 d.p.   (c) 3 d.p.
   (i)   6.12 × 7.54   (ii)   89.1 × 0.67   (iii)   90.53 × 6.29
   (iv)   98.6 ÷ 5.78   (v)   67.2 ÷ 101.45

7 In each of these short problems decide upon the most suitable accuracy for the answer.
   Then calculate the answer.
   Give a reason for your degree of accuracy.
   (a) One gallon is 4.54596… litres.
        How many litres is 9 gallons?
   (b) What is the cost of 0.454 kg of cheese at £5.21 per kilogram?
   (c) The total length of 7 equal sticks, lying end to end, is 250 cm.
        How long is each stick?
   (d) A packet of 6 bandages costs £7.99.
        How much does one bandage cost?
   (e) Petrol costs 91.4 pence a litre. I buy 15.6 litres.
        How much will I have to pay?

Rounding using significant figures

Consider the calculation   600.02 × 7500.97 = 4500732.0194
To 1 d.p. it is 4500732.0,   to 2 d.p. it is 4500732.02.
The answers to either 1 or 2 d.p. are very close to the actual answer and are almost as long.
There is little advantage in using either of these two roundings.
The point of a rounding is that it is a more convenient number to use.

Another kind of rounding uses **significant figures**.
The **most** significant figure in a number is the figure which has the greatest place value.

Consider the number 237.
The figure 2 has the greatest place value. It is worth 200.
So, 2 is the most significant figure.

Noughts which are used to locate the decimal point and preserve the place value of other figures are not significant.

In the number 0.00328, the figure 3 has the greatest place value.
So, 3 is the most significant figure.

To round a number to a given number of significant figures

When rounding a number to one, two or more significant figures:
   1. Start from the most significant figure and count the required number of figures.
   2. Look at the next figure to the right of this and
      ● if the figure is 5 or more round up,
      ● if the figure is less than 5 round down.
   3. Add noughts, as necessary, to locate the decimal point and preserve the place value.
   4. When answering a problem remember to include any units and state the degree of approximation used.

EXAMPLES

1 Round the numbers 75,   135,   1478 and 2500 to one significant figure.

Number	Rounded to 1 sig. fig.
75	80
135	100
1478	1000
2500	3000

Notation:
Often significant figure
is shortened to sig. fig.

2 Write   4 500 732.0194   to 2 significant figures.

The figure after the first 2 significant figures **45** is 0.
This is less than 5, so round down, leaving 45 unchanged.
Add noughts to 45 to locate the decimal point and preserve place value.
So,   4 500 732.0194 = 4 500 000   to 2 sig. fig.

3 Write   0.000364907   to 1 significant figure.

The figure after the first significant figure 3 is 6.
This is 5 or more, so round up, 3 becomes 4.
So,   0.000364907 = 0.0004   to 1 sig. fig.

Notice that the noughts before the 4 locate the decimal point and preserve place value.

Choosing a suitable degree of accuracy

In some calculations it would be wrong to use the complete answer from the calculator.
The result of a calculation involving measurement should not be given to a greater degree of accuracy
than the measurements used in the calculation.

EXAMPLE

What is the area of a rectangle measuring 4.6 cm by 7.2 cm?

Note:
To find the area of a rectangle:
multiply length by breadth.

$4.6 \times 7.2 = 33.12$
Since the measurements used in the calculation (4.6 cm and 7.2 cm) are given to 2 significant figures
the answer should be as well.
33 cm^2 is a more suitable answer.

Exercise 3.4

1 Write these numbers correct to one significant figure.
   (a) 17   (b) 523   (c) 350   (d) 1900   (e) 24.6

2 Copy and complete this table.

Number	456 000	454 000	7 981 234	7 981 234	1290	19 602
sig. fig.	2	2	3	2	2	1
Answer	460 000					

3 Write these numbers correct to one significant figure.
   (a) 0.083   (b) 0.086   (c) 0.00948   (d) 0.0095

4 Copy and complete this table.

Number	0.000567	0.093748	0.093748	0.093748	0.010245	0.02994
sig. fig.	2	2	3	4	2	2
Answer						

5 This display shows the result of $3400 \div 7$.

$$\boxed{485.7142857}$$

What is the result correct to two significant figures?

6 Carry out these calculations giving the answers correct to
(a)   1 sig. fig.   (b)   2 sig. fig.   (c)   3 sig. fig.

   (i) 672×123 (ii) 6.72×12.3 (iii) 78.2×12.8
   (iv) $7.19 \div 987.5$ (v) $124 \div 65300$

7 A rectangular field measures 18.6 m by 25.4 m.
Calculate the area of the field, giving your answer to a suitable degree of accuracy.

8 In each of these short problems decide upon the most suitable accuracy for the answer.
Then work out the answer, remembering to state the units.
Give a reason for your degree of accuracy.
   (a)   The area of a rectangle measuring 13.2 cm by 11.9 cm.
   (b)   The area of a football pitch measuring 99 m by 62 m.
   (c)   The total length of 13 tables placed end to end measures 16 m.
         How long is each table?
   (d)   The area of carpet needed to cover a rectangular floor measuring 3.65 m by 4.35 m.

Estimation

It is always a good idea to find an **estimate** for any calculation.
An estimate is used to check that the answer to the actual calculation is of the right magnitude (size).
If the answer is very different to the estimate then a mistake has possibly been made.

Estimation is done by approximating every number in the calculation to one significant figure.
The calculation is then done using the approximated values.

EXAMPLES

1 Estimate 421×48.

Round 421 to one significant figure: 400
Round 48 to one significant figure: 50
$400 \times 50 = 20\,000$

Use long multiplication to calculate 421×48.   Comment on your answer.

2 Estimate $608 \div 19$.

Round the numbers in the calculation to one significant figure.
$600 \div 20 = 30$

Use long division to calculate $608 \div 19$.   Comment on your answer.

Exercise 3.5   Do not use a calculator for this exercise.

1 Last year, Hannah paid these four telephone bills:

> £48   £89   £62   £103

By using approximations to one significant figure, **estimate** her total telephone bill for the year.

2 Bernard plans to buy a conservatory costing £8328 and furniture costing £984.
By using suitable approximations, **estimate** the total amount Bernard plans to spend.
Show all your working.

3 Make estimates to these calculations by using approximations to one significant figure.
  (a) (i) 39×21 (ii) 115×18 (iii) 797×53 (iv) 913×59
  (b) (i) $76 \div 18$ (ii) $597 \div 29$ (iii) $889 \div 61$ (iv) $3897 \div 82$

4 Lilly ordered 39 prints of her holiday photographs to give to her friends.
Each print cost 52 pence.
Use suitable approximations to **estimate** the total cost of the prints. Show your working.

5 Chairs are arranged in rows.
There are 18 chairs in each row.
Estimate the total number of chairs in 27 rows.

6 A book contains 576 pages which are grouped into 32 chapters of equal length.
Estimate the number of pages in each chapter.

7 Last year Alex used 964 litres of petrol to drive 15 209 kilometres.
Use suitable approximations to estimate her petrol consumption in kilometres per litre.

8 (a) When estimating the answer to 29×48 the approximations 30 and 50 are used.
How can you tell that the estimation must be bigger than the actual answer?
  (b) When estimating the answer to $182 \div 13$ the approximations 200 and 10 are used.
Will the estimate be bigger or smaller than the actual answer? Explain your answer.

Using a calculator

A calculator is a very useful piece of equipment.
But you must know how to use it properly.

Calculators vary but almost all do calculations in the right order.
Use your calculator to work out
> $3 + 4 \times 5$

by entering the following key sequence:

> [3] [+] [4] [×] [5] [=]

You should get the answer 23.

If you wish to work out
> $(3 + 4) \times 5$

then you must use the brackets on your calculator.
Try entering the following key sequence:

> [( ] [3] [+] [4] [ )] [×] [5] [=]

You should get the answer 35.

Scientific calculator

If your calculator works in a different way
refer to the instruction booklet supplied
with the calculator or ask someone for help.

Approximation and Estimation

1 Work out the following without using a calculator.
Then use a calculator to check your answers.

Question	Without a calculator	Using a calculator
(a) $\dfrac{5 \times 6}{1 + 2}$	$\dfrac{5 \times 6}{1 + 2} = \dfrac{30}{3} = 10$	 Answer: 10
(b) $\dfrac{50}{22 - 12}$	$\dfrac{50}{22 - 12} = \dfrac{50}{10} = 5$	$\boxed{5}\ \boxed{0}\ \boxed{\div}\ \boxed{(}\ \boxed{2}\ \boxed{2}$ $\boxed{-}\ \boxed{1}\ \boxed{2}\ \boxed{)}\ \boxed{=}$ Answer: 5
(c) $\dfrac{8 \times 9}{6 \times 6}$	$\dfrac{8 \times 9}{6 \times 6} = \dfrac{72}{36} = 2$	$\boxed{(}\ \boxed{8}\ \boxed{\times}\ \boxed{9}\ \boxed{)}\ \boxed{\div}$ $\boxed{(}\ \boxed{6}\ \boxed{\times}\ \boxed{6}\ \boxed{)}\ \boxed{=}$ Answer: 2

2 Use estimation to show that $\dfrac{78.5 \times 0.51}{18.7}$ is close to 2.

Approximating: $78.5 = 80$ to 1 sig. fig.
$0.51 = 0.5$ to 1 sig. fig.
$18.7 = 20$ to 1 sig. fig.

$$\frac{80 \times 0.5}{20} = \frac{40}{20} = 2 \text{ (estimate)}$$

Remember:
When you are asked to estimate, write each number in the calculation to one significant figure.

Using a calculator $\dfrac{78.5 \times 0.51}{18.7} = \dfrac{40.035}{18.7} = 2.140909\ldots$

Is 2.140909 reasonably close to 2?   Yes.

Exercise **3.6**

1 Work out the following without using a calculator.
Then use a calculator to check your answers.
Write down the key sequence you pressed on your calculator.

(a) $\dfrac{20}{22 - 18}$ (b) $\dfrac{4 \times 6}{7 + 5}$ (c) $\dfrac{21 - 9}{3 \times 2}$ (d) $\dfrac{48}{10 + 6}$ (e) $\dfrac{56 - 30}{13}$ (f) $\dfrac{12 + 18}{11 - 5}$

2 John estimated 43×47 to be about 2000. Explain how he did it.

3 (a) Write down the numbers you could use to get an approximate answer to 196×311.
(b) Write down your approximate answer.
(c) Use a calculator to find the difference between your approximate answer and the exact answer.

4 By using approximations to one significant figure find estimates to these products.
Then carry out the calculations with the original figures.
Compare your estimate to the actual answer.

(a) 32×41 (b) 12×66 (c) 58×34 (d) 72×45
(e) 4.2×1.8 (f) 8.9×3.1 (g) 48.1×4.2 (h) 103.4×2.9

5 Find estimates to these divisions by using approximations to one significant figure.
Then carry out the calculations with the original figures.
Compare your estimate to the actual answer.

(a) 594 ÷ 18   (b) 609 ÷ 21   (c) 256 ÷ 16   (d) 840 ÷ 35
(e) 10.78 ÷ 4.9   (f) 19.68 ÷ 4.1   (g) 30.4 ÷ 3.2   (h) 203.49 ÷ 5.1

6 Make estimates of the following by rounding each number in the calculation to one significant figure.
Then use a calculator to carry out an accurate calculation.
Compare the answer given by the calculator to the estimate.

(a) $\dfrac{51 \times 199}{22}$ (b) $\dfrac{581}{12 \times 29}$ (c) $\dfrac{18 \times 57}{38 \div 4}$ (d) $\dfrac{62}{86 \div 9} + 48$

7 Find estimates to these calculations by using approximations to one significant figure.
Then carry out the calculations with the original figures.
Compare your estimate to the actual answer.

(a) $\dfrac{9.9 \times 4.1}{4.8}$ (b) $\dfrac{11.6 + 49}{6.2}$ (c) $\dfrac{400 \times 0.29}{6.2}$ (d) $\dfrac{81.7 \times 4.9}{1.9 \times 10.3}$

Accuracy in measurement

No measurement is ever exact.
Measures which can lie within a range of possible values are called **continuous measures**.
The value of a continuous measure depends on the accuracy of whatever is making the measurement.

Jane is 160 cm tall to the nearest 10 cm.
What are the limits between which her actual height lies?

Height is a continuous measure.
When rounding to the nearest 10 cm:
The minimum value that rounds to 160 cm is 155 cm.
155 cm is the minimum height that Jane can be.

The maximum value that rounds to 160 cm is 164.999… cm.
164.999… cm is the maximum height that Jane can be.
For ease the value 164.999… cm is normally called 165 cm.

So, Jane's actual height is any height from 155 cm to 165 cm.
This can be written as the inequality:
155 cm ≤ Jane's height < 165 cm

All possible heights for Jane can be shown on a number line.

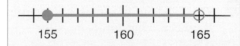

The hollow circle indicates that 165 is **not** included

EXAMPLES

1 The length of a pencil is 17 cm to the nearest centimetre.
What are the limits between which the actual length of the pencil lies?

When rounding to the nearest centimetre:

The smallest value that rounds to 17 cm is 16.5 cm.
The largest value that rounds to 17 cm is 17.4999… cm.
For ease 17.4999… is called 17.5.

So, the actual length of the pencil lies between 16.5 cm and 17.5 cm.
16.5 cm ≤ length of pencil < 17.5 cm

2 A concrete block weighs 1.8 kg, correct to the nearest tenth of a kilogram.
What is the minimum possible weight of the concrete block?

Minimum weight = 1.8 kg − 0.05 kg = 1.75 kg

Approximation and Estimation

41

1. A girl's height is 168 cm, correct to the nearest centimetre.
   What is the minimum possible height of the girl?

2. Harry weighs 49 kg, correct to the nearest kilogram.
   What is the maximum possible weight he could be?

3. The height of a building is 9 m, correct to the nearest metre.
   Copy and complete the inequality: …… ⩽ height of building < ……

4. A running track is 100 m in length, correct to the nearest metre.
   What is the minimum length of the track?

5. A brick weighs 840 g, correct to the nearest 10 g.
   What is the minimum and maximum possible weight of the brick?

6. A piece of cheese weighs 285 grams, correct to the nearest 5 grams.
   What is the maximum possible weight of the piece of cheese?

7. An athlete completed a race in 11.6 seconds, correct to the nearest tenth of a second.
   What is the minimum possible time the athlete could have taken?

8. A pane of glass weighs 9.4 kg, correct to one decimal place.
   What is the minimum possible weight of the pane of glass?

9. The length of a table is 2.7 m, correct to the nearest tenth of a metre.
   Write down the least and greatest possible length of the table.

10. A glass contains 24 ml of milk, correct to the nearest millilitre.
    Find the minimum possible number of millilitres in four glasses.

11. Loaves of bread each weigh 0.8 kg, correct to the nearest 100 g.
    Write down the minimum and maximum possible weight of a loaf of bread.

12. The capacity of a tank is 220 litres, correct to the nearest 5 litres.
    Between what limits does the actual capacity lie?

What you need to know

- How to round numbers to the nearest 10, 100, 1000.
- In real-life problems a rounding must be used which gives a sensible answer.
- How to approximate using **decimal places**.
    1. Write the number using one more decimal place than asked for.
    2. Look at the last decimal place and
        - if the figure is 5 or more round up,
        - if the figure is less than 5 round down.
- How to approximate using **significant figures**.
    When rounding a number to one, two or more significant figures:
    1. Start from the most significant figure and count the required number of figures.
    2. Look at the next figure to the right of this and
        - if the figure is 5 or more round up,
        - if the figure is less than 5 round down.
    3. Add noughts, as necessary, to locate the decimal point and preserve the place value.
- When answering a problem, include any units and state the degree of approximation used.
- Use approximation to **estimate** that the actual answer to a calculation is of the right magnitude.
- How to use a calculator to check answers to calculations.
- You should be able to recognise limitations on the accuracy of data and measurements.

1 Round 8475 to the nearest   (a) 10,   (b) 100,   (c) 1000.

2 (a) The exact number of people who watched the cup final on TV was 6 732 125.
   (i) Copy and complete this newspaper headline with a sensible rounded number.
   "......Watch Cup Final on TV"
   (ii) Copy and complete this sentence. My number is rounded to the nearest .... people.
 (b) **"110 000 Listen To New Radio Station"**
   The number is given to the nearest thousand.
   What is the **smallest** possible number of listeners?   AQA

3 The summit of Mount Everest is at a height of 29 078 feet.
 (a) What is the height of Mount Everest to the nearest ten feet?
 (b) In a newspaper report the height of Mount Everest is given as 30 000 feet.
   To what accuracy has the height been given?   AQA

4 Baljinder's grandmother will be 90 next week.
 Baljinder has made a large birthday cake for a celebration party and wants to put 90 candles
 on it. Her local shop sells birthday cake candles in packets of 12.
 (a) How many packets must Baljinder buy?

 Each packet costs 29 pence.
 (b) How much does Baljinder pay?   AQA

5 Round these numbers to 1 significant figure.
 (a) 72   (b) 138   (c) 754   (d) 650   (e) 78   (f) 987

6 Bob is set this problem.   782 + 32 − 292
 Bob works out the answer as 810.
 (a) Use estimation to show that Bob's answer is wrong.
 (b) Work out the correct answer to Bob's problem.   AQA

7 Rachel plans to buy a house costing £174 015 and a car costing £12 988.
 Rachel wants an estimate of the total she will
 spend on the two items.
 (a) Which two numbers must she add together?

 Rachel estimates that the total cost of these
 two items is £300 000.
 (b) (i) Is she correct?
   (ii) Give a reason for your answer.

    AQA

8 (a) Mohini **estimated**   21 × 29   to be about 600.
   Explain how she did it.
 (b) Show how you would find an **estimate** for   1980 ÷ 43.
   Write down your estimate for   1980 ÷ 43.   AQA

9 Find **estimates** to these calculations by using approximations to one significant figure.
 (a) 86.5 × 1.9   (b) 2016 ÷ 49.8

10 Alun has a part-time job. He is paid £28 each day he works.
 Last year he worked 148 days.
 Estimate Alun's total pay for last year.
 Write down your calculation and answer.   AQA

11 Aimee uses her calculator to multiply 18.7 by 0.96. Her answer is 19.752.
 Without finding the exact value of   18.7 × 0.96,   explain why her answer must be wrong.

12 Angela needs to buy 132 chocolate biscuits for a Christmas party.
The biscuits are sold in packs of nine.
How many packs must Angela buy?

AQA

13 Estimate the value of $\dfrac{399 \times 78}{201}$

14 Jonathan uses his calculator to work out the value of 42.2×0.027.
The answer he gets is 11.394.
Use approximation to show that his answer is wrong.

AQA

15 Clive said that the weight of his apple is 129.625 grams.
Explain why the weight given by Clive is not sensible.

16 The number of people at a beach party is 90, to the nearest 10.
What is the smallest and largest possible number of people at the party?

17 The display shows the result of $179 \div 7$.
What is the result correct to
(a) two decimal places,
(b) one decimal place,
(c) one significant figure?

$$\boxed{25.57142857}$$

18 Jo has to calculate $\dfrac{481 + 97}{32}$
She calculates the answer to be 180.625.
By rounding each number to one significant figure estimate whether her answer is of the right order of magnitude. Show your working.

19 (a) A rectangular lawn measures 27 metres by 38 metres.
A firm charges £9.95 per square metre to turf the lawn.
Estimate the charge for turfing the lawn.
(b) Is your estimate too large or too small?
Give a reason for your answer.

20 What is the smallest possible length of a piece of string which measures 25 cm to the nearest centimetre?

21 A bus is 18 m in length, correct to the nearest metre.
(a) What is the minimum possible length of the bus?
(b) Copy and complete this inequality: …… ⩽ length of bus < ……

22 Use your calculator to find the value of $\dfrac{2.971 + 21.42}{17.12 - 1.469}$
Give your answer to one decimal place.

AQA

23 Use your calculator to find the value of $\dfrac{3.81 \times 17.21}{8.49 + 7.38}$
Give your answer to two decimal places.

AQA

24 Work out $\dfrac{4.7 \times 20.1}{5.6 - 1.8}$
(a) Write down your full calculator display.
(b) Use estimation to check your answer. Show each step of your working.

AQA

25 The floor of a lounge is a rectangle which measures 5.23 m by 3.62 m.
The floor is to be carpeted.
(a) Calculate the area of carpet needed.
Give your answer to an appropriate degree of accuracy.
(b) Explain why you chose this degree of accuracy.

26 A book weighs 3.2 kg, given to the nearest 100 g.
Find the minimum possible weight of the book.

CHAPTER 4

Negative Numbers

In Chapter 1 we used a number line to show whole numbers.
This number line can be extended to include **negative whole numbers**.

Negative whole numbers, zero and positive whole numbers are called **integers**.
−5 can be read as "minus five" or "negative five".
A number written without a sign before it is assumed to be positive. +5 has the same value as 5.
Real-life situations which use negative numbers include temperature, bank accounts and depths below sea level.
Can you think of any other situations where negative numbers are used?

Ordering numbers

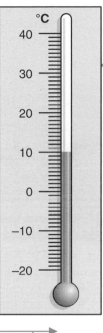

The thermometer

−5°C is colder than −1°C.	2°C is warmer than −3°C.
−3°C is colder than 1°C.	4°C is warmer than −5°C.
−4°C is colder than 0°C.	0°C is warmer than −3°C.
2°C is colder than 4°C.	5°C is warmer than 2°C.

As you move up the thermometer the temperatures become warmer.

As you move down the thermometer the temperatures become colder.

The number line

−5 is less than −1.	2 is more than −3.
−3 is less than 1.	4 is more than −5.
−4 is less than 0.	0 is more than −3.
2 is less than 4.	5 is more than 2.

As you move from left to right along the number line the numbers become bigger.

As you move from right to left along the number line the numbers become smaller.

EXAMPLES

1 List these temperatures from coldest to hottest:
3°C, 5°C, −2°C, 0°C, −4°C.

−4°C, −2°C, 0°C, 3°C, 5°C.

2 List these numbers in ascending order (from lowest to highest):
50, −41, −18, −11, 28, 9.

−41, −18, −11, 9, 28, 50.

1 Copy and complete these sentences using the words 'colder' or 'warmer' as appropriate.
   (a)   −2°C is ...... than −5°C.   (b)   −1°C is ...... than   4°C.
   (c)   2°C is ...... than −4°C.   (d)   −10°C is ...... than −5°C.

2 Copy and complete these sentences using the words 'less' or 'more' as appropriate.
   (a)   −3 is ...... than   2.   (b)   1 is ...... than   −5.
   (c)   −4 is ...... than −1.   (d)   −4 is ...... than −10.

3 At midnight on New Year's Day the temperatures in some cities were as shown:

Edinburgh	−7°C	London	0°C	Moscow	−22°C
New York	−17°C	Rome	3°C	Colombo	21°C
Cairo	15°C				

   (a)   Which city recorded the highest temperature?
   (b)   Which city recorded the lowest temperature?
   (c)   List the temperatures from coldest to hottest.

4 List these temperatures from coldest to hottest.
   (a)   23°C, −28°C, −3°C, 19°C, −13°C.   (b)   −9°C, −11°C, 12°C, 10°C, −7°C, 0°C.
   (c)   27°C, 18°C, −29°C, −15°C, 2°C.   (d)   20°C, −15°C, −20°C, 0°C, −5°C, 10°C.

5 List these numbers from lowest to highest.
   (a)   31, −78, 51, −39, −16, −9, 11.   (b)   5, 1, −1, −3, −5, −2, 0, 2, 4.
   (c)   99, −103, 104, 5, −3, 52, −63, −19.   (d)   30, 10, −30, −50, −20, 0, 40.
   (e)   27, −30, 17, 0, −15, −10, 8.

Subtracting a larger number from a smaller number

Work out $3 - 5$.
To work out smaller number − larger number:
   ● Do the calculation the other way round. $3 - 5$ becomes $5 - 3$.
   ● Put a minus sign in front of the answer. So, $3 - 5 = -2$.

This is the same as starting at 3 on a number line and going 5 places to the left, to get to -2.

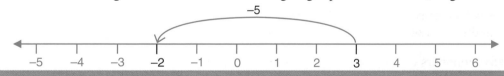

EXAMPLES

1 Work out $7 - 13$.
Do $13 - 7 = 6$.   Then $7 - 13 = -6$.

2 Calculate $21 - 34$.
Do $34 - 21 = 13$.   Then $21 - 34 = -13$.

3 Alec has £50 in his bank account.   He writes a cheque for £80.   What is his new balance?

His new balance is given by the calculation £50 − £80.
$80 - 50 = 30$.
So, $50 - 80 = -30$.
The new balance is −£30.   This means that Alec's account is overdrawn by £30.

1 Use the number line to work out the following.

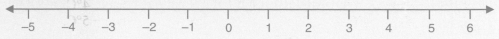

(a) $4 - 3$ (b) $1 - 3$ (c) $2 - 4$ (d) $5 - 9$ (e) $3 - 6$

2 Draw a number line to show each of these statements.
   (a) $7 - 10 = -3$ (b) $4 - 6 = -2$ (c) $3 - 4 = -1$

3 Work out the following.
   (a) $1 - 4$ (b) $3 - 5$ (c) $5 - 8$
   (d) $10 - 12$ (e) $4 - 7$ (f) $8 - 12$
   (g) $13 - 20$ (h) $24 - 36$ (i) $10 - 20$
   (j) $23 - 30$ (k) $29 - 50$ (l) $20 - 21$

4 What number should be put in the box to make each of these statements correct?

   (a) $7 - \boxed{} = -2$ (b) $\boxed{} - 6 = -5$ (c) $9 - \boxed{} = -3$

   (d) $-3 - 7 = \boxed{}$ (e) $\boxed{} - 50 = -20$ (f) $10 - \boxed{} = -5$

5 Mr Armstrong has £25 in the bank.
He writes a cheque for £100.
What is his new balance?

6 The temperature inside a fridge is 6°C above zero.
The temperature inside a freezer is 5°C below zero.
By how many degrees is the temperature inside the freezer below the temperature inside the fridge?

7 At midnight, the temperature in York is 3°C below freezing and in Bath the temperature is 2°C above freezing.
What is the difference in temperature between York and Bath?

8 Brad is 8 cm shorter than Alex and Cath is 9 cm taller than Alex.
By how many centimetres is Cath taller than Brad?

9 Adrian is 5 kg heavier than Tim. Matt is 3 kg lighter than Tim.
What is the difference in weight between Matt and Adrian?

10 Negative numbers can be used for depths below sea level.

Use negative numbers to answer the following.
   (a) At what depth is the diver?
   (b) At what depth is the treasure chest?

What is the difference in height between
   (c) the helicopter and the parachutist,
   (d) the diver and the jellyfish,
   (e) the diver and the treasure chest,
   (f) the bird and the jellyfish,
   (g) the parachutist and the treasure chest,
   (h) the kite and the jellyfish,
   (i) the helicopter and the kite,
   (j) the diver and the helicopter,
   (k) the bird and the treasure chest?

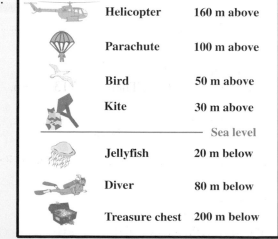

Helicopter	160 m above
Parachute	100 m above
Bird	50 m above
Kite	30 m above
	Sea level
Jellyfish	20 m below
Diver	80 m below
Treasure chest	200 m below

Negative Numbers

Addition and subtraction using negative numbers

Think of the number line as a series of stepping stones.

← -10 -9 -8 -7 -6 -5 -4 -3 -2 -1 0 1 2 3 4 5 6 7 8 9 10 →

What is $-4-5$?

Using the number line: Start at -4 and move 5 to the **left**.
The answer is -9.
$-4-5 = -9$

$-4-5$ can be written as: $-4 + (-5)$ or $-4 - (+5)$.
So, $-4 + (-5) = -9$ and $-4 - (+5) = -9$.

What is $-3 - (+7)$? $-3 - (+7)$ is the same as $-3 - 7$.

Using the number line: Start at -3 and move 7 to the **left**.
$-3 - (+7) = -10$

$-3 - (-7)$ must start at -3 and move 7 to the **right**.
$-3 - (-7)$ is the same as $-3 + 7$.
So, $-3 - (-7) = 4$ and $-3 + 7 = 4$.

What is $-5 + (+7)$?
$+7$ can be written as 7.
$-5 + (+7)$ is the same as $-5 + 7 = 2$.

To add or subtract negative numbers:
Replace double signs with a single sign.
Start on the number line with the first number.
Then move left or right according to the single sign.

> $+ \ +$ can be replaced by $+$
> $- \ -$ can be replaced by $+$
> $+ \ -$ can be replaced by $-$
> $- \ +$ can be replaced by $-$

EXAMPLES

1 Work out $2 + (-6)$.

$+-$ can be replaced with $-$.
Start at 2 and move 6 to the left.
$2 + (-6) = 2 - 6 = -4$

2 Work out $-2 - (-8)$.

$--$ can be replaced with $+$.
Start at -2 and move 8 to the right.
$-2 - (-8) = -2 + 8 = 6$

3 Work out $-4 - (+6)$.

$-+$ can be replaced with $-$.
Start at -4 and move 6 to the left.
$-4 - (+6) = -4 - 6 = -10$

4 Work out $-4 + (-3) + 6 - (-5) - (+3)$.

Replace signs.
$= -4 - 3 + 6 + 5 - 3$
$= 1$

Exercise **4.3**   Do not use a calculator for this exercise.

1 Work out.
(a) $-3 + (+5)$
(b) $5 + (-4)$
(c) $-2 + (-7)$
(d) $-1 + (+9)$
(e) $7 + (-3)$
(f) $15 + (-20)$
(g) $-11 + (+4)$
(h) $11 + (-4)$
(i) $8 + (-7)$
(j) $-8 + (-7)$
(k) $3 + (+3) + (-9)$
(l) $-7 + (-5) + 6$

2 Work out.
(a) $8 - (-5)$
(b) $-4 - (-10)$
(c) $10 - (+3)$
(d) $6 - (-1)$
(e) $-5 - (-10)$
(f) $-4 - (+8)$
(g) $-7 - (-6)$
(h) $7 - (-6)$
(i) $-2 - (+9)$
(j) $2 - (-9)$
(k) $5 - (+5) + 9$
(l) $-10 - (-6) + 4$

3 Work out.

(a) $-3 - (-8)$ (b) $5 + (-2)$ (c) $7 - (+4)$

(d) $-9 - (-5) + (-3)$ (e) $7 + (-8) - (+5)$ (f) $-2 - (-7) - 6$

4 Work out.

(a) $10 + 5 - 8 + 6 - 7$ (b) $12 + 8 - 15 + 7 - 20$ (c) $30 - 20 + 12 - 50$

(d) $6 + 12 - 14 - 4$ (e) $37 - 23 - 24 - 25$ (f) $12 + 13 + 14 - 20$

5

| Edinburgh $-7°C$ Moscow $-22°C$ New York $-17°C$ Rome $3°C$ Cairo $15°C$ |

What is the difference in temperature between

(a) Edinburgh and Rome,   (b) Edinburgh and New York,

(c) Moscow and New York,   (d) Moscow and Cairo?

6 The temperature inside a freezer was $-23°C$.
After two hours the temperature had risen by $8°C$.
What is the temperature in the freezer then?

7 The temperature inside an igloo is $-5°C$.
The temperature outside the igloo is $17°C$ cooler.
What is the temperature outside the igloo?

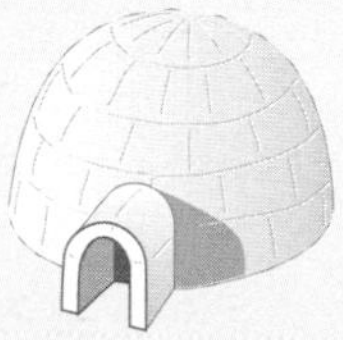

8 The temperature of an iceberg is $-13°C$.
The temperature of the sea is $15°$ warmer than the iceberg.
What is the temperature of the sea?

Multiplying and dividing negative numbers

You will need to know these rules for multiplying and dividing negative numbers:

When multiplying:
$+ \times + = +$
$- \times - = +$
$+ \times - = -$
$- \times + = -$

When dividing:
$+ \div + = +$
$- \div - = +$
$+ \div - = -$
$- \div + = -$

The diagram shows the multiplication table extended to include negative numbers.

Second number

×	−5	−4	−3	−2	−1	0	1	2	3	4	5
−5	25	20	15	10	5	0	−5	−10	−15	−20	−25
−4	20	16	12	8	4	0	−4	−8	−12	−16	−20
−3	15	12	9	6	3	0	−3	−6	−9	−12	−15
−2	10	8	6	4	2	0	−2	−4	−6	−8	−10
−1	5	4	3	2	1	0	−1	−2	−3	−4	−5
0	0	0	0	0	0	0	0	0	0	0	0
1	−5	−4	−3	−2	−1	0	1	2	3	4	5
2	−10	−8	−6	−4	−2	0	2	4	6	8	10
3	−15	−12	−9	−6	−3	0	3	6	9	12	15
4	−20	−16	−12	−8	−4	0	4	8	12	16	20
5	−25	−20	−15	−10	−5	0	5	10	15	20	25

First number

Describe any patterns you can see in the table.

Division is the opposite (inverse) operation to multiplication.

If $a \times b = c$,
then $c \div b = a$
and $c \div a = b$.

If $(+5) \times (-2) = -10$,
then $(-10) \div (-2) = +5$
and $(-10) \div (+5) = -2$.

EXAMPLES

1 Work out $(+7) \times (-5)$.

Signs: $+ \times - = -$
Numbers: $7 \times 5 = 35$
So, $(+7) \times (-5) = -35$.

2 Work out $(-4) \times (-8)$.

Signs: $- \times - = +$
Numbers: $4 \times 8 = 32$
So, $(-4) \times (-8) = 32$.

3 Work out $(+8) \div (-2)$.

Signs: $+ \div - = -$
Numbers: $8 \div 2 = 4$
So, $(+8) \div (-2) = -4$.

4 Work out $(-36) \div (-3)$.

Signs: $- \div - = +$
Numbers: $36 \div 3 = 12$
So, $(-36) \div (-3) = 12$.

> **Work logically:**
> Work out the sign first.
> Then work out the numbers.

Exercise 4.4

Do not use a calculator for this exercise.

1
(a) $(+7) \times (+5)$
(b) $(-7) \times (+5)$
(c) $(-7) \times (-5)$
(d) $5 \times (+2)$
(e) $(+5) \times (-2)$
(f) $(-5) \times (-2)$
(g) $(-1) \times (-1)$
(h) $8 \times (-3)$
(i) $(-8) \times (+3)$
(j) $(-5) \times 9$
(k) $(-8) \times (-8)$
(l) $(-7) \times 6$
(m) $(-7) \times (-6)$
(n) $8 \times (-10)$
(o) $(-8) \times (+10)$
(p) $(-4) \times (-8)$

2
(a) $(+5) \times (-2) \times (+2)$
(b) $(+4) \times (-3) \times (-5)$
(c) $(-3) \times (-2) \times (-5)$
(d) $(-5) \times (+3) \times (-4)$
(e) $(-5) \times (+3) \times (+4)$
(f) $(-5) \times (-4) \times (-5)$

3
(a) $(-8) \div (+2)$
(b) $(-8) \div (-2)$
(c) $(+20) \div (+4)$
(d) $(+20) \div (-4)$
(e) $(-20) \div (+4)$
(f) $(-20) \div (-4)$
(g) $(+18) \div (+3)$
(h) $(-18) \div (+3)$
(i) $(-24) \div (-6)$
(j) $(+24) \div (-3)$
(k) $(-30) \div (-5)$
(l) $(-30) \div (+6)$

4 In a multiple choice test there are 5 marks for a right answer and -3 marks for a wrong answer.

For example: A student has 8 questions right and 17 wrong.
What is his overall mark?
$8 \times 5 + 17 \times -3 = 40 + -51 = 40 - 51 = -11$

In a 25-question test these students have the following right and wrong answers.
They can leave a question out rather than guess a wrong answer.

	Ahmed	Bridget	Chris	Dileep	Evan
Right	10	12	6	7	9
Wrong	8	13	17	18	16

(a) How many marks did each student score?
(b) Put the students in order from first to fifth.

What you need to know

You should be able to:
- Use **negative numbers** in context such as temperatures and bank accounts.
- Realise where negative numbers come on a **number line**.
- Put numbers in order (including negative numbers).
- Add $(+)$, subtract $(-)$, multiply $(\times)$ and divide $(\div)$ with negative numbers.

1 Write these temperatures in order, starting with the coldest.

$-11°C$ $-15°C$ $7°C$ $11°C$ $-21°C$ $-3°C$

AQA

2 The table shows the midday temperatures in these towns one day.

Town	Selby	Poole	Woking
Temperature (°C)	-8	-2	-5

(a) Which town has the highest midday temperature?
(b) Which town has the lowest midday temperature?

3 Calculate.
(a) $-7 - 11$ (b) $-7 + 11$ (c) $-7 - (-11)$

4 Complete the following.
(a) $-4 - \Box = -9$ (b) $6 - \Box = 8$

AQA

5 The temperatures at Athens and Moscow are taken at the same time.

Athens $8°C$ Moscow $-5°C$

(a) How many degrees colder is Moscow than Athens?
(b) At the same time, Oslo is $3°C$ colder than Moscow.
     What is the temperature in Oslo?

6 At midnight the temperature was $2°C$. At 8 am the next day the temperature was $-3°C$.
By how many degrees did the temperature fall?

7 A diver is 6 metres below the surface of the water. A bird is flying 27 metres above the water.
How many metres is the bird above the diver?   AQA

8 Carry out these multiplications.
(a) $(+6) \times (+4)$ (b) $(+6) \times (-4)$ (c) $(-6) \times (+4)$ (d) $(-6) \times (-4)$
(e) 8×5 (f) $(-8) \times 5$ (g) $8 \times (-5)$ (h) $(-8) \times (-5)$

9 Carry out these divisions.
(a) $(+50) \div (-10)$ (b) $(-12) \div (-6)$ (c) $(-18) \div 3$ (d) $24 \div 6$

10 (a) Complete the boxes.

(i) $\Box \times \Box = \boxed{-10}$ (ii) $\Box \div \Box = \boxed{-1}$

(b) Work out $\dfrac{(-1) \times (-7) \times (+8)}{(-2)}$ AQA

11 The temperature inside a fridge is $3°C$. The temperature inside a freezer is $-18°C$.
(a) How much colder is it inside the freezer than inside the fridge?

(b) Calculate $\dfrac{9 \times (-18)}{5} + 32$ to find the temperature of the freezer in degrees Fahrenheit.

AQA

12 Ama and Bob take part in a quiz. The rules are:

> For each correct answer $+4$ points
>
> For each wrong answer -3 points

After the first round Ama has $+6$ points and Bob has $+13$ points.
In the second round each person answers 5 more questions.
Bob scores -1 point in this round. This is added to his previous points.
What is the least number of questions Ama must answer correctly in the second round to have
more points than Bob **in total**?

AQA

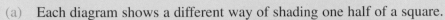

Activity

(a)   Each diagram shows a different way of shading one half of a square.

   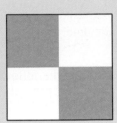

Find some more ways of shading one half of a square.

(b)   Each diagram shows a different way of dividing the square into quarters.

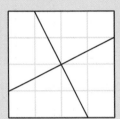 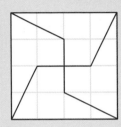 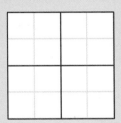 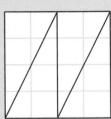

Find some more ways of dividing a square into quarters.

Shaded fractions

What fraction of this rectangle is shaded?

The rectangle is divided into **eight** squares.
The squares are all the same size.
Three of the squares are shaded.

$\frac{3}{8}$ of the rectangle is shaded.

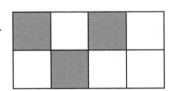

In a fraction:
The top number is called
the **numerator**.
The bottom number
is called the **denominator**.

EXAMPLES

1   What fraction of this rectangle is shaded?

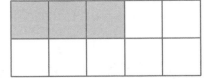

The rectangle is divided into 10 equal parts.
Three of the equal parts are shaded.

So, $\frac{3}{10}$ of the rectangle is shaded.

2   Shade $\frac{2}{5}$ of a rectangle.

Draw a rectangle and divide it into
5 equal parts.
Then shade two of the parts.

① What fraction of each of these rectangles is shaded?

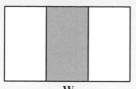 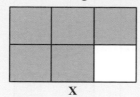 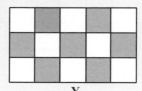 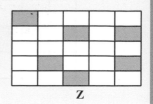

      W   **X**   **Y**   **Z**

② Make two copies of each rectangle.

(a) Shade $\frac{1}{4}$ of rectangle A.

(b) Shade $\frac{3}{4}$ of rectangle A.

(c) Shade $\frac{1}{5}$ of rectangle B.

(d) Shade $\frac{3}{5}$ of rectangle B.

(e) Shade $\frac{1}{2}$ of rectangle C.

(f) Shade $\frac{3}{4}$ of rectangle C.

(g) Shade $\frac{2}{3}$ of rectangle D.

(h) Shade $\frac{6}{9}$ of rectangle D.

(i) Shade $\frac{1}{3}$ of rectangle E.

(j) Shade $\frac{3}{12}$ of rectangle E.

(k) Shade $\frac{3}{7}$ of rectangle F.

(l) Shade $\frac{12}{28}$ of rectangle F.

Rectangle A   **Rectangle B**

  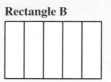

Rectangle C   **Rectangle D**

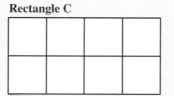 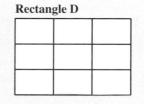

Rectangle E   **Rectangle F**

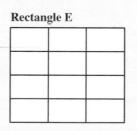 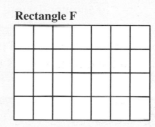

③ What fraction of each of these diagrams is shaded?

(a)   (b)   (c)   (d)

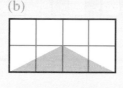 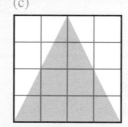 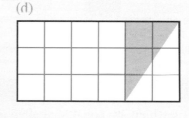

④ What fraction of each of these shapes is shaded?

(a)   (b)   (c)   (d)   (e)

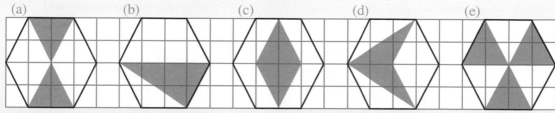

5 Look at these diagrams.

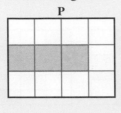 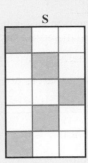

(a) Which diagram has $\frac{3}{12}$ shaded?

(b) Which diagram has $\frac{1}{4}$ shaded?

(c) Which diagram has $\frac{6}{15}$ shaded?

(d) Which diagram has $\frac{2}{5}$ shaded?

(e) Which diagram has $\frac{1}{2}$ shaded?

(f) Which diagram has $\frac{1}{3}$ shaded?

6 You are asked to design a flag.

(a) In one design, $\frac{1}{2}$ of the flag is red and $\frac{1}{3}$ of the flag is blue.

The rest of the flag is white.
(i) On squared paper, draw some 2 by 3 rectangles and design some possible flags.
(ii) What fraction of the flag is white?

(b) In another design, $\frac{2}{3}$ of the flag is red and $\frac{1}{4}$ of the flag is blue.

The rest of the flag is white.
(i) On squared paper, draw some 3 by 4 rectangles and design some possible flags.
(ii) What fraction of the flag is white?

Activity

What fraction of each of these rectangles is shaded?

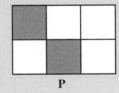 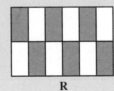 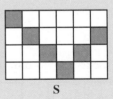

Which rectangles have the same fraction shaded?

Equivalent fractions

Fractions which are equal are called **equivalent fractions**.

Rectangle Q has $\frac{3}{12}$ shaded, $\frac{3}{12} = \frac{1}{4}$.

Rectangle S has $\frac{6}{24}$ shaded, $\frac{6}{24} = \frac{1}{4}$.

Each of the fractions $\frac{1}{4}$, $\frac{3}{12}$, $\frac{6}{24}$,

is the same fraction written in different ways.

These fractions are all equivalent to $\frac{1}{4}$.

Write down two more fractions equivalent to $\frac{1}{4}$.

To write an equivalent fraction:
Multiply the numerator and denominator by the **same** number.

For example. $\frac{1}{4} = \frac{1 \times 3}{4 \times 3} = \frac{3}{12}$

$\frac{1}{4} = \frac{1 \times 6}{4 \times 6} = \frac{6}{24}$

EXAMPLES

1 Write down three fractions equivalent to $\frac{5}{7}$.

The numerators are any multiples of 5. For example: 5, 10, 15, 20, …

The denominators are the same multiples of 7: 7, 14, 21, 28, …

This gives the fractions: $\frac{5}{7}$, $\frac{10}{14}$, $\frac{15}{21}$, $\frac{20}{28}$, …

2 The fraction $\frac{2}{3}$ is equivalent to the fraction $\frac{?}{12}$. Find the value of the unknown numerator.

3 has been multiplied by 4 to get 12.
So, 2 must also be multiplied by 4 to get the unknown numerator.
The unknown numerator is 8.

3 Write the fractions $\frac{3}{4}$, $\frac{2}{5}$ and $\frac{7}{10}$ in ascending order.

Find equivalent fractions for $\frac{3}{4}$, $\frac{2}{5}$ and $\frac{7}{10}$ with the same denominators.

$\frac{3}{4} = \frac{6}{8} = \frac{9}{12} = \frac{12}{16} = \mathbf{\frac{15}{20}}$ $\qquad \frac{2}{5} = \frac{4}{10} = \frac{6}{15} = \mathbf{\frac{8}{20}}$ $\qquad \frac{7}{10} = \mathbf{\frac{14}{20}}$

The fractions in ascending order are: $\frac{2}{5}$, $\frac{7}{10}$, $\frac{3}{4}$.

Simplifying fractions

Fractions can be **simplified** if both the numerator and denominator can be divided by the **same number**.

To write a fraction in its **simplest form** divide both the numerator and denominator by the largest number that divides into them both.

This is sometimes called **cancelling** a fraction.

Remember:
Multiplication and division are inverse (opposite) operations. Equivalent fractions can also be made by dividing the numerator and denominator of a fraction by the same number.

EXAMPLES

1 Write the fraction $\frac{25}{30}$ in its simplest form.

The largest number that divides into both the numerator and denominator of $\frac{25}{30}$ is 5.

$\frac{25}{30} = \frac{25 \div 5}{30 \div 5} = \frac{5}{6}$ $\qquad \frac{25}{30} = \frac{5}{6}$ in its simplest form.

2 In a class of 28 pupils there are 12 boys.
What fraction of the pupils are boys?
Write this fraction in its simplest form.
12 out of the 28 pupils in the class are boys.

The fraction of boys $= \frac{12}{28}$ $\qquad \frac{12}{28} = \frac{12 \div 4}{28 \div 4} = \frac{3}{7}$ $\qquad \frac{12}{28} = \frac{3}{7}$ in its simplest form.

3 Simplify $\frac{24}{30}$.

The largest number that divides into both 24 and 30 is 6.

$\frac{24}{30} = \frac{24 \div 6}{30 \div 6} = \frac{4}{5}$

4 Write 42 as a fraction of 70.
Give your answer in its simplest form.

42 as a fraction of 70 is $\frac{42}{70}$. 2 divides into both 42 and 70. $\frac{42}{70} = \frac{42 \div 2}{70 \div 2} = \frac{21}{35}$

7 divides into both 21 and 35. $\frac{21}{35} = \frac{21 \div 7}{35 \div 7} = \frac{3}{5}$

$\frac{42}{70} = \frac{3}{5}$ in its simplest form.

Do not use a calculator.

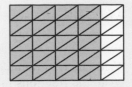

1 Write three equivalent fractions for the shaded part of this rectangle.
What is the simplest form of the shaded fraction?

2 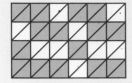   Write three equivalent fractions for the shaded part of this rectangle.
What is the simplest form of the fraction for the shaded part?

3 The diagrams show that: $\frac{1}{2} = \frac{2}{4} = \frac{3}{6}$

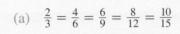

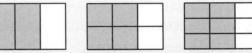

Copy the diagrams.
Add two more diagrams to show that: $\frac{1}{2} = \frac{2}{4} = \frac{3}{6} = \frac{4}{8} = \frac{5}{10}$

4 Copy and draw more diagrams to show that:

(a) $\frac{2}{3} = \frac{4}{6} = \frac{6}{9} = \frac{8}{12} = \frac{10}{15}$

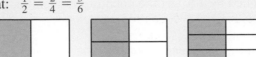

(b) $\frac{3}{4} = \frac{6}{8} = \frac{9}{12} = \frac{12}{16} = \frac{15}{20}$

(c) $\frac{5}{6} = \frac{10}{12} = \frac{15}{18} = \frac{20}{24} = \frac{25}{30}$

5 The diagram shows a 6 by 4 rectangle with a fraction of the rectangle shaded.

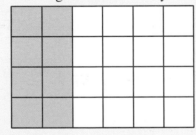

(a) How many $\frac{1}{3}$'s are shaded?

(b) How many $\frac{1}{6}$'s are shaded?

(c) How many $\frac{1}{24}$'s are shaded?

6 Write down three fractions equivalent to:
(a) $\frac{1}{3}$ (b) $\frac{2}{9}$ (c) $\frac{5}{8}$ (d) $\frac{4}{5}$ (e) $\frac{3}{10}$ (f) $\frac{7}{12}$

7 Each of these pairs of fractions are equivalent.
In each case find the value of the unknown numerator.
(a) $\frac{1}{3}$ and $\frac{?}{6}$ (b) $\frac{?}{8}$ and $\frac{6}{16}$ (c) $\frac{?}{4}$ and $\frac{12}{16}$

8 Each of these pairs of fractions are equivalent.
In each case find the value of the unknown denominator.
(a) $\frac{5}{?}$ and $\frac{15}{18}$ (b) $\frac{24}{64}$ and $\frac{3}{?}$ (c) $\frac{7}{?}$ and $\frac{56}{96}$

9 Write these fractions in ascending order. $\frac{5}{8}$ $\frac{3}{4}$ $\frac{7}{16}$

10 Write these fractions in descending order. $\frac{2}{3}$ $\frac{3}{5}$ $\frac{7}{10}$ $\frac{8}{15}$

11 Which of these fractions is the largest? $\frac{4}{5}$ $\frac{13}{20}$ $\frac{7}{10}$ $\frac{3}{4}$

12 Write each of these fractions in its simplest form.

(a) $\frac{6}{8}$ (b) $\frac{12}{15}$ (c) $\frac{18}{27}$ (d) $\frac{22}{99}$ (e) $\frac{50}{75}$ (f) $\frac{16}{40}$ (g) $\frac{12}{50}$ (h) $\frac{52}{65}$

13 Write the first number as a fraction of the second.
Write the fractions in their simplest form.

(a) 4, 20 (b) 3, 12 (c) 8, 12 (d) 24, 60 (e) 60, 105

Questions 14 to 18. Give your answers as fractions in their simplest form.

14 In a class of 32 there are 4 left-handed students.
What fraction of the students are left-handed?

15 A box of 50 chocolates includes 30 soft-centred chocolates.
What fraction of the chocolates are soft-centred?

16 In each hour a television channel shows:
        programmes for 48 minutes and adverts for 12 minutes.
For what fraction of an hour are
(a) programmes shown, (b) adverts shown?

17 Mr Jones plans a car journey.
(a) The journey is 50 km long. Mr Jones plans to stop after 35 km.
    What fraction of the total distance is this?
(b) The journey takes 60 minutes which includes a 12-minute stop.
    For what fraction of the total time does Mr Jones stop on his journey?

18 A group of students were asked some questions about how they travelled to school.
$\frac{1}{2}$ of the students said they walked. $\frac{1}{3}$ of the students said they travelled by bus.
(a) The rest of the group came by car.
    What fraction of the group came by car?
(b) In the group there were more than 20 students and less than 30.
    How many students were in the group?

Types of fractions

This diagram shows that when 5 cakes are shared equally between 2 people they get $2\frac{1}{2}$ cakes each.

This diagram shows that when 5 cakes are shared equally among 4 people they get $1\frac{1}{4}$ cakes each.

Numbers like $2\frac{1}{2}$ and $1\frac{1}{4}$ are called **mixed numbers** because they are a mixture of whole numbers and fractions.
Mixed numbers can be written as **improper** or '**top heavy**' fractions.
These are fractions where the numerator is larger than the denominator.

EXAMPLES

1 Write $3\frac{4}{7}$ as an improper fraction.

$$3\frac{4}{7} = \frac{(3 \times 7) + 4}{7} = \frac{21 + 4}{7} = \frac{25}{7}$$

2 Write $\frac{32}{5}$ as a mixed number.

$32 \div 5 = 6$ remainder 2.
$\frac{32}{5} = 6\frac{2}{5}$

Finding fractions of quantities

1 Find $\frac{2}{5}$ of £65.

Divide £65 into 5 equal parts.
$$£65 \div 5 = £13.$$

13	13	13	13	13

Each of these parts is $\frac{1}{5}$ of £65.

Two of these parts is $\frac{2}{5}$ of £65.

13	13	13	13	13

So, $\frac{2}{5}$ of £65 = 2 × £13 = £26.

2 A coat costing £138 is reduced by $\frac{1}{3}$.
What is the reduced price of the coat?

Find $\frac{1}{3}$ of £138.

$\frac{1}{3}$ of £138 = £138 ÷ 3 = £46

So, reduced price = £138 − £46
                  = £92

Exercise 5.3   Do not use a calculator in questions 1 to 3.

1 Change the following improper fractions to mixed numbers:

(a) $\frac{13}{10}$ (b) $\frac{3}{2}$ (c) $\frac{17}{8}$ (d) $\frac{15}{4}$ (e) $\frac{23}{5}$ (f) $\frac{34}{7}$ (g) $\frac{7}{2}$ (h) $\frac{11}{3}$ (i) $\frac{16}{9}$

2 Change the following mixed numbers to improper fractions:

(a) $2\frac{7}{10}$ (b) $1\frac{3}{5}$ (c) $5\frac{5}{6}$ (d) $3\frac{3}{20}$ (e) $4\frac{5}{9}$ (f) $7\frac{4}{7}$ (g) $3\frac{1}{4}$ (h) $4\frac{2}{3}$ (i) $2\frac{3}{8}$

3 Calculate:

(a) $\frac{1}{4}$ of 12   (b) $\frac{1}{5}$ of 20   (c) $\frac{1}{10}$ of 30   (d) $\frac{1}{6}$ of 48   (e) $\frac{2}{5}$ of 20

(f) $\frac{3}{10}$ of 30   (g) $\frac{2}{7}$ of 42   (h) $\frac{5}{9}$ of 36   (i) $\frac{5}{6}$ of 48   (j) $\frac{3}{8}$ of 32

4 Richard has 30 marbles. He gives $\frac{1}{5}$ of them away.

(a) How many marbles does he give away?
(b) How many marbles has he got left?

5 Stella has collected 48 tokens. She needs $\frac{1}{6}$ more to claim a prize.
What is the total number of tokens she needs to claim a prize?

6 Aisha has 36 balloons. She sells $\frac{2}{9}$ of them.
How many balloons has she got left?

7 Alfie collects £12.50 for charity. He gives $\frac{3}{5}$ of it to Oxfam.
How much does he give to other charities?

8 Ken has saved £5.60. Paula has saved $\frac{3}{8}$ more than Ken.
How much has Paula saved?

9 In a sale all prices are reduced by $\frac{3}{10}$.
What is the sale price of a microwave which was originally priced at £212?

10 Lauren and Amelia share a bar of chocolate. The chocolate bar has 24 squares.
Lauren eats $\frac{3}{8}$ of the bar. Amelia eats $\frac{5}{12}$ of the bar.

(a) How many squares has Lauren eaten?
(b) How many squares has Amelia eaten?
(c) What fraction of the bar is left?

There are 12 sweets in a packet.   Alena eats $\frac{2}{3}$ of the sweets and Sead eats $\frac{1}{4}$ of the sweets.
What fraction of the packet of sweets have they eaten altogether?

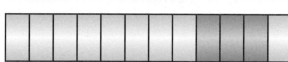

Alena eats $\frac{2}{3}$ of the sweets in the packet.

$$\frac{2}{3} \text{ of } 12 = 8 \qquad \frac{2}{3} = \frac{8}{12}$$

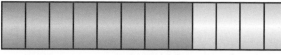

Sead eats $\frac{1}{4}$ of the sweets in the packet.

$$\frac{1}{4} \text{ of } 12 = 3 \qquad \frac{1}{4} = \frac{3}{12}$$

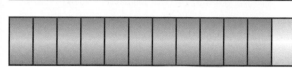

Together Alena and Sead eat $\frac{2}{3} + \frac{1}{4}$ of the packet.

$$\frac{2}{3} + \frac{1}{4} = \frac{8}{12} + \frac{3}{12} = \frac{11}{12}$$

Together Alena and Sead eat $\frac{11}{12}$ of the packet.

How to add (and subtract) fractions

Calculate $1\frac{3}{4} + \frac{5}{6}$

Change mixed numbers to improper ('top heavy') fractions. $1\frac{3}{4} = \frac{7}{4}$

The calculation then becomes $\frac{7}{4} + \frac{5}{6}$

Find the smallest number into which both 4 and 6 will divide.

4 divides into: 4,   8,   12,   16,   …
6 divides into: 6,   12,   18,   …
So, 12 is the smallest.

> Fractions must have the **same denominator** before addition (or subtraction) can take place.
>
> *What happens when you use a denominator that is **not** the smallest?*

Change the original fractions to equivalent fractions using the smallest number as the new denominator. $\frac{7}{4} = \frac{21}{12}$ and $\frac{5}{6} = \frac{10}{12}$

Add the new numerators.
Keep the new denominator the same. $\frac{21}{12} + \frac{10}{12} = \frac{21 + 10}{12} = \frac{31}{12}$

Write the answer in its simplest form. $\frac{31}{12} = 2\frac{7}{12}$

EXAMPLES

1   Work out $\frac{5}{8} - \frac{7}{12}$.

8 divides into: 8,   16,   24,   …   and   12 divides into: 12,   24,   …
24 is the smallest number into which both 8 and 12 divide.

$$\frac{5}{8} = \frac{5 \times 3}{8 \times 3} = \frac{15}{24} \quad \text{and} \quad \frac{7}{12} = \frac{7 \times 2}{12 \times 2} = \frac{14}{24}$$

$$\frac{5}{8} - \frac{7}{12} = \frac{15}{24} - \frac{14}{24} = \frac{1}{24}$$

2   Work out $\frac{3}{4} + \frac{2}{3}$.

$$\frac{3}{4} + \frac{2}{3} = \frac{9}{12} + \frac{8}{12} = \frac{17}{12} = 1\frac{5}{12}$$

> **Remember:**
> ● Add the numerators only.
> ● When the answer is an improper fraction change it into a mixed number.

Fractions . . . . Fractions . . . . Fractions . . . . Fractions

Do not use a calculator in this exercise.

1 Work out:

(a) $\frac{1}{4} + \frac{1}{8}$ (b) $\frac{1}{3} + \frac{1}{4}$ (c) $\frac{1}{2} + \frac{1}{5}$ (d) $\frac{1}{3} + \frac{1}{5}$ (e) $\frac{1}{2} + \frac{1}{7}$

2 Work out:

(a) $\frac{1}{4} - \frac{1}{8}$ (b) $\frac{1}{3} - \frac{1}{4}$ (c) $\frac{1}{2} - \frac{1}{5}$ (d) $\frac{1}{3} - \frac{1}{5}$ (e) $\frac{1}{2} - \frac{1}{7}$

3 Work out:

(a) $\frac{1}{2} + \frac{3}{4}$ (b) $\frac{2}{3} + \frac{5}{6}$ (c) $\frac{3}{4} + \frac{4}{5}$ (d) $\frac{5}{7} + \frac{2}{3}$ (e) $\frac{3}{8} + \frac{5}{6}$

4 Calculate:

(a) $\frac{5}{8} - \frac{1}{2}$ (b) $\frac{13}{15} - \frac{1}{3}$ (c) $\frac{5}{6} - \frac{5}{24}$ (d) $\frac{7}{15} - \frac{2}{5}$ (e) $\frac{3}{4} - \frac{5}{12}$

5 Calculate:

(a) $2\frac{3}{4} + 1\frac{1}{2}$ (b) $1\frac{1}{2} + 2\frac{1}{3}$ (c) $1\frac{3}{4} + 2\frac{5}{8}$ (d) $2\frac{1}{4} + 3\frac{3}{5}$ (e) $4\frac{3}{5} + 1\frac{5}{6}$

6 Calculate:

(a) $2\frac{1}{2} - 1\frac{2}{5}$ (b) $1\frac{2}{3} - 1\frac{1}{4}$ (c) $3\frac{3}{4} - 2\frac{3}{8}$ (d) $5\frac{2}{5} - 2\frac{1}{10}$ (e) $4\frac{5}{12} - 2\frac{1}{6}$

7 Calculate:

(a) $\frac{1}{4} - \frac{1}{5}$ (b) $\frac{2}{5} + \frac{3}{4}$ (c) $\frac{7}{8} - \frac{2}{3}$ (d) $3\frac{3}{10} + 2\frac{3}{20}$ (e) $3\frac{3}{8} - 2\frac{5}{16}$

8 Colin buys a bag of flour.

He uses $\frac{1}{3}$ to bake a cake and $\frac{1}{2}$ to make a loaf.

(a) What fraction of the bag of flour has he used?

(b) What fraction of the bag of flour is left?

9 Kathryn and Matt share a bottle of cola.

Kathryn drinks $\frac{1}{4}$ of the cola.   Matt drinks $\frac{1}{5}$ of the cola.

What fraction of the bottle of cola is left?

10 Both Lee and Mary have a packet of the same sweets.

Mary eats $\frac{1}{3}$ of her packet.   Lee eats $\frac{3}{4}$ of his packet.

(a) Find the difference between the fraction Mary eats and the fraction Lee eats.

Lee gives his remaining sweets to Mary.

(b) What fraction of a packet does Mary now have?

11 Jon, Billy and Cathy are the only candidates in a school election.

Jon got $\frac{7}{20}$ of the votes.   Billy got $\frac{2}{5}$ of the votes.

(a) What fraction of the votes did Cathy get?

(b) Which candidate won the election?

12 A school has pupils in Years 7 to 13.

$\frac{7}{12}$ of its pupils are in Years 7 to 9 and $\frac{3}{10}$ of its pupils are in Years 10 and 11.

What fraction of the pupils in the school are in Years 12 and 13?

13 A bag of sweets contains chocolates, toffees and mints.

$\frac{2}{5}$ are chocolates and there are an equal number of toffees and mints.

What fraction of the sweets are toffees?

Multiplying fractions

How to multiply fractions

Calculate $\frac{3}{8} \times \frac{1}{9}$

Simplify, where possible, by cancelling. $\frac{\overset{1}{\cancel{3}}}{8} \times \frac{1}{\underset{3}{\cancel{9}}}$

Multiply the numerators.
Multiply the denominators. $\frac{1 \times 1}{8 \times 3} = \frac{1}{24}$

Write the answer in its simplest form. $\frac{3}{8} \times \frac{1}{9} = \frac{1}{24}$

To simplify:
Divide a numerator **and** a denominator by the **same number**.

In this case:
3 and 9 can be divided by 3.
$3 \div 3 = 1$ and $9 \div 3 = 3$.

EXAMPLES

1 Calculate $\frac{3}{8} \times 12$.

$\frac{3}{8} \times \frac{12}{1}$

Simplify by cancelling.

$= \frac{3}{\underset{2}{\cancel{8}}} \times \frac{\overset{3}{\cancel{12}}}{1}$

Any whole number can be written as a fraction with denominator 1.

$12 = \frac{12}{1}$

Multiply out.

$= \frac{3 \times 3}{2 \times 1} = \frac{9}{2}$

Write the answer in its simplest form.

$\frac{9}{2} = 4\frac{1}{2}$

2 Work out $\frac{5}{6} \times \frac{3}{4}$

$\frac{5}{6} \times \frac{3}{4}$

Simplify by cancelling.

$= \frac{5}{\underset{2}{\cancel{6}}} \times \frac{\overset{1}{\cancel{3}}}{4}$

Multiply out.

$= \frac{5 \times 1}{2 \times 4} = \frac{5}{8}$

Exercise 5.5

Do not use a calculator in this exercise.

1 Work out. Give your answers as mixed numbers.
(a) $\frac{1}{2} \times 7$ (b) $\frac{1}{3} \times 8$ (c) $\frac{3}{5} \times 3$ (d) $\frac{5}{8} \times 10$ (e) $\frac{6}{7} \times 8$
(f) $\frac{2}{3} \times 9$ (g) $\frac{3}{5} \times 10$ (h) $\frac{3}{4} \times 12$ (i) $\frac{3}{8} \times 12$ (j) $\frac{7}{10} \times 15$

2 Work out:
(a) $\frac{1}{2} \times \frac{1}{3}$ (b) $\frac{1}{4} \times \frac{1}{5}$ (c) $\frac{3}{5} \times \frac{1}{6}$ (d) $\frac{5}{7} \times \frac{1}{3}$ (e) $\frac{2}{3} \times \frac{1}{4}$
(f) $\frac{1}{2} \times \frac{3}{4}$ (g) $\frac{1}{4} \times \frac{2}{5}$ (h) $\frac{2}{3} \times \frac{1}{2}$ (i) $\frac{1}{10} \times \frac{5}{8}$ (j) $\frac{3}{10} \times \frac{1}{6}$

3 Calculate:
(a) $\frac{2}{3} \times \frac{3}{4}$ (b) $\frac{3}{4} \times \frac{2}{5}$ (c) $\frac{2}{5} \times \frac{5}{6}$ (d) $\frac{2}{3} \times \frac{1}{2}$ (e) $\frac{3}{10} \times \frac{5}{8}$
(f) $\frac{2}{5} \times \frac{5}{7}$ (g) $\frac{3}{4} \times \frac{2}{3}$ (h) $\frac{3}{10} \times \frac{5}{6}$ (i) $\frac{4}{5} \times \frac{3}{8}$ (j) $\frac{7}{12} \times \frac{4}{5}$

4 Calculate:
(a) $\frac{4}{5} \times 20$ (b) $\frac{3}{4} \times 7$ (c) $\frac{1}{3} \times \frac{1}{4}$ (d) $\frac{2}{3} \times \frac{1}{8}$ (e) $\frac{3}{10} \times \frac{5}{9}$

5 Work out:
(a) $1\frac{1}{2} \times 5$ (b) $1\frac{1}{4} \times 6$ (c) $1\frac{1}{5} \times 4$ (d) $3\frac{2}{3} \times 2$ (e) $2\frac{3}{5} \times 3$

Fractions . . . . Fractions . . . . Fractions . . . .

6 A packet of butter weighs 250 g.
Adrian uses $\frac{2}{5}$ of the packet to make sandwiches.
How many grams did he use?

7 Bradley needs $\frac{3}{8}$ of a kilogram of flour to make one cake.
How many kilograms of flour does he need to make 4 cakes?

8 A class has 30 pupils.
$\frac{2}{5}$ of the pupils are boys and $\frac{1}{4}$ of the boys wear glasses.
How many boys in the class wear glasses?

9 Spencer has $\frac{1}{4}$ of a pint of milk.   He uses $\frac{1}{2}$ of the milk.
(a)   What fraction of a pint does he use?
(b)   What fraction of a pint is left?

10 Kylie has $\frac{2}{3}$ of a litre of orange.   She drinks $\frac{2}{5}$ of the orange.
(a)   What fraction of a litre does she drink?
(b)   What fraction of a litre is left?

11 Tony eats $\frac{1}{5}$ of a bag of sweets.
He shares the remaining sweets equally among Bob, Jo and David.
(a)   What fraction of the bag of sweets does Bob get?
(b)   What is the smallest possible number of sweets in the bag?

Dividing fractions

How to divide fractions

The method normally used when one fraction is divided by another is to change the division to a multiplication. The fractions can then be multiplied in the usual way.

$$\text{Calculate} \quad \frac{7}{15} \div \frac{1}{5}$$

Change the division to a multiplication. $\quad \frac{7}{15} \div \frac{1}{5} = \frac{7}{15} \times \frac{5}{1}$

Simplify, where possible, by cancelling. $\quad \frac{7}{\overset{}{\underset{3}{15}}} \times \frac{\overset{1}{5}}{1}$

Multiply the numerators.
Multiply the denominators. $\quad \frac{7 \times 1}{3 \times 1} = \frac{7}{3}$

Write the answer in its simplest form. $\quad \frac{7}{3} = 2\frac{1}{3}$

EXAMPLES

1   Work out $\frac{2}{3} \div 5$.

$\frac{2}{3} \div 5$

$= \frac{2}{3} \times \frac{1}{5}$

$= \frac{2}{15}$

> Divide by 5 is the same as multiply by $\frac{1}{5}$.

2   Calculate $\frac{2}{5} \div \frac{4}{9}$.

$\frac{2}{5} \div \frac{4}{9}$

$= \frac{2}{5} \times \frac{9}{4}$

> Divide by $\frac{4}{9}$ is the same as multiply by $\frac{9}{4}$.

$= \frac{\overset{1}{2}}{5} \times \frac{9}{\underset{2}{4}}$

$= \frac{9}{10}$

Exercise 5.6   Do not use a calculator in this exercise.

1 Work out. Give your answers in their simplest form.

(a) $\frac{1}{2} \div 5$ (b) $\frac{1}{3} \div 2$ (c) $\frac{1}{5} \div 4$ (d) $\frac{4}{5} \div 2$ (e) $\frac{9}{10} \div 3$

(f) $\frac{3}{4} \div 2$ (g) $\frac{2}{3} \div 2$ (h) $\frac{2}{5} \div 4$ (i) $\frac{3}{4} \div 6$ (j) $\frac{6}{7} \div 3$

2 Work out:

(a) $\frac{1}{2} \div \frac{1}{4}$ (b) $\frac{1}{5} \div \frac{1}{2}$ (c) $\frac{1}{4} \div \frac{1}{2}$ (d) $\frac{1}{8} \div \frac{1}{4}$ (e) $\frac{1}{10} \div \frac{1}{4}$

(f) $\frac{3}{4} \div \frac{1}{2}$ (g) $\frac{2}{9} \div \frac{1}{3}$ (h) $\frac{7}{8} \div \frac{1}{3}$ (i) $\frac{2}{3} \div \frac{1}{5}$ (j) $\frac{3}{4} \div \frac{1}{8}$

3 Calculate:

(a) $\frac{2}{3} \div \frac{4}{5}$ (b) $\frac{3}{8} \div \frac{2}{3}$ (c) $\frac{3}{5} \div \frac{3}{4}$ (d) $\frac{2}{5} \div \frac{3}{10}$ (e) $\frac{3}{8} \div \frac{9}{16}$

(f) $\frac{7}{12} \div \frac{7}{18}$ (g) $\frac{4}{9} \div \frac{2}{3}$ (h) $\frac{7}{10} \div \frac{3}{5}$ (i) $\frac{9}{20} \div \frac{3}{10}$ (j) $\frac{21}{25} \div \frac{7}{15}$

4 Calculate:

(a) $\frac{1}{5} \div 3$ (b) $\frac{1}{5} \div \frac{1}{4}$ (c) $\frac{3}{5} \div 3$ (d) $\frac{3}{5} \div \frac{1}{4}$ (e) $\frac{3}{5} \div \frac{3}{4}$

5 One-fifth of a pint of milk is used to make one cup of coffee.
How many cups of coffee can be made with 4 pints of milk?

6 Neil uses $\frac{1}{2}$ of a block of paté to make 5 sandwiches.
What fraction of the block of paté does he put on each sandwich?

7 Lauren uses $\frac{2}{3}$ of a bag of flour to make 6 muffins.
What fraction of the bag of flour is used for each muffin?

8 (a) A shelf is three-quarters of a metre in length.
How many books of width $\frac{3}{4}$ cm can stand on the shelf?

(b) Another shelf is half the length and the books on the shelf are twice the width.
How many books can stand on this shelf?

Fractions on a calculator

Fraction calculations can be done quickly using the fraction button on a calculator.

On most calculators the fraction button looks like this … $\boxed{a^b/_c}$

EXAMPLES

1 Use a calculator to work out $1\frac{4}{5} + \frac{2}{3}$.

This can be calculated with this calculator sequence.

$\boxed{1}\ \boxed{a^b/_c}\ \boxed{4}\ \boxed{a^b/_c}\ \boxed{5}\ \boxed{+}\ \boxed{2}\ \boxed{a^b/_c}\ \boxed{3}\ \boxed{=}$

This gives the answer $2\frac{7}{15}$.

2 Calculate $\frac{7}{12}$ of 32.

This can be calculated with this calculator sequence.

$\boxed{7}\ \boxed{a^b/_c}\ \boxed{1}\ \boxed{2}\ \boxed{\times}\ \boxed{3}\ \boxed{2}\ \boxed{=}$

This gives the answer $18\frac{2}{3}$.

Use a calculator to check your answers to some of the questions in Exercises 5.4, 5.5 and 5.6.

All fractions can be written as decimals and vice versa.

Changing decimals to fractions

$0.7 = \frac{7}{10}$ $0.03 = \frac{3}{100}$ $0.009 = \frac{9}{1000}$

0.35 can be written as a fraction.
Using place value:

$0.35 = \frac{3}{10} + \frac{5}{100} = \frac{30}{100} + \frac{5}{100} = \frac{35}{100}$

This can be written as a fraction in its simplest form.

$\frac{35}{100} = \frac{35 \div 5}{100 \div 5} = \frac{7}{20}$

$0.35 = \frac{35}{100} = \frac{7}{20}$

Write equivalent fractions with denominator 100.

$$\frac{3}{10} = \frac{3 \times 10}{10 \times 10} = \frac{30}{100}$$

EXAMPLES

Change the following decimals to fractions in their simplest form.

1 0.02 $0.02 = \frac{2}{100} = \frac{2 \div 2}{100 \div 2} = \frac{1}{50}$

2 0.225 $0.225 = \frac{225}{1000} = \frac{225 \div 25}{1000 \div 25} = \frac{9}{40}$

Changing fractions to decimals

EXAMPLES

Change the following fractions to decimals.

1 $\frac{1}{5} = 1 \div 5 = 0.2$

2 $\frac{11}{20} = 11 \div 20 = 0.55$

$\frac{1}{5}$ means $1 \div 5$.

$1 \div 5$ can be worked out using:
short division, long division or a calculator.

Remember:

$11 \div 20 = 11.00 \div 20$

$$20)\overline{1\ 1.0^{11}0^{10}0}^{\ 0.5\ 5}$$

Recurring decimals

Some decimals have recurring digits.
These are shown by:
    a single dot above a single recurring digit,
    a dot above the first digit and the last digit of a set of recurring digits.

For example:

$\frac{1}{3} = 0.3333333\ldots$ $= 0.\dot{3}$

$\frac{123}{999} = 0.123123123\ldots$ $= 0.\dot{1}2\dot{3}$

$\frac{41}{70} = 0.5857142857142\ldots = 0.5\dot{8}5714\dot{2}$

$\frac{3}{11} = 0.27272727\ldots$ $= 0.\dot{2}\dot{7}$

Exercise **5.7**   Do not use a calculator for questions 1 to 4.

(5)

1 Match the decimals in Box A with the fractions in Box B.

Box A				
0.5	0.2	0.75	0.7	0.01

Box B				
$\frac{3}{4}$	$\frac{1}{2}$	$\frac{1}{100}$	$\frac{1}{5}$	$\frac{7}{10}$

2 Change the following decimals to fractions in their simplest form.

(a) 0.12   (b) 0.6   (c) 0.32   (d) 0.175   (e) 0.45   (f) 0.65
(g) 0.22   (h) 0.202   (i) 0.28   (j) 0.555   (k) 0.625   (l) 0.84

3 Change the following fractions to decimals.

(a) (i) $\frac{1}{4}$ (ii) $\frac{1}{2}$ (iii) $\frac{3}{4}$

(b) (i) $\frac{1}{10}$ (ii) $\frac{3}{10}$ (iii) $\frac{7}{10}$

(c) (i) $\frac{2}{5}$ (ii) $\frac{3}{5}$ (iii) $\frac{4}{5}$

4 Change the following fractions to decimals.

(a) (i) $\frac{3}{20}$ (ii) $\frac{7}{20}$ (iii) $\frac{19}{20}$

(b) (i) $\frac{4}{25}$ (ii) $\frac{9}{25}$ (iii) $\frac{23}{25}$

(c) (i) $\frac{7}{100}$ (ii) $\frac{23}{100}$ (iii) $\frac{106}{200}$

5 Change these fractions to decimals.

(a) $\frac{1}{8}$ (b) $\frac{5}{8}$ (c) $\frac{9}{40}$ (d) $\frac{29}{40}$

6 Write these decimals using dots to represent recurring digits.

(a) 0.77777...   (b) 0.363636...   (c) 0.135135...   (d) 0.166666...

7 Write each of these fractions as recurring decimals, using dots to represent recurring digits.

(a) $\frac{2}{3}$ (b) $\frac{4}{9}$ (c) $\frac{5}{6}$ (d) $\frac{8}{11}$ (e) $\frac{4}{15}$

8 Change these fractions to decimals. Give your answers correct to two decimal places.

(a) $\frac{1}{3}$ (b) $\frac{1}{6}$ (c) $\frac{3}{7}$ (d) $\frac{5}{11}$ (e) $\frac{7}{9}$

What you need to know

- The top number of a fraction is called the **numerator**, the bottom number is called the **denominator**.
- To write **equivalent fractions**, the numerator and denominator of a fraction are multiplied (or divided) by the **same** number.   e.g. $\frac{3}{8} = \frac{3 \times 4}{8 \times 4} = \frac{12}{32}$
- In its **simplest form**, the numerator and denominator of a fraction have no common factor, other than 1.
- $2\frac{1}{2}$ is an example of a **mixed number**. It is a mixture of whole numbers and fractions.
- $\frac{5}{2}$ is an **improper** (or 'top heavy') fraction.
- Fractions must have the **same denominator** before **adding** or **subtracting**.
- All fractions can be written as decimals.
  Some decimals have **recurring digits**.
  These are shown by:
    a single dot above a single recurring digit,   e.g. $\frac{2}{3} = 0.6666... = 0.\dot{6}$
    a dot above the first digit and
    the last digit of a set of recurring digits,   e.g. $\frac{5}{11} = 0.454545... = 0.\dot{4}\dot{5}$

Do not use a calculator for questions 1 to 11.

1 (a) What fraction of the rectangle is shaded? (b) Copy and shade $\frac{1}{3}$ of this rectangle.

AQA

2 Which of these two fractions is the bigger? $\frac{3}{4}$ or $\frac{2}{3}$ Give a reason for your answer. AQA

3 Which of these fractions is equivalent to two-fifths? $\frac{3}{9}$ $\frac{4}{12}$ $\frac{8}{20}$ $\frac{12}{25}$

4 (a) Rachael has a bar of chocolate. It has 20 pieces.
Rachael breaks off five pieces.
What fraction of the bar is this?

(b) Sally also has a bar of chocolate with 20 pieces.
She has eaten $\frac{1}{5}$ of it. How many pieces has she eaten? AQA

5 Rob has 40 sunbeds for hire on a beach in Portugal.
On Tuesday he rents out $\frac{3}{5}$ of his sunbeds.

(a) How many sunbeds does he rent out?

On Wednesday he rents out 30 sunbeds.
(b) Express this as a fraction of the total, giving your fraction in its simplest form. AQA

6 Brenda's dog eats $\frac{5}{8}$ of a tin of dog food each day.
How many tins of food does the dog eat each week?

7 Work out. (a) $2\frac{1}{4} + 1\frac{2}{3}$ (b) $\frac{2}{5} - \frac{1}{8}$ (c) $\frac{3}{5} \times \frac{5}{7}$ (d) $\frac{5}{8} \div \frac{3}{4}$

8

All types of cabins for hire			
Season	Summer	Spring	Winter
Price	Summer Price	Pay $\frac{3}{4}$ of Summer Price	Discount $\frac{2}{3}$ off Summer Price

(a) The Summer price of a Standard Cabin is £360.
How much do you pay for a Standard Cabin in Spring?
(b) The Summer price of a deluxe cabin is £438.
How much do you pay for a deluxe Cabin in Winter? AQA

9 A necklace is made from 60 beads.
$\frac{3}{10}$ of the beads are red. $\frac{9}{20}$ of the beads are blue. The rest of the beads are white.
What fraction of the beads are white? Give this fraction in its simplest form.

10 In a school $\frac{8}{15}$ of the pupils are girls. $\frac{3}{16}$ of the girls are left-handed.
What fraction of the pupils in the school are left-handed girls?

11 36 girls and 24 boys applied to go on a rock climbing course.
$\frac{2}{3}$ of the girls and $\frac{3}{4}$ of the boys went on the course.
What fraction of the 60 students who applied went on the course?
Write the fraction in its simplest form.

12 (a) Convert $\frac{1}{7}$ to a decimal, giving your answer correct to three decimal places.
(b) Place the following numbers in order of size, starting with the smallest.
11.14 $1\frac{1}{7}$ 1.14   1.41   1.014

AQA

CHAPTER 6

Working with Number ●●●●●

Multiples

Numbers in the 4 times table are called **multiples** of 4.
Numbers in the 10 times table are called **multiples** of 10.

A table of multiples

×	1	2	3	4	5	6	7	8	9	10
1	1	2	3	4	5	6	7	8	9	10
2	2	4	6	8	10	12	14	16	18	20
3	3	6	9	12	15	18	21	24	27	30
4	4	8	12	16	20	24	28	32	36	40
5	5	10	15	20	25	30	35	40	45	50
6	6	12	18	24	30	36	42	48	54	60
7	7	14	21	28	35	42	49	56	63	70
8	8	16	24	32	40	48	56	64	72	80
9	9	18	27	36	45	54	63	72	81	90
10	10	20	30	40	50	60	70	80	90	100

EXAMPLES

1 Write down the first five multiples of 5.

$$1 \times 5 = 5$$
$$2 \times 5 = 10$$
$$3 \times 5 = 15$$
$$4 \times 5 = 20$$
$$5 \times 5 = 25$$

The first five multiples of 5 are:
5, 10, 15, 20 and 25.

2 What is the eighth multiple of 9?

The eighth multiple of 9 is $8 \times 9 = 72$.

3 The fifth multiple of a number is 30.
What is the number?

$5 \times 6 = 30$.
So, the number is 6.

The shaded numbers in the table are the
multiples of 2.
Multiples of 2 are called **even numbers** and
end in 0, 2, 4, 6 or 8.
Odd numbers end in 1, 3, 5, 7 or 9.
6, 12, 18, 24, … are **multiples** of 6.
The 8th **multiple** of 7 is $8 \times 7 = 56$.
3×8 has the same value as 8×3.

Exercise **6.1**   Do not use a calculator.

1 Write down the first five multiples of:
  (a) 10   (b) 3   (c) 7   (d) 6   (e) 9   (f) 20

2 Copy and complete the following.
  (a) The fifth multiple of 4 is ……
  (b) The seventh multiple of 6 is ……
  (c) The …… multiple of 6 is 18.
  (d) The …… multiple of 8 is 56.
  (e) The sixth multiple of …… is 60.
  (f) The eighth multiple of …… is 72.

3 (a) What multiple of 6 is the third multiple of 4?
  (b) What multiple of 8 is the fourth multiple of 4?
  (c) What multiple of 20 is the tenth multiple of 10?
  (d) What multiple of 3 is the sixth multiple of 4?
  (e) What multiple of 12 is the fourth multiple of 9?

4 (a) Write down a multiple of 7 between 30 and 40.
  (b) Write down a multiple of 8 between 40 and 50.

(a) Look at the table of multiples on Page 67.
What can you say about the following?
- (i) Even multiples of an even number.
- (ii) Even multiples of an odd number.
- (iii) Odd multiples of an even number.
- (iv) Odd multiples of an odd number.

(b) Using **O** for an odd number and **E** for an even number copy and complete these multiplication tables.

(i)

×	2	3	6	7	9
2	E				
3		O	E		
6					
7					
9					

(ii)

×	O	E
O		
E		

(c) Why are there more even numbers than odd numbers in the table of multiples?

Activity

The product of 1 and 12 is $1 \times 12 = 12$.
Write down **all** the other pairs of **whole numbers** that have a product of 12.
Write down all the pairs of whole numbers that have a product of 6.
Write down all the pairs of whole numbers that have a product of 5.
Write down all the pairs of whole numbers that have a product of 48.

> When numbers are multiplied together the answer is called the **product** of the numbers.

Factors

Pairs of **whole numbers** which have a product of 6 are 1×6 and 2×3.
1,   2,   3,   and   6   are called **factors** of 6.

EXAMPLES

1 Find all the factors of 30.

Find **all** the pairs of whole numbers that have a product of 30.
$$30 \times 1 = 30 \qquad 15 \times 2 = 30$$
$$10 \times 3 = 30 \qquad 6 \times 5 = 30$$
1,   2,   3,   5,   6,   10,   15   and   30 are all factors of 30.

2 Find all the factors of 7.

Only one pair of numbers has a product of 7.
$$1 \times 7 = 7$$
7 has just two factors,   1   and   7.

Prime numbers

Numbers like 7 are called **prime numbers**.
A prime number has exactly **two** factors, 1 and the number itself.
The first few prime numbers are: 2,   3,   5,   7,   11,   13,   …
The number 1 is not a prime number because it has only one factor.

Common factors

The factors of 20 are: **1**,   **2**,   4,   **5**,   **10**,   20.   The factors of 50 are: **1**,   **2**,   **5**,   **10**,   25,   50.

1, 2, 5 and 10 are factors of both 20 **and** 50.
They are called the **common factors** of 20 and 50.

Exercise 6.2   Do not use a calculator.

1. These pairs of numbers have a product of 12. 1×12 2×6 3×4
   (a) List all the factors of 12.
   (b) Explain why 8 is not a factor of 12.

2. (a) Find all the pairs of whole numbers that have a product of 18.
   (b) Write down all the factors of 18.

3. (a) Find all the pairs of whole numbers that have a product of 20.
   (b) Write down all the factors of 20.

4. Find all the factors of:
   (a) 16   (b) 28   (c) 36   (d) 45   (e) 48   (f) 50   (g) 60   (h) 80

5. (a) Find all the factors of: (i) 2   (ii) 3   (iii) 5   (iv) 7   (v) 11   (vi) 13
   (b) Find two more numbers with only two factors.

6. (a) Find all the factors of: (i) 4   (ii) 9   (iii) 25   (iv) 49
   (b) Find two more numbers with only three factors.

7. (a) Find all the factors of: (i) 6   (ii) 10   (iii) 14   (iv) 26   (v) 55   (vi) 38
   (b) Find two more numbers with only four factors.

8. Which of these numbers have common factors of 1, 2 and 3? 6,   16,   26,   36,   46.

9. Find the common factors of:
   (a) 10 and 15,   (b) 12 and 20,   (c) 16 and 18,   (d) 24 and 36,   (e) 12, 18 and 36.

10. Consider these numbers. 3,   4,   5,   14,   20,   27,   35,   60.
    (a) Which number is a factor of 10?
    (b) Which number is a multiple of 9?
    (c) Which numbers are prime numbers?

11. (a) How many multiples of 6 are factors of 36?
    (b) How many multiples of 5 are factors of 120?
    (c) How many factors of 100 are multiples of 2?
    (d) How many factors of 96 are multiples of 4?

12. Draw a 100 square on squared paper.

1	2	3	4	5	6	7	8	9	10
11	12	13	14	15	16	17	18	19	20

   (a) On your 100 square shade all the multiples of 2 **except** 2.

1	2	3	4	5	6	7	8	9	10
11	12	13	14	15

   (b) Next, shade all the multiples of 3 **except** 3.

1	2	3	4	5	6	7	8	9	10
11	12	13	14	15

   (c) All the multiples of 4 are already shaded. Explain why.
   (d) Shade all the multiples of 5 **except** 5.
   (e) Why have all the multiples of 6 **already** been shaded?
   (f) Shade all the multiples of 7 **except** 7.
   (g) Explain why 11 is the next unshaded number. Shade all the multiples of 11 **except** 11.
   (h) Continue to shade multiples of unshaded numbers (except the unshaded number).
   (i) Write a list of all the unshaded numbers less than 50 (except 1).
       How many factors has each of the numbers in your list?
       What is the special name for these numbers?

Powers

Products of the same number, like

$$3 \times 3, \quad 5 \times 5 \times 5, \quad 10 \times 10 \times 10 \times 10 \times 10,$$

can be written in a shorthand form using **powers**.

For example:

$3 \times 3 = 3^2$	This is read as '3 to the power of 2'.	3^2 has the value 9.
$5 \times 5 \times 5 = 5^3$	This is read as '5 to the power of 3'.	5^3 has the value 125.
$10 \times 10 \times 10 \times 10 \times 10 = 10^5$	This is read as '10 to the power of 5'.	10^5 has the value 100 000.

Index form

Numbers written in shorthand form like 3^2, 5^3 and 10^5 are said to be in **index form**.
This is sometimes called **power** form.

An expression of the form $a \times a \times a \times a \times a$ can be written in index form as a^5.
a^5 is read as 'a to the **power** 5'. a is the **base** of the expression. 5 is the **index** or **power**.

Exercise 6.3   Do not use a calculator.

1. Copy and complete.
   (a) $5^2 = \ldots \times \ldots$
   (b) $2^3 = \ldots \times \ldots \times \ldots$
   (c) $8^3 = \ldots \times \ldots \times \ldots$

2. Write each of the following as a power.
   (a) $4 \times 4 \times 4$
   (b) 8×8
   (c) $10 \times 10 \times 10 \times 10 \times 10 \times 10$

3. Copy and complete this table of the powers of 10.

	Expression	Index form	Value
	$10 \times 10 \times 10 \times 10 \times 10$	10^5	100 000
(a)		10^4	
(b)	$10 \times 10 \times 10$		
(c)			100
(d)	10		

4. Work out the value of:
   (a) 3^3
   (b) 6^2
   (c) 4^3
   (d) 12^2
   (e) 5^3
   (f) 10^6

5. Write the following as products of powers.
   For example: $2 \times 2 \times 2 \times 3 \times 3 \times 5 = 2^3 \times 3^2 \times 5$
   (a) $2 \times 2 \times 3 \times 3$
   (b) $2 \times 3 \times 3 \times 3 \times 5$
   (c) $2 \times 3 \times 5 \times 5$
   (d) $2 \times 2 \times 2 \times 3 \times 5 \times 5$
   (e) $5 \times 3 \times 3 \times 3 \times 5 \times 5$

Prime factors

The factors of 18 are   1,   2,   3,   6,   9   and   18.
Two of these factors, 2 and 3, are prime numbers.
The **prime factors** of 18 are 2 and 3.

Those factors of a number which are prime numbers are called **prime factors**.

Products of prime factors

All numbers can be written as the product of their prime factors.
For example: $6 = 2 \times 3$ $20 = 2 \times 2 \times 5$ $168 = 2 \times 2 \times 2 \times 3 \times 7$

Powers can be used to write numbers as the product of their prime factors in a shorter form.
For example: $20 = 2^2 \times 5$ $168 = 2^3 \times 3 \times 7$

A **factor tree** can be used to help write numbers as the product of their prime factors.

For example, this factor tree shows that:

$40 = 2 \times 20$
$40 = 2 \times 2 \times 10$
$40 = 2 \times 2 \times 2 \times 5$
$40 = 2^3 \times 5$

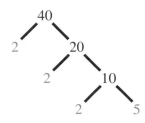

The branches of a factor tree stop when a prime factor is obtained.

So, 40 written as the product of its prime factors is $2^3 \times 5$.

EXAMPLES

1 Find the prime factors of 42.

First find the factors of 42.
$42 \times 1 = 42$
$21 \times 2 = 42$
$14 \times 3 = 42$
$7 \times 6 = 42$

Factors of 42 are:
1, 2, 3, 6, 7, 14, 21 and 42.
2, 3 and 7 are prime numbers.
The prime factors of 42 are 2, 3 and 7.

2 Write 50 as the product of its prime factors.

The factor tree shows that:
$50 = 2 \times 25$
$50 = 2 \times 5 \times 5$

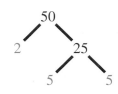

50 written as the product of its prime factors is 2×5^2.

Exercise 6.4   Do not use a calculator for questions 1 to 3.

1 Find the prime factors of:
   (a) 12   (b) 20   (c) 28   (d) 45   (e) 66   (f) 108

2 Write the following numbers as products of their prime factors.
   (a) 12   (b) 20   (c) 28   (d) 45   (e) 66   (f) 108

3 Write 128 as a product of its prime factors.

4 Write 1000 as a product of its prime factors.

5 A number written in terms of its prime factors is $2^3 \times 5^5$.
   (a) Calculate the number.
   (b) Write, in terms of its prime factors, a number which is 5 times the size.

Least common multiples

The first few multiples of three are:
          3, 6, 9, 12, **15**, 18, 21, 24, 27, **30**, 33, 36, 39, 42, **45**, …

The first few multiples of five are:
          5, 10, **15**, 20, 25, **30**, 35, 40, **45**, 50, …

15, 30, 45, … are multiples of both 3 **and** 5.
They are called **common multiples** of 3 and 5.

The smallest number that is a multiple of both 3 and 5 is 15.
The **least common multiple** of 3 and 5 is 15.

The least common multiple (**LCM**) of two numbers is the smallest number that is a multiple of them both.

EXAMPLE

Find the least common multiple of 20 and 45.
Start with the multiples of 45: 45, 90, 135, 180, 225, 270, …
Which is the lowest multiple of 45 which is also a multiple of 20?
Multiples of 20: 20, 40, 60, 80, 100, 120, 140, 160, **180**, …

So, the least common multiple of 20 and 45 is 180.

Highest common factors

The factors of 20 are: **1, 2,** 4, **5, 10,** 20.
The factors of 50 are: **1, 2, 5, 10,** 25, 50.

1, 2, 5 and 10 are factors of both 20 **and** 50.
They are called the **common factors** of 20 and 50.

The largest number that is a factor of both 20 and 50 is 10.
The **highest common factor** of 20 and 50 is 10.

The highest common factor (**HCF**) of two numbers is the largest number that is a factor of them both.

$20 \times 1 = 20$ $50 \times 1 = 50$
$10 \times 2 = 20$ $25 \times 2 = 50$
$5 \times 4 = 20$ $10 \times 5 = 50$

EXAMPLE

The factors of 18 are: 1, 2, 3, 6, 9, 18.
The factors of 45 are: 1, 3, 5, 9, 15, 45.

The common factors of 18 and 45 are: 1, 3 and 9.

The highest common factor of 18 and 45 is 9.

$18 \times 1 = 18$ $45 \times 1 = 45$
$9 \times 2 = 18$ $15 \times 3 = 45$
$6 \times 3 = 18$ $9 \times 5 = 45$

Exercise 6.5   Do not use a calculator.

1 Find the least common multiple of:
(a) 8 and 12   (b) 5 and 32   (c) 10 and 20   (d) 15 and 18
(e) 30 and 45   (f) 4, 6 and 8   (g) 5, 8 and 10   (h) 45, 90 and 105

2 Find the highest common factor of:
(a) 12 and 66   (b) 8 and 24   (c) 16 and 18   (d) 20 and 36
(e) 33 and 88   (f) 16, 20 and 28   (g) 15, 39 and 45   (h) 45, 90 and 105

3 (a) Write 24 as a product of prime factors.
(b) Write 54 as a product of prime factors.
(c) What is the highest common factor of 24 and 54?
(d) What is the least common multiple of 24 and 54?

4 (a) Find the value of x when $2^2 \times 3^x = 108$.
(b) Write 162 as a product of prime factors.
(c) What is the highest common factor of 108 and 162?
(d) What is the least common multiple of 108 and 162?

5 The bell at St. Gabriel's church rings every 6 minutes.
At St. Paul's, the bell rings every 9 minutes.
Both bells ring together at 9.00 am.
When is the next time both bells ring together?

72

Square numbers

Whole numbers raised to the power 2 are called **square numbers**.

$1^2 = 1 \times 1 = 1$ 1^2 is read as '1 squared'.   1 is a square number.
$2^2 = 2 \times 2 = 4$ 2^2 is read as '2 squared'.   4 is a square number.
$3^2 = 3 \times 3 = 9$ 3^2 is read as '3 squared'.   9 is a square number.

> To **square a number** multiply it by itself.

Square numbers can be shown as square patterns of dots.

$1^2 = 1$ $2^2 = 4$ $3^2 = 9$

> **Squaring on a calculator**
> 1.6^2 is read as '1.6 squared'.
>
> To calculate 1.6^2 use this sequence of buttons: $\boxed{1}$ $\boxed{.}$ $\boxed{6}$ $\boxed{x^2}$ $\boxed{=}$
>
> If there is no x^2 button try this:
> $\boxed{1}$ $\boxed{.}$ $\boxed{6}$ $\boxed{\times}$ $\boxed{=}$

Find the next two square numbers.
Numbers that are not whole numbers can also be squared.
For example:

$1.6^2 = 1.6 \times 1.6 = 2.56$
2.56 is **not** a square number. *Why not?*

Cube numbers

Whole numbers raised to the power 3 are called **cube numbers**.

$1^3 = 1 \times 1 \times 1 = 1$ 1^3 is read as '1 cubed'.   1 is a cube number.
$2^3 = 2 \times 2 \times 2 = 8$ 2^3 is read as '2 cubed'.   8 is a cube number.
$3^3 = 3 \times 3 \times 3 = 27$ 3^3 is read as '3 cubed'.   27 is a cube number.

Cube numbers can be shown using small cubes.

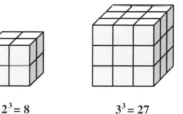

$1^3 = 1$ $2^3 = 8$ $3^3 = 27$

Draw a diagram to show 4^3.
What is the value of 4^3?

Numbers that are not whole numbers can also be cubed.
For example:

$1.6^3 = 1.6 \times 1.6 \times 1.6 = 4.096$
4.096 is **not** a cube number. *Why not?*

> 1.6^3 is read as '1.6 cubed'.

Using a calculator

Powers

The **squares** and **cubes** of numbers can also be calculated using the $\boxed{x^y}$ button on a calculator.

The $\boxed{x^y}$ button can be used to calculate the value of a number x raised to the power of y.

Reciprocals

The **reciprocal** of a number is the value obtained when the number is divided into 1.

The reciprocal of a number x is $\frac{1}{x}$.

A number times its reciprocal equals 1.

For example: the reciprocal of 2 is $\frac{1}{2}$, and $2 \times \frac{1}{2} = 1$.

> To find the reciprocal of a number on a calculator use the $\boxed{\frac{1}{x}}$ button.
> 0 (zero) has no reciprocal.

Working with Number

1 Calculate the value of 2.6^3.

To do the calculation enter the following sequence into your calculator:

$$\boxed{2}\ \boxed{.}\ \boxed{6}\ \boxed{x^y}\ \boxed{3}\ \boxed{=}$$

This gives $2.6^3 = 17.576$

2 Find the reciprocal of 5.

The reciprocal of 5 is $\frac{1}{5}$.

$1 \div 5 = 0.2$

The reciprocal of 5 is 0.2

To find the reciprocal of 5 on your calculator use the sequence: $\boxed{5}\ \boxed{\frac{1}{x}}$

Exercise 6.6   Do not use a calculator for questions 1 to 6.

1 (a) What is the square of 7? (b) What is the cube of 5? (c) What is the reciprocal of 2?

2 (a) Complete this list of square numbers from 1^2 to 20^2.

$$1^2 = 1 \times 1 = 1 \qquad 2^2 = 2 \times 2 = 4 \qquad 3^2 = 3 \times 3 = 9$$

(b) Copy and continue the difference pattern shown below for your list of square numbers.

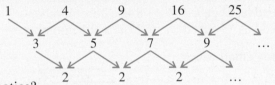

What do you notice?

(c) Use the pattern to find 21^2 from 20^2.

3 Complete this list of cube numbers from 1^3 to 10^3.

$$1^3 = 1 \times 1 \times 1 = 1 \qquad 2^3 = 2 \times 2 \times 2 = 8 \qquad 3^3 = 3 \times 3 \times 3 = 27$$

4 Consider the numbers: $\boxed{8 \quad 16 \quad 27 \quad 36 \quad 64 \quad 100}$

Which of these numbers is both a square number **and** a cube number?

5 Connie says that $2^2 + 3^2 = 5^2$. Is she right? Explain your answer.

6 (a) Calculate the value of: (i) $(-3)^2$ (ii) $(-2)^3$ (iii) $(-4)^2$ (iv) $(-5)^3$

(b) What do you notice about the signs of your answers?

7 Use the $\boxed{x^y}$ button on your calculator to find the value of:

(a) (i) 13^2 (ii) 17^2 (iii) 2.5^2 (iv) 0.8^2 (v) 9.7^2

(b) (i) 6^3 (ii) 15^3 (iii) 2.4^3 (iv) 0.7^3 (v) 5.6^3

8 (a) Find the reciprocals of these numbers without using a calculator,
then use a calculator to check your answers.

(i) 2 (ii) 5 (iii) 10 (iv) 0.5 (v) 0.1 (vi) 0.2

(b) Use the $\boxed{\frac{1}{x}}$ button on your calculator to find the reciprocals of:

(i) 4 (ii) 20 (iii) 25 (iv) 0.25 (v) 0.4 (vi) 0.16

9 Show by means of an example, that a number times its reciprocal is equal to 1.

10 Calculate the value of:

(a) $3^3 \times 10^3$ (b) $10^5 \div 5^3$ (c) $2.6 + \frac{1}{2.6}$ (d) 2.2^3

(e) $8.5^2 - 1.3^2$ (f) $\frac{5}{(0.4)^2}$ (g) $(1.9 + 2.2)^2 \times 1.5$ (h) $0.8^2 \times \frac{1}{0.5}$

Square roots

The opposite of squaring a number is called finding the **square root**.

For example:

The square root of 16 is 4 because $4^2 = 16$.

$$4 \xrightarrow{\text{square}} 16$$
$$4 \xleftarrow{\text{square root}} 16$$

The square root of 3.24 is 1.8 because $1.8^2 = 3.24$.

$$1.8 \xrightarrow{\text{square}} 3.24$$
$$1.8 \xleftarrow{\text{square root}} 3.24$$

$\sqrt{}$ This special symbol stands for the square root.

For example: $\sqrt{9} = 3$ $\sqrt{2.56} = 1.6$

Note: $-3 \times -3 = 9$ and $-1.6 \times -1.6 = 2.56$

So, the square root of a number can be **positive** or **negative**.
In most cases we only use the positive square root.

Square roots on a calculator

To calculate $\sqrt{2.56}$ use this sequence of buttons:

Square roots can be worked out on a calculator without using special buttons.
A method called **trial and improvement** can be used.

EXAMPLE

You are asked to find the square root of 18.6 but your calculator does not have a square root button.
Use trial and improvement to find the square root of 18.6 to an accuracy of one decimal place.
Show your method clearly.

4^2 $= 4 \times 4$ $= 16$	so, the square root of 16 is 4		
5^2 $= 5 \times 5$ $= 25$	so, the square root of 25 is 5	So, try 4.5	*Why?*
$4.5^2 = 4.5 \times 4.5 = 20.25$	so, the square root of 20.25 is 4.5	So, try 4.3	*Why?*
$4.3^2 = 4.3 \times 4.3 = 18.49$	so, the square root of 18.49 is 4.3	So, try 4.4	*Why?*
$4.4^2 = 4.4 \times 4.4 = 19.36$	so, the square root of 19.36 is 4.4	So, try 4.35	*Why?*
$4.35^2 = 4.35 \times 4.35 = 18.9225$	so, the square root of 18.9225 is 4.35		

This shows that the square root of 18.6 lies between 4.3 and 4.35.
So, correct to one decimal place, the square root of 18.6 is 4.3.

When using trial and improvement:
- Work methodically using trials first to the nearest whole number, then to one decimal place etc.
- Do at least one trial to one more decimal place than the required accuracy to be sure of your answer.

Cube roots

The opposite of cubing a number is called finding the **cube root**.
For example: the cube root of 27 is 3 because $3^3 = 27$.

Exercise 6.7

Do not use a calculator for questions 1 to 4.

1. What is (a) the square root of 36, (b) the cube root of 8?

2. Write down the value of: (a) $\sqrt{25}$ (b) $\sqrt{100}$ (c) $\sqrt{64}$ (d) $\sqrt{49}$

3. Calculate the value of: (a) $\sqrt{3^2 + 4^2}$ (b) $\sqrt{13^2 - 12^2}$

4. Jake says, "The square root of 55 lies between 7 and 8."
   Is he right? Explain your answer.

5. Use your calculator to find $\sqrt{96}$. Give your answer to one decimal place.

6 Calculate. (a) $\sqrt{5.3}$ (b) $\sqrt{300}$ (c) $\sqrt{4.8^2 - 2.7^2}$
Give your answers correct to one decimal place.

7 (a) Use the method of trial and improvement to find the square roots of:
(i) 20 (ii) 108 (iii) 7.6
Give your answers to an accuracy of one decimal place.
(b) Check each of your answers using the square root button.

8 Use the method of trial and improvement to find the length of the side of a square carpet of area 3.5 square metres.
Give your answer to an accuracy of one decimal place.

9 The floor of a garage is a square and has an area of 55 m².
Use the method of trial and improvement to find the length of one side of the floor.
Give your answer to an accuracy of one decimal place.

Multiplying and dividing numbers with powers

Remember: $3 \times 3 = 3^2$ and $5 \times 5 \times 5 = 5^3$
Both 3^2 and 5^3 are examples of numbers with powers.

EXAMPLES

This example introduces a method for multiplying powers of the same number.

1 Calculate the value of $6^5 \times 6^4$ in power form.
$6^5 = 6 \times 6 \times 6 \times 6 \times 6$ and $6^4 = 6 \times 6 \times 6 \times 6$
$6^5 \times 6^4 = (6 \times 6 \times 6 \times 6 \times 6) \times (6 \times 6 \times 6 \times 6)$
$= 6 \times 6 \times 6 \times 6 \times 6 \times 6 \times 6 \times 6 \times 6$
This gives: $6^5 \times 6^4 = 6^9$

Can you see a quick way of working out the power of the answer?

This example introduces a method for dividing powers of the same number.

2 Calculate the value of $6^7 \div 6^4$ in power form.
$6^7 \div 6^4 = \dfrac{6^7}{6^4} = \dfrac{6 \times 6 \times 6 \times \cancel{6} \times \cancel{6} \times \cancel{6} \times \cancel{6}}{\cancel{6} \times \cancel{6} \times \cancel{6} \times \cancel{6}}$
$= 6 \times 6 \times 6$
$= 6^3$
This gives: $6^7 \div 6^4 = 6^3$

Can you see a quick way of working out the power of the answer?

Rules for multiplying and dividing powers of the same number

When **multiplying**: powers of the same base are **added**.	In general: $a^m \times a^n = a^{m+n}$
When **dividing**: powers of the same base are **subtracted**.	In general: $a^m \div a^n = a^{m-n}$

EXAMPLES

Simplify each of these expressions.

(a) $3^3 \times 3^2 = 3^{3+2} = 3^5$

(b) $4^5 \div 4^2 = 4^{5-2} = 4^3$

Exercise 6.8

Do not use a calculator in this exercise. Leave your answers in power form.

1 Simplify each of these expressions.
  (a) $2^5 \times 2^2$ (b) $4^3 \times 4^6$ (c) $6^2 \times 6$ (d) $8^4 \times 8^3$
  (e) $9^3 \times 9^2$ (f) $2^3 \times 2^5$ (g) $5^5 \times 5^7$ (h) $3^0 \times 3^3$

2 Simplify.
  (a) $2^5 \div 2^2$ (b) $4^7 \div 4^5$ (c) $6^2 \div 6$ (d) $8^4 \div 8^3$
  (e) $3^{11} \div 3^5$ (f) $2^3 \div 2^2$ (g) $5^5 \div 5^3$ (h) $11^3 \div 11^0$

3 Simplify.
  (a) $8^5 \times 8^2$ (b) $2^6 \times 2^3$ (c) $7^7 \div 7^2$ (d) $5^2 \div 5^2$
  (e) $4^2 \div 4$ (f) $6^5 \times 6^4$ (g) $10^6 \div 10^2$ (h) $3^8 \times 3^4$

4 Simplify.
  (a) $3 \times 3^2 \times 3^3$ (b) $\dfrac{10 \times 10^3}{10^2}$ (c) $\dfrac{4^3 \times 4^3}{4}$ (d) $\dfrac{5^5 \times 5^2}{5^4}$
  (e) $\dfrac{2 \times 2^5}{2^3}$ (f) $\dfrac{5 \times 5^2}{5^3}$ (g) $\dfrac{7^3 \times 7^5}{7^2}$ (h) $\dfrac{3^5 \times 3^3}{3^4}$

5 Write, as a single power of 10.
  (a) $10^3 \times 10^2$ (b) $10^6 \div 10^2$ (c) $\dfrac{10^7 \times 10^2}{10^3}$ (d) $\dfrac{10^{10}}{10^4 \times 10}$

Large numbers and your calculator

Scientists who study the planets and the stars work with very large numbers.

Approximate distances from the Sun to some planets are:

Earth 149 000 000 km   Mars 228 000 000 km   Pluto 5 898 000 000 km

A calculator displays very large numbers in **standard index form**.
To represent large numbers in standard index form you need to use powers of 10.

Number	1 000 000	100 000	10 000	1000	100	10
Power of 10	10^6	10^5	10^4	10^3	10^2	10^1

For example: $2\,600\,000 = 2.6 \times 1\,000\,000 = 2.6 \times 10^6$
Therefore $2\,600\,000 = 2.6 \times 10^6$ in **standard index form**.

Calculator displays

Work out $3\,000\,000 \times 25\,000\,000$ *on your calculator.*
Write down the display.

Most calculators will show the answer as: | 7.5   13 |

In **standard index form** the answer is 7.5×10^{13}, which is $75\,000\,000\,000\,000$.

EXAMPLES

1 Write 5.6×10^7 as an ordinary number.
  $5.6 \times 10^7 = 5.6 \times 10\,000\,000 = 56\,000\,000$

2 Write the calculator display: | 7.3   05 | as an ordinary number.
  | 7.3   05 | means 7.3×10^5 and $7.3 \times 10^5 = 7.3 \times 100\,000 = 730\,000$

Use a calculator for question 3 only.

1   Write each of these numbers as an ordinary number.
   (a) 6×10^5 (b) 2×10^3 (c) 5×10^7 (d) 9×10^8
   (e) 3.7×10^9 (f) 2.8×10^1 (g) 9.9×10^{10} (h) 7.1×10^4

2   Write these calculator displays as ordinary numbers.

   (a) | 4.5   03 | (b) | 7.8   07 | (c) | 5.3   05 | (d) | 3.25   04 |

3   Use your calculator to work out each of the following.
   Write each of the answers: (i) as on your calculator display,   (ii) as an ordinary number.
   (a) $300\,000 \times 200\,000\,000$ (b) $120\,000 \times 80\,000\,000$
   (c) $15\,000 \times 700\,000\,000$ (d) $65\,000 \times 2\,000\,000\,000$
   (e) $480\,000 \times 500\,000\,000$ (f) $50\,000 \times 50\,000\,000$

Small numbers and your calculator

Scientists who study microbiology work with numbers that are very small.
The smallest living cells are bacteria which have a diameter of about 0.000 025 cm.
Blood cells have a diameter of about 0.000 75 cm.

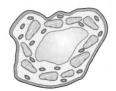

A calculator displays small numbers in **standard index form**.
It does this in the same sort of way that it does for large numbers.
To represent very small numbers in standard index form you need to use powers of 10 for numbers
less than 1.

Number	1000	100	10	1	0.1	0.01	0.001	0.000 1
Power of 10	10^3	10^2	10^1	10^0	10^{-1}	10^{-2}	10^{-3}	10^{-4}

For example: $0.000\,037 = 3.7 \times 0.000\,01 = 3.7 \times 10^{-5}$
Therefore $0.000\,037 = 3.7 \times 10^{-5}$ in **standard index form**.

Calculator displays

Work out $0.000\,007 \times 0.000\,9$ *on your calculator.*
Write down the display.

Most calculators will show the answer as: | 6.3   −09 |

In standard index form the answer is 6.3×10^{-9},   which is   0.000 000 006 3.

EXAMPLES

1   Write 2.9×10^{-6} as an ordinary number.
        $2.9 \times 10^{-6} = 2.9 \times 0.000\,001 = 0.000\,002\,9$

2   Write the calculator display: | 1.5   −03 | as an ordinary number.

   | 1.5   −03 | means 1.5×10^{-3} and $1.5 \times 10^{-3} = 1.5 \times 0.001 = 0.001\,5$

Use a calculator for question 3 only.

1   Write each of these numbers as an ordinary number.
   (a) 3.5×10^{-1} (b) 5×10^{-4} (c) 7.2×10^{-5} (d) 6.1×10^{-3}
   (e) 1.17×10^{-10} (f) 8.135×10^{-7} (g) 6.462×10^{-2} (h) 4.001×10^{-9}

2 Write these calculator displays as ordinary numbers.

(a) | 3.4   −03 | (b) | 5.65   −05 | (c) | 7.2   −04 | (d) | 9.13   −01 |

3 Use your calculator to work out each of the following.
Write each of the answers: (i) as on the calculator display, (ii) as an ordinary number.
(a) 0.000 03 × 0.000 02
(b) 0.000 045 ÷ 3000
(c) 0.000 75 × 0.000 000 04
(d) 0.002 3 ÷ 5 000 000
(e) 0.053 × 0.000 000 08
(f) 0.000 006 4 × 0.000 015

What you need to know

- **Multiples** of a number are found by multiplying the number by 1, 2, 3, 4, …
  For example: the multiples of 8 are $1 \times 8 = 8$, $2 \times 8 = 16$, $3 \times 8 = 24$, $4 \times 8 = 32$, …
- **Factors** of a number are found by listing all the products that give the number.
  For example: $1 \times 6 = 6$ and $2 \times 3 = 6$. So, the factors of 6 are: 1, 2, 3 and 6.
- The **common factors** of two numbers are the numbers which are factors of **both**.
- A **prime number** is a number with only two factors, 1 and the number itself.
  The first few prime numbers are: 2, 3, 5, 7, 11, 13, 17, 19, 23, 29, 31, …
- The **prime factors** of a number are those factors of the number which are prime numbers.
- The **Least Common Multiple** of two numbers is the smallest number that is a multiple of both.
- The **Highest Common Factor** of two numbers is the largest number that is a factor of both.
- An expression such as $3 \times 3 \times 3 \times 3 \times 3$ can be written in a shorthand way as 3^5.
  This is read as '3 to the power 5'. The number 3 is the **base** of the expression. 5 is the **power**.
- Numbers raised to the power 2 are **squared**.
  Whole numbers squared are called **square numbers**.
  Squares can be calculated using the x^2 button on a calculator.
  The opposite of squaring a number is called finding the **square root**.
  Square roots can be calculated using the $\sqrt{\ }$ button on a calculator.
- The square root of a number can be positive or negative.
  For example: the square root of 4 can be $+2$ or -2.
- Numbers raised to the power 3 are **cubed**.
  Whole numbers cubed are called **cube numbers**.
  The opposite of cubing a number is called finding the **cube root**.
  For example: the cube root of 27 is 3 because $3^3 = 27$.
- **Powers**
  The squares and cubes of numbers can be worked out on a calculator by using the x^y button.
  The x^y button can be used to calculate the value of a number x raised to the power of y.
- **Reciprocals**
  The reciprocal of a number is the value obtained when the number is divided into 1.
  The reciprocal of a number can be found on a calculator by using the $\frac{1}{x}$ button.
  A number times its reciprocal equals 1. Zero has no reciprocal.
- Square roots can be found using a method called **trial and improvement**.
- Powers of the same base are **added** when terms are **multiplied**.   e.g. $5^2 \times 5^3 = 5^5$
  Powers of the same base are **subtracted** when terms are **divided**.   e.g. $7^5 \div 7^3 = 7^2$

You should be able to:

- Use the x^2, x^y, $\sqrt{\ }$ and $\frac{1}{x}$ buttons on a calculator to solve a variety of problems.
- Interpret a calculator display showing very large and very small numbers in **standard index form**.

Do not use a calculator for questions 1 to 25.

1 (a) Find all the factors of 12.
(b) Write down the factors of 12 which are also factors of 18.

2 (a) Write down three multiples of 7.
(b) (i) Which of these numbers are both a multiple of 2 **and** a multiple of 5?

| 2 | 5 | 10 | 15 | 18 | 20 | 25 |

(ii) Explain how you know that a number is both a multiple of 2 **and** a multiple of 5.

3 Find the common factors of 16 and 24.

4 (a) Explain why 19 is a prime number, but 15 is **not** a prime number.
(b) What is the first prime number larger than 20?

5 Part of a sequence of numbers is shown.   3   6   9   12   15   18
(a) Which of these numbers are multiples of 9?
(b) Which of the numbers are factors of 6?
(c) Which of these numbers is a prime number?   AQA

6 (a) What is the square of 6?
(b) What is the cube of 4?

7

| 3 | 9 | 20 | 25 | 29 | 75 | 92 | 100 |

Which of the numbers in the box are:
(a) square numbers,   (b) factors of 100,   (c) prime numbers?   AQA

8 Consider the numbers: 8, 9, 11, 17 and 121.
(a) Write down all the factors of these numbers.
(b) (i) Which of the numbers have only two factors?
(ii) What special name is given to these numbers?
(c) (i) Which of these numbers have exactly 3 factors?
(ii) What special name is given to these numbers?

9 (a) State which of these numbers are multiples of 4.
2, 3, 8, 15, 22, 25, 29, 39.
(b) Jane tries to find a cube number in this list.
She says, "39 is a cube number because $13^3 = 39$."
Explain why Jane is wrong.   AQA

10 (a) What is the square root of 64?
(b) What is the cube root of 64?

11 (a) What is the value of $2^3 - \sqrt{25}$?
(b) Work out the value of 10^4.

12 Which is greater, $\sqrt{625}$ or 3^3? Show working to explain your answer.   AQA

13 Cameron states that the sum of four consecutive numbers is always a multiple of 4.
Give an example to show that this statement is not true.

14 (a) Write 1 000 000 as a power of 10.
(b) What is the reciprocal of 4?

15 (a) Write 3.45×10^{10} as an ordinary number.
(b) Write 5.43×10^{-7} as an ordinary number.

16 Write down the value of: (a) 9^2,   (b) $\sqrt{49}$,   (c) 5^3.

17 Write 24 as a product of its prime factors.

18 Find the value of $\sqrt{2^4 \times 3^2}$.

19 (a) Find the value of p when $2^p \times 3 = 48$.
(b) Write 72 as a product of prime factors.
(c) What is the highest common factor of 48 and 72?
(d) What is the least common multiple of 48 and 72?   AQA

20 A blue light flashes every 18 seconds and a green light flashes every 30 seconds.
The two lights flash at the same time.
After how many seconds will the lights next flash at the same time?

21 Write 120 as a product of its prime factors.

22 Look at the numbers shown on these calculator displays.

| 8.3   08 | 3.9   10 | 6.7   −05 | 9.3   −03 | 5.6   09 |

(a) Write the largest number as an ordinary number.
(b) Write the smallest number as an ordinary number.

23 Calculate the exact value of $8^3 \times 2^3$.

24 Look at this number pattern.
This pattern continues.
(a) Write down the next line of the pattern.
(b) Use the pattern to work out 6666667^2.

$$7^2 = 49$$
$$67^2 = 4489$$
$$667^2 = 444889$$
$$6667^2 = 44448889$$

AQA

25 Simplify. Leave each answer as a single power of 5.

(a) $5^4 \times 5^7$ (b) $5^7 \div 5^3$ (c) $\dfrac{5^5 \times 5^3}{5^4}$

26 Use your calculator to work out 0.4^5.

27 Calculate. (a) $\sqrt{5.76}$ (b) $\sqrt{5.76} + 3.5^2$

28 What is the value of $\sqrt{50}$?
Give your answer correct to one decimal place.

29 Calculate the value of $0.7^3 + \sqrt{30}$.
Give your answer correct to one decimal place.

30 Kim is trying to find the square root of 31 without using the square root button on her calculator.
She tries 5 and gets $5 \times 5 = 25$.
She tries 6 and gets $6 \times 6 = 36$.
(a) What number should she try next?
(b) Continue this method to find the square root of 31 correct to one decimal place.   AQA

31 (a) Find the square root of 1296.

(b) Find the value of $\dfrac{1}{\sqrt{1296}}$, correct to 3 decimal places.

32 Calculate: $2.6^3 \times \sqrt{4.3 + 2.8}$
Give your answer correct to one decimal place.

33 (a) Calculate the value of $\dfrac{1}{(5.3 - 4.8)^2}$.

(b) Calculate $\dfrac{25.94 - 9.27}{12.43 + 5.16}$.
Give your answer to an appropriate degree of accuracy.

Working with Number

The meaning of a percentage

A fraction with denominator 100 has the special name - **percentage**.
'Per cent' means 'out of 100'.
The symbol for per cent is %.
A percentage can be written as a fraction with denominator 100.

> 10% means 10 out of 100.
>
> 10% can be written as $\frac{10}{100}$.
>
> 10% is read as '10 percent'.

EXAMPLE

What percentage of this diagram is shaded?

The large square is divided into 100 smaller squares.

5 of the smaller squares are shaded.

$\frac{5}{100}$ of the diagram is shaded. $\frac{5}{100} = 5\%$

So, 5% of the diagram is shaded.

What percentage of each of these diagrams is shaded?

Changing percentages to decimals and fractions

> To change a percentage to a decimal or a fraction: **divide by 100**

EXAMPLES

1 Write 38% as a fraction in its simplest form.

38% means '38 out of 100'.

This can be written as $\frac{38}{100}$.

$\frac{38}{100} = \frac{38 \div 2}{100 \div 2} = \frac{19}{50}$

$38\% = \frac{19}{50}$

Remember:
To write a fraction in its **simplest form**
divide both the numerator and
denominator of the fraction by the **largest**
number that divides into them both.

2 Write 38% as a decimal.

$38\% = \frac{38}{100} = 38 \div 100 = 0.38$

3 Write 2.5% as a decimal.

$2.5\% = \frac{2.5}{100} = 2.5 \div 100 = 0.025$

Remember:
To change a fraction to a
decimal divide the numerator
by the denominator.

Exercise 7.1

Do not use a calculator for questions 1 to 3.

1 What percentage of each diagram is shaded?

(a)   (b)   (c)   (d)

(e)   (f)   (g)   (h)

2 (a) Draw a 10 by 10 square on squared paper.

  (i) Calculate $\frac{2}{5}$ of 100.

  (ii) Shade $\frac{2}{5}$ of a 10 by 10 square.
    What percentage of the square is shaded?

  (b) Repeat (a) for the following fractions.
    (i) $\frac{3}{5}$ (ii) $\frac{7}{10}$ (iii) $\frac{9}{20}$ (iv) $\frac{6}{25}$ (v) $\frac{23}{50}$ (vi) $\frac{17}{25}$

3 Copy and complete this table to show the percentages given as:
(a) fractions in their simplest form,
(b) decimals.

Percentage	10%	20%	25%	50%	75%	80%
Fraction	$\frac{1}{10}$					
Decimal	0.1					

4 Change these percentages to fractions in their simplest form.
(a) 15% (b) 5% (c) 18% (d) 52% (e) 23% (f) 12.5%

5 Change these percentages to decimals.
(a) 15% (b) 5% (c) 47% (d) 72% (e) 87.5% (f) 150%

Changing decimals and fractions to percentages

To change a decimal or a fraction to a percentage: **multiply by 100**

EXAMPLES

1 Change 0.3 to a percentage.

$0.3 \times 100 = 30$
So, 0.3 as a percentage is 30%.

2 Change 0.875 to a percentage.

$0.875 \times 100 = 87.5$
So, 0.875 as a percentage is 87.5%.

3 Change $\frac{7}{10}$ to a percentage.

$\frac{7}{10} \times 100 = 7 \times 100 \div 10$
$= 700 \div 10 = 70\%$

4 Change $\frac{11}{25}$ to a percentage.

$\frac{11}{25} \times 100 = 11 \times 100 \div 25$
$= 1100 \div 25 = 44\%$

Comparing fractions

Fractions can be compared by first writing them as percentages.

Exercise 7.2

Do not use a calculator for questions 1 to 3.

1 Copy and complete this table to work out the percentage equivalents of the fractions given.

Fraction	$\frac{3}{10}$	$\frac{2}{5}$	$\frac{3}{25}$	$\frac{7}{20}$
Percentage				

2 Copy and complete this table to work out the percentage equivalents of the decimals given.

Decimal	0.7	0.45	0.05	1.2
Percentage				

3 What is $\frac{1}{3}$ as a percentage?

4 Change these fractions to percentages.

(a) $\frac{17}{50}$ (b) $\frac{12}{25}$ (c) $\frac{30}{200}$ (d) $\frac{4}{5}$ (e) $\frac{135}{500}$ (f) $\frac{13}{20}$ (g) $\frac{2}{3}$ (h) $\frac{2}{9}$

5 Change these decimals to percentages.

(a) 0.15   (b) 0.32   (c) 0.125   (d) 0.07   (e) 1.12   (f) 0.015

6 Write in order of size, lowest first:

(a) $\frac{1}{2}$ 60% $\frac{2}{5}$ 0.55   (b) 43% $\frac{9}{20}$ 0.42 $\frac{11}{25}$ (c) $\frac{23}{80}$ 28% $\frac{57}{200}$ 0.2805

7 Peter scores 96 out of 120.
What percentage did he get?

8 Change each of these marks to a percentage.

(a) Maths: 27 out of 30.   (b) French: 34 out of 40.
(c) Science: 22 out of 25.   (d) Art: 48 out of 60.

9 Which rectangle has the greater percentage shaded?

 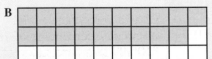

10 In an ice hockey competition Team A won 8 out of the 11 games they played whilst Team B won 5 of their 7 games.

Which team has the better record in the competition?

Expressing one quantity as a percentage of another

To work out one number as a percentage of another there are two steps.

Step 1 Write the numbers as a fraction. **Step 2** Change the fraction to a percentage.

EXAMPLES

The numbers in the fraction must be in the **same** units.

1 What is 30p as a percentage of £2?

£2 = 200p

Step 1
30p as a fraction of 200p is $\frac{30}{200}$.

Step 2
$\frac{30}{200} \times 100 = 30 \times 100 \div 200 = 15\%$

So, 30p as a percentage of £2 is 15%.

2 A newspaper contains 48 pages, 6 of which are Sports pages.
What percentage of the pages are Sports pages?

Step 1
6 out of 48 pages are Sports pages.
$\frac{6}{48} = 6 \div 48 = 0.125$

Step 2
$0.125 \times 100 = 12.5$
12.5% of the pages are Sports pages.

Exercise 7.3

Do not use a calculator for questions 1 to 6.

1 What is (a) 30 as a percentage of 50,
(b) 7 as a percentage of 10,
(c) 42 as a percentage of 200?

2 What is (a) 6 minutes as a percentage of 1 hour,
(b) 30 mm as a percentage of 5 cm,
(c) 150 g as a percentage of 1 kg?

3 There are 8 yellow fruit drops in a packet of 25 fruit drops.
What percentage of the fruit drops are yellow?

4 James saved £30 and then spent £9.
What percentage of his savings did he spend?

5 A Youth Club has 200 members. 80 of the members are boys.
(a) What percentage of the members are boys?
(b) What percentage of the members are girls?

6 240 people took part in a survey. 30 of them were younger than 18.
What percentage were younger than 18?

7 A bar of chocolate has 32 squares. Jane eats 12 of the squares.
What percentage of the bar does she eat?

8 Billy earns £9 per hour. He gets a wage rise of 27 pence per hour.
What is his percentage wage rise?

9 What is (a) £2 as a percentage of £6,
(b) 20 cm as a percentage of 160 cm,
(c) £105.09 as a percentage of £186?

10 A new car costs £13 500. The dealer gives a discount of £1282.50.
What is the percentage discount?

11 There are 600 pupils in Years 9 to 13 of a High school.
360 pupils are in Years 10 and 11. 15% of pupils are in Years 12 and 13.
What percentage of pupils are in Year 9?

Percentages

Finding a percentage of a quantity

EXAMPLES

1 Find 20% of £56.

Step 1 Divide by 100.
£56 ÷ 100 = £0.56

Step 2 Multiply by 20.
£0.56 × 20 = £11.20
So, 20% of £56 is £11.20.

To find 1% of a quantity divide the quantity by 100.
To find 20% of a quantity multiply 1% of the quantity by 20.
This is the same as the method you would use to
find $\frac{20}{100}$ of a quantity.

2 The price of a car is £12 500.
A car salesman offers a 7% discount.
How much is the discount?

Discount = 7% of £12 500
1% of £12 500 = £12 500 ÷ 100 = £125
7% of £12 500 = £125 × 7 = £875
The discount is £875.

3 David invests £500 in a building society.
He earns 6% interest per year.
How much interest does he get after one year?

Interest = 6% of £500
1% of £500 = £500 ÷ 100 = £5
6% of £500 = £5 × 6 = £30
The interest is £30.

Exercise 7.4

Do not use a calculator in this exercise.

1 Find:
(a) 10% of 500
(b) 5% of 800
(c) 20% of 700
(d) 30% of 200
(e) 65% of 30
(f) 85% of 20
(g) 12% of 500
(h) 32% of 200

2 Find:
(a) 20% of £80
(b) 75% of £20
(c) 30% of £220
(d) 15% of £350
(e) 30% of 80 kg
(f) 35% of 800 m
(g) 45% of £25
(h) 60% of 20 pence

3 Garry has 300 marbles. 20% of the marbles are blue. 35% of the marbles are red.
The rest of the marbles are white.
(a) How many marbles are (i) blue, (ii) red?
(b) What percentage of the marbles are white?

4 David invests £400 in a building society. He earns 5% interest per year.
How much interest does he get in one year?

5 There are 450 seats in a theatre. 60% of the seats are in the stalls.
How many seats are in the stalls?

6 A salesman earns a bonus of 3% of his weekly sales.
How much bonus does the salesman earn in a week when his sales are £1400?

7 Jenny gets a 15% discount on a theatre ticket. The normal cost is £18.
How much does she save?

8 Dipak earns £350 per week. He gets a wage rise of 3%.
How much extra does he earn each week?

9 (a) In a school of 1200 pupils 45% are boys. How many are girls?
(b) 30% of the girls at this school are under 13. How many girls are under 13?

10 A dozen biscuits weigh 720 g.
The amount of flour in a biscuit is 40% of the weight of a biscuit.
What is the weight of flour in **each** biscuit?

Percentage change

EXAMPLES

1 Increase £25 by 30%.

First find 30% of £25.
$\frac{30}{100} \times 25 = 0.3 \times 25 = 7.5$
30% of £25 is £7.50.
£25 increased by 30% = £25 + £7.50
$\qquad\qquad\qquad = £32.50$

3 A shirt normally priced at £24 is reduced by 15% in a sale.
How much does it cost in the sale?

Reduction in price = 15% of £24
$15 \div 100 \times 24 = 0.15 \times 24 = 3.6$
15% of £24 = £3.60
The shirt costs £24 − £3.60 = £20.40.

2 Decrease £600 by 12%.

First find 12% of £600.
$\frac{12}{100} \times 600 = 0.12 \times 600 = 72$
12% of £600 is £72.
£600 decreased by 12% = £600 − £72
$\qquad\qquad\qquad = £528$

4 A packet of cereals weighs 440 g.
A special offer packet contains 30% more.
What is the weight of a special offer packet?

Extra contents = 30% of 440 g
$\qquad\qquad = 440 \div 100 \times 30$
$\qquad\qquad = 132\,g$
$440 + 132 = 572$

A special offer packet weighs 572 g.

Exercise 7.5

Do not use a calculator for questions 1 to 5.

1 Increase:
(a) £400 by 20%   (b) £300 by 40%   (c) £2000 by 40%   (d) £600 by 80%
(e) £3000 by 15%   (f) £900 by 40%   (g) £50 by 60%   (h) £10 by 30%
(i) £15 by 10%   (j) £50 by 15%

2 Decrease:
(a) £600 by 30%   (b) £800 by 25%   (c) £2500 by 20%   (d) £250 by 40%
(e) £12 000 by 15%   (f) £7000 by 35%   (g) £600 by 15%   (h) £55 by 90%
(i) £42 by 20%   (j) £63 by 35%

3 A mobile telephone company offers a 20% discount on calls made in March.
The normal cost of a peak time call is 50 pence per minute.
How much does a peak time call cost in March?

4 Abdul earns £200 per week. He gets a wage rise of 7.5%. What is his new weekly wage?

5 A packet of breakfast cereal contains 660 g. A special offer packet contains an extra 15%.
How many grams of breakfast cereal are in the special offer packet?

6 Prices in a sale are reduced by 18%. The normal price of a shirt is £22.50.
Calculate its sale price.

7 The price of a gold watch is £278. What does it cost with a 12% discount?

8 The price of a used car is £5200. What does it cost with a 9.5% discount?

9 The price of a new kitchen is £8650. What does it cost with a 35% discount?

10 Jane's salary of £14 000 is increased by 4%. Calculate her new salary.

11 A car was valued at £13 500 when new. After one year it lost 22% of its value.
What was the value of the car after one year?

12 Louisa puts £480 into a bank account.
At the end of one year interest at 2.5% is added to her account.
How much is in her account at the end of one year?

More complicated percentage problems

Use a calculator in this exercise.
Where appropriate give your answers to 3 significant figures.

1 James wins a lottery prize of £1 764 000.
He pays £529 000 for a house.
What percentage of his prize did he spend on the house?

2 Milk costs 35 pence a pint. How much does it cost after a 14% increase?

3 Petrol costs 89.9 pence a litre. What does it cost after a 2.4% decrease?

4 One can of paint covers an area of 28 m². Harry buys 3 cans of paint to cover 70 m².
What percentage of the paint is used?

5 There are 633 pupils in a school.
230 of the pupils walk to school, 212 travel by bus and 150 come by car.
(a) What percentage walk to school?
(b) What percentage come by car?

18% of the pupils who normally come by car start to travel on a new bus route.
(c) What percentage of the pupils now travel by bus?

6 A car was valued at £13 500 when new. After one year it lost 22% of its value.
At the end of two years it was sold for £8200.
What percentage of its original value did the car lose in its second year?

7 In 2001 house prices increased by 19.6%. In 2002 house prices increased by 17.4%.
A house was valued at £78 000 at the beginning of 2001.
What was the value of the house at the end of 2002?

Percentage increase and decrease

Sadik and Chandni took Maths tests in October and June.

My mark went up from 54% to 72%.

My mark went up from 42% to 60%.

Who has made the most improvement?

They have both improved by a score of 18%, so by one measure they have both improved equally.
Another way of comparing their improvement is to use the idea of a percentage increase.

$$\text{Percentage increase} = \frac{\text{actual increase}}{\text{initial value}} \times 100\%$$

Remember:
To calculate
% increase or % decrease
always use the initial value.

Comparing percentage increases is the best way to decide
whether Sadik or Chandni has made the most improvement.
Explain why.

For Sadik
% increase = $\frac{18}{54} \times 100\% = 33.3\%$

For Chandni
% increase = $\frac{18}{42} \times 100\% = 42.9\%$

Both calculations are correct to one decimal place.

A percentage decrease can be calculated in a similar way.

$$\text{Percentage decrease} = \frac{\text{actual decrease}}{\text{initial value}} \times 100\%$$

EXAMPLES

1 A shop buys pens for 15 pence and sells them for 21 pence.
What is their percentage profit?

Actual profit = 21 pence − 15 pence = 6 pence

$$\% \text{ profit} = \frac{\text{actual profit}}{\text{initial value}} \times 100$$

$$= \frac{6}{15} \times 100 = 40\%$$

2 Pam buys a micro-scooter for £24.
She sells the micro-scooter for £15.
What is her percentage loss?

Actual loss = £24 − £15 = £9

$$\% \text{ loss} = \frac{\text{actual loss}}{\text{initial value}} \times 100$$

$$= \frac{9}{24} \times 100 = 37.5\%$$

Exercise 7.7

Do questions 1 to 5 without a calculator.

1 A shop buys calculators for £5 and sells them for £6.
Find the percentage profit.

2 John's weekly wage rises from £250 to £265.
What is John's percentage wage rise?

3 On Monday peaches cost 15p each. On Tuesday peaches cost 12p each.
What is the percentage reduction in price?

4 The price of a book increases from £8 to £9.
What is the percentage increase in price?

5 A man buys a boat for £25 000 and sells it for £18 000.
Find his percentage loss.

6 The rent on Karen's flat increased from £80 to £90 per week.
(a) Find the percentage increase in her rent.

At the same time Karen's wages increased from £250 per week to £280 per week.
(b) Find the percentage increase in her wages.
Comment on your answers.

7 In October, Sam scored 50% in an English test. In January he improved to 66%.
In the same tests, Becky scored 40% and 56%.
Who has made the most improvement? Explain your answer.

8 A sample of soil is dried in an oven. Its mass reduces from 65 g to 45 g.
Find the percentage decrease in the mass.

9 The value of car A when new was £13 000. The value of car B when new was £16 500.
After one year the value of car A is £11 200 and the value of car B is £13 500.
Calculate the percentage loss in the values of cars A and B after one year.

10 A rectangle has length 12 cm and width 8 cm.
The length is increased by 7.5% and the width decreased by 12.5%.
Find the change in area as a percentage of the original area of the rectangle.

11 In 2001 Miles bought £2000 worth of shares.
In 2002 the value of his shares decreased by 10%.
In 2003 the value of his shares increased by 20%.
By what percentage has the value of his shares changed from 2001 to 2003?

12 The price of a micro-scooter is reduced by 10%.
In a sale, the new price is reduced by a further 10%.
By what percentage has the original price of the micro-scooter been reduced in the sale?

Percentages . . . . Percentages . . . . Percentages . . . .

What you need to know

- 'Per cent' means 'out of 100'. The symbol for per cent is %.
- A percentage can be written as a fraction with denominator 100.

  For example: 10% can be written as $\frac{10}{100}$.

- To change a decimal or a fraction to a percentage - **multiply by 100**.

  For example: 0.12 as a percentage is $0.12 \times 100 = 12\%$.

  $\frac{3}{25}$ as a percentage is $\frac{3}{25} \times 100 = 3 \times 100 \div 25 = 12\%$.

- To change a percentage to a decimal or a fraction - **divide by 100**.

  For example: 18% as a decimal is $18 \div 100 = 0.18$.

  18% as a fraction is $\frac{18}{100}$ which in its simplest form is $\frac{9}{50}$.

- Percentage increase $= \dfrac{\text{actual increase}}{\text{initial value}} \times 100\%$ Percentage decrease $= \dfrac{\text{actual decrease}}{\text{initial value}} \times 100\%$

Increasing and decreasing

200 increased by 10% and then decreased by 10%

$200 \;\rightarrow\; +10\% \;\rightarrow\; 220 \;\rightarrow\; -10\% \;\rightarrow\; 198$

Actual change $= 200 - 198 =$ decrease of 2

$\frac{2}{200} = \frac{2}{200} \times 100 = 2 \times 100 \div 200 = 1\%$

2 as a percentage of 200 is 1%.
The combined result of a 10% increase followed by a 10% decrease
is a 1% decrease.

Investigate increasing and decreasing different quantities by different percentages.
You might find a spreadsheet useful.

Review Exercise 7

Do not use a calculator for questions 1 to 13.

1 Copy and complete the following.

(a) Fraction Decimal Percentage
$\frac{1}{2}$ = ☐ = 50%

(b) Fraction Decimal Percentage
☐ = 0.3 = 30%

(c) Fraction Decimal Percentage
$\frac{29}{100}$ = 0.29 = ☐

AQA

2 (a) What percentage of these squares is shaded?

(b) Copy and shade 15% of this diagram.

3 Toby's last ten holidays were spent in the following countries.

England   France   Spain   England   Italy

France   England   England   Spain   Spain

What percentage of Toby's holidays were spent in England?

90

4 Here are some numbers:  $\frac{3}{5}$ 0.6 $\frac{6}{20}$ $\frac{6}{10}$ 0.06

Which numbers are equal to 60%?

5 Lee eats five sweets from a packet of twenty.
What percentage has he eaten?

AQA

6 (a) Write 0.47 as a percentage.
(b) Write $\frac{7}{20}$ as a percentage.

7 In a sale, prices are reduced by 25%.
What is the sale price of a puzzle which normally costs £1?

8 Write in order of size, lowest first:

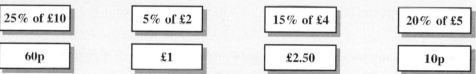

0.41 $\frac{2}{5}$ 39% $\frac{21}{50}$

9 A packet contains 20 biscuits.
(a) Joe eats 20% of the biscuits.
How many biscuits does he eat?
(b) Sylvia eats 9 of the biscuits.
What percentage of the biscuits does she eat?

10 Match the pairs.

25% of £10	5% of £2	15% of £4	20% of £5
60p	£1	£2.50	10p

AQA

11 A bag contains 60 beads.
(a) Emily uses 30% of the beads to make a necklace.
How many beads does she use?
(b) Laura uses 12 beads to make a bracelet.
What percentage of the beads does she use?

12 A recent survey shows there are 20 000 different types of fish in the world.
People catch only 9000 different types.
What percentage of the different types of fish do people catch?

AQA

13 60% of a garden is a rockery. $\frac{1}{4}$ of the garden is a patio.
The rest of the garden is lawn.
What percentage of the garden is lawn?

AQA

14 Jane wants to buy this car.
The deposit is $\frac{2}{5}$ of the price of the car.
Jane's father gives her 30% of the price.
Will this be enough for her deposit?
You must explain your answer fully.

£1800

AQA

15 Tim invests £650 in a building society. He earns 5% interest in the first year.
How much interest does he earn?

16 There are 800 houses on an estate. 45% of the houses are detached.
(a) How many detached houses are on the estate?

520 of the houses on the estate have garages.
(b) What percentage of the houses on the estate have garages?

17 Joe earns £650 in May. In June he earns 20% more.
How much does he earn in June?

AQA

18 A roll of carpet is 20 m long. Beryl buys 18 m of carpet from the roll.
What percentage of the roll did she buy?

19 A train has 1200 seats. 85% of the seats are occupied.
How many seats are unoccupied?

20 In a nine carat gold ring $\frac{9}{24}$ of the weight is pure gold.
What percentage of the weight of the ring is pure gold?

AQA

21 In a local election 3750 people could have voted. 2150 people actually voted.
What percentage of the people voted?

AQA

22 An estate agent makes the following charge for the sale of a house.

Sale price of house	Up to £50 000	Over £50 000
Charge by estate agent	3% of the sale price	3% of the first 50 000 plus 2% on the remainder

Calculate the charge made by the estate agent for a house sold for £229 000.

AQA

23 (a) A year ago Martin was 1.60 m tall. He is now 4% taller.
Calculate his height now.
(b) Martin now weighs 58 kg. A year ago he weighed 51 kg.
Calculate the percentage increase in his weight.
Give your answer to an appropriate degree of accuracy.

AQA

24 (a) In the men's long jump, Leroy improved on his first jump of 8.15 m by 8%.
What was the length of this improved jump?
(b) In the women's long jump, Sabrina improved her jump from 6.40 m to 7.36 m.
What was the percentage increase in her jump?

AQA

25 Emma is organising a Christmas party and prints 400 tickets.
She sells 40% of the tickets to her friends at College, and divides the remainder equally
between Ben, Lisa, Kirsty and Frank.
(a) (i) How many does Kirsty receive?
(ii) Write 40% as a decimal.
(b) Kirsty gives $\frac{1}{3}$ of her tickets away to her friends.
What percentage, of the original 400 tickets, does Kirsty give away?

AQA

26 Jason bought an old bicycle for £36.
He repaired it and resold it for £52.
What was his percentage profit?

AQA

27 Brian buys 300 CDs for £1400. He sells $\frac{3}{4}$ of them at £9 each.
He then reduces the price of the remaining CDs by 30%.
When he has sold 290 CDs, he gives the last 10 to a charity.
(a) How much money does he receive from selling the CDs?
(b) Find the percentage profit which Brian made on these CDs.

AQA

28 In a sale, all prices are reduced by 20%.
On blue cross day, all sale prices are reduced by 50%.
Calculate the prices on blue cross day as a percentage of prices before the sale.

29 In 1998 a house is valued at £240 000.
In 2005 the house is valued at £325 000.
Calculate the percentage increase in the value of the house.

Time and Money

The time of day is given in terms of hours, minutes and sometimes seconds.
There are 24 hours in each day, 60 minutes in each hour and 60 seconds in each minute.

24-hour clock and 12-hour clock times

The time can be given using the 12-hour clock or the 24-hour clock.

The watch and the digital clock both show the same time.
The time on the watch is 5.45 pm using 12-hour clock time.
The digital clock shows 5.45 pm as 1745 using 24-hour clock time.

> **12-hour clock times**
> Times before midday are given as am.
> Times after midday are given as pm.

> **24-hour clock times**
> The first two figures give the hours.
> The last two figures give the minutes.

EXAMPLES

1 A video recorder uses 24-hour clock times.
  (a) What time is shown by the video recorder at 6.30 pm?
  (b) The video is set to record programmes from 1120 to 1645.
      What are these times in 12-hour clock time?

  (a) 6.30 pm is equivalent to 1830.   (b) 1120 is equivalent to 11.20 am.
                                                  1645 is equivalent to 4.45 pm.

> **12-hour to 24-hour clock times**
> Times before midday:
>   use the same figures.
> Times after midday:
>   add 12 to the hours.

> **24-hour to 12-hour clock times**
> Times before midday:
>   use the same figures and
>   include **am**.
> Times after midday:
>   subtract 12 from the hours and
>   include **pm**.

2 A motorist left Liverpool at 10.50 am and arrived in Birmingham at 1.20 pm.
How long did the journey take?

Method 1 (subtraction)

```
1 3.2 0      1.  Write the times as 24-hour clock times.
1 0.5 0      2.  Subtract the minutes.
─────            20 − 50 cannot be done.
                 Exchange 1 hour for 60 minutes.
─────            60 + 20 − 50 = 30 minutes.
  2 6 0
1 3̶.2̶ 0     3.  Subtract the hours.
1 0.5 0          12 − 10 = 2.
─────
  2.3 0
```

Method 2 (adding on)

10.50 to 11.00 =	10 minutes
11.00 to 13.00 = 2 hours	
13.00 to 13.20 =	20 minutes
Total time	= 2 hours 30 minutes

The journey took 2 hours 30 minutes.

1. Write these 12-hour clock times in 24-hour clock time.
   (a) 10.30 am   (b) 10.30 pm   (c) 1.45 am   (d) 1.45 pm   (e) 11.50 pm

2. Write these 24-hour clock times in 12-hour clock time.
   (a) 1415   (b) 0525   (c) 2320   (d) 1005   (e) 1705

3.
   The clock shows the time an alarm goes off in the morning.
   What time does the alarm go off
   (a) in 12-hour clock time,
   (b) in 24-hour clock time?

4. The clocks show the time a school starts in the
   morning and finishes in the afternoon.
   (a) Write these times in 12-hour clock time.
   (b) Write these times in 24-hour clock time.
   (c) How long is the school day?

   Start   **Finish**

5. START: 11:54
   FINISH: 13:35
   A video is set to record a film using 24-hour clock time.
   The start and finish times are shown.
   (a) Write these times in 12-hour clock time.
   (b) How long did the film last?

6. The times of some Monday afternoon ITV programmes are shown.

12 30 News	**1 20** Three minutes
12 55 Shortland Street	**1 25** Home and Away

   (a) Give the times of these programmes using 24-hour clock time.
   (b) How many minutes does Shortland Street last?

7. A coach left Poole at 1340 and arrived in Swanage at 1428.
   (a) What was the arrival time in 12-hour clock time?
   (b) How many minutes did the journey take?

8. A train left Paddington at 1315 and arrived in Exeter at 1605.
   (a) At what time did the train leave in 12-hour clock time?
   (b) How long did the journey take?

9. A train left Manchester at 9.10 am and arrived in Reading at 1.25 pm.
   (a) What was the arrival time in 24-hour clock time?
   (b) How long did the journey take?

10. A coach leaves Bournemouth at 10.50 am to travel to London.
    The journey takes 2 hours 40 minutes.
    At what time does the coach reach London?
    Give your answer in   (a) 12-hour clock time,   (b) 24-hour clock time.

11. A plane flies from Southampton to Jersey.
    The plane leaves Southampton at 1255. The flight takes 48 minutes.
    At what time does the plane arrive in Jersey?
    Give your answer in   (a) the 24-hour clock,   (b) the 12-hour clock.

12. Mrs Hill took 3 hours 56 minutes to drive from Bath to Blackpool.
    She left Bath at 1045. At what time did she arrive in Blackpool?
    Give your answer in   (a) the 24-hour clock,   (b) the 12-hour clock.

Timetables

Bus and rail timetables are usually given in 24-hour clock time.
Here is part of a rail timetable.

Kidderminster	1035	1115	1155	1240	1325	1410
Bewdley	1050	1130	—	1300	—	1430
Arley	1105	1148	—	1318	1403	1448
Highley	1114	1158	—	1328	—	1458
Hampton Loade	1125	1210	—	1340	1425	1510
Bridgnorth	1140	1225	1310	1355	1440	1525

Some trains do not stop at every station. This is shown by a dash on the timetable.

How many minutes does the journey take on the 1035 train from Kidderminster to Arley?

Jean catches the 1155 train from Kidderminster to Bridgnorth.
What is her arrival time in 12-hour clock time?
How long does the journey take?

Alex lives in Bewdley.
What is the time of the last train he can catch to keep an appointment in Bridgnorth at 1.15 pm?

Exercise 8.2

1 The times of rail journeys from Guildford to Reading are shown.

Guildford	1415	1428	1515	1528
North Camp	1427	1444	1527	1544
Wokingham	1441	1508	1541	1608
Reading	1450	1524	1550	1624

(a) Richard catches the 1415 from Guildford to Reading.
  (i) How many minutes does the journey take?
  (ii) What is his arrival time in 12-hour clock time?
(b) Kate catches the 1444 from North Camp to Wokingham.
  (i) How many minutes does the journey take?
  (ii) What is her arrival time in 12-hour clock time?

2 The times of some trains from Hastings to Charing Cross are shown.

Hastings	0702	0802	0857	0900	0957	1102	1257
Crowhurst	—	—	0908	—	1008	—	1308
Battle	0715	0815	0912	—	1012	—	1312
Tunbridge Wells	0745	0845	0943	0940	1043	1141	1342
Sevenoaks	0805	0905	1003	—	1103	1201	1403
Charing Cross	0834	0934	1032	1025	1132	1230	1432

(a) John catches the 0745 from Tunbridge Wells to Charing Cross.
  How many minutes does the journey take?
(b) Aimee catches the 0857 from Hastings to Charing Cross.
  How long does the journey take?
(c) Sarah catches the 1257 from Hastings to Tunbridge Wells.
  What is her arrival time using the 12-hour clock?
(d) Keith wants to be in Charing Cross by 1030.
  What is the latest train he can catch from Battle?

Time and Money . . . Time and Money . . .

3 The times of some bus journeys are shown.

Poole	1210	1225	1240	1255
Ashley Cross	1218	1233	1248	1303
Branksome	1228	1243	1258	1313
Westbourne	1233	1248	1303	1318
Bournemouth	1240	1255	1310	1325

(a) Terry catches the bus at 1258 from Branksome to Westbourne.
    (i) How many minutes does the journey take?
    (ii) What is his arrival time in 12-hour clock time?

(b) Adrian has to be in Westbourne by 1 pm.
    What is the time of the latest bus he can catch from Ashley Cross?

4 The table shows the train services from Oxford to Birmingham.

Oxford	1109	1204	1313	1413	1503	1628	1736
Banbury	1142	1239	1337	—	1521	1652	1800
Leamington	1201	1300	1359	1454	1542	1713	1822
Coventry	1218	1317	1416	1512	1559	1730	1839
Birmingham	1247	1345	1445	1540	1635	1758	1911

(a) Carol catches the 1142 from Banbury to Coventry.
    How long does the journey take?

(b) Arnold needs to be in Birmingham before ten to five in the afternoon.
    What is the latest train he can catch from Oxford?

(c) Debbie arrives at Leamington station at 4.30 pm.
    What time is the next train to Birmingham?

5 The timetable shows some rail journeys from Waterloo to Brookwood.

Waterloo	1510	1520	1523	1538	1540
Clapham Junction	1516	—	1529	—	—
Surbiton	1528	—	—	—	1558
Woking	1542	1546	1549	1603	1612
Brookwood	1547	—	—	—	1617

(a) Nick catches the 1538 from Waterloo to Woking.
    (i) What is his time of arrival in 12-hour clock time?
    (ii) How long does the journey take?

(b) Anne-Marie arrives at Waterloo station at 3.30 pm.
    What time is the next train to Surbiton?

6 Some of the coach services from Woking to Heathrow airport are shown.

Woking	0610	0650	0720	0750	0820	Then	1830	1900	2000
Terminal 1	0650	0730	0800	0830	0900	every	1900	1930	2030
Terminal 2	0655	0735	0805	0835	0905	30	1905	1935	2035
Terminal 3	0700	0740	0810	0840	0910	mins	1910	1940	2040
Terminal 4	0710	0750	0820	0850	0920	until	1920	1950	2050

(a) Helen arrives at Woking at 3 pm.
    She catches the next coach to Heathrow.
    (i) At what time does it leave Woking?
    (ii) At what time does it arrive at Terminal 3?

(b) Leroy needs to be at Terminal 2 at 6 pm.
    What is the latest time he can catch a coach from Woking?

Spending

Spending money is part of daily life.
Every day people have to deal with many different situations involving money.
Money is needed to buy fares for journeys, for purchases at shops, for hiring cars and equipment and for buying large items such as furniture.

When a large sum of money is needed to make a purchase, **credit** may be arranged.
This involves paying for the goods over a period of time by agreeing to make a number of weekly or monthly repayments.
It may also involve paying a **deposit**. The cost of credit may be more than paying cash.

EXAMPLES

1. Hamish pays £2.73 for 1 kg of pears and 2 kg of apples.
   Pears cost 89p per kilogram.
   How much per kilogram are apples?

   2 kg of apples cost   273 − 89 = 184p
   1 kg of apples costs   184 ÷ 2 = 92p
   Apples cost 92p per kilogram.

   > Use the same units:
   > £2.73 is 273p

2. A motor home costs £19 950.
   It can be bought on credit by paying a deposit of £7000 and 36 monthly payments of £395.
   How much more is paid for the motor home when it is bought on credit?

Deposit:	£ 7 000
Payments: £395 × 36 =	£14 220
Credit Price:	£21 220

   Difference: £21 220 − £19 950 = £1270
   Credit price is £1270 more.

Exercise 8.3

Do not use a calculator for questions 1 to 7.

1. Esther pays £30.47 for a wheelbarrow, a fork and a spade.
   The fork costs £7.49. The spade costs £6.99.
   How much does the wheelbarrow cost?

2. Gail pays £2.32 for a packet of cereal, a bag of sugar and a carton of milk.
   The packet of cereal costs £1.38. The sugar costs 65p.
   How much does the carton of milk cost?

3. Mr Grey pays £6 for 2 adult fares and 3 child fares on the bus. The fare for an adult is £1.65.
   How much is the fare for a child?

4. Amy pays £2.12 for a celery, a cucumber and a lettuce.
   The lettuce costs 67p. The cucumber costs 59p.
   How much does the celery cost?

5. Mrs Connor pays £3.27 for 2 kg of bananas and 1.5 kg of apples. Apples cost 90p per kilogram.
   How much per kilogram are bananas?

6. A family pay £6.35 for 4 cups of coffee and 3 cups of tea. A cup of coffee costs 95p.
   How much is a cup of tea?

7. Sam pays £66.40 for 200 bricks and 9 paving slabs. The bricks cost 26p each.
   How much is a paving slab?

8 Mushrooms cost £2.10 per kilogram. Tomatoes cost 96p per kilogram.
James buys 200 g of mushrooms and 500 g of tomatoes.
How much does James have to pay?

9 Mr Jones pays £4.14 for 400 g of Brie and 250 g of Stilton. Stilton costs £7.60 per kilogram.
How much per kilogram is Brie?

10 A building supplier hires out cement mixers.
There is a delivery charge of £15 and a hire charge of £8 per day.
(a) How much would it cost for the delivery and hire of a cement mixer for 4 days?
(b) A builder pays £95 for the delivery and hire of a cement mixer.
For how many days did he hire it?

11 The cost of hiring a carpet cleaner is £7 per day plus a delivery charge of £10.
(a) What is the cost of hiring a carpet cleaner for 2 days, including delivery?
(b) Gus pays a total of £66 to hire a carpet cleaner, including delivery.
For how many days did he hire the cleaner?

12 (a) Alex hires a van for one day and drives 45 miles.
How much is the total hire charge?
(b) Bob hires a van for 3 days.
The total hire charge is £114.
How many miles did Bob drive?

£30 per day
+
20p per mile driven

VAN FOR HIRE

13 The price of a pram is £299.
It can be bought on credit by paying a deposit of £50 and 10 monthly payments of £27.50.
How much more is paid for the pram when it is bought on credit?

14 A car costs £4950.
It can be bought on credit by paying a deposit of £2000 and 24 monthly payments of £149.50.
How much more is paid for the car when it is bought on credit?

15 The cash price of a settee is £900.
It can be bought on credit by paying a deposit of 10% of the cash price and 30 monthly payments of £32.50.
How much more is paid for the settee when it is bought on credit?

16 A washing machine costs £475.
It can be bought on credit by paying a deposit of 10% of the cash price and 24 monthly payments of £19.50.
How much more is paid for the washing machine when it is bought on credit?

Best buys

When shopping we often have to make choices between products which are packed in various sizes and priced differently. If we want to buy the one which gives the better value for money we must compare prices using the same units.

EXAMPLE

Peanut butter is available in small or large jars, as shown.
Which size is the better value for money?

Compare the number of grams per penny for each size.
Small: 250 ÷ 58 = 4.31... grams per penny.
Large: 454 ÷ 106 = 4.28... grams per penny.

The small size gives more grams per penny and is better value.

SMALL
250 g
58p

LARGE
454 g
£1.06

Exercise 8.4 In each question you must show all your working.

1. Milk is sold in 1 pint, 2 pint and 4 pint containers. The cost of a 1 pint container is 35p, the cost of a 2 pint container is 62p and the cost of a 4 pint container is £1.08.
   (a) How much per pint is saved by buying a 2 pint container instead of two 1 pint containers?
   (b) How much per pint is saved by buying a 4 pint container instead of two 2 pint containers?

2. Mushroom soup is sold in two sizes.
   A small tin costs 43p and weighs 224 g. A large tin costs 89p and weighs 454 g.
   Which size gives more grams per penny?

3. Strawberry jam is sold in two sizes.
   A small pot costs 52p and weighs 454 g. A large pot costs 97p and weighs 822 g.
   Which size gives more grams per penny?

4. Jars of pickled onions are sold at the following prices: 460 g at 65p or 700 g at 98p.
   Which size is better value for money?

5. Honey is sold in two sizes.
   A large pot costs £1.98 and weighs 454 g. A small pot costs 86p and weighs 185 g.
   Which size is better value for money?

6. Cottage cheese costs 85p for 120 g, £1.55 for 250 g and £6 for 1 kg.
   Which size is the best value for money?

7. Two bottles of sauce are shown.
   Which size gives better value for money?

Small	Medium
285 g	567 g
38p	74p

8.  Which of these two bottles of "Active" drink is better value for money?

1.5 litre ACTIVE 90p

2 litre ACTIVE £1.30

9. Toothpaste is sold in small, medium and large sizes.
   The small size contains 75 ml and costs 85p.
   The medium size contains 125 ml and costs £1.45.
   The large size contains 180 ml and costs £2.05.
   Which size is the best value for money?

10. Oscar wants to buy a camcorder. He looks at two different advertisements.

DAISY'S

OUR PRICE 30% OFF

Recommended price £640

ALFIE'S

OUR PRICE 1/3 OFF

Recommended price £657

   (a) Find the actual selling price of each camcorder.
   (b) Which camcorder has the bigger discount?

Some goods and services are subject to a tax called **value added tax**, or **VAT**, which is calculated as a percentage of the price or bill. Total amount payable = cost of item or service + VAT

For most purchases the rate of VAT is 17.5%.   For gas and electricity the rate of VAT is 5%. Some goods are exempt from VAT.

EXAMPLE

A bill at a restaurant is   £24 + VAT at 17.5%.
What is the total bill?

VAT: £24 × 0.175 = £4.20

Total bill: £24 + £4.20 = £28.20

The total bill is £28.20.

Remember:
$17.5\% = \frac{17.5}{100} = 0.175$

Exercise 8.5

Do not use a calculator for questions 1 and 2.

1 Naomi's gas bill is £120 plus VAT at 5%. How much VAT does she have to pay?

2 Joe receives an electricity bill for £70 plus VAT at 5%.
   (a) Calculate the amount of VAT charged.   (b) What is the total bill?

3 A washing machine costs £340 plus VAT at 17.5%.
   (a) Calculate the amount of VAT charged.
   (b) What is the total cost of the washing machine?

£340 + VAT

4 A car service costs £90 plus VAT at 17.5%.
   (a) Calculate the amount of VAT charged.   (b) What is the total cost of the service?

5 Mrs Swan receives a gas bill for £179.53. VAT at 5% is added to the bill.
   (a) How much VAT does she have to pay?   (b) What is the total bill?

6 A bike costs £248 plus VAT at 17.5%. What is the total cost of the bike?

7 A ladder costs £145 plus VAT at 17.5%. What is the total cost of the ladder?

8 Joyce buys a greenhouse for £184 plus VAT.
   VAT is charged at 17.5%. What is the total cost of the greenhouse?

9 James receives a telephone bill for £66 plus VAT at 17.5%. How much is the total bill?

10 A loft conversion costs £23 000 plus VAT at 17.5%. What is the total cost?

11 George buys vertical blinds for his windows.
   He needs three blinds at £65 each and two blinds at £85 each.
   VAT at 17.5% is added to the cost of the blinds.
   How much do the blinds cost altogether?

HIRE A CAR

12 A car is hired for two days and driven 90 miles.
   VAT at 17.5% is added to the hire charges.
   How much does it cost to hire the car altogether?

£35 per day
plus
10 pence per mile

Foreign currency

When we go abroad we have to pay for goods and services in the currency of the country we are visiting. We therefore need to change pounds (£) into other currencies.

The rate of exchange varies from day to day.

The table below shows the exchange rates on one day.

EXCHANGE RATE	
Each £ will buy	
European currency	1.55 euros
Japan	173 yen
Malta	0.63 liri
Norway	12.48 krone
Switzerland	2.30 francs
USA	1.42 dollars

EXAMPLE

What is the value, in £s and pence, of 500 Norwegian krone?

$$12.48 \text{ krone} = £1$$
$$500 \text{ krone} = 500 \div 12.48$$
$$= £40.0641\ldots$$
$$500 \text{ krone} = £40.06, \text{ to the nearest penny.}$$

Exercise 8.6

Use the table of exchange rates above to answer these questions.

1 How much will I receive if I change £200 into
   (a) European euros,
   (b) Japanese yen,
   (c) Maltese liri,
   (d) Norwegian krone,
   (e) Swiss francs,
   (f) United States dollars?

2 How much would each item cost in £s? Give your answers to the nearest penny.
   (a) A vase in Switzerland for 90 francs.
   (b) A radio in Japan for 5000 yen.
   (c) A pair of shoes in Italy for 75 euros.
   (d) A meal in Norway for 225 krone.
   (e) A pair of jeans in the United States for 35 dollars.

3 (a) A tourist changes £25 into euros. How many euros does she receive?
   (b) She pays 23.25 euros for a gift. What is the cost of the gift in £s?

4 Norman travels to Switzerland. He changes £120 into francs.
   (a) How many francs does he receive?
   (b) He pays 24.50 francs for a box of chocolates.
        What is the cost of the chocolates in £s?

5 Dolores travels to England from Spain. She changes 600 euros into £.
   (a) How much, in £s and pence, does she receive?
   (b) She buys a theatre ticket for £30.
        What is the cost of the theatre ticket in euros?

6 Marcel travels to England from Norway. He changes 3000 krone into £.
   (a) How much, in £s, does he receive?
   (b) He pays £45 for bed and breakfast.
        What is the cost of bed and breakfast in krone?

7 Sue changes £500 into dollars for a trip to the USA.
   (a) How many dollars does she receive?
   (b) On holiday she spends 680 dollars. She changes the remaining dollars back into £s.
        There is a £3 charge for changing the money.
        How much, in £s, will she receive?

8 Jeff has just returned from Malta. He changes 85 liri back into £s.
   There is a £3 charge for changing the money.
   How much, in £s, will he receive?

9 In France a car costs 9000 euros. In Japan the same car costs 1 million yen.
   In which country is the car cheaper? By how much?

Time and Money . . . . Time and Money . . . .

- Time can be given using either the **12-hour clock** or the **24-hour clock**. When using the 12-hour clock:

  times **before** midday are given as am,

  times **after** midday are given as pm.

- **Timetables** are usually given using the 24-hour clock.

- When considering a **best buy**, compare quantities by using the same units. For example, find which product gives more grams per penny.

- **Value added tax**, or **VAT**, is a tax on some goods and services and is added to the bill.

- **Exchange rates** are used to show what £1 will buy in foreign currencies.

Review Exercise 8

Do not use a calculator for questions 1 to 7.

1 A film starts at 7.50 p.m. It lasts for 145 minutes.
What time does it finish?

AQA

2 (a) A taxi left Russell Square at 1643. It arrived at Waterloo Station 22 minutes later.
At what time did the taxi arrive at Waterloo Station?
Give your answer in 12 hour clock time.

(b) Part of a railway timetable is shown.

London Waterloo	1630	1645	1715	1745	1830	1850
Southampton	1739	1810	1825	1859	1940	2018
Bournemouth	1812	1831	1856	1929	2011	2101
Poole	1825	1905	1907	1942	2023	2116
Weymouth	1913	—	1953	2028	2111	—

George catches the 1745 from London Waterloo to Bournemouth.
How long does the journey take?

AQA

3 Copy and complete the following shopping bill and work out the total.

8 oranges at 32p each	£2.56
7 bananas at 17p each	....
2 punnets of strawberries at £1.12p each	....

AQA

4 The prices of a bottle of red wine and a kilogram of Cheddar Cheese are shown.

What is the total cost of twelve bottles of red wine
and a quarter of a kilogram of Cheddar Cheese?

RED WINE £3.29

CHEDDAR CHEESE £4.96 per kg

AQA

5 Edward pays 81p for 2 pencils and 3 pens.
A pencil costs 12p. How much does a pen cost?

6 Packets of chocolate biscuits are sold in two sizes.

Standard Packet £1.09
Contents: 12 biscuits

Large Packet £3.07
Contents: 36 biscuits

Which size is better value for money? Show your working.

AQA

7 (a) While in the USA, John pays $30 for a pair of trousers. The exchange rate is $1.50 to £1. Calculate the cost of the pair of trousers in £.

(b) John also buys an ink cartridge which costs $25.
He has to pay a tax of 6% which is added on to this cost.
Calculate the total cost of the ink cartridge in $.

AQA

8 David buys 0.6 kg of grapes and 0.5 kg of apples. He pays £1.36 altogether.
The grapes cost £1.45 per kilogram.
How much per kilogram are apples?

AQA

9 (a) Anita hires a van for one day.
She drives 68 miles.
How much is the hire charge?

(b) John hires a van for two days.
The total hire charge is £90.72.
How many miles did he drive?

SOUTHERN RENTAL

Van Hire Charges

£36 per day
plus
12 pence per mile

AQA

10 Karen buys the television on credit.

WIDE SCREEN TELEVISION

Cash Price £950

Credit Price
£50 Deposit
Plus 10 payments of £99.50

(a) How much is the credit price?
(b) How much more does she pay than the cash price?

11 Josie decides to buy a motorcycle on credit.
The credit terms are:

Cash price	£1680.
No deposit,	
Credit charge	12.5% of cash price.

The total cost is the credit charge plus the cash price.
She pays by 12 equal monthly payments.
How much does she pay each month?

AQA

12   Jam is sold in two sizes.
A large pot of jam costs 88p and weighs 822 g.
A small pot of jam costs 47p and weighs 454 g.
Which pot of jam is better value for money?
You must show all your working.

AQA

13 An electricity bill is £73.28 plus VAT at 5%. Calculate the VAT charged.

AQA

14 A PC costs £499 + VAT at $17\frac{1}{2}\%$.
What is the total cost of the PC?

15 A bottle of lemonade costs 95 pence in London.
In Rome, the same size bottle costs 1.20 euros.
There are 1.60 euros to £1.
In which city does the lemonade cost more? You should show all your working.

AQA

CHAPTER 9

Personal Finance

Wages

Hourly pay

Many people are paid by the hour for their work. In most cases they receive a **basic hourly rate** for a fixed number of hours and an **overtime rate** for any extra hours worked.

EXAMPLE

A car-park attendant is paid £6.20 per hour for a basic 40-hour week.
Overtime is paid at time and a half.
One week an attendant works 48 hours.
How much does he earn?

Basic Pay: £6.20 × 40 = £248.00
Overtime: 1.5 × £6.20 × 8 = £74.40
                  Total pay = £322.40

Overtime paid at 'time and a half' means 1.5 × normal hourly rate.
In this example, the hourly overtime rate is given by:
1.5 × £6.20

Common overtime rates are 'time and a quarter', 'double time', etc.

Commission

As an incentive for their employees to work harder some companies pay a basic wage (fixed amount) plus commission.
The amount of commission is usually expressed as a percentage of the value of the sales made by the employee.

EXAMPLE

An estate agent is paid a salary of £18 000 per year plus commission of 0.5% on the sales of all houses.
Last year the estate agent sold houses to the value of £2 640 500.
How much did the estate agent earn last year?

Annual salary: £18 000
Commission: 0.005 × £2 640 500 = £13 202.50
                  Total pay = £31 202.50

Remember:
$0.5\% = \frac{0.5}{100} = 0.005$

Exercise 9.1

Do not use a calculator for questions 1 and 2.

1. Helen is paid £6.50 per hour.
   She works 20 hours a week.
   How much does she earn each week?

2. John does a part-time job for 7 hours a week.
   He is paid £52.50 a week.
   What is his hourly rate of pay?

3 Mike earns £364.80 a week.
He is paid £9.60 per hour.
How many hours a week does Mike work?

4 Jean earns a basic £7.20 an hour.
When she works overtime she is paid at time and a half.
How much is she paid for 1 hour of overtime?

5 On Saturday, Ros works for 3 hours 20 minutes.
She is paid £6.90 per hour.
How much does Ros earn on Saturday?

6 Burt starts work at 0845 and finishes work at 1300 every day.
He is paid £6.20 per hour.
How much does Burt earn each day?

7 Amrit is paid at double time for working on a Bank Holiday.
His basic rate of pay is £7.50 per hour.
How much is Amrit paid for working 8 hours on a Bank Holiday?

8 A secretary is paid a basic £192 for working 30 hours a week.
If she works overtime she is paid at one and a half times her basic hourly rate.
How much is she paid for 1 hour of overtime?

9 Tom is paid at time and a half for overtime.
His overtime rate of pay is £9.90 per hour.
What is his basic rate of pay per hour?

10 A chef is paid £12.40 per hour for a basic 38-hour week.
Overtime is paid at time and a half.
How much does the chef earn in a week in which she works 50 hours?

11 A mechanic is paid £9.80 per hour for a basic 40-hour week.
Overtime is paid at time and a quarter.
One week the mechanic works 42 hours.
How much does he earn?

12 A hairdresser is paid £8.20 per hour for a basic 35-hour week.
One week she works two hours overtime at time and a half and $3\frac{1}{2}$ hours overtime at
time and a quarter.
How much is she paid that week?

13 A driver is paid £68.85 for $4\frac{1}{2}$ hours of overtime.
Overtime is paid at time and a half.
What is his basic hourly rate of pay?

14 A furniture salesperson is paid an annual salary of £14 000 plus commission of 2% on sales.
Last year the salesperson sold £500 000 worth of furniture.
How much did the salesperson earn last year?

15 A car salesperson is paid an annual salary of £12 500 plus commission of 2.5% on sales.
How much does the salesperson earn in a year in which cars to the value of £868 000 are sold?

16 A double glazing salesperson is paid £620 per month plus commission of 5% on sales.
How much does he earn in a month in which he makes sales of £22 600?

17 An estate agent is paid 1.5% commission on the sales of houses.
How much commission does he earn for selling a house for one million pounds?

Income tax

The amount you earn for your work is called your **gross pay**.
Your employer will make deductions from your gross pay for income tax, National Insurance, etc.
Pay after all deductions have been made is called **net pay**.

The rates of tax and the bands (ranges of income) to which they apply vary.

The amount of **income tax** you pay will depend on how much you earn.
Everyone is allowed to earn some money which is not taxed, this is called a **tax allowance**.
Any remaining money is your **taxable income**.

EXAMPLE

George earns £6310 per year.
His tax allowance is £4895 per year and he pays tax at 10p in the £ on his taxable income.
How much income tax does George pay per year?

Taxable income: £6310 − £4895 = £1415
   Income tax: £1415 × 0.10   = £141.50

George pays income tax of £141.50 per year.

An income tax rate of 10% is often expressed as '10p in the pound (£)'.

Exercise 9.2

1   Joan earns £12 680 per year.
   Her tax allowance is £4895 per year.
   What is her taxable income?

2   Tony is paid £294 per week for 52 weeks a year.
   His tax allowance is £6505 per year.
   What is his annual taxable income?

3   Lyn earns £6080 per year.
   Her tax allowance is £4895 per year and she pays tax at 10p in the £ on her taxable income.
   (a)   What is her annual taxable income?
   (b)   How much income tax does she pay per year?

4   Brian earns £684 per month.
   His tax allowance is £6505 per year and he pays tax at 10p in the £ on his taxable income.
   (a)   What is his annual taxable income?
   (b)   How much income tax does he pay per year?
   (c)   How much income tax does he pay per month?

5   Kay has an annual salary of £23 980.
   Her tax allowance is £4895 per year.
   She pays tax at 10p in the £ on the first £2090 of her taxable income and 22p in the £ on the remainder.
   How much income tax does she pay per year?

6   Julie earns £990 per month.
   Her tax allowance is £4895 per year and she pays tax at 10p in the £ on the first £2090 of her taxable income and 22p in the £ on the remainder.
   How much income tax does she pay per month?

7   Jim is paid £292 per week for 52 weeks a year.
   His tax allowance is £4895 per year.
   He pays tax at 10p in the £ on the first £2090 of his taxable income and 22p in the £ on the remainder.
   How much income tax does he pay per week?

8 Reg has an annual salary of £49 880. His tax allowance is £4895 per year.
He pays tax at 10p in the £ on the first £2090 of his taxable income, 22p in the £ on the next £30 310 and 40p in the £ on the remainder.
Calculate how much income tax he pays per year.

9 Alex has an annual salary of £41 240. Her tax allowance is £4895 per year.
She pays tax at 10p in the £ on the first £2090 of her taxable income, 22p in the £ on the next £30 310 and 40p in the £ on the remainder. She is paid monthly.
How much income tax does she pay per month?

Household bills

The cost of living includes many bills for services provided to our homes. Electricity, gas and telephone charges are all examples of **quarterly bills** which are sent out four times a year.
Some bills are made up of two parts:
    A fixed (standing) charge, for providing the service.
    A charge for the quantity of the service used (amount of electricity, duration of telephone calls, etc.)

Other household bills include taxes payable to the local council, water charges and the cost of the insurance of the house (structure) and its contents.

EXAMPLE

Mrs Davis receives a quarterly bill for electricity.
The standing charge for the quarter is £10.40 and she has used 264 units of electricity at 6.85p per unit.
How much is her electricity bill?

Cost of units used: 264 × 6.85p = 1808.4p
                                            = 1808p, to the nearest penny.

Electricity bill = standing charge + cost of units used
                       = £10.40 + £18.08
                       = £28.48

Her electricity bill is £28.48

> ○○○○○○○○○○
> The actual cost is rounded down to the nearest penny.
>
> Change the cost of units to £.

Exercise 9.3

1 Last year the Evans family received four quarterly gas bills.
    March: £134.26   June: £52.00   September: £33.49   December: £80.25
  (a) What was their total bill for the year?
  (b) The family can pay for their gas by 12 equal monthly instalments.
       How much would each instalment be?

2 Last year the Price family used 2734 units of electricity. Each unit costs 6.85 pence.
Calculate the cost of the units used in £ and pence.

3 Mrs Cotton uses 1064 units of electricity during one quarter.
Find the cost of her electricity bill if each unit costs 6.85 pence and the quarterly charge is £10.40.

4 During one quarter Mr Singh uses 5934 kWh of gas.
Calculate the cost of the gas used if each kWh costs 1.714 pence.

5 Mr Jones receives an electricity bill for £56.84.
The bill includes a quarterly charge of £10.40 and the cost per unit is 6.85 pence.
Calculate to the nearest whole number, the number of units he has used.

6 Mrs Madan has to pay £121.83 for the gas used during one quarter.
The cost of the gas used is 1.714 pence per kWh.
How many kWh of gas has she used?
Give your answer to a suitable degree of accuracy.

7 John pays his council tax by 10 instalments.
His first instalment is £98.65 and the other 9 instalments are £97 each.
How much is his total council tax?

8 Mr Peters has an annual council tax of £1982.28.
He pays the council tax in 10 instalments.
The first instalment is £200.28 and the remaining amount is payable in 9 instalments of equal value.
How much is the second instalment?

9 Mrs Dear checks her water bill. She has used 58 cubic metres of water at 97.04 pence per cubic metre and there is a standing charge of £11.
How much is her bill?

10 The table shows the premiums charged by an insurance company to insure a house and its contents.

	Buildings and Contents Insurance	
	Buildings	Contents
Annual premium for each £1000 insured	£1.50 Minimum £20 per year	£5.00

(a) Jim has bought a house valued at £154 000.
How much would he pay to insure the house?
(b) The cost for Mr Brown to insure his house is £315.
What is the value of his house?
(c) Mrs Crow insures the contents of her house for £24 000.
What is the annual premium?
(d) Andy insures his flat valued at £84 000 and its contents valued at £19 000.
Calculate the total cost of the insurance premium.

11 George insures his house valued at £284 000 and its contents valued at £27 500.
The annual premiums for the insurance are:

> Buildings: £1.35 per £1000 of cover,   Contents: 56p per £100 of cover.

Calculate the total cost of the insurance premium.

12 Naomi rents a flat and pays £69.44 to insure its contents.
Contents insurance costs 56p for each £100 insured.
For how much are the contents insured?

13 The table shows the monthly payments for loans.

	12 MONTHS	24 MONTHS
LOAN, £	Monthly repayment	Monthly repayment
5000	492.95	287.20
3000	295.79	172.32
2000	197.15	114.88

Marc takes out a loan for £3000 over a period of 12 months.
Holly takes out a loan for £5000 over a period of 24 months.
(a) What is the difference in their monthly repayments?
(b) What is the total amount that Holly is charged for her loan?

Savings

Money invested in a savings account or a bank or building society earns **interest**, which is usually paid once a year. When the interest is paid out each year and not added to your account it is called **Simple Interest**.

The amount of Simple Interest an investment earns can be calculated using:

Simple Interest $= \dfrac{\text{Amount}}{\text{invested}} \times \dfrac{\text{Time in}}{\text{years}} \times \dfrac{\text{Rate of interest}}{\text{per year}}$

Banks and building societies advertise the **yearly rates** of interest payable.

For example, 6% per year.

Interest, usually calculated annually, can also be calculated for shorter periods of time.

EXAMPLE

Find the Simple Interest paid on £600 invested for 6 months at 8% per year.

Simple Interest
$= 600 \times \dfrac{6}{12} \times \dfrac{8}{100}$
$= 600 \times 0.5 \times 0.08$
$= £24$

The Simple Interest paid is £24.

Note: Interest rates are given 'per year'. The length of time for which an investment is made is also given in years.
6 months $= \dfrac{6}{12}$ years. *Explain why.*

Exercise 9.4

Do not use a calculator for questions 1 and 2.

1 Find the simple interest paid on £200 for 1 year at 5% per year.

2 Calculate the simple interest on £500 invested at 6% per year after
(a) 1 year, (b) 6 months.

3 A school savings account pays interest at 8% per year.
Find the simple interest paid on savings of £50 after 6 months.

4 Calculate the simple interest paid on an investment of £6000 at 7.5% per year after 6 months.

5 Find the simple interest on £800 invested for 9 months at 8% per year.

What you need to know

- **Hourly pay** is paid at a **basic rate** for a fixed number of hours.
  Overtime pay is usually paid at a higher rate such as time and a half, which means each hour's work is worth 1.5 times the basic rate.
- Everyone is allowed to earn some money which is not taxed. This is called a **tax allowance**.
- Tax is only paid on income earned in excess of the tax allowance. This is called **taxable income**.
- Gas, electricity and telephone bills are paid **quarterly**.
  Some bills consist of a standing charge plus a charge for the amount used.
- Money invested in a savings account at a bank or building society earns **interest**.
  Simple Interest is when the interest is paid out each year and not added to your account.
  Simple Interest = Amount invested × Time in years × Rate of interest per year.

Review Exercise 9

Do not use a calculator for questions 1 to 5.

1 Mr Blaney gets four telephone bills each year.
He pays for his bills by 12 equal monthly payments.
Last year his telephone bills were: £46.48, £49.36, £45.65 and £50.51.
How much was each monthly payment?

2 David works 30 hours at £6.60 per hour. How much does he earn?

3 Edward has an annual income of £5795. He has a tax allowance of £4895.
(a) Calculate Edward's taxable income.

He pays tax at the rate of 10p in the £ on his taxable income.
(b) How much tax does Edward pay per year?   AQA

4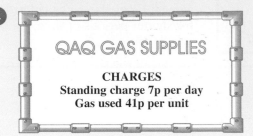

Rachel's gas meter was read on 1 March.

The reading was | 1 | 7 | 4 | 2 |

92 days later the meter was read again.

The reading was | 1 | 9 | 5 | 6 |

Calculate the total gas bill that Rachel will have to pay for the 92 days from 1 March.   AQA

5 Noreen insures her house for £91 000 and the contents for £18 000.
The annual premiums for the insurance are:

> Buildings: 21p per £100 of cover,   Contents: 98p per £100 of cover.

Noreen calculates the total annual cost of the premiums to be £3675.
Use **estimation** to check Noreen's calculation to see if she is correct.
You **must** show all your working.   AQA

6 Andrew works for 3 hours 20 minutes. He is paid £8.40 per hour.
How much does Andrew earn?   AQA

7 Yasmine earns £7.20 per hour for working a basic week of 38 hours.
Overtime is paid at time and a half. The table shows the hours Yasmine worked last week.

Day	Monday	Tuesday	Wednesday	Thursday	Friday
Hours	10	8	7	9	12

(a) How many hours overtime did she work last week?
(b) Calculate her total pay for last week.

8 Bernadette has to pay tax at the rate of 10p in the £ on the first £2090 of her taxable income and 22p in the £ on the remainder. Her taxable income is £2967.
How much tax does she have to pay?

9 The table shows the monthly repayments for different loans.

Amount of loan	Period of loan (months)		
	24	**36**	**60**
£5000	242.05	168.14	107.29
£7500	363.08	252.21	160.94
£10 000	484.10	336.28	214.58

Davina borrows £7500. She repays the loan over a period of 60 months.
Calculate the amount of interest she has to pay.

10 Sophie invests £1550 in a building society at 9% interest per year.
How much money will she have after 6 months?   AQA

11 Hannah invests £360 in a building society account at 4.8% per year.
Find the simple interest paid on her investment after 4 months.

CHAPTER 10

Ratio and Proportion

Ratio

 Some faces are
SMILERS.

Some faces are
GLUMS.

For the ratio 3 : 2 say 3 to 2.

In a group of 10 faces the **ratio** of SMILERS to GLUMS is 3 : 2.
This means that for every three SMILERS there are two GLUMS.

In the group there are 6 SMILERS and 4 GLUMS.

EXAMPLES

1 In a group of 16 faces there are 10 SMILERS and 6 GLUMS.
What is the ratio of SMILERS to GLUMS?

For every 5 SMILERS there are 3 GLUMS. So the ratio of SMILERS to GLUMS is 5 : 3.

2 Draw 15 faces where the ratio of SMILERS to GLUMS is 4 : 1.

For every 4 SMILERS there is 1 GLUM.
So draw sets of 4 SMILERS and 1 GLUM until there
are a total of 15 faces.

How many sets of 4 SMILERS and 1 GLUM *are there?*

Exercise 10.1

1 In a group of 14 faces there are 10 SMILERS and 4 GLUMS.

Copy and complete the following.
(a) For every 5 SMILERS there are ...... GLUMS.
(b) The ratio of SMILERS to GLUMS is 5 : ....

2 In a group of 12 faces there are 9 SMILERS and 3 GLUMS.

Copy and complete the following.
(a) For every ...... SMILERS there is 1 GLUM.
(b) The ratio of SMILERS to GLUMS is ... : 1.

3 In a group of 24 faces there are 15 SMILERS and 9 GLUMS.
Copy and complete the following.
(a) For every 5 SMILERS there are …… GLUMS.
(b) The ratio of SMILERS to GLUMS is 5 : ….

4 In a group of 30 faces there are 21 SMILERS and 9 GLUMS.
Copy and complete the following.
The ratio of SMILERS to GLUMS is … : 3.

5 (a) Draw 10 faces where the ratio of SMILERS to GLUMS is 4 : 1.
(b) Draw 12 faces where the ratio of SMILERS to GLUMS is 1 : 3.

6 The ratio of SMILERS to GLUMS is 5 : 2.

(a) How many SMILERS are there when there are 30 GLUMS?
(b) How many GLUMS are there when there are 30 SMILERS?
(c) How many FACES are there when there are …
(i) 40 SMILERS, (ii) 40 GLUMS?

7 The ratio of SMILERS to GLUMS is 4 : 3.

(a) How many SMILERS are there when there are 12 GLUMS?
(b) How many GLUMS are there when there are 12 SMILERS?
(c) How many FACES are there when there are …
(i) 48 SMILERS, (ii) 48 GLUMS?

8 How many SMILERS and how many GLUMS are there when …
(a) the ratio of SMILERS to GLUMS is 7 : 3 and there are 20 faces?
(b) the ratio of SMILERS to GLUMS is 3 : 2 and there are 15 faces?

9 (a) Look at this group of faces.

(i) What fraction of the faces are GLUMS?
(ii) What fraction of the faces are SMILERS?
(iii) What is the ratio of GLUMS to SMILERS?
(b) In another group of faces $\frac{1}{4}$ are GLUMS.
What is the ratio of SMILERS to GLUMS?

10 (a) In this group of faces the ratio of SMILERS to GLUMS is 7 : 3.

(i) What percentage of the faces are GLUMS?
(ii) What percentage of the faces are SMILERS?
(b) In another group of faces 40% are GLUMS.
What is the ratio of SMILERS to GLUMS?

11 In a group of faces the ratio of SMILERS to GLUMS is 2 : 3.
What fraction of the faces are SMILERS?

12 In a group of faces the ratio of SMILERS to GLUMS is 3 : 1.
What percentage of the faces are SMILERS?

Equivalent ratios

Ratios are used only to **compare** quantities. They do not give information about actual values.

For example.
A necklace is made using red beads and white beads in the ratio **3 : 4**.
This gives no information about the actual numbers of beads in the necklace.
The ratio **3 : 4** means that for every 3 red beads in the necklace there are 4 white beads.
The **possible** numbers of beads in the necklace are shown in the table.

Red beads	3	6	9	12
White beads	4	8	12	16
Total beads	7	14	21	28

Make similar tables when the ratio of red beads to white beads in the necklace is:

(a) 4 : 5   (b) 2 : 3   (c) 3 : 1

The ratios 3 : 4, 6 : 8, 9 : 12, …
are different forms of the **same** ratio.
They are called **equivalent** ratios.

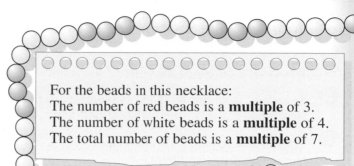

For the beads in this necklace:
The number of red beads is a **multiple** of 3.
The number of white beads is a **multiple** of 4.
The total number of beads is a **multiple** of 7.

Simplifying ratios

To simplify a ratio divide both of the numbers in the ratio by the **same** number.
A ratio with whole numbers which cannot be simplified is in its **simplest form**.

EXAMPLES

1 Find 3 ratios that are equivalent to the ratio 2 : 1.
$2 \times 2 : 1 \times 2 = 4 : 2$
$2 \times 3 : 1 \times 3 = 6 : 3$
$2 \times 4 : 1 \times 4 = 8 : 4$
3 ratios equivalent to the ratio 2 : 1 are 4 : 2, 6 : 3 and 8 : 4.

Finding equivalent ratios:
To find equivalent ratios multiply or divide each number in the ratio by the **same** number.

2 The ratio of boys to girls in a school is 3 : 4.
There are 72 boys. How many girls are there?

$72 \div 3 = 24$
To find a ratio equivalent to 3 : 4 where the first number in the ratio is 72,
multiply each number in the ratio by 24.
$3 \times 24 : 4 \times 24 = 72 : 96$
The number of girls = 96.

3 Write the ratio 15 : 9 in its simplest form.

15 and 9 can both be divided by 3.
$15 \div 3 : 9 \div 3 = 5 : 3$
5 : 3 cannot be simplified.
The ratio 15 : 9 in its simplest form is 5 : 3.

In its simplest form a ratio contains **only** whole numbers.
There are **no** units.
In order to simplify the ratio both quantities in the ratio must be in the **same units**.

4 Write the ratio 2 cm : 50 mm in its simplest form.

$2\,cm : 50\,mm = 20\,mm : 50\,mm = 20 : 50$
Divide both parts of the ratio by 10.
$20 \div 10 : 50 \div 10 = 2 : 5$
The ratio 2 cm : 50 mm in its simplest form is 2 : 5.

Do not use a calculator in this exercise.

1 Give three ratios equivalent to each ratio. (a) 6 : 1 (b) 7 : 2 (c) 3 : 5

2 Give the simplest form of each of these ratios.
(a) 3 : 6 (b) 9 : 27 (c) 9 : 12 (d) 10 : 25 (e) 30 : 40 (f) 22 : 55
(g) 9 : 21 (h) 18 : 8 (i) 36 : 81 (j) 35 : 15

3 Each of these pairs of ratios are equivalent.
(a) 3 : 4 and 9 : n. (b) 2 : 7 and 8 : n. (c) 8 : n and 2 : 25. (d) 25 : n and 5 : 4.
In each case calculate the value of n.

4 The heights of two friends are in the ratio 7 : 9. The shorter of the friends is 154 cm tall.
What is the height of the taller of the friends?

5 Sugar and flour are mixed in the ratio 2 : 3.
How much sugar is used with 600 g of flour?

6 The ratio of boys to girls in a school is 4 : 5. There are 80 girls.
How many boys are there?

7 I earn £300 per week.

I earn £270 per week.

The amounts Jenny and James earn
is in the ratio of their ages.
Jenny is 20 years old. How old is James?

8 A necklace contains 30 black beads and 45 gold beads.
What is the simplest form of the ratio of black beads to gold beads on the necklace?

9 On Monday a hairdresser uses 800 ml of shampoo and 320 ml of conditioner.
Write in its simplest form the ratio of shampoo : conditioner used on Monday.

10 Denise draws a plan of her classroom. On her plan Denise uses 2 cm to represent 5 m.
Write the scale as a ratio in its simplest form.

11 On a map a pond is 3.5 cm long. The pond is actually 52.5 m long.
Write the scale as a ratio in its simplest form.

12 Write each of these ratios in its simplest form.
(a) £2 : 50p (b) 20p : £2.50 (c) £2.20 : 40p (d) 6 m : 240 cm
(e) 2 kg : 500 g (f) 1 kg : 425 g (g) 90 cm : 2 m (h) 5 km : 200 m
(i) 20 seconds : 5 minutes (j) $\frac{1}{2}$ minute : 15 seconds

13 Sam spends 90p a week on comics. Tom spends £4 a week on comics.
Write the ratio of the amounts Tom and Sam spend on comics in its simplest form.

14 The cards in a pack are marked ☒ or ⊡   25% of the cards are marked ☒

What is the ratio of cards marked ☒ : ⊡ in its simplest form?

15 An alloy is made of tin and zinc. 40% of the alloy is tin.
What is the ratio of tin : zinc in its simplest form?

16 A box contains blue biros and red biros. $\frac{1}{3}$ of the biros are blue.
What is the ratio of blue biros to red biros in the box?

17 A necklace is made from 40 beads. $\frac{2}{5}$ of the beads are white. The rest of the beads are red.
Find the ratio of the number of red beads to the number of white beads in its simplest form.

EXAMPLES

1 James and Sally share £20 in the ratio 3 : 2.
How much do they each get?

Add the numbers in the ratio.   3 + 2 = 5
For every £5 shared: James gets £3,   Sally gets £2.
20 ÷ 5 = 4   There are 4 shares of £5 in £20.
James gets   £3 × 4 = £12.   Sally gets   £2 × 4 = £8.
So, James gets £12 and Sally gets £8.

2 A necklace is made using red beads and gold beads in the ratio 7 : 3.
A total of 30 beads are used in the necklace.

Calculate the number of red beads and gold beads used to make the necklace.

Add the numbers in the ratio.   7 + 3 = 10
Number of shares.   30 ÷ 10 = 3
Red beads   7 × 3 = 21.   Gold beads   3 × 3 = 9.
There arc 21 red beads and 9 gold beads.

Exercise 10.3   Do not use a calculator for questions 1 to 6.

1 (a) Share 9 in the ratio 2 : 1.   (b) Share 20 in the ratio 3 : 1.
   (c) Share 35 in the ratio 1 : 4.   (d) Share 100 in the ratio 9 : 1.

2 A bag contains 12 toffees. Bruce and Emily eat them at the ratio 1 : 2.
How many toffees does Emily eat?

3 A box contains gold coins and silver coins. The ratio of gold coins to silver coins is 1 : 9.
There are 20 coins in the box. How many silver coins are in the box?

4 Sunny and Chandni share £48 in the ratio 3 : 1. How much do they each get?

5 Anouska spent £35 on travel and admission to a pop concert.
The cost of travel to the cost of admission was 1 : 4.
What was the cost of admission?

6 Copy and complete.

		Quantity				
		40 marbles	20 sweets	80 kg	200 g	£1200
Shared in the ratio	4 : 1					
	3 : 2					

7 (a) Share £35 in the ratio 2 : 3.   (b) Share £56 in the ratio 4 : 3.
   (c) Share £5.50 in the ratio 7 : 4.   (d) Share £4.80 in the ratio 3 : 5.

8 A necklace contains 72 beads. The ratio of red beads to blue beads is 5 : 3.
How many red beads are on the necklace?

9 £480 is shared in the ratio 7 : 3.
What is the difference between the larger share and the smaller share?

10 In 1901, the total population of England and Wales was 32 528 000.
The ratio of the population of England to the population of Wales was 15 : 1.
What was the population of Wales in 1901?

11 A bag contains red beads and black beads in the ratio 1 : 3. What fraction of the beads are red?

12 A box contains red biros and black biros in the ratio 1 : 4.
What percentage of the biros are black?

13 In a school the ratio of the number of boys to the number of girls is 3 : 5.
What fraction of the pupils in the school are girls?

14 The ratio of non-fiction books to fiction books in a library is 2 : 3.
Find the percentage of fiction books in the library.

15 John is 12 years old and Sara is 13 years old. They share some money in the ratio of their ages.
What percentage of the money does John get?

16 In the UK there are 240 939 km² of land.
The ratio of agricultural land to non-agricultural land is approximately 7 : 3.
Estimate the area of land used for agriculture.

17 At the start of a game Jenny and Tim have 40 counters each.
At the end of the game the number of counters that Jenny and Tim each have is in the ratio 5 : 3.
   (a)   How many counters do Jenny and Tim have at the end of the game?
   (b)   How many counters did Jenny win from Tim in the game?

18 On a necklace, for every 10 black beads there are 4 red beads.
   (a)   What is the ratio of black beads to red beads in its simplest form?
   (b)   If the necklace has 15 black beads how many red beads are there?
   (c)   If the necklace has a total of 77 beads how many black beads are there?
   (d)   Why can't the necklace have a total of 32 beads?

19 The lengths of the sides of a triangle are in the ratio 4 : 6 : 9.
The total length of the sides is 38 cm. Calculate the length of each side.

20 To make concrete a builder mixes gravel, sand and cement in the ratio 4 : 2 : 1.
The builder wants 350 kg of concrete. How much gravel does the builder need?

21 The angles of a triangle are in the ratio 2 : 3 : 4. Calculate each angle.

22 A bag contains some red, green and black sweets. 30% of the sweets are red.
The ratio of the numbers of green sweets to black sweets is 5 to 9.
What percentage of the total number of sweets are black?

23 Alan, Beth and Catrina share some money in the ratio 1 : 3 : 4.
   (a)   What percentage of the money do they each receive?
   (b)   What fraction of Beth's share is Alan's share?
   (c)   What fraction of Alan and Catrina's share is Beth's share?

Proportion

Some situations involve comparing **different** quantities.
For example, when a motorist buys fuel the more he buys the greater the cost.
In this situation the quantities can change but the ratio between the quantities stays the same.
When two different quantities are in the **same ratio** they are said to be in **direct proportion**.

EXAMPLE

4 cakes cost £1.20.
Find the cost of 7 cakes.

4 cakes cost £1.20
1 cake costs   £1.20 ÷ 4 = 30p
7 cakes cost   30p × 7 = £2.10   So, 7 cakes cost £2.10.

This is sometimes called the **unitary method**.
   (a)   **Divide** by 4 to find the cost of **1** cake.
   (b)   **Multiply** by 7 to find the cost of 7 cakes.

1 5 candles cost 80 pence.
   (a) What is the cost of 1 candle?
   (b) What is the cost of 8 candles?

2 Georgina works for 4 hours and earns £28.
   (a) How much does she earn in 1 hour?
   (b) How much does she earn in 10 hours?

3 5 bananas cost £1.50.
   (a) What is the cost of 1 banana?
   (b) What is the cost of 8 bananas?

£1.50

4 Gina works for 10 hours and earns £65.
   (a) How much does Gina earn in 1 hour?
   (b) How much does Gina earn in 20 hours?

5 Alistair pays £1.90 for 2 cups of tea.
   How much would he pay for 3 cups of tea?

6 Jean pays £168 for 10 square metres of carpet.
   How much would 12 square metres of carpet cost?

7 Alfie is paid £48.80 for working 4 hours overtime.
   How much would he be paid for 5 hours overtime?

8 Aimee pays £1.14 for 3 kg of potatoes.
   How much would 7.5 kg of potatoes cost?

3 kg   **7.5 kg**

9 5 litres of petrol costs £4.50.
   How much would 18 litres of petrol cost?

10 These ingredients are needed to make an apple crumble for 6 people.

540 g apples	150 g flour
75 g butter	75 g sugar

   (a) How much sugar is needed to make an apple crumble for 4 people?
   (b) How much apple is needed to make an apple crumble for 8 people?

11 9 metres of stair carpet cost £83.70.
   How much does 9.6 metres cost?

12 29 euros is about the same as £20. Sue spends 47 euros.
   How many pounds is this?

13 Mary phones her uncle in New York.
   Phone calls to New York are charged at the rate of £2.20 for a 5-minute call.
   (a) How much would a 7-minute call to New York cost?
   (b) Mary's call cost £5.28.
      How long was her call?

14 These ingredients are needed to make macaroni cheese for 4 people.

Macaroni … 120 g	Cheese … 72 g	Flour … 30 g	Milk … 850 ml

   (a) How much cheese is needed to make macaroni cheese for 10 people?
   (b) How much milk is needed to make macaroni cheese for 3 people?
   (c) How much macaroni is needed to make macaroni cheese for 7 people?

10

Ratio and Proportion

15 A car travels 6 miles in 9 minutes. If the car travels at the same speed:
    (a)   how long will it take to travel 8 miles,   (b)   how far will it travel in 24 minutes?

16 5 litres of paint cover an area of 30 m².
    (a)   What area will 2 litres of paint cover?   (b)   How much paint is needed to cover 72 m²?

17 A school is organising three trips to the zoo.

Our trip is on Monday.
45 students are going.
The total cost is £468.

Our trip is on Tuesday.
25 students are going.

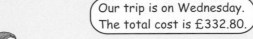

Our trip is on Wednesday.
The total cost is £332.80.

    (a)   How much does Tuesday's trip cost?
    (b)   How many students are going to the zoo on Wednesday?

18 A 42-litre paddling pool is filled at the rate of 12 litres of water every 5 minutes.
How long will it take to fill the pool?

19 50 g of flour and 90 ml of milk are needed to make 8 biscuits.
    (a)   How much flour is needed to make 27 biscuits?
    (b)   How many biscuits can be made with 225 g of flour?

Some biscuits are made with 300 g of flour.
    (c)   How much milk is needed?

20 A piece of beef weighs 1.5 kg and costs £5.22.
How much would a piece of beef weighing 2.4 kg cost?
Give your answer to a suitable degree of accuracy.

What you need to know

- The ratio 3 : 2 is read '3 to 2'.
- A ratio is used only to **compare** quantities.
  A ratio does not give information about the exact values of quantities being compared.
- To simplify a ratio divide both of the numbers in the ratio by the **same** number.
  A ratio with whole numbers which cannot be simplified is in its s**implest form**.
  All quantities in a ratio must have the **same units** before the ratio can be simplified.
  For example, £2.50 : 50p = 250p : 50p = 5 : 1.
- When two different quantities are in the **same ratio** the two quantities are in **direct proportion.**
  For example, the amount and cost of fuel bought by a motorist are in proportion.

Review Exercise 10
Do not use a calculator for questions 1 to 6.

1 A packet contains 5 white balloons and 15 red balloons.
What is the ratio of white balloons to red balloons in its simplest form?

2 A tin contains 21 nuts and 28 bolts.
What is the ratio of nuts to bolts in its simplest form?

3 A sewing box contains pins and needles in the ratio 4 : 1. There are 36 pins in the box.
How many needles are in the box?

4 In a class the ratio of students with dark hair to those with light hair is 3 : 2.
There are 18 students with dark hair.
How many students have light hair?

5 A bag contains white buttons and red buttons in the ratio 1 : 2. The bag contains 24 buttons. How many white buttons are in the bag?

6 These ingredients are needed to make 20 scones.

| 500 g flour   100 g dried fruit   250 g butter   water to mix |

(a) How much dried fruit would you need for 5 scones?
(b) How much flour would you need for 30 scones?

AQA

7 Yesterday a shop sold 300 yogurts.
The number of natural yogurts sold to the number of fruit yogurts sold was in the ratio 1 : 5.
How many natural yogurts were sold?

8 A recipe for 18 buns needs: **150 g flour   100 g butter   50 g sugar**
(a) Calculate the amount of flour needed for 45 buns. Give your answer in kilograms.
(b) Lucy has plenty of flour and butter but only one kilogram of sugar.
How many buns can she make with this?

AQA

9 Fiona is given £24 for her birthday. She spends 5 times as much as she saves.
How much does she save?

AQA

10 Max shares £420 with a friend in the ratio 5 : 3. How much does each receive?

AQA

11 A test is taken by 28 pupils. The ratio of the number of boys to the number of girls is 3 : 4.
How many boys take the test?

AQA

12 (a) A bag contains 3 black counters and 2 white counters.
Another bag contains 30 counters.
There is the same **ratio** of black counters to white counters.
How many black counters are in the bag?
(b) Another bag contains black counters and white counters in the ratio 7 to 13.
What percentage of the counters are black?

AQA

13 (a) A wood contains fir trees and yew trees in the ratio 3 : 4.
There are 18 fir trees in the wood.
How many yew trees are in the wood?
(b) Another wood contains 20 beech trees and 12 ash trees.
Write, in its simplest form, the ratio of beech trees to ash trees.

14 Alison pays £483 for a package holiday.
The cost of travel to the cost of accommodation is in the ratio 3 : 4.
What is the cost of the accommodation?

AQA

15 A plan is drawn using a scale of 2 centimetres to represent 5 metres.
Write the scale as a ratio in the form 1 : n, where n is a whole number.

16 The cost of 30 bottles of water is £21. What is the cost of 45 bottles of water?

AQA

17 Two pints of milk cost 58p. What is the cost of 5 pints of milk at the same price per pint?

AQA

18 The cost of 8 metres of material is £18.40. What is the cost of 5 metres of the same material?

19 On a map the distance between two houses is 19 mm.
The actual distance between the houses is 3.8 km.
What is the scale of the map?

20 Jack is building a brick wall for a garage. He decides to use a mortar mix given by:
1 part cement,   1 part lime,   6 parts sand.
(a) What fraction of the mix will be sand?
(b) What percentage of the mix will be cement?
(c) He uses 1200 cm³ of sand for the mix. What volume of lime will he need?

AQA

CHAPTER 11

Speed and Other Compound Measures

Speed

Speed is a measurement of how fast something is travelling.
It involves two other measures, **distance** and **time**.
Speed can be worked out using this formula.

$$\text{Speed} = \frac{\text{Distance}}{\text{Time}}$$

○○○○○○○○○○○○○○○○○○○○○○○○○○○○
Speed can be thought of as the **distance** travelled in
one unit of time (1 hour, 1 second, …)

Speed can be measured in:

kilometres per hour (km/h), metres per second (m/s), miles per hour (mph), and so on.

Average speed

When the speed of an object is **constant** it means that the object doesn't slow down or go faster.
However, in many situations, speed is not constant.
For example:

A sprinter needs time to start from the starting blocks
and is well into the race before running at top speed.
A plane changes speed as it takes off and lands.

In situations like this the idea of **average speed** can be used.
The formula for average speed is:

$$\text{Average speed} = \frac{\text{Total distance travelled}}{\text{Total time taken}}$$

The formula linking speed, distance and time can be rearranged and remembered as:

| (average) **speed** = (total) **distance** ÷ (total) **time** |
| (total) **distance** = (average) **speed** × (total) **time** |
| (total) **time** = (total) **distance** ÷ (average) **speed** |

| S = D ÷ T |
| D = S × T |
| T = D ÷ S |

EXAMPLES

1 Robert drives a distance of 260 km. His journey takes 5 hours.
What is his average speed on the journey?

$$\text{Speed} = \frac{\text{Distance}}{\text{Time}} = \tfrac{260}{5} = 260 \div 5 = 52 \text{ km/h}$$

2 Lisa drives at an average speed of 80 kilometres per hour on a journey that takes 3 hours.
What distance has she travelled?

Distance = Speed × Time = 80 × 3 = 240 km   *So in 3 hours she travels 240 km.*

3 Lucy cycles at an average speed of 7 km/h on a journey of 28 km.
How long does she take?

Time = Distance ÷ Speed = 28 ÷ 7 = 4   *So her journey takes 4 hours.*

4 A cheetah takes 4 seconds to travel 100 m.
What is the speed of the cheetah?

$$\text{Speed} = \frac{\text{Distance}}{\text{Time}} = \tfrac{100}{4} = 100 \div 4 = 25 \text{ m/s}$$

Do not use a calculator.

1 John cycles 16 miles in 2 hours. What is his average speed in miles per hour?

2 Sue runs 21 km in 3 hours. What is her average speed in kilometres per hour?

3 Joe swims 100 m in 4 minutes. What is his average speed in metres per minute?

4 Calculate the average speed for each of the following journeys in kilometres per hour.

	Total distance travelled	Total time taken
(a)	60 km	3 hours
(b)	100 km	2 hours
(c)	10 km	$2\frac{1}{2}$ hours

5 Jackie runs 15 km in $1\frac{1}{2}$ hours.   What is her average speed in kilometres per hour?

6 Beverley walks for 2 hours at an average speed of 4 km/h.
How many kilometres does she walk?

7 Howard cycles for 5 hours at an average speed of 6 miles per hour. How far does he cycle?

8 Aubrey runs at 6 km/h for $\frac{1}{2}$ hour.   How far does he run?

9 Calculate the total distance travelled on each of the following journeys.

	Total time taken	Average speed
(a)	3 hours	50 km/h
(b)	2 hours	43 km/h
(c)	$\frac{1}{2}$ hour	80 km/h

10 Ahmed drives 30 miles at an average speed of 60 miles per hour.
How long does the journey take?

11 Lauren cycles 100 m at 5 metres per second. How long does she take?

12 A coach travels 75 km at an average speed of 50 km/h. How long does the journey take?

13 Calculate the total time taken on each of the following journeys.

	Total distance travelled	Average speed
(a)	30 km	10 km/h
(b)	80 km	40 km/h
(c)	210 km	60 km/h

14 Aimee cycles 27 km at 12 km/h. How long does she take?

15 Penny cycles to work at 18 km/h. She takes 20 minutes. How far does she cycle to work?

16 A train travels 100 km at an average speed of 80 km/h. How long does the journey take?

17 Bristol is 40 miles from Gloucester.
(a) How long does it take to cycle from Bristol to Gloucester at 16 miles per hour?
(b) How long does it take to drive from Bristol to Gloucester at 32 miles per hour?

18 A car travels 120 km in 2 hours.
(a) What is its average speed in kilometres per hour?
(b) How many hours would the car take to travel 120 km if it had gone twice as fast?

19 Liam drives for 60 km at an average speed of 40 km/h. He starts his journey at 9.50 am.
At what time does his journey end?

EXAMPLE

The Scottish Pullman travels from London to York, a distance of 302.8 km in 1 hour 45 minutes.
It then travels from York to Edinburgh, a distance of 334.7 km in 2 hours 30 minutes.
Calculate the average speed of the train between London and Edinburgh.

Total distance travelled = 302.8 + 334.7 = 637.5 km
Total time taken = 1 hr 45 mins + 2 hr 30 mins = 4 hr 15 mins = 4.25 hours
Average speed = $\frac{637.5}{4.25}$ = 150 km/h

Exercise 11.2

1 On the first part of a journey a car travels 140 km in 3 hours.
On the second part of the journey the car travels 160 km in 2 hours.
(a) What is the total distance travelled on the journey?
(b) What is the total time taken on the journey?
(c) What is the average speed of the car over the whole journey?

2 Lisa runs two laps of a 400 m running track.
The first lap takes 70 seconds. The second lap takes 90 seconds.
What is her average speed over the two laps?

3 Jenny sets out on a journey at 10.20 am.
She completes her journey at 1.05 pm. She travels a total distance of 27.5 km.
Calculate her average speed in kilometres per hour.

4 Harry drives for 40 km at an average speed of 60 km/h. He starts his journey at 9.50 am.
At what time does his journey end?

5 Sally cycles 38 km at an average speed of 23 km/h. She starts her journey at 9.30 am.
At what time does she finish? Give your answer to the nearest minute.

6 Chandni runs from Newcastle to Whitley Bay and then from Whitley Bay to Blyth.

| Newcastle to Whitley Bay | Time taken: 1 hr 20 min | Distance: 20 km |
| Whitley Bay to Blyth | Average speed: 0.2 km/min | Distance: 12 km |

(a) Calculate Chandni's average speed over the whole journey.
(b) Chandni left Newcastle at 10.50 am.
At what time did she arrive in Blyth?

7 Angela, Ben and Cathy drive from London to Glasgow.
Angela takes 12 hours 30 minutes driving at an average speed of 64 km/h.
Ben drives at an average speed of 100 km/h.
(a) How long does Ben take?

Cathy takes 7 hours 12 minutes.
(b) What is Cathy's average speed?

8 Ron runs 400 m in 1 minute 23.2 seconds.
Calculate his average speed in (a) metres per second, (b) kilometres per hour.

9 A cheetah runs at a speed of 90 km/h for 6 seconds.
How many metres does the cheetah run?

10 The distance from the Sun to the Earth is about 150 million kilometres.
It takes light from the Sun about 500 seconds to reach the Earth.
Calculate the speed of light in metres per second.

Other compound measures

Density

Density is a compound measure because it involves two other measures, **mass** and **volume**.
The formula for density is:

$$\text{Density} = \frac{\text{Mass}}{\text{Volume}}$$

The formula linking density, mass and volume can be rearranged and remembered as:

$$\text{Volume} = \frac{\text{Mass}}{\text{Density}}$$

$$\text{Mass} = \text{Density} \times \text{Volume}$$

For example, if a metal has a density of 2500 kg/m³ then 1 m³ of the metal weighs 2500 kg.

EXAMPLES

1 A block of metal has mass 500 g and volume 400 cm³.
Calculate the density of the metal.
Density $= \frac{500}{400} = 1.25$ g/cm³

2 The density of a certain metal is 3.5 g/cm³. A block of the metal has volume 1000 cm³.
Calculate the mass of the block.
Mass $= 3.5 \times 1000 = 3500$ g.

Population density

Population density is a measure of how populated an area is. The formula for population density is:

$$\text{Population density} = \frac{\text{Population}}{\text{Area}}$$

EXAMPLE

Cumbria has a population of   489 700   and an area of   6824 km².
Surrey has a population of   1 036 000   and an area of   1677 km².
Which county has the greater population density?

The population densities are:

Cumbria $\frac{489\,700}{6824} = 71.8$ people/km².   Surrey $\frac{1\,036\,000}{1677} = 617.8$ people/km².

Surrey has the greater population density.

Exercise 11.3

1 A metal bar has a mass of 960 g and a volume of 120 cm³.
Find the density of the metal in the bar.

2 A block of copper has a mass of 2160 g. The block measures 4 cm by 6 cm by 10 cm.
What is the density of copper?

3 A paperweight is made of glass. It has a volume of 72 cm³ and a mass of 180 g.
What is the density of the glass?

4 A silver necklace has a mass of 300 g. The density of silver is 10.5 g/cm³.
What is the volume of the silver?

5 A can in the shape of a cuboid is full of oil.
It measures 30 cm by 15 cm by 20 cm. The density of oil is 0.8 g/cm³.
What is the mass of the oil?

6 A bag of sugar has a mass of 1 kg.
The average density of the sugar in the bag is 0.5 g/cm³.
Find the volume of sugar in the bag.

7 A block of concrete has dimensions 15 cm by 25 cm by 40 cm.
The block has a mass of 12 kg. What is the density of the concrete?

8 A rectangular pane of glass measures 60 cm by 120 cm by 0.5 cm.
The density of glass is 2.6 g/cm³. What is the mass of the glass?

9 The population of Northern Ireland is 1 595 000.
The area of Northern Ireland is 13 483 km².
Calculate the population density of Northern Ireland.

10 The table shows the total population, land area and the population densities for some
countries in Europe.

	Country	Area km²	Population	Population density people/km²
(a)	Belgium	?	9 970 000	326.6
(b)	France	543 960	56 700 000	?
(c)	UK	244 090	?	235.2

Calculate the missing figures in the table.

What you need to know

- **Speed** is a compound measure because it involves **two** other measures.

- **Speed** is a measure of how fast something is travelling. It involves the measures **distance** and **time**.

  $$\text{Speed} = \frac{\text{Distance}}{\text{Time}}$$

- In situations where speed is not constant, **average speed** is used.

  $$\text{Average speed} = \frac{\text{Total distance travelled}}{\text{Total time taken}}$$

- The formula linking speed, distance and time can be rearranged and remembered as:

  (average) **speed** = (total) **distance** ÷ (total) **time**
  (total) **distance** = (average) **speed** × (total) **time**
  (total) **time** = (total) **distance** ÷ (average) **speed**

- Two other commonly used compound measures are **density** and **population density**.

- **Density** is a compound measure which involves the measures **mass** and **volume**.

  $$\text{Density} = \frac{\text{Mass}}{\text{Volume}}$$

- **Population density** is a measure of how populated an area is.

  $$\text{Population density} = \frac{\text{Population}}{\text{Area}}$$

Review Exercise 11

Do not use a calculator for questions 1 to 6.

1 The chart shows the distances in kilometres
between some towns.
Mrs Hill drove from Manchester to Southampton.
She completed the journey in 4 hours.
What was her average speed for the journey in
kilometres per hour?

London
326	Manchester
270	60	Sheffield
129	376	334	Southampton

2 A train travels 120 kilometres at an average speed of 80 kilometres per hour.
How long will the journey take?

124

3 Mr Mogg took 5 hours to drive from Cardiff to Leeds.
His average speed was 48 miles per hour.
What is the distance from Cardiff to Leeds?

4 Lorraine has a jet-ski ride. She travels at an average speed of 28 km/h for half an hour.
How far does she travel?

5 A coach takes a quarter of an hour to travel 20 km on a motorway.
What is the average speed of the coach in kilometres per hour?

6 Alex cycles at an average speed of 12 miles per hour for 15 minutes.
How far does she cycle?

7 A train travels 150 miles in 2 hours 30 minutes. Find its average speed in miles per hour. AQA

8 Simon drives 9 miles to work. His journey takes 20 minutes.
What is Simon's average speed in miles per hour?

9 A lorry travels from Middlesbrough to York in $1\frac{1}{4}$ hours. The distance is 50 miles.
Find the average speed of the lorry in miles per hour.

10 Kath has to drive 17 km to work. Calculate her average speed for the journey when she leaves home at 0750 and gets to work at 0820.

11 Matt drove 120 miles at an average speed of 45 mph.
Calculate the time Matt took for the journey. Give your answer in hours and minutes.

12 Nick started his journey at 0945. He travelled 175 km at an average speed of 50 km/h.
At what time did his journey end?

13 Kelly cycles 36 kilometres in 4 hours 30 minutes.
Calculate her average speed in kilometres per hour.

14 Flik lives 2.4 km from school.
How many minutes does she take to walk to school if her average walking speed is 4 km/h?

15 An aeroplane flies from New York to Los Angeles, a distance of 2475 miles, at an average speed of 427 miles per hour.
How long does the flight take in hours and minutes? AQA

16 Jane cycles from *A* to *B* and then from *B* to *C*.
Details of each stage of her journey are given below.

A to *B* Distance 55 km.	Average speed 22 km/h.	
B to *C* Time taken 1 hour 30 minutes.	Average speed 30 km/h.	

Calculate Jane's average speed over the whole of her journey from *A* to *C*. AQA

17 Arnold runs 400 m in 1 minute. What is his average speed in km/h?

18 A tiger runs at a speed of 50 km/h for 9 seconds. How many metres does the tiger run? AQA

19 A glass ornament has a mass of 350 g and a volume of 120 cm³.
Calculate the density of the glass.

20 A gold bracelet has a mass of 84 g. The density of gold is 19.3 g/cm³.
What is the volume of the bracelet?

21 Light travels at 186 284 miles per second.
The planet Jupiter is 483.6 million miles from the Sun.
Using suitable approximations, estimate the number of **minutes** light takes to travel from the Sun to Jupiter. AQA

Speed and Other Compound Measures

Do not use a calculator for this exercise.

1 (a) Write the number 5423 in words.
  (b) (i) Place these numbers in order, starting with the smallest.
     316, 38, 139, 31, 1310
   (ii) Which of these numbers are odd numbers?

2 (a) What numbers are needed to complete these sums?
   (i) $100 - 13 = \square$ (ii) $44 + \square = 100$ (iii) $100 - \square = 39$
  (b) (i) Work out 70×100.   (ii) Work out $950 \div 100$.

3 How much bigger is $72 \div 3$ than $72 \div 4$?

4 (a) The table shows the distances between towns in miles.
   (i) How far is it from Penzance to Liverpool?
   (ii) Rachel drives from York to Newcastle.
     She then drives from Newcastle to Edinburgh.
     Later, she returns directly from Edinburgh to York.
     How far does she drive altogether?

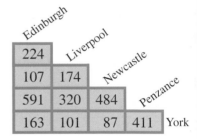

  (b) Brian's mileage readings at the start and at the end of a journey are shown.

Start   **End**

   Work out how far Brian travelled.

  (c) Sue organises a trip for 53 people.
    They use mini-buses for the trip. A mini-bus holds 12 people.
    How many mini-buses are needed for the trip?   AQA

5 Adam sells magazines at 98 pence per copy. He sells 100 magazines.
  How much money does Adam collect?
  (a) Give your answer in pence.
  (b) Give your answer to (a) in pounds.   AQA

6 Place these numbers in order of size, **smallest** first.
     3.5, −4, 0, 4, 3.05   AQA

7 Tony buys these four items.

 £1.28   £3.45   75p   £1.39

  (a) What is the total cost of these items?
  (b) Tony pays with a £10 note. How much change does he get?   AQA

8 (a) Richard has a dental appointment at 1425.
    What time is his appointment using the 12-hour clock?
  (b) Richard leaves home at 1345. He gets to the dentist at 1417.
    How long did his journey take?

9 Five numbers are shown:

$$23 \quad 26 \quad 56 \quad 73 \quad 74$$

  (a) Which two of these numbers have a total of 100?
  (b) Which two of these numbers have a difference of 50?
  (c) What is the highest number you can make by multiplying two of these numbers together?

AQA

10 (a) Write $\frac{1}{5}$ as a decimal.
  (b) Write 0.7 as a fraction.
  (c) Write 50% as a fraction.

AQA

11 Work out. (a) $528 + 273$ (b) $16 - 4 \times 3$ (c) $12 \div 4 + 2$

12 Work out. (a) $20\,000 \times 600$ (b) $\frac{40\,000}{800}$ (c) $\frac{20 \times 90}{60^2}$

AQA

13 (a) Copy and shade $\frac{1}{4}$ of this shape.

  (b) Copy and shade 10% of this shape.

  (c) What fraction of this shape is shaded?
       Give your answer in its simplest form.

AQA

14 On Monday morning a dentist treated 20 patients.

  (a) $\frac{1}{4}$ of the patients had a check-up.
       How many of the patients had a check-up?

  (b) 10 patients had a filling.
       What fraction of the patients had a filling?
       Give your answer in its simplest form.

  (c) $\frac{3}{10}$ of the patients were children.
       How many patients were children?

15 The table shows the midday temperature in two cities on one day.

London	3°C
Moscow	−8°C

  (a) How much colder is Moscow than London?
  (b) Paris is 5°C colder than London.
       What is the temperature in Paris?

16 Ali buys a number of CDs for £392. Each CD costs £7.
Work out how many CDs Ali has bought.

AQA

17 A cement mixer costs £5.95 per day to hire.
There is a delivery charge of £7.50.
How much does it cost altogether to hire the cement mixer for 3 days?

AQA

18 (a) Write 0.3 as a fraction.
  (b) Find the value of 3^2.
  (c) Write 34.219 correct to (i) 1 decimal place,   (ii) 2 decimal places.
  (d) Find the value of 3×4^2.

AQA

19 Paul writes down five numbers.

| 2 | 5 | 6 | 7 | 9 |

(a) Paul uses these numbers to make the five-digit number 92576.
    (i) What is the value of the 5 in Paul's number?
    (ii) What is Paul's number correct to the nearest thousand?
(b) Paul uses all five numbers again.
    What is the smallest **even** number that Paul can make?   AQA

20 (a) Write 10% as a fraction.
    (b) Find 10% of £50.

21 Steven pays 96 pence for 3 oranges and 2 grapefruit. A grapefruit costs 27 pence.
How much is an orange?

22 Work out.   (a) $32.4 - 14.9$ (b) $23.6 \div 8$ (c) 4×0.4

23 Work out.   (a) $600 - 327$ (b) $768 \div 24$ (c) $2^3 \times 5^2$ AQA

24 Sarmad swims 80 lengths of a swimming pool every day.
The swimming pool is 20 metres long.
(a) (i) How far does Sarmad swim each day?
    (ii) What is this distance in kilometres?
(b) Sarmad swims lengths of front crawl and lengths of backstroke in the ratio 1 : 3.
    How many lengths of front crawl does he swim each day?   AQA

25 A garage charges £36 per hour.
What is the charge for a job which takes 1 hour 20 minutes?

26 Find the value of:
(a) 10^5 (b) $\sqrt{49}$ (c) $8^2 \div 2^3$ (d) 0.1×0.9 (e) $\frac{9}{20}$ as a decimal

27 Anna is taking the written part of her driving test.
In the room there are fifty seats for candidates.
Candidates are sitting on 80% of these seats.
(a) How many candidates are in the room?

$\frac{2}{5}$ of the candidates in the room are female.
(b) How many candidates are female?

Twelve people pass the written test.
(c) What fraction of the candidates in the room passed the test?
    Give your fraction in its simplest form.   AQA

28 Every day, Ron drives a bus between Cambridge and Oxford.
His bus can carry 40 passengers.
On Monday, $\frac{5}{8}$ of the seats for passengers are used.
(a) How many passengers are on the bus on Monday?

On Tuesday, 30 passengers are on the bus.
(b) What percentage of the seats are used?   AQA

29 (a) Find $\frac{3}{5}$ of 60.
(b) Work out. (i) $\frac{2}{3} + \frac{3}{4}$ (ii) $\frac{1}{2} - \frac{2}{5}$
(c) Write down a decimal that lies between $\frac{1}{4}$ and $\frac{1}{3}$.
(d) Work out $\frac{3}{4} \times \frac{2}{5}$.

30 A box contains red candles and white candles in the ratio 1 : 4.
The box contains 30 candles.
How many red candles are there?

31 Raspberry jam is sold in two sizes.

Which size is the better value for money?
You **must** show all your working.

£2.69   £1.39

AQA

32 Estimate the value of   97 × 7.1

AQA

33 Debbie wants to calculate $\frac{70.24}{9.8 - 3.08}$.

(a)   Write each of the numbers in Debbie's calculation to the nearest whole number.
(b)   Hence, find an estimate of the answer for Debbie.

AQA

34 A lorry travels 100 miles in 2 hours 30 minutes.
Find its average speed in miles per hour.

35 There are 400 people at a fair.
Of these 400 people, $\frac{1}{4}$ are over 60   and $\frac{1}{5}$ are under 10.

(a)   How many people are **not** over 60 and are **not** under 10?
(b)   Of the 400 people, 240 are female.
What percentage are female?

AQA

36 (a)   The newspaper heading is given to the nearest thousand.
What is the smallest possible size of the audience?
(b)   Given that   37 × 249 = 9213,   find the exact value of $\frac{92130}{37}$.

37 Which is bigger, 4^3 or 7^2?
Show all your working.

38 A gas bill is £112.40 plus VAT at 5%.
Calculate the VAT charged.

39 A turkey should be cooked for 40 minutes per kilogram.
A turkey weighs 4 kilograms. It is placed in the oven at 1030.
At what time of day will it be cooked?

AQA

40 (a)   The cost of 20 grapefruit is £7.
Find the cost of one grapefruit.
(b)   Pears cost £1.20 per kilogram. Apples cost 30% less than pears.
What is the cost of 1 kg of apples?

41 The number of spaces is given to the nearest ten.
What is the minimum and maximum possible number of spaces?

Parking
400 spaces

42 The cash price of a computer is £960.
The computer can be bought on credit by paying a deposit of 10% of the cash price and
12 monthly payments of £79.50.
How much **more** is paid when the computer is bought on credit?

AQA

43 (a)   Write down the value of 7^2.   (b)   What is the value of $\sqrt{81}$?

(c)   What is the reciprocal of $\frac{3}{2}$?   (d)   Simplify $\frac{9^2 \times 9^3}{9^4}$.

(e)   Write 6.5×10^5 as an ordinary number.

44 An examination in history is marked out of 60 marks.
Reg gets 54 marks.
What percentage of the marks does he get?

45 Use the calculation $487 \times 3.53 = 1719.11$ to find the value of:
(a) 487×0.0353 (b) 48700×0.00353 AQA

46 A cycle route is 30 miles long. A leaflet states that this route can be covered in $2\frac{1}{2}$ hours.
(a) Calculate the average speed required to complete the route in the time stated.
(b) A cyclist completes the route in $2\frac{1}{2}$ hours.
He averages 8 miles an hour for the first $\frac{1}{2}$ hour.
Calculate his average speed for the remainder of the journey.   AQA

47 Work out.   (a) $\frac{3}{4}$ of 5.6   (b) 0.2×0.4 (c) $3\frac{1}{4} - 1\frac{2}{5}$

48 Red tulips and yellow tulips are used for a flower display.
The display uses 40 tulips. The ratio of red tulips to yellow tulips is 3 : 5.
How many tulips are red?

49 A sports pitch has a length of 75 metres, correct to the nearest metre.
Write down the least and the greatest possible length of this pitch.   AQA

50 The volume of a metal prism is 30 cm³. The mass of the prism is 210 g.
What is the density of the metal in g/cm³?

51 (a) Express 72 as a product of its prime factors.
(b) Find the highest common factor of 48 and 72.

52 A racing car travels 3 miles in 45 seconds.
Find the average speed of the car in miles per hour.

53 Paul invests £500 for two years, at 4% simple interest, paid yearly.
Paul says that the interest will be £40.
Is Paul correct? Explain clearly how you obtained your answer.

54 Andi tells Cato:
"When driving your car, if you increase your speed by 10% and then decrease your speed by 10% you will be travelling at a slower speed than you were to begin with."
Is he correct? Explain your answer.

55 (a) Use approximations to estimate the value of $\frac{9.67^2}{0.398}$. You **must** show all your working.
(b) (i) p and q are prime numbers. Find the values of p and q when $p^3 \times q = 24$.
(ii) Write 18 as a product of prime factors.
(iii) What is the least common multiple of 24 and 18?   AQA

56 Are these statements true or false?

> **A:** The product of two prime numbers is always a prime number.
> **B:** The sum of two consecutive numbers is always odd.

Explain each answer.

57 Carol is paid £6.30 per hour for a basic 30-hour week.
Overtime is paid at time and a third.
Last week Carol's total pay was £218.40.
How many hours and minutes overtime did Carol work last week?   AQA

Number
Calculator Paper

You may use a calculator for this exercise.

1   (a)   Write the number 5624 in words.
     (b)   Write these numbers in order, starting with the smallest.

$$39 \qquad -4 \qquad 3 \qquad 120 \qquad -15$$

2   Jainil orders a bouquet for his mother's birthday.

BEAUTIFUL BLOOMS		
Basic bouquet	£9	50p
6 Roses @ 40p each		
4 Carnations @ 30p each		
Delivery		
Total		

He chooses a bouquet which has a basic cost of £9.50.
He asks for 6 roses and 4 carnations to be added to the flowers.
He agrees to pay for the flowers to be delivered at a cost of £3.50.
Copy and complete the bill.

AQA

3   Rosie gets on a bus at 1745 and gets off the bus at 1806.
     (a)   How long was she on the bus?
     (b)   What time did she get off the bus using the 12-hour clock?

4   (a)   Write 937   (i)   to the nearest 10,   (ii)   to the nearest 100.
     (b)   Calculate 937 × £2.35. Give your answer to the nearest pound.
     (c)   Write 0.3   (i)   as a fraction,   (ii)   as a percentage.
     (d)   Work out $\frac{3}{7}$ of 42.

AQA

5   A ball of wool costs 69p. Bill buys 19 balls of wool.
     (a)   How much do the balls of wool cost?

He pays for the wool with a £20 note.
     (b)   How much change is Bill given?

This change is given using the least number of notes and coins.
     (c)   How is the change given?

AQA

6   The temperatures in six cities one day in February were:

$$7°C \qquad 11°C \qquad -3°C \qquad 5°C \qquad 0°C \qquad -4°C$$

     (a)   Which temperature is the coldest?
     (b)   Which temperature is the warmest?

7   Bob's basic rate of pay is £6.20 per hour.
Overtime is paid at time and a half.
One week Bob works 3 hours overtime.
How much is Bob paid for his overtime?

8   Write 86.739   (a)   to the nearest whole number,
                   (b)   to one decimal place.

9 (a) The number 3482 is multiplied by 100.
What is the value of 8 in the answer?

(b) The number 3482 is divided by 10.
What is the value of 4 in the answer?   AQA

10 Nisha wants to buy some reprints of her holiday photos.
Nisha has a £5 note. The reprints cost 41 pence each.

(a) How many reprints can Nisha buy?

(b) How much change does Nisha receive?   AQA

11 (a) Write $2\frac{1}{4}$ as a decimal.

(b) Place the following numbers in order of size, starting with the smallest.

$$2\frac{1}{4} \qquad 1.53^2 \qquad 2.21 \qquad 2.035 \qquad \sqrt{4.78}$$ AQA

12 A packet of butter is shown.
The butter is kept in a freezer at −15°C.
How many degrees is this below the storage temperature?

> BEST BUTTER
> Store at −9°C
> Weight 4 kg
                                                                         AQA

13 A tall office block has 60 floors.

(a) One lift stops at $\frac{1}{4}$ of the floors.
At how many floors does the lift stop?

(b) 20% of the floors are not occupied.
How many floors **are** occupied?   AQA

14 A school party of 500 people is going on a coach trip.
Each coach can seat 44 passengers.
Each coach costs £95 to hire.

(a) Work out how many coaches are needed.

(b) Work out the total cost of all the coaches that are needed.   AQA

15 Every Saturday Amy works in a tea shop.
She is paid £6 per hour and 20p for every customer which she serves.
On one Saturday, Amy works 4 hours and serves 30 customers.
How much is she paid for that Saturday?   AQA

16 (a) Susan goes on a motoring holiday to France.
At the start of her holiday, the milometer in her car shows:

$$27849$$

(i) What is the number 27849 to the nearest thousand?

(ii) On her holiday, Susan drives a total of 1753 miles.
What will the milometer show at the end of the holiday?

(b) Petrol costs £0.85 per litre in England.
Calculate the cost of 23 litres of petrol.   AQA

17 (a) Write $\frac{3}{8}$ as a decimal.

(b) Write 0.8 as a fraction. Give your answer in its lowest terms.   AQA

18 A youth club holds a raffle.
The raffle prize is a CD player costing £97.95.
Raffle tickets are sold at 20 pence each.
If 817 tickets are sold, how much profit is made?   AQA

19 The land area of a farm is 385 acres.

(a) $\frac{1}{5}$ of the land is used to grow barley. How many acres is this?

(b) 15% of the land is not used. How many acres is this?

AQA

20 Calculate $\frac{594}{19 \times 4}$ Give your answer to the nearest whole number.

21 Laura buys 18 cartons of juice.
She pays with a £10 note. She gets £2.98 change.
How much is each carton of juice?

22 A ladder costs £126 plus VAT at $17\frac{1}{2}\%$.
Calculate the VAT charged.

23 Dan writes down five numbers.   3   6   8   9   24

(a) Which of these numbers are factors of 12?

(b) Which of these numbers is a prime number?

24 Two melons cost £2.34.
How much will three melons cost?

AQA

25 (a) (i) Copy and complete the following number pattern:

$$11 \quad = \quad 11$$
$$11 \times 11 \quad = \quad 121$$
$$11 \times 11 \times 11 \quad = \quad \boxed{}$$
$$\boxed{} \quad = \quad \boxed{}$$

(ii) Look at the numbers in the right hand column.
Write down what you notice about these numbers.

(b) Use your calculator to work out the next line of the pattern.
What do you notice now?

AQA

26 Mary buys a new kitchen for £5390.

(a) Mary pays 10% deposit.
How much is the deposit?

(b) After paying the deposit, Mary pays the rest of the cost in 12 equal instalments.
How much is each instalment?

AQA

27 Dave drives 15 miles to work. The journey takes 20 minutes.
What is Dave's average speed in miles per hour?

AQA

28

> **Exchange Rate:** £1 = 1.48 euros

Calculate the difference in £s between 300 euros and £300.

29 Carl buys 1.2 kg of potatoes and 0.4 kg of carrots. He pays 98p in total.
The potatoes cost 70p per kilogram.
What is the cost of 1 kg of carrots?

AQA

30 Use your calculator to find:

(a) 2.3^2 (b) $(285 - 198) \div 2.9$ (c) $\sqrt{38}$, correct to 1 d.p.

31 (a) John is 8.4 kg heavier than Alan. The sum of their weights is 133.2 kg.
How heavy is Alan?

(b) Before starting a diet Derek weighed 80 kg. He now weighs 8 kg less.
Calculate his weight loss as a percentage of his previous weight.

AQA

32 (a) The weights and prices of two tins of pineapple are shown.

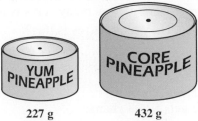

227 g
27 pence

432 g
52 pence

Which tin of pineapple gives more grams per penny?
You **must** show all your working.

(b) A greengrocer sold 16 cauliflowers and 24 cabbages.
The total cost was £18.16. The cost of a cauliflower was 55p.
What was the cost of a cabbage?

AQA

33 How much more is paid when the motorboat is bought on credit terms instead of cash?

MOTOR
BOAT
Cash price...£12 800
Credit terms...
Deposit of 30% of cash price
plus 36 monthly payments of £295

34 Mr Trim earns £5970 per year.
He has a tax allowance of £4895 per year and tax is paid at 10p in the £ on his taxable income.
How much tax does he pay per year?

35 A suitcase costs £56 plus VAT at $17\frac{1}{2}\%$. What is the total price of the suitcase?

36 A sponge cake for eight people needs 120 g of sugar.
John makes a sponge cake for five people.
(a) Calculate the weight of sugar he needs.

John cuts his cake into five equal slices. He gives three slices to his mother.
(b) What percentage of the cake does John have left?

AQA

37 Steff buys 5.1 kg of nails for £3.06.
How much would she pay for 3.2 kg of the same nails?

AQA

38 Zoe chooses three consecutive numbers.
Show that the sum of the first and the last of these three numbers is always an even number.

AQA

39 Scooters can be hired at a cost of £21 per day plus 5.3 pence per mile.
Mustapha hires a scooter for three days and rides it 136 miles.
Find the total cost of hiring the scooter.

AQA

40 Give examples to show that the sum of two prime numbers can be an even number or an odd number.

41 A cycle route is 45 km long.
A cyclist completes the route in 3 hours.
He averages 16 kilometres per hour for the first $1\frac{1}{2}$ hours.
Calculate his average speed for the remainder of the journey.

42 Tom buys 200 tomato plants at a total cost of £40 to sell at a school fair.

He sells $\frac{3}{4}$ of them at 50p each.

He then reduces the price of the remaining plants to 40p.

At the end of the day there are 18 plants left which have **not** been sold.

(a) How much money does he receive from selling the tomato plants at the school fair?

(b) Find the percentage profit which Tom made on these tomato plants.   AQA

43 (a) Use your calculator to find the value of $\dfrac{19.8^2}{7.19 + 2.73}$

(b) Show how you can use approximation to check your answer.

44 Calculate the simple interest on £480 invested at 7% for 6 months.

45 Wyn has 600 yo-yo's to sell.   He sells $\frac{3}{5}$ of them at £2 each.

He then reduces the price of the remaining yo-yo's by 40%.

In the end he is left with 24 damaged yo-yo's which he cannot sell.

Wyn paid £500 for the yo-yo's.

Find the percentage profit which he made on the sale of these yo-yo's.

46 (a) Calculate the exact value of 5^9.

(b) Find the reciprocal of 7. Give your answer correct to 3 decimal places.

(c) Work out $\dfrac{0.36 \times 0.89}{47 - 3.9}$ Give your answer correct to 2 significant figures.

47 Calculate $\dfrac{5.8 \times \sqrt{21.48}}{\sqrt{12}}$ Give your answer to one decimal place.

48 The table shows the amount of foreign currency you can buy with £1.

EXCHANGE RATES - *£1 will buy:*	France 1.56 euros	USA 1.37 dollars

An American has 400 000 dollars to spend on a holiday villa in the south of France.

She sees a villa for sale for 450 000 euros.

Does she have enough money to buy it?

You **must** show all your working.

49 Greg drove 115 miles at an average speed of 43 mph.

Calculate the time Greg took for the journey. Give your answer in hours and minutes.   AQA

50 Work out $2.37^2 - \sqrt{5.8}$ Give your answer correct to two decimal places.

51 A farm has 427 acres of land. 135 acres are used for grazing.

What percentage of the land is used for grazing?

52 Three musicians received £100 between them for playing in a concert.

They divided their pay in the ratio of the number of minutes for which each played.

Angela played for 8 minutes, Fran played for 14 minutes and Dan played for 18 minutes.

How much did each receive?   AQA

53 A bar of Fruit & Nut chocolate normally weighs 200 g.

The ratio by weight of a special offer bar to a normal bar is 5 : 4.

What is the weight of a special offer bar?   AQA

54 Starting with $4^3 = 64$,   use trial and improvement to find the cube root of 104,

correct to one decimal place. Show all your working.

55 A gold bar has a mass of 1 kilogram. The density of gold is 19.3 g/cm³.

What is the volume of the gold?

CHAPTER 12

Introduction to Algebra

Algebra is sometimes called the language of Mathematics.
Algebra uses letters in place of numbers.

A class of children line up. We cannot see how many
children there are altogether because of a tree.
We can say there are n children in the line.
The letter n is used in place of an unknown number.

Three more children join the line.
There are now $n + 3$ children in the line.

This picture shows two lines of n children.
So, there are $n + n$ or $2 \times n$ children altogether.
The simplest way to write this is $2n$.

Both $n + 3$ and $2n$ are examples of **algebraic expressions**.

Exercise **12.1**

Write algebraic expressions for each of the following questions.

1 There are n children in a queue. 4 more children join the queue.
How many children are in the queue now?

2 There are n children in a queue. 3 children leave the queue.
How many children are left in the queue?

3 There are 3 classes with n children in each class.
How many children are there altogether?

4 I have m marbles in a bag. I put in another 6 marbles.
How many marbles are now in the bag?

5 I have m marbles. I lose 12 marbles.
How many marbles do I have left?

6 I have 8 bags of marbles. Each bag contains m marbles.
How many marbles do I have altogether?

7 There are p pencils in a pencil case. I take one pencil out.
How many pencils are left in the pencil case?

8 There are p pencils in a pencil case. I put in another 5 pencils.
How many pencils are now in the pencil case?

9 I have 25 pencil cases. There are p pencils in each pencil case.
How many pencils do I have altogether?

10 I have 6 key rings. There are k keys on each key ring.
    How many keys do I have altogether?

11 What is the cost of b biscuits costing 5 pence each?

12 Three cakes cost a total of c pence.
    What is the cost of one cake?

13 Five kilograms of apples cost a pence.
    What is the cost of one kilogram of apples?

14 A group of 36 students are split into g groups.
    How many students are in each group?

15 There are t toffees in a tin.
    How many toffees are there in   (a) 2 tins,   (b) 10 tins?

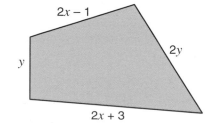

Expressions and terms

Consider this situation:
      $2n$ students start a typing course.
      3 of the students leave the course.
      How many students remain on the course?

$2n - 3$ students remain.
$2n - 3$ is an **algebraic expression**, or simply an **expression**.

An expression is just an answer made up of letters and numbers.
$+2n$ and -3 are **terms** of the expression.

Note:
A term includes the sign, + or −.
$2n$ has the same value as $+2n$.

Simplifying expressions

Adding and subtracting terms

You can add and subtract terms with the same letter.
This is sometimes called **simplifying an expression**.

$a + a = 2a$ $6a - 2a = 4a$
$5k + 3k = 8k$ $2d - 3d = -d$
$3p + 5 + p - 1 = 4p + 4$ $4x - 4x = 0$

  $6 + a$ cannot be simplified.
$5p - 2q$ cannot be simplified.
  $x^2 + x$ cannot be simplified.
$ab + ba = 2ab$

Note that:
A simpler way to write $1d$ is just d.

$-1d$ can be written as $-d$.

$0d$ is the same as 0.

Just as with ordinary numbers,
you can add terms in any order.
$a - 2a + 5a = a + 5a - 2a - 4a$

EXAMPLE

Write down an expression for the perimeter of this shape. Give your answer in its simplest form.

Perimeter is the total distance round the outside of the shape.
$y + 2x - 1 + 2y + 2x + 3$
Imagine that each term is written on a separate card.

| $+y$ | $+2x$ | -1 | $+2y$ | $+2x$ | $+3$ |

The cards can be arranged in any order.

| $+2x$ | $+2x$ | $+y$ | $+2y$ | $+3$ | -1 |

Simplify this expression to get: $4x + 3y + 2$
The perimeter of the shape is $4x + 3y + 2$.

1 Write simpler expressions for the following.

(a) $y + y$ (b) $c + c + c$ (c) $x + x + x + x + x$

(d) $p + p + p + p + p + p + p$ (e) $t + t + t - t$ (f) $d + d + d - d + d$

(g) $2n + n$ (h) $2y + 3y$ (i) $5g + g + 4g$

(j) $2m + 5m + m$ (k) $5z + 4z + z + 3z$ (l) $5r - 3r$

(m) $7t - 2t$ (n) $5y - y$ (o) $5j + 2j - 4j$

(p) $9c - 2c - 3c$ (q) $3x - x + 5x$ (r) $12w - 7w - 4w$

(s) $5d + 7d - 12d$ (t) $-2y - 3y$ (u) $3x - 8x$

(v) $2a - 5a - 12a + a$ (w) $3b + 5b - 4b + 2b$ (x) $m - 2m + 3m$

2 Write an expression for the perimeter of each shape.
Give each answer in its simplest form.

(a)   (b)   (c)   (d)

  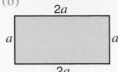 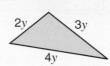 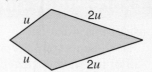

3 Which of these expressions cannot be simplified?
Give a reason for each of your answers.

(a) $v + v$ (b) $v + 4$ (c) $2v + v + 4$ (d) $v + w$

4 Simplify where possible.

(a) $5x + 3x + y$ (b) $w + 3v - v$ (c) $2a + b - 3b$

(d) $2x + 3y + 3x$ (e) $5 + 7u - 2$ (f) $p + 3q + q$

(g) $3d - 5c - 2c$ (h) $3y + 1 - y$ (i) $-a + b + 2a$

(j) $3m + n + m$ (k) $5c + 4c - d$ (l) $2x + y - x$

(m) $-p + 4p + 3p$ (n) $5 - 9k + 4k$ (o) $2a - a + 3$

5 Simplify where possible.

(a) $3a + 5a + 2b + b$ (b) $p + 2q + 2p + q$ (c) $m + 2m - n + 3n$

(d) $2x + 3y - x - 5y$ (e) $3x - x + 5y - 2y$ (f) $2d + 5 - d - 2$

(g) $3a - 5a + 2b + b$ (h) $a - 2a + 7 + a$ (i) $2a - b + 3b - a$

(j) $-f + g - f - g$ (k) $2v - w - 3w - v$ (l) $7 - 2t - 9 - 3t$

(m) $-p + 3q - 3p + q$ (n) $5 - 9k - 4 + 2k$ (o) $2c + d + 4 - c - 2d + 7$

6 Write down an expression for the perimeter of each shape.
Give each answer in its simplest form.

(a)   (b)   (c)   (d)

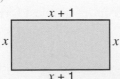 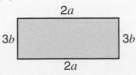 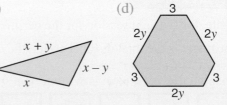

7 Simplify.

(a) $xy + yx$ (b) $3pq - qp$ (c) $5ab - 2ba$

(d) $3x^2 - x^2$ (e) $5y^2 + 4y^2$ (f) $a^2 + 5a^2 - 2a^2$

(g) $d^2 - 2g^2 - g^2 + d^2$ (h) $3t^2 + t + 2t^2 - 2t$ (i) $3m^2 - 4m^2 + 7m - m$

(j) $p^2 - 2p + p^2 + p$

Multiplying and dividing terms

$6 \times a = 6a$	$a \times b = ab$	$x \times 2 = 2x$	$b \times b \times b = b^3$	$8a \div 2 = 4a$
$5 \times 2a = 10a$	$3x \times y = 3xy$	$x \times x = x^2$	$5c \times 4c = 20c^2$	$9x \div x = 9$

EXAMPLE

Find an expression for the area of this rectangle.

Area = length × breadth
$\qquad$ = $3d \times d$

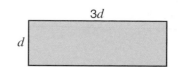

The simplest way to write an expression for the area is $3d^2$.

Exercise **12.3**

1 Write these expressions in a simpler form.

(a) $3 \times a$ $\qquad$ (b) $7 \times b$ $\qquad$ (c) $2 \times 4 \times c$ $\qquad$ (d) $3 \times 3 \times d$

(e) $e \times 4$ $\qquad$ (f) $f \times 8$ $\qquad$ (g) $3 \times 2p$ $\qquad$ (h) $3q \times 5$

(i) $r \times r$ $\qquad$ (j) $g \times g$ $\qquad$ (k) $2g \times g$ $\qquad$ (l) $2g \times 3g$

(m) $t \times 4t$ $\qquad$ (n) $3t \times 4t$ $\qquad$ (o) $5u \times 3u$ $\qquad$ (p) $3m \times 5m$

(q) $3d \times 3d$ $\qquad$ (r) $5x \times 3x$ $\qquad$ (s) $4y \times 3y$ $\qquad$ (t) $3k \times 2k$

2 Simplify.

(a) $3 \times (-y)$ $\qquad$ (b) $y \times (-5)$

(c) $(-2) \times (-y)$ $\qquad$ (d) $3 \times (-2y)$

(e) $t \times (-t)$ $\qquad$ (f) $2t \times (-t)$

(g) $(-2t) \times 5t$ $\qquad$ (h) $(-2t) \times (-5t)$

Remember:
$$2 \times (-x) = -2x$$
$$(-2) \times (-x) = 2x$$

3 Simplify.

(a) $10a : 2$ $\qquad$ (b) $16b \div 4$ $\qquad$ (c) $12x \div x$ $\qquad$ (d) $20y \div y$

(e) $8y \div 4$ $\qquad$ (f) $8y \div y$ $\qquad$ (g) $8y \div 4y$ $\qquad$ (h) $18p \div p$

(i) $18p \div 6$ $\qquad$ (j) $18p \div 6p$ $\qquad$ (k) $18k \div 2k$ $\qquad$ (l) $18a \div 3a$

(m) $28g \div 7g$ $\qquad$ (n) $10m \div 2m$ $\qquad$ (o) $20t \div 5t$ $\qquad$ (p) $27x \div 3x$

4 Simplify.

(a) $6y \div (-3)$ $\qquad$ (b) $(-6y) \div 2$

(c) $(-5m) \div 5$ $\qquad$ (d) $(-5m) \div (-5)$

(e) $3a \div (-1)$ $\qquad$ (f) $(-10d) \div 2$

(g) $6g \div (-2)$ $\qquad$ (h) $(-3k) \div -3$

Remember:
$$2x \div (-2) = -x$$
$$(-2x) \div 2 = -x$$
$$(-2x) \div (-2) = x$$

5 Simplify.

(a) $a \times b$ $\qquad$ (b) $x \times y$ $\qquad$ (c) $y \times y$ $\qquad$ (d) $2 \times p \times q$

(e) $2 \times a \times a$ $\qquad$ (f) $3 \times x \times y$ $\qquad$ (g) $3 \times a \times 2 \times b$ $\qquad$ (h) $3 \times g \times 4 \times h$

(i) $2 \times d \times 3 \times d$ $\qquad$ (j) $3g \times g$ $\qquad$ (k) $a \times 5b$ $\qquad$ (l) $2g \times 3h$

(m) $a \times b \times c$ $\qquad$ (n) $m \times m \times m$ $\qquad$ (o) $2 \times d \times d \times d$ $\qquad$ (p) $g \times g \times g \times 3$

(q) $2x \times 3x \times x$ $\qquad$ (r) $5m \times m \times 2n$ $\qquad$ (s) $3a \times b \times c$ $\qquad$ (t) $2p \times 3q \times 3r$

6 Write an expression for the area of each shape. Give each answer in its simplest form.

(a) $\qquad$ (b) $\qquad$ (c) $\qquad$ (d)

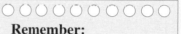

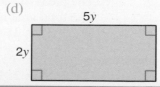

Multiplying and dividing algebraic expressions with powers

a^2 and b^3 are examples of **algebraic expressions with powers**.

The rules for multiplying and dividing powers of the same number can be used to simplify algebraic expressions involving powers.

EXAMPLES

1 Simplify $a^2 \times a^3$

a^2 means $a \times a$ $\quad a^3$ means $a \times a \times a$

$a^2 \times a^3 = (a \times a) \times (a \times a \times a) = a^5$

$a^2 \times a^3 = a^{2+3} = a^5$

When **multiplying**:
powers of the same base are **added**.
In general: $a^m \times a^n = a^{m+n}$

2 Simplify $b^5 \div b^3$

$b^5 \div b^3 = \dfrac{b^5}{b^3} = \dfrac{b \times b \times \cancel{b} \times \cancel{b} \times \cancel{b}}{\cancel{b} \times \cancel{b} \times \cancel{b}} = b^2$

$b^5 \div b^3 = b^{5-3} = b^2$

When **dividing**:
powers of the same base are **subtracted**.
In general: $a^m \div a^n = a^{m-n}$

Exercise 12.4

1 Simplify.

(a) $y^2 \times y$ 　　(b) $t^3 \times t^2$ 　　(c) $a^3 \times a^3$ 　　(d) $g^7 \times g^3$

(e) $x^3 \times x^8$ 　　(f) $m^2 \times m^5$ 　　(g) $k \times k^4$ 　　(h) $h^3 \times h^5$

2 Simplify.

(a) $y^3 \div y$ 　　(b) $a^4 \div a^3$ 　　(c) $x^5 \div x^5$ 　　(d) $t^7 \div t^3$

(e) $g \div g^2$ 　　(f) $h^3 \div h^5$ 　　(g) $x^8 \div x^5$ 　　(h) $m^3 \div m^4$

3 Simplify.

(a) $a \times b \times a^2$ 　　(b) $m \times m^3 \times n^2$ 　　(c) $2y \times y^2$ 　　(d) $3d^2 \times 2d^3$

(e) $a^2b^2 \times a^3$ 　　(f) $6b^3 \div b$ 　　(g) $10m^3 \div 2m^3$ 　　(h) $12a^5 \div 3a$

4 Simplify.

(a) $\dfrac{t^3}{t^2}$ 　　(b) $\dfrac{g^2}{g^3}$ 　　(c) $\dfrac{m^2 \times m}{m}$ 　　(d) $\dfrac{y^2 \times y^3}{y^4}$

(e) $\dfrac{y \times y^3}{y^2}$ 　　(f) $\dfrac{m^2 \times m^3}{m^6}$ 　　(g) $\dfrac{2t^3 \times t}{t^2}$ 　　(h) $\dfrac{6g^5 \times g}{2g^3}$

Brackets

Some expressions contain brackets. $2(a + b)$ means $2 \times (a + b)$.

You can multiply out brackets in an expression either by using a diagram or by expanding.

To multiply out $2(x + 3)$ using the **diagram method**:

$2(x + 3)$ means $2 \times (x + 3)$.
This can be shown using a rectangle.
The areas of the two parts are $2x$ and 6.
The total area is $2x + 6$.
$2(x + 3) = 2x + 6$

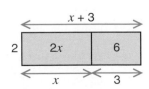

To multiply out $2(x + 3)$ by **expanding**:

$2(x + 3) = 2 \times x + 2 \times 3$
$= 2x + 6$

EXAMPLES

1 Multiply out the bracket $3(4a + 5)$.

Diagram method
$3(4a + 5) = 12a + 15$

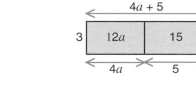

Expanding
$3(4a + 5) = 3 \times 4a + 3 \times 5$
$= 12a + 15$

2 Expand $x(x - 5)$.

$x \times x = x^2$ and $x \times -5 = -5x$
$x(x - 5) = x^2 - 5x$

Exercise 12.5

1 Use the diagrams to multiply out the brackets.

(a)

$2(x + 5) = \ldots$

(b)

$3(a + 6) = \ldots$

(c)

$4(y + 3) = \ldots$

(d)

$2(2a + 1) = \ldots$

(e)

$2(3y + 2) = \ldots$

(f)

$3(a + b) = \ldots$

2 Draw your own diagrams to multiply out these brackets.

(a) $3(x + 2)$ (b) $2(y + 5)$ (c) $2(2x + 1)$ (d) $3(p + q)$

3 Match the pairs of cards.

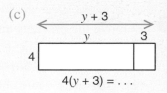

$2(q + 2)$ $2(q - 1)$ $2(2q + 1)$ $2(2 \quad q)$

$4q + 2$ $4 - 2q$ $2q + 4$ $2q - 2$

4 Use the diagrams to multiply out the brackets.

(a)

$a(a + 1) = \ldots$

(b)

$d(2 + d) = \ldots$

(c)

$x(2x + 1) = \ldots$

5 Multiply out the brackets by expanding.

(a) $2(x + 4)$ (b) $3(t - 2)$ (c) $4(5 - a)$ (d) $5(3 - 2d)$

(e) $6(b + 2c)$ (f) $3(2m - 5n)$ (g) $x(x + 3)$ (h) $t(t - 3)$

(i) $g(2g + 3)$ (j) $m(2 - 3m)$ (k) $t(3t + 5)$ (l) $m(m - n)$

To simplify an expression involving brackets:
- Remove the brackets.
- Simplify by collecting like terms together.

$$3(t + 4) + 2$$ — Remove the brackets
$$= 3t + 12 + 2$$ — Simplify
$$= 3t + 14$$

6 Multiply out the brackets and simplify.

(a) $2(x + 1) + 3$ (b) $3(a + 2) + 5$ (c) $6(w - 4) + 7$

(d) $4 + 2(p + 3)$ (e) $3 + 3(q - 1)$ (f) $1 + 3(2 - t)$

(g) $4(z + 2) + z$ (h) $5(t + 3) + 3t$ (i) $3(c - 2) - c$

(j) $2a + 3(a - 3)$ (k) $y + 2(y - 5)$ (l) $5x + 3(2 - x)$

(m) $4(2a + 5) + 3$ (n) $-2x + 4(3x - 3)$ (o) $3(p - 5) - p + 4$

(p) $3a + 2(a + b)$ (q) $3(x + y) - 2y$ (r) $2(p - q) - 3q$

(s) $2x + x(3 - x)$ (t) $a(a - 3) + a$ (u) $y(2 - y) + y^2$

7 Remove the brackets and simplify.

(a) $2(x + 1) + 3(x + 2)$ (b) $3(a + 1) + 2(a + 5)$ (c) $4(y + 2) + 5(y + 3)$

(d) $2(3a + 1) + 3(a + 1)$ (e) $3(2t + 5) + 5(4t + 3)$ (f) $3(z + 5) + 2(z - 1)$

(g) $7(q - 2) + 5(q + 6)$ (h) $5(x + 3) + 6(x - 3)$ (i) $8(2e - 1) + 4(e - 2)$

(j) $2(5d + 4) + 2(d - 1)$ (k) $m(m - 2) + m(2m - 1)$ (l) $a(3a + 2) + a(a - 3)$

8 Multiply out the brackets and simplify.

(a) $-3(x + 2)$ (b) $-3(x - 2)$

(c) $-2(y - 5)$ (d) $-2(3 - x)$

(e) $-3(5 - y)$ (f) $-4(1 + a)$

(g) $5 - 2(a + 1)$ (h) $5d - 3(d - 2)$

(i) $4b - 2(3 + b)$ (j) $-3(2p + 3)$

(k) $5m - 2(3 + 2m)$ (l) $2(3d - 1) - d + 3$

(m) $-a(a - 2)$ (n) $2d - d(1 + d)$

(o) $x^2 - x(1 - x) + x$ (p) $-3g(2g + 3)$

(q) $t^2 - 2t(3 - 3t)$ (r) $2m - 2m(m - 3)$

Remember:

$$(-2) \times (+3) = -6$$
$$(-2) \times (-3) = +6$$

so, $-2(x + 3) = -2x - 6$
and $-2(x - 3) = -2x + 6$

9 Expand the brackets and simplify.

(a) $2(a + 1) - (3 - a)$ (b) $3(y - 1) - 2(y + 1)$ (c) $2(1 + m) - 3(1 - m)$

(d) $5(x - 2) - 2(x - 3)$ (e) $2(5 + d) - 3(3 + d)$ (f) $4(t - 1) - 3(t + 2)$

(g) $3(2m + 1) - 2(m - 3)$ (h) $4(2x - 3) - 3(3x + 2)$ (i) $2(4 + 3a) - 3(4a - 5)$

More brackets

This method can be extended to multiply out $(x + 2)(x + 3)$.

The areas of the four parts are: x^2, $3x$, $2x$ and   6.

$(x + 2)(x + 3) = x^2 + 3x + 2x + 6$

Collect like terms and simplify   (i.e. $3x + 2x = 5x$)

$$= x^2 + 5x + 6$$

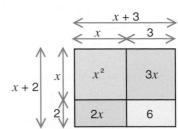

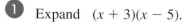
1 Expand $(x + 3)(x - 5)$.

The diagram method works with negative numbers.

$(x + 3)(x - 5) = x^2 - 5x + 3x - 15$
$\qquad\qquad\qquad = x^2 - 2x - 15$

	x	-5
x	x^2	$-5x$
3	$3x$	-15

2 Expand $(x - 1)(x + 3)$.

As you become more confident you may not need a diagram to expand the brackets.

$$(x - 1)(x + 3)$$

1   $x \times x = x^2$
2   $x \times 3 = 3x$
3   $-1 \times x = -x$
4   $-1 \times 3 = -3$

$(x - 1)(x + 3) = x^2 + 3x - x - 3$
$\qquad\qquad\qquad = x^2 + 2x - 3$

Exercise 12.6

Questions 1 to 9.   Use diagrams to multiply out the brackets.

1 $(x + 3)(x + 4)$ **2** $(x + 1)(x + 5)$ **3** $(x - 5)(x + 2)$

4 $(x + 1)(x - 2)$ **5** $(x - 2)(x - 6)$ **6** $(x + 1)(x + 2)$

7 $(x + 2)(x + 3)$ **8** $(x + 2)(x - 3)$ **9** $(x - 2)(x - 3)$

Questions 10 to 24.   Expand the following brackets. Only draw a diagram if necessary.

10 $(x + 8)(x - 2)$ **11** $(x + 5)(x - 2)$ **12** $(x - 1)(x + 3)$

13 $(x - 3)(x - 2)$ **14** $(x - 4)(x - 1)$ **15** $(x - 7)(x + 2)$

16 $(x + 3)(x - 1)$ **17** $(x - 1)(x + 5)$ **18** $(x - 2)(x + 5)$

19 $(x + 3)(x - 3)$ **20** $(x + 5)(x - 5)$ **21** $(x + 7)(x - 7)$

22 $(x - 10)(x + 10)$ **23** $(x + 3)^2$ **24** $(x - 3)^2$

Factorising

Factorising is the opposite operation to removing brackets.
For example: to remove brackets

$$2(x + 5) = 2x + 10$$

To factorise $3x + 6$ we can see that $3x$ and 6 have a **common factor** of 3, so,

$$3x + 6 = 3(x + 2)$$

A **common factor** can also be a **letter**.
Both y^2 and $5y$ can be divided by y.
To factorise $y^2 - 5y$ we take y as the common factor,
   so, $y^2 - 5y = y(y - 5)$

Common factors:
The **factors** of a number are all the numbers that will divide exactly into the number.
Factors of 6 are 1, 2, 3 and 6.

A **common factor** is a factor which will divide into two or more terms.

1 Factorise $4x - 6$.

Each term has a factor of 2.
So, the common factor is 2.
$4x - 6 = 2(2x - 3)$

2 Factorise $x^2 + 3x$.

Each term has a factor of x.
So, the common factor is x.
$x^2 + 3x = x(x + 3)$

Exercise 12.7

1 Copy and complete.
   (a) $2x + 2y = 2(\ldots + \ldots)$
   (b) $3a - 6b = 3(\ldots - \ldots)$
   (c) $6m + 8n = 2(\ldots + \ldots)$
   (d) $x^2 - 2x = x(\quad)$
   (e) $ab + a = a(\quad)$
   (f) $2x - xy = x(\quad)$
   (g) $2b - 4a = 2(\quad)$
   (h) $2x^2 + 3x = x(\quad)$
   (i) $g - g^2 = g(\quad)$

2 Factorise.
   (a) $2a + 2b$
   (b) $5x - 5y$
   (c) $3d + 6e$
   (d) $4m - 2n$
   (e) $6a + 9b$
   (f) $6a - 8b$
   (g) $8t + 12$
   (h) $5a - 10$
   (i) $4d - 2$
   (j) $3 - 9g$
   (k) $5 - 20m$
   (l) $4k + 4$

3 Factorise.
   (a) $xy - xz$
   (b) $fg + gh$
   (c) $ab - 2b$
   (d) $3q + pq$
   (e) $a + ab$
   (f) $gh - g$
   (g) $a^2 + 3a$
   (h) $5t - t^2$
   (i) $d - d^2$
   (j) $m^2 + m$
   (k) $5r^2 - 3r$
   (l) $3x^2 + 2x$

What you need to know

You should be able to:
- Write simple algebraic expressions.
- Simplify expressions and rules by collecting like terms together.
   e.g. $2d + 3d = 5d$ and $3x + 2 - x + 4 = 2x + 6$
- Multiply simple expressions together. e.g. $2a \times a = 2a^2$ and $y \times y \times y = y^3$
- Recall and use these properties of powers:
   Powers of the same base are **added** when terms are **multiplied**.
   Powers of the same base are **subtracted** when terms are **divided**.

   $a^m \times a^n = a^{m+n}$
   $a^m \div a^n = a^{m-n}$

- Multiply out brackets. e.g. $2(x - 5) = 2x - 10$, $x(x - 5) = x^2 - 5x$ and
   $(x + 2)(x + 5) = x^2 + 5x + 2x + 10 = x^2 + 7x + 10$
- Factorise expressions. e.g. $3x - 6 = 3(x - 2)$ and $x^2 + 5x = x(x + 5)$

Review Exercise 12

1 A lollipop costs t pence.
   Write an expression for the cost of 6 lollipops.

2 Tom is x years old. Naomi is 3 years older than Tom.
   How old is Naomi in terms of x?

3 Mary has p sweets.
   Write down the number of sweets each person has in terms of p.
   (a) Carol has five more than Mary.
   (b) Abdul has twice as many as Mary.
   (c) Tina has three fewer than Abdul.

AQA

4 (a) Aimee is *n* years old. Her brother Ben is two years younger.
Write down Ben's age in terms of *n*.

(b) Aimee's mother is three times as old as Aimee.
Write down her mother's age in terms of *n*.

(c) Aimee's father is four years older than her mother.
Write down her father's age in terms of *n*.

(d) Write an expression for the combined ages of all four members of the family.
Simplify your answer. AQA

5 Cakes cost 25 pence each.

(a) Barry buys 6 cakes for his friends.
How much do the cakes cost altogether?

(b) Jane buys *n* cakes.
Write down an expression for the cost of *n* cakes. AQA

6 Simplify (a) $w + w + w$,   (b) $2w + 5 - w - 3$,   (c) $w \times w$.

7 (a) Simplify. $5p + 2q - q + 2p$ (b) Multiply out. $4(r - 3)$ AQA

8 Write an expression in terms of *d* for the perimeter of this shape.

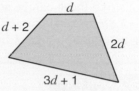

9 (a) Simplify   (i) $x + 3x + x$,   (ii) $6x \div 2$,   (iii) $y \times 3y$.

(b) Multiply out   (i) $4(x + y)$,   (ii) $5(x - 2y)$. AQA

10 Simplify. (a) $ab + 2ba$ (b) $a^2 - a + 3a$ (c) $3(x - 2) - x$ (d) $3x + 2(x + 1)$

11 (a) Multiply out and simplify as much as possible $5(a - 2b) + 7b$.

(b) A loaf of bread costs *y* pence. Write down an expression for the cost of 3 loaves. AQA

12 In the triangle *ABC*, the side *AB* has length *x* units.
AC is twice the length of *AB*.
BC is three units shorter than *AC*.

(a) Write expressions in terms of *x*, for
(i) *AC*,
(ii) *BC*.

(b) Write an expression for the perimeter of the triangle, in terms of *x*.
Give your answer in its simplest form. AQA

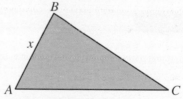

13 Multiply out and simplify where possible.   (a) $y(y - 4)$ (b) $4(3y + 1) - 5y$

14 Multiply out   (a) $x(x + 3)$,   (b) $x(x^2 - 3x)$.

15 Simplify.   (a) $p \times p \times p$ (b) $2a \times 3b \times 4c$ (c) $x^3 \div x^3$ AQA

16 (a) Simplify. $4x + 3x + 7y - 2x + 3y$

(b) Expand and simplify. $x(2x - 3) + 4(x^2 + 1)$ AQA

17 (a) Factorise. (i) $6x - 15$ (ii) $y^2 + 7y$

(b) Multiply out and simplify. $3(y + 2) - 2(y - 3)$

18 Simplify.   (a) $t^3 \times t^5$ (b) $p^6 \div p^2$ (c) $\dfrac{a^3 \times a^2}{a}$ AQA

19 (a) Expand and simplify. $(x + 1)(x - 2)$ (b) Simplify. $a^2b^3 \times \dfrac{a^3}{b}$

CHAPTER 13

Solving Equations ● ● ● ● ●

Activity

Can you solve these puzzles?

● **Nueve** is a Spanish number.
If you add 1 to **nueve** you get 10.
What is **nueve**?

● What number must be put in each shape to make the statements correct?

$\Box + 3 = 8$ $\qquad$ ⬡ $\times 3 = 30$ $\qquad$ $2 \times$ ◯ $- 3 = 7$

These are all examples of **equations.**
Equations like these can be solved using a method known as **inspection**.
Instead of words or boxes, equations are usually written using letters for the unknown numbers.
Solving an equation means finding the numerical value of the letter which fits the equation.

EXAMPLES

Solve these equations by inspection.

1 $\quad x - 2 = 6$

$\quad x = 8$

$\quad$ Reason: $\mathbf{8} - 2 = 6$

2 $\quad 2y = 10$

$\quad y = 5$

$\quad$ Reason: $2 \times \mathbf{5} = 10$

Remember:
A letter or a symbol stands for an unknown number.

$2y$ means $2 \times y$.

Exercise 13.1

1 What number must be put in the box to make each of these statements true?

(a) $\Box + 4 = 7$ $\qquad$ (b) $15 - \Box = 11$ $\qquad$ (c) $13 = \Box + 4$ $\qquad$ (d) $11 = \Box - 5$

2 Solve these equations by inspection.

(a) $x + 2 = 6$ $\qquad$ (b) $a + 7 = 10$ $\qquad$ (c) $y - 4 = 4$

(d) $6 + t = 12$ $\qquad$ (e) $h - 15 = 7$ $\qquad$ (f) $d + 4 = 5$

(g) $z - 5 = 25$ $\qquad$ (h) $p + 7 = 7$ $\qquad$ (i) $c + 1 = 100$

3 What number must be put in the box to make each of these statements true?

(a) $3 \times \Box = 15$ $\qquad$ (b) $\Box \times 4 = 20$ $\qquad$ (c) $\Box \div 2 = 9$ $\qquad$ (d) $7 = \Box \div 3$

4 Solve these equations by inspection.

(a) $3a = 12$ $\qquad$ (b) $5e = 30$ $\qquad$ (c) $8 = 2p$ $\qquad$ (d) $4 = 8y$

(e) $\frac{d}{2} = 5$ $\qquad$ (f) $\frac{t}{3} = 3$ $\qquad$ (g) $\frac{m}{7} = 4$ $\qquad$ (h) $\frac{x}{5} = 20$

5 What number must be put in the box to make each of these statements true?

(a) $2 \times \Box + 3 = 5$ $\qquad$ (b) $\Box \times 3 + 5 = 17$ $\qquad$ (c) $3 + \Box \times 2 = 11$

(d) $5 \times \Box - 1 = 9$ $\qquad$ (e) $4 \times \Box - 5 = 7$ $\qquad$ (f) $\Box \times 3 - 6 = 9$

Solving equations by working backwards

I think of a number and then subtract 3.
The answer is 5.
What is the number I thought of?

Imagine that x is the number I thought of.
The steps of the problem can be shown in a diagram.

x ⟶ | subtract 3 | ⟶ Answer 5

Now work backwards, doing the opposite calculation.

8 ⟵ | add 3 | ⟵ 5

The opposite of 'subtracting 3' is 'adding 3'.

The number I thought of is 8.

EXAMPLES

1. Ken thinks of a number.
   He multiplies it by 5.
   His answer is 30.
   What number did Ken think of?

   x ⟶ | multiply by 5 | ⟶ 30

   6 ⟵ | divide by 5 | ⟵ 30

   Ken's number is 6.

   Remember:

Forwards	Backwards
add	subtract
subtract	add
multiply	divide
divide	multiply

2. I think of a number, multiply it by 3 and add 4.
   The answer is 19.
   What is my number?

   x ⟶ | multiply by 3 | ⟶ | add 4 | ⟶ 19

   5 ⟵ | divide by 3 | ⟵ 15 ⟵ | subtract 4 | ⟵ 19

   The number I thought of is 5.

Exercise 13.2   Solve these equations by working backwards.

1. I think of a number and then add 4.   The answer is 7.
   What is my number?

2. Jan thinks of a number and then subtracts 5.   Her answer is 9.
   What is her number?

3. I think of a number and then multiply it by 2.   The answer is 10.
   What is my number?

4. Lou thinks of a number. He multiplies it by 2 and then subtracts 5.   The answer is 7.
   What is his number?

5. I think of a number, subtract 5 and then multiply by 2.   The answer is 12.
   What is my number?

6. Beth thinks of a number. She multiplies it by 3 and adds 4.
   If the answer is 19, what is her number?

Solving Equations . . . Solving Equations . . . Solving Equations . . .

7 I think of a number, add 4 then multiply by 3.   The answer is 24.
What is my number?

8 Steve thinks of a number. He multiplies it by 5 and then adds 2.   The answer is 17.
What is his number?

9 I think of a number, multiply it by 3 and then subtract 5.   The answer is 7.
What is my number?

10 Solve this puzzle.

> Begin with x.   Double it and then add 3.   The result is equal to 17.   What is the value of x?

11 Kathryn thinks of a number. She adds 3 and then doubles the result.
(a)   What number does Kathryn start with to get an answer of 10?
(b)   Kathryn starts with x. What is her answer in terms of x?

12 Sarah thinks of a number. She subtracts 2 and multiplies by 3.
(a)   What number does Sarah start with to get an answer of 21?
(b)   Sarah starts with x. What is her answer in terms of x?

13 Ali thinks of a number. He multiplies it by 2 and then adds 3.
(a)   What number does Ali start with to get an answer of 15?
(b)   Ali starts with x. What is his answer in terms of x?

The balance method

It is not always easy to solve equations by inspection.
To solve harder equations a better method has to be used.
Here is a method that works a bit like a balance.

①

②

These scales are balanced.

You can add the same amount to both sides
and they still balance.

③

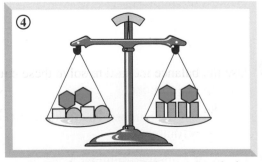

④

You can subtract the same amount from both sides
and they still balance.

You can double (or halve) the amount on both
sides and they still balance.

Equations work in the same way.
If you do the same to both sides of an equation, it is still true.

EXAMPLES

Use the balance method to solve these equations. Explain what you are doing.

1 Solve $d - 13 = 5$.

$d - 13 = 5$
Add 13 to both sides.
$d = 18$

2 Solve $x + 7 = 16$.

$x + 7 = 16$
Subtract 7 from both sides.
$x = 9$

3 Solve $4a = 20$.

$4a = 20$
Divide both sides by 4.
$a = 5$

The aim is to find out what number the letter stands for, by ending up with **one letter** on one side of the equation and a **number** on the other side.

4 Solve $4n + 5 = 17$.

$4n + 5 = 17$
Subtract 5 from both sides.
$4n = 12$
Divide both sides by 4.
$n = 3$

Look at the examples carefully.
The steps taken to solve the equations are explained.
Notice that, doing the same to both sides means: **adding** the **same number** to both sides.
subtracting the **same number** from both sides.
dividing both sides by the **same number**.
multiplying both sides by the **same number**.

Exercise 13.3

1 Use the balance method to solve these equations.
Write down the steps that you use to solve each equation.
(a) $y + 4 = 7$ (b) $x + 5 = 11$ (c) $a + 10 = 20$
(d) $e + 9 = 24$ (e) $d + 6 = 17$ (f) $c + 15 = 35$
(g) $9 + x = 11$ (h) $2 + y = 21$ (i) $8 + m = 15$

2 Use the balance method to solve these equations. Explain each step of your working.
(a) $q - 5 = 2$ (b) $m - 2 = 8$ (c) $n - 7 = 9$
(d) $p - 6 = 12$ (e) $x - 11 = 20$ (f) $y - 3 = 14$
(g) $a - 1 = 1$ (h) $g - 3 = 1$ (i) $h - 5 = 7$

3 Use the balance method to solve these equations.
(a) $28 + x = 42$ (b) $t - 15 = 13$ (c) $f + 16 = 34$
(d) $y - 12 = 7$ (e) $14 + b = 21$ (f) $x - 9 = 20$
(g) $7 + m = 11$ (h) $k - 2 = 3$ (i) $5 + y = 12$

4 Use the balance method to solve these equations. Write down the steps that you use.
(a) $3c = 12$ (b) $5a = 20$ (c) $4f = 12$
(d) $8p = 24$ (e) $6h = 30$ (f) $10u = 20$
(g) $\frac{d}{3} = 10$ (h) $\frac{e}{2} = 7$ (i) $\frac{f}{4} = 5$

5 Use the balance method to solve these equations. Show each step of your working.
(a) $2p + 1 = 9$ (b) $4t - 1 = 11$ (c) $3h - 7 = 14$
(d) $3 + 4b = 11$ (e) $5d - 8 = 42$ (f) $2x + 3 = 15$
(g) $2 + 3c = 17$ (h) $3n - 1 = 8$ (i) $4x + 3 = 11$

6 Solve these equations.
There is no need to explain your working if you are confident of what you are doing.

(a) $a - 5 = 7$ (b) $7x = 28$ (c) $\frac{a}{3} = 9$

(d) $3x + 5 = 11$ (e) $4b + 8 = 32$ (f) $6x - 9 = 15$

(g) $6k - 7 = 5$ (h) $7b + 4 = 25$ (i) $9c - 12 = 6$

(j) $5c + 7 = 42$ (k) $8y - 5 = 27$ (l) $2x + 5 = 17$

7 Solve these equations.

(a) $6a = 18$ (b) $4x - 7 = 29$ (c) $8a + 7 = 7$

(d) $1 + 6p = 7$ (e) $8y - 14 = 26$ (f) $12 + 5p = 32$

(g) $3x - 5 = 22$ (h) $5 + 2k = 13$ (i) $5m + 3 = 18$

More equations

All the equations you have solved so far have had whole number solutions,
but the solutions to equations can include negative numbers and fractions.

EXAMPLES

1 Solve $5x = 2$.

$5x = 2$
Divide both sides by 5.
$x = \frac{2}{5}$

2 Solve $-4a = 20$.

$-4a = 20$
Divide both sides by -4.
$a = -5$

3 Solve $6m - 1 = 2$.

$6m - 1 = 2$
Add 1 to both sides.
$6m = 3$
Divide both sides by 6.
$m = \frac{3}{6}$
$m = \frac{1}{2}$

4 Solve $5 - 4n = -1$.

$5 - 4n = -1$
Subtract 5 from both sides.
$-4n = -6$
Divide both sides by -4.
$n = 1.5$

Exercise 13.4

1 Solve these equations. Explain each step of your working.

(a) $4k = 2$ (b) $2a = -6$ (c) $-3d = 12$

(d) $-8n = 4$ (e) $t + 3 = -2$ (f) $n - 3 = -2$

(g) $2m + 1 = 4$ (h) $3x - 2 = 5$ (i) $2y + 5 = 4$

2 Solve these equations.

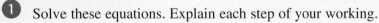

(a) $5x = -10$ (b) $2y + 7 = 1$ (c) $4t + 10 = 2$

(d) $5 - a = 7$ (e) $2 - d = 5$ (f) $3 - 2g = 9$

(g) $4t = 2$ (h) $2x = 15$ (i) $5d = 7$

(j) $4a - 5 = 1$ (k) $3 + 5g = 4$ (l) $2b - 5 = 4$

3 Solve these equations.

(a) $x - 1 = -3$ (b) $3 + 2n = 2$ (c) $2 - x = 3$

(d) $4 - 3y = 13$ (e) $2x - 1 = -3$ (f) $3 - 5x = 18$

(g) $4x + 1 = -5$ (h) $-2 - 3x = 10$ (i) $2 - 4x = 8$

What you need to know

- The solution of an equation is the value of the unknown letter that fits the equation.

You should be able to:
- Solve simple equations by inspection.   e.g. $x + 2 = 5$, $x - 3 = 7$, $2x = 10$
- Solve simple equations by working backwards.
- Use the balance method to solve equations which are difficult to solve by inspection.

Review Exercise 13

1 What number must be put in the box to make each of these statements correct?

    (a) $\square + 5 = 9$ (b) $7 - \square = 4$ (c) $3 \times \square = 18$ (d) $\dfrac{\square}{3} = 6$

2 Petra thinks of a number, then doubles it. Her answer is 16.
What number did she think of?

3

Think of a number. Multiply it by three. Now add seven.

The number I think of is 14.

My answer is 94.

   Teacher   **Tina**   **Abel**

    (a) What answer does Tina get?
    (b) What is the number Abel thinks of?   AQA

4 Jacob uses this rule.

       "Start with a number, divide it by 2 and then add 3.
                  Write down the result."

    (a) What is the result when Jacob starts with 8?
    (b) What number did Jacob start with when the result is 5?   AQA

5 Bob uses this rule:

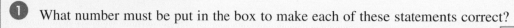
    Start with a number. Multiply it by 3. Take away 5. Write down the answer.

    (a) What is the answer if Bob starts with x?
    (b) What is the answer if Bob starts with -1?
    (c) What number must Bob start with to get an answer of 16?

6 Solve these equations.   (a) $x - 2 = 5$ (b) $5x = 20$ AQA

7 Solve. (a) $\dfrac{x}{5} = 4$ (b) $3x + 5 = 17$ (c) $5n - 3 = 7$

8 Solve these equations.
    (a) $y + 3 = 5$ (b) $2t + 8 = 2$ (c) $4g = 2$ (d) $5x - 1 = 2$

9 Solve. (a) $3a = 15$ (b) $b + 8 = 5$ (c) $7c - 3 = 11$

10 Solve the equation $6x + 27 = 99$.

11 Solve these equations.   (a) $2x + 3 = -5$ (b) $5 + 4y = 7$

12 Solve these equations.   (a) $-5x = 4$ (b) $6 - 2y = 7$

CHAPTER 14 Further Equations

Solving equations was first covered in Chapter 13.
Here are some reminders.

EXAMPLES

Solve the following equations.

1. $5x = 15$
   $x = 3$

2. $m - 5 = 3$
   $m = 8$

3. $2a + 4 = 7$
   $2a = 3$
   $a = 1\frac{1}{2}$

4. $3k + 5 = 2$
   $3k = -3$
   $k = -1$

The aim is to find the numerical value of the letter, by ending up with **one letter** on one side of the equation and a **number** on the other side of the equation.

Exercise 14.1

Solve these equations. The solution will not always be a whole number.

1. $4a = 12$
2. $x + 5 = 7$
3. $2m = -6$
4. $6y = 3$
5. $-3y = 15$
6. $\frac{k}{3} = 5$
7. $2x + 1 = 7$
8. $3 + 4w = 5$
9. $5n + 7 = 3$
10. $2m + 3 = -1$
11. $8 + 2g = 5$
12. $2p + 9 = 18$
13. $-5n - 6 = 19$
14. $4y + 5 = 11$
15. $5 - 2d = 10$

Equations with brackets

Equations can include brackets.
Before using the balance method any brackets must be removed by multiplying out.
This is called **expanding**.

Remember:
$2(x + 3)$ means $2 \times (x + 3)$
$2(x + 3) = 2 \times x + 2 \times 3$
$= 2x + 6$

Once the brackets have been removed the balance method can be used as before.

EXAMPLES

1. Solve $3(x + 2) = 12$.

   $3(x + 2) = 12$
   Expand the brackets.
   $3x + 6 = 12$
   $3x = 6$
   $x = 2$

2. Solve $5(3y - 7) = 25$.

   $5(3y - 7) = 25$
   Expand the brackets.
   $15y - 35 = 25$
   $15y = 60$
   $y = 4$

Exercise 14.2

1 Solve.
(a) $2(x + 3) = 12$ (b) $4(a + 1) = 12$ (c) $5(t + 4) = 30$
(d) $2(y + 4) = 8$ (e) $3(e + 2) = 21$ (f) $6(3 + x) = 30$

2 Solve.
(a) $3(p - 2) = 9$ (b) $6(c - 2) = 24$ (c) $2(x - 1) = 4$
(d) $4(y - 3) = 24$ (e) $2(g - 3) = 16$ (f) $8(q - 3) = 40$

3 Solve.
(a) $3(a + 1) = 15$ (b) $2(b - 2) = 8$ (c) $4(c + 2) = 12$
(d) $6(d - 3) = 36$ (e) $7(2 + e) = 49$ (f) $5(f + 2) = 30$

4 Solve.
(a) $3(2w + 1) = 15$ (b) $2(4s + 5) = 34$ (c) $4(1 + 3x) = 28$
(d) $6(3g - 7) = 12$ (e) $4(2q - 1) = 28$ (f) $8(3t - 5) = 32$
(g) $3(2w + 1) = 27$ (h) $4(7 - 2x) = 4$ (i) $5(3y - 10) = 25$

5 Solve these equations. The solution will not always be a whole number.
(a) $3(p + 2) = 3$ (b) $2(3 - d) = 10$ (c) $2(1 - 3g) = 14$
(d) $2(x - 5) = 7$ (e) $5(y + 1) = 7$ (f) $2(1 + 3t) = 5$
(g) $2(2t - 1) = 5$ (h) $3(2a - 3) = 6$ (i) $5(m - 2) = 3$

Equations with letters on both sides

In some questions letters appear on both sides of the equation.

EXAMPLES

1 Solve $3x + 1 = x + 7$.

$3x + 1 = x + 7$
Subtract 1 from both sides.
$\quad 3x = x + 6$
Subtract x from both sides.
$\quad 2x = 6$
Divide both sides by 2.
$\quad\quad x = 3$

2 Solve $2a - 3 = 9 - a$.

$2a - 3 = 9 - a$
Add 3 to both sides.
$\quad 2a = 12 - a$
Add a to both sides.
$\quad 3a = 12$
Divide both sides by 3.
$\quad\quad a = 4$

Exercise 14.3

1 Solve the following equations. Write down the steps that you use.
(a) $3x = 20 - x$ (b) $5q = 12 - q$ (c) $2t = 15 - 3t$
(d) $5e - 9 = 2e$ (e) $3g - 8 = g$ (f) $y + 3 = 5 - y$
(g) $4x + 1 = x + 7$ (h) $7k + 3 = 3k + 7$ (i) $3a - 1 = a + 7$
(j) $3p - 1 = 2p + 5$ (k) $6m - 1 = m + 9$ (l) $3d - 5 = 5 + d$
(m) $2y + 1 = y + 6$ (n) $3 + 5u = 2u + 12$ (o) $4q + 3 = q + 3$

2 Solve.
(a) $3d = 32 - d$ (b) $3q = 12 - q$ (c) $3c + 2 = 10 - c$
(d) $4t + 2 = 17 - t$ (e) $4w + 1 = 13 - 2w$ (f) $2e - 3 = 12 - 3e$
(g) $2g + 5 = 25 - 2g$ (h) $2z - 6 = 14 - 3z$ (i) $5m + 2 = 20 + 2m$
(j) $5a - 4 = 3a + 6$ (k) $3 + 4x = 15 + x$ (l) $6y - 11 = y + 4$

3 Solve these equations. The solution will not always be a whole number.

(a) $3m + 8 = m$ (b) $2 - 4t = 12 + t$ (c) $5p - 3 = 3p - 7$

(d) $5x - 7 = 3x$ (e) $3 + 5a = a + 5$ (f) $2b + 7 = 11 - 3b$

(g) $4 - 4y = y$ (h) $7 + 3d = 10 - d$ (i) $f - 6 = 3f + 1$

4 Solve these equations.

(a) $2(x + 3) - 5 = 9$ (b) $2(a - 1) + a = 3$ (c) $4(3 - 2m) - 7 = 2m$

(d) $3(a + 4) = 2 + a$ (e) $3(y - 5) = y - 4$ (f) $4(n + 2) = 2n + 5$

(g) $4d + 3 = 2(d - 3)$ (h) $7k + 2 = 5(k - 4)$ (i) $2(4t + 5) = t - 18$

(j) $5q - 2(q + 1) = 4$ (k) $x = 8 - 2(x + 3)$ (l) $4 - 3(a - 2) = a$

Using equations to solve problems

So far, you have been given equations and asked to solve them.
The next step is to **form an equation** first using the information given in a problem.
The equation can then be solved in the usual way.

EXAMPLE

The triangle has sides of length: x cm, $2x$ cm and 7 cm.

(a) Write an expression, in terms of x,
for the perimeter of the triangle.
Give your answer in its simplest form.

(b) The triangle has a perimeter of 19 cm.
By forming an equation find the value of x.

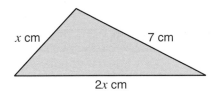

(a) The perimeter of the triangle is: $x + 2x + 7$ cm
In its simplest form, the perimeter is: $(3x + 7)$ cm

(b) The perimeter of the triangle is 19 cm, so, $3x + 7 = 19$
$$3x = 12$$
$$x = 4$$

Exercise 14.4

1 (a) Write an expression, in terms of x, for the sum of the angles of the triangle.
Give your answer in its simplest form.

(b) The sum of the angles is 180°.
By forming an equation find the value of x.

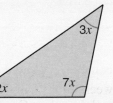

2 The weights of three packages are shown.

k kilograms

$2k$ kilograms

$3k$ kilograms

(a) Write an expression, in terms of k, for the total weight of the packages.

(b) The packages weigh 15 kilograms altogether.
By forming an equation find the weight of the lightest package.

3 Bernadette pays 96 pence for a newspaper and a magazine.
The magazine costs twice as much as the newspaper.

  (a) The newspaper costs x pence.
  Write an expression, in terms of x,
  for the price of the magazine.

  (b) By forming an equation find the
  price of the magazine.

4

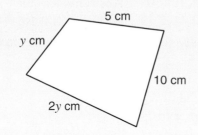

  (a) Write an expression, in terms of y,
  for the perimeter of this shape.
  Give your answer in its simplest form.

  (b) The shape has a perimeter of 39 cm.
  By forming an equation find the value of y.

5 A bag contains the following balls.

  a yellow balls
  $2a + 1$ red balls
  $3a + 2$ blue balls

  (a) Write an expression, in terms of a,
  for the total number of balls in the bag.
  (b) The bag contains 45 balls.
  How many yellow balls are in the bag?

6 Dominic is 7 years younger than Marcie.
  (a) Dominic is n years old.
  Write an expression, in terms of n, for Marcie's age.
  (b) The sum of their ages is 43 years.
  By forming an equation find the ages of Dominic and Marcie.

7 The diagram shows the lengths of three rods.

  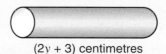

  $(y - 5)$ centimetres y centimetres $(2y + 3)$ centimetres

  (a) Write an expression, in terms of y, for the total length of the rods.
  (b) The total length of the rods is 30 centimetres.
  What is the length of the longest rod?

8 Grace is given a weekly allowance of £p.
  Aimee is given £4 a week **more** than Grace.
  Lydia is given £3 a week **less** than Grace.
  (a) Write an expression, in terms of p, for the amount given to
      (i) Aimee,   (ii) Lydia,   (iii) all three girls.
  (b) The three girls are given a total of £25 a week altogether.
  By forming an equation find the weekly allowance given to each girl.

9 The cost of a pencil is x pence. The cost of a pen is 10 pence more than a pencil.
  (a) Write an expression, in terms of x, for the cost of a pen.
  (b) Write an expression, in terms of x, for the total cost of a pencil and two pens.
  (c) The total cost of a pencil and two pens is 65 pence.
  Form an equation in x and solve it to find the cost of a pencil.

Trial and improvement

Some equations cannot be solved directly.
Numerical solutions can be found by making a guess and improving the accuracy of the guess by **trial and improvement**.
This can be a time-consuming method of solving equations and is often used only as a last resort for solving equations which cannot be easily solved by algebraic or graphical methods.

A solution to the equation
$x^3 - 4x = 7$ lies between 2 and 3.
Find this solution to 1 decimal place.

Because the solution lies between 2 and 3, notice that:

when $x = 2$ $2^3 - 4(2) = 0$ Answer is less than 7.
when $x = 3$ $3^3 - 4(3) = 15$ Answer is greater than 7.

We are trying to find a value for x which produces the answer 7.

First guess: $x = 2.5$ $2.5^3 - 4(2.5) = 5.625$ Too small.
Second guess: $x = 2.6$ $2.6^3 - 4(2.6) = 7.176$ Too big.

The solution lies between 2.5 and 2.6, but 2.6 gives an answer closer to 7.
The answer is probably 2.6, correct to 1 d.p.
To be certain, try $x = 2.55$. $2.55^3 - 4(2.55) = 6.381375$ Too small.

The solution lies between 2.55 and 2.6.
2.6 is the solution, correct to 1 d.p.

Exercise 14.5

You will need a calculator for this exercise.

1   Use trial and improvement to solve $x^3 = 54$,   correct to one decimal place.
The working can be shown in a table.

x	x^3	
3	27	Too small
4	64	Too big
3.5		

2   Use trial and improvement to solve these equations.
(a) $w^3 = 72$ (b) $4x^3 = 51$

3   A solution to the equation $x^3 + 2x = 40$ lies between 3 and 4.
Find this solution to 1 decimal place.

x	$x^3 + 2x$	
3	33	Too small
3.5	49.875	Too big

4   A solution to the equation $x^3 + 5x = 880$ lies between 9 and 10.
Use trial and improvement to find this solution to one decimal place. Show your trials.

5   The volume, V, of a cuboid is given as $V = x^3 - x$.
Use trial and improvement to find the value of x when $V = 100\,cm^3$.
Give your answer to one decimal place.

What you need to know

You should be able to:
- Solve equations with brackets.   e.g. $4(3 + 2x) = 36$
- Solve equations with unknowns on both sides of the equals sign.   e.g. $3x + 1 = x + 7$
- Use equations to solve problems.
- Use **trial and improvement** to solve equations. The accuracy of the value of the unknown letter is improved until the required degree of accuracy is obtained.

① Solve the equations   (a) $2x + 3 = 11,$ (b) $3y + 7 = 1 - y.$ AQA

② Solve these equations: (a) $2x + 10 = 29$ (b) $5x - 4 = 8 - x$ AQA

③ Solve the equations.   (a) $5x + 4 = 39$ (b) $4(x + 4) = 26$ AQA

④ Solve these equations.
(a) $7x = 56$ (b) $5x + 7 = 15$ (c) $8x - 1 = 4x + 19$ AQA

⑤ Solve.   (a) $3(x - 2) = 9$ (b) $6x + 3 = x - 2$

⑥ Solve these equations.   (a) $4x + 6 = 11$ (b) $2(5 + 2x) = 16$ AQA

⑦ Solve these equations.   (a) $5x - 12 = 23$ (b) $3x + 4 = 5x + 8$ AQA

⑧ Solve the equation. $7x - 13 = 5(x - 3)$ AQA

⑨ Solve.   (a) $3 - 4q = 11$ (b) $4(2t - 3) + 4t = 6$

⑩ Hassan is twice as old as Ali. Their ages add up to 39 years.
How old is: (a)   Ali,   (b)   Hassan?   AQA

⑪ (a)   Write an expression, in terms of x,
      for the perimeter of this shape.
(b)   The perimeter is 58 cm.
      By forming an equation find the value of x.
(c)   What is the length of the longest side of the shape?

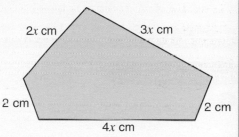

2x cm   3x cm

2 cm   2 cm

4x cm

⑫ A drink costs x pence. A cake costs 7 pence more than a drink.
(a)   Write down, in terms of x, an expression for
     (i)   the cost of a cake,   (ii)   the total cost of two drinks and a cake.
(b)   The total cost of two drinks and a cake is 97 pence.
     Form an equation in x and solve it to find the cost of a cake.

⑬

n litres $(3n + 1)$ litres

The diagram shows two cans of oil.
The cans hold a total of 3 litres of oil.
By forming an equation find the amount of oil in the larger can.

⑭ Use a trial and improvement method to solve the equation $x^3 - x = 300.$
Give your answer to one decimal place.
Show all your trials in a table, as shown.

Trial x	$x^3 - x$	Too high/too low
8	504	Too high

AQA

⑮ The equation $x^3 + x = 20$ has a solution between 2 and 3.
Use a trial and improvement method to find the solution correct to two decimal places.
Show all your working.

CHAPTER 15

Formulae ●●●●●●●●●●●●

Most people at some time make use of **formulae** to carry out routine calculations.
A **formula** represents a rule written using numbers, letters and mathematical signs.
When using a formula you will need to **substitute** your own values for the letters in order to carry out your calculation.

Substitution

Substituting whole numbers

EXAMPLE

Find the value of
(a) $a + 5$, (b) $a - 3$, (c) $3a$, (d) $a \times a$, when $a = 4$.

(a) $a + 5$ (b) $a - 3$ (c) $3a$ (d) $a \times a$
$= 4 + 5$ $= 4 - 3$ $= 3 \times 4$ $= 4 \times 4$
$= 9$ $= 1$ $= 12$ $= 16$

Exercise 15.1 Do not use a calculator.

1 $m = 3$. Find the value of
(a) $m + 2$ (b) $m - 1$ (c) $4m$ (d) $m \times m$

2 $t = 5$. Find the value of
(a) $5 + t$ (b) $3 - t$ (c) $2t$ (d) $t \times t$

3 $x = 4$. Find the value of
(a) $x + x$ (b) $x - 4$ (c) $3x$ (d) $x \times x \times 2$

4 $a = 6$ and $b = 3$. Find the value of
(a) $3a + 2b$ (b) $b - a$ (c) $\dfrac{a}{b}$ (d) $a \times b$ (e) $2(a + b)$

5 $p = 10$ and $q = 5$. Find the value of
(a) $p + q$ (b) $p - 2q$ (c) $\dfrac{p}{q}$ (d) $p \times q$ (e) $3(p - q)$

6 $x = 15$ and $y = 6$. Find the value of
(a) $x + 2y$ (b) $x - 3y$ (c) $\dfrac{x}{y}$ (d) $x \times y$ (e) $6(x - y)$

Substituting negative numbers

EXAMPLE

Find the value of
(a) $a + 5$, (b) $a - 3$, (c) $3a$, when $a = -4$.

(a) $a + 5$ (b) $a - 3$ (c) $3a$
$= -4 + 5$ $= -4 - 3$ $= 3 \times -4$
$= 1$ $= -7$ $= -12$

Exercise 15.2 — Do not use a calculator.

1 $m = -3$. Find the value of
   (a) $m + 2$ (b) $m - 1$ (c) $4m$ (d) $2m + 9$

2 $t = -5$. Find the value of
   (a) $5 + t$ (b) $t - 3$ (c) $2t$ (d) $3t - 1$

3 $x = -4$. Find the value of
   (a) $x + x$ (b) $x - 4$ (c) $3x$ (d) $10 + 2x$

4 $a = 6$ and $b = -3$. Find the value of
   (a) $3a + 2b$ (b) $b - a$ (c) $\frac{a}{b}$ (d) $a \times b$ (e) $2(a + b)$

5 $p = -10$ and $q = 5$. Find the value of
   (a) $p + q$ (b) $p - 2q$ (c) $\frac{p}{q}$ (d) $p \times q$ (e) $3(p - q)$

6 $x = 15$ and $y = -6$. Find the value of
   (a) $x + 2y$ (b) $y - x$ (c) $\frac{x}{y}$ (d) $x \times y$ (e) $6(x + y)$

Writing expressions and formulae

A lollipop costs 15 pence.
How much will n lollipops cost?
Write a formula for the cost, C, in pence, of n lollipops.

Each lollipop costs 15 pence.

So, n lollipops cost $15 \times n$ pence $= 15n$ pence.

> $15n$ is an **algebraic expression**.

If the cost of n lollipops is C pence, then $C = 15n$.

> $C = 15n$ is a **formula**.

Formulae can be used in lots of situations.

The grid shows the numbers from 1 to 50.
An **L** shape has been drawn on the grid.
It is called $\mathbf{L_{14}}$ because the lowest number is 14.

What is the sum of the numbers in $\mathbf{L_{14}}$?

The **L** shape can be moved to different parts of the grid.
We can find the sum of the numbers for each shape.

1	2	3	4	5	6	7	8	9	10
11	12	13	14	15	16	17	18	19	20
21	22	23	24	25	26	27	28	29	30
31	32	33	34	35	36	37	38	39	40
41	42	43	44	45	46	47	48	49	50

A formula for the sum of the numbers, $\mathbf{S_n}$,
can be written in terms of n for shape $\mathbf{L_n}$.
$$S_n = n + (n + 10) + (n + 20) + (n + 21)$$
$$S_n = 4n + 51$$

An **expression** is just an answer using letters and numbers.
A **formula** is an algebraic rule. It always has an equals sign.

EXAMPLES

1 A hedge is l metres long. A fence is 50 metres longer than the hedge.
Write an **expression**, in terms of l, for the length of the fence.

The fence is $(l + 50)$ metres long.

2 Boxes of matches each contain 48 matches.
Write down a **formula** for the number of matches, m, in n boxes.

$m = 48 \times n$ This could be written as $m = 48n$.

1 A pencil costs *y* pence.
  (a) What is the cost of 5 pencils?
  (b) A ruler costs 8 pence more than a pencil.
      What is the cost of a ruler?

2 Egg boxes hold 12 eggs each.
  How many eggs are there in *e* boxes?

3 I am *a* years old.
  (a) How old will I be in 1 years time?
  (b) How old was I four years ago?
  (c) How old will I be in *n* years time?

4 A child is making a tower with toy bricks.
  He has *b* bricks in his tower.
  Write an expression for the number of bricks in the tower after he takes 3 bricks from the top.

5 Paul is *h* cm tall.
  Sue is 12 cm taller than Paul.
  Write down an expression for Sue's height in terms of *h*.

6 John has *d* CDs.
  (a) Carol has twice as many CDs as John.
      Write down an expression for the number of CDs that Carol has in terms of *d*.
  (b) Fred has 5 more CDs than Carol.
      Write down an expression for the number of CDs that Fred has in terms of *d*.

7 A packet of biscuits costs *y* pence.
  Write down a formula for the cost, *P* pence, of another packet which costs
  (a) five pence more than the first packet,
  (b) two pence less than the first packet,
  (c) twice the cost of the first packet.

8 David is *d* years old.
  Copy and complete this table to show the ages, *A*, of these people.

Name	Clue	Age
Alec	3 years older than David.	$A = d + 3$
Ben	2 years younger than David.	
Charlotte	Twice as old as David.	
Erica	Half David's age.	

9 Write a formula for the perimeter, *P*, for each of these shapes in terms of the letters given.

(a)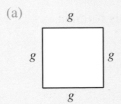
(b)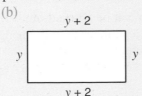
(c)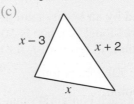
(d)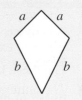

10 A caravan costs £25 per day to hire.
  Write a formula for the cost, *C*, in £s, to hire the caravan for *d* days.

11 The cost of hiring a ladder is given by: £12 per day, plus a delivery charge of £8

(a) Bill hired a ladder for 3 days. How much did he pay?
(b) Sam hired a ladder for 6 days. How much did he pay?
(c) Fred hired a ladder for x days.
   Write down a formula for the total cost, £C, in terms of x.

12 The grid shows the numbers from 1 to 50.
A **T** shape has been drawn on the grid.
It is called $\mathbf{T_{23}}$ because the lowest number is 23.

1	2	3	4	5	6	7	8	9	10
11	12	13	14	15	16	17	18	19	20
21	22	23	24	25	26	27	28	29	30
31	32	33	34	35	36	37	38	39	40
41	42	43	44	45	46	47	48	49	50

Calculate the sum of the numbers in:
(a) $\mathbf{T_{16}}$ (b) $\mathbf{T_{28}}$ (c) $\mathbf{T_2}$

(d) The diagram on the right shows $\mathbf{T_n}$.
   Copy and complete the **T** shape in terms of n.

(e) Write a formula for the sum of the numbers, $\mathbf{S_n}$,
   in terms of n, for shape $\mathbf{T_n}$.
   Write your answer in its simplest form.

Using formulae

The formula for the perimeter of a rectangle is $P = 2L + 2W$.
By **substituting** values for the length, L, and the width, W, you can calculate the value of P.

For example.
To find the perimeter of a rectangle 5 cm in length and 3 cm in width,
substitute $L = 5$ and $W = 3$ into $P = 2L + 2W$.

$P = 2 \times 5 + 2 \times 3$
$ = 10 + 6$
$ = 16$ The perimeter of the rectangle is 16 cm.

EXAMPLES

1 Here is a formula for the area of a rectangle.

   Area = length × width

Use the formula to find the area of a
rectangle 8 cm in length and 3 cm in width.
Area = length × width
$ = 8 \times 3$
$ = 24 \, \text{cm}^2$

2 $G = 4t - 1$.
Find the value of G when $t = \frac{1}{2}$.

$G = 4t - 1$
$ = 4 \times \frac{1}{2} - 1$
$ = 2 - 1$
$ = 1$

Exercise **15.4**   Do not use a calculator for questions 1 to 10.

1 The wages earned by an hourly paid person can be worked out using this formula.

   Wages earned = hours worked × pay per hour

Work out the wages earned by a person who works 8 hours at £6 per hour.

2 The number of points scored by a soccer team can be worked out using this formula.

   Points scored = 3 × games won + games drawn

A team has won 5 games and drawn 2 games. How many points have they scored?

Formulae . . . . Formulae . . . . Formulae . . . .

3 This formula is used to work out the profit, in £s, made on a coach journey.

$$\text{Profit (£)} = 12 \times \text{number of passengers} - 50$$

How much profit is made on a coach journey with 20 passengers?

4 Here is a formula for the perimeter of a rectangle.

$$\textbf{Perimeter} = \textbf{2} \times \textbf{(length + width)}$$

A rectangle is 9 cm in length and 4 cm in width.
Use the formula to work out the perimeter of this rectangle.

5 $T = 5a - 3$.　Find the value of T when　$a = 20$.

6 $X = 3y + 5$.　Work out the value of X when　$y = 1$.

7 $M = 4n + 1$.　(a)　Work out the value of M when　$n = -2$.
　　　　　　　　(b)　Work out the value of n when　$M = 19$.

8 $H = 3g - 5$.　(a)　Find the value of H when　$g = 2.5$.
　　　　　　　　(b)　Find the value of g when　$H = 13$.

9 The number of matches, M, needed to make a pattern of P pentagons is given by the
　　formula:　$M = 4P + 1$.
Find the number of matches needed to make 8 pentagons.

10 The distance, d metres, travelled by a lawn mower in t minutes is given by the
　　formula:　$d = 24t$.
Find the distance travelled by the lawn mower in 4 minutes.

11 Convert 30° Centigrade to Fahrenheit using the formula:　$F = C \times 1.8 + 32$

12 $T = 45W + 30$　is used to calculate the time in minutes needed to cook a joint of beef
weighing W kilograms. How many minutes are needed to cook a joint of beef weighing 2.4 kg?

Further substitution into formulae

EXAMPLES

1 $H = 3(4x - y)$
Find the value of H when　$x = 5$　and　$y = 7$.

$H = 3(4x - y)$
$\quad = 3(4 \times 5 - 7)$
$\quad = 3(20 - 7)$
$\quad = 3(13)$
$\quad = 39$

2 $W = x^2 + 2$
Find the value of W when　$x = 3$.

$W = x^2 + 2$
$\quad = 3 \times 3 + 2$
$\quad = 9 + 2$
$\quad = 11$

Remember:
x^2 means $x \times x$.

Exercise 15.5　　　Do not use a calculator for questions 1 to 15.

1 $F = 5(v + 6)$.　What is the value of F when　(a)　$v = 1$,　(b)　$v = 9$,　(c)　$v = -9$?

2 $V = 2(7 + 2x)$.　What is the value of V when　(a)　$x = 3$,　(b)　$x = -3$,　(c)　$x = \frac{1}{2}$?

3 $P = 3(5 - 2d)$.　What is the value of P when　(a)　$d = 2$,　(b)　$d = 4$,　(c)　$d = 0.5$?

4 $C = 8(p + q)$.　What is the value of C when
　　(a)　$p = 5$　and　$q = 8$,　(b)　$p = 6$　and　$q = -2$,　(c)　$p = 5$　and　$q = -8$?

5 $S = ax + 4$. What is the value of S when
(a) $a = 12$ and $x = 3$, (b) $a = 3$ and $x = -2$, (c) $a = 5$ and $x = 0.4$?

6 $T = a(x + 4)$. What is the value of T when
(a) $a = 5$ and $x = 3$, (b) $a = 2$ and $x = -5$, (c) $a = -3$ and $x = 2$?

7 $K = ab + c$. Work out the value of K when
(a) $a = 3$, $b = 2$ and $c = 5$, (b) $a = 5$, $b = 3$ and $c = -2$.

8 $L = xy - z$. Work out the value of L when
(a) $x = 2$, $y = 3$ and $z = 4$, (b) $x = -4$, $y = 2$ and $z = 3$.

9 $S = a^2$. Find the value of S when (a) $a = 3$, (b) $a = -3$, (c) $a = 10$.

10 $R = p^2 + 2p$. Find the value of R when (a) $p = 3$, (b) $p = -3$.

11 $K = m^2 - 5m$. Work out the value of K when $m = -4$.

12 $S = 2a^2$. Find the value of S when (a) $a = 3$, (b) $a = -3$, (c) $a = 10$.

13 $S = (2a)^2$. Find the value of S when (a) $a = 3$, (b) $a = -3$, (c) $a = 10$.

14 $T = 3a^2 - 9$. Work out the value of T when $a = 4$.

15 $A = x^3$. Find the value of A when (a) $x = 2$, (b) $x = 3$, (c) $x = 4$.

16 $S = 2t^3$. Find the value of S when (a) $t = 2$, (b) $t = 3$, (c) $t = 4$.

17 Convert 77 degrees Fahrenheit to Centigrade using the formula: $C = (F - 32) \div 1.8$

18 The voltage, V volts, in a circuit with resistance, R ohms, and current, I amps, is given by the
formula: $V = IR$.
Find the voltage in a circuit when $I = 12$ and $R = 20$.

19 A simple formula for the motion of a car is $F = ma + R$.
Find F when $m = 500$, $a = 0.2$ and $R = 4000$.

20 The formula $F = \dfrac{mv^2}{r}$ describes the motion of a cyclist rounding a corner.
Find F when $m = 80$, $v = 6$ and $r = 20$.

Writing and using formulae

EXAMPLE

(a) Nick has a birthday party for 10 people.
How much does it cost?
(b) Tony has a birthday party for x people.
Write a formula for the cost £T, in terms of x.
(c) Jean pays £140 for her birthday party.
How many people went to the party?

Birthday Party Specials
£20, plus £8 per person

(a) Nick's party costs: £20 + 10 × £8
= £100

(b) Cost for x people in £ = $x \times 8 = 8x$
Total cost in £   = $20 + 8x$
Total cost is £T
So, formula is $T = 20 + 8x$

(c) Using the formula $T = 20 + 8x$.
Jean's party costs £140, so, $T = 140$.
$$140 = 20 + 8x$$
$$8x = 120$$
$$x = 15$$
15 people went to Jean's party.

1 The cost of a taxi journey is:

> £3 plus £2 for each kilometre travelled

(a) Alex travels 5 km by taxi.
How much does it cost?

(b) A taxi journey of k kilometres costs £C.
Write a formula for the cost, C, in terms of k.

(c) Adrian paid £7 for a taxi journey.
Use your formula to find the number of kilometres he travelled.

2 A rule to find the cooking time, C minutes, of a chicken which weighs k kilograms, is:

> multiply the weight of the chicken by 40 and then add 20

(a) Find the cooking time for a chicken which weighs 3 kg.

(b) Write a formula for C in terms of k.

(c) Use your formula to find the weight of a chicken which has a cooking time of 100 minutes.

3 An approximate rule for changing temperatures in degrees Celsius, C, to temperatures in degrees Fahrenheit, F, is given by the rule:

> double C and add on 30

(a) Find the value of F when $C = 6$.

(b) Write down a formula for F in terms of C.

(c) Use your formula to find the value of C when $F = 58$.

4 A teacher uses this rule to work out the number of exercise books he needs for Year 11 students.

> 3 books per student, plus 50 extra books

(a) This year there are 120 students in Year 11.
How many books are needed?

(b) Using b for the number of books and n for the number of students, write down the teacher's rule for b in terms of n.

(c) For the next Year 11, he will need 470 books.
How many students will be in Year 11 next year?

5

(a) How much does it cost to hire the carpet cleaner for 3 days?

(b) Using T for the total cost in £, and d for the number of days hired, write a formula for T in terms of d.

(c) Sarah paid a total of £96 to hire the carpet cleaner.
For how many days did she hire the carpet cleaner?

6 Scaffolding can be hired.
The hire charge is calculated using this formula:

> forty-five pounds per day plus a fixed charge of seventy pounds

(a) How much would it cost to hire scaffolding for 5 days?

(b) Using C for the total cost in £, and n for the number of days, write a formula for C in terms of n.

(c) A builder paid £475 altogether to hire some scaffolding.
For how many days did he hire the scaffolding?

Rearranging formulae

Sometimes it is easier to use a formula if you **rearrange** it first.

EXAMPLES

1 $k = \frac{8m}{5}$

Rearrange the formula to give m in terms of k.

$$k = \frac{8m}{5}$$

Multiply both sides by 5.

$$5k = 8m$$

Divide both sides by 8.

$$\frac{5k}{8} = m$$

We say we have **rearranged the formula** $k = \frac{8m}{5}$ to make m the **subject** of the formula.

2 $y = 2x + 8$

Make x the subject of the formula.

$$y = 2x + 8$$

Subtract 8 from both sides.

$$y - 8 = 2x$$

Divide both sides by 2.

$$\tfrac{1}{2} y - 4 = x$$

y is the subject of $y = 2x + 8$, x is the subject of $x = \tfrac{1}{2} y - 4$.

3 A cuboid has length 8 cm and breadth 5 cm.
The volume of the cuboid is 140 cm³.
Calculate the height of the cuboid.

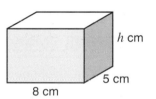

The formula for the volume of a cuboid is $V = lbh$,

so, $h = \frac{V}{lb}$ (by dividing both sides of $V = lbh$ by lb)

Substitute $V = 140$, $l = 8$ and $b = 5$ in $h = \frac{V}{lb}$.

$$h = \frac{140}{8 \times 5} = 3.5$$

The height of the cuboid is 3.5 cm.

Exercise **15.7**

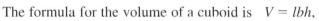

1 Make m the subject of these formulae.
   (a) $a = m + 5$
   (b) $a = x + m$
   (c) $a = m - 2$
   (d) $a = m - b$

2 Make x the subject of these formulae.
   (a) $y = 4x$
   (b) $y = ax$
   (c) $y = \frac{x}{2}$
   (d) $y = \frac{x}{a}$
   (e) $y = \frac{3x}{5}$

3 Make p the subject of these formulae.
   (a) $y = 2p + 6$
   (b) $t = 5p + q$
   (c) $m = 3p - 2$
   (d) $r = 4p - q$

4 The cost, £C, of hiring a car for n days is given by $C = 35 + 24n$.
Make n the subject of the formula.

5 $V = IR$. Rearrange the formula to give R in terms of V and I.

6 The perimeter of a square is $P = 4d$.
  (a) Rearrange the formula to give d in terms of P.
  (b) Find d when $P = 2.8$ cm.

7 The area of a rectangle is $A = lb$.
  (a) Rearrange the formula to give l in terms of A and b.
  (b) Find l when $A = 27$ cm² and $b = 4.5$ cm.

8 The speed of a car is $S = \dfrac{D}{T}$.
  (a) (i) Change the subject to D.
      (ii) Find D when $S = 48$ km/h and $T = 2$ hours.
  (b) (i) Change the subject to T.
      (ii) Find T when $S = 36$ km/h and $D = 90$ km.

9 The perimeter of a rectangle is $P = 2(l + b)$.
  (a) Change the subject to b.
  (b) Find b when $P = 18$ cm and $l = 4.8$ cm.

10 $y = mx + c$
  (a) Rearrange the formula to give x in terms of y, m and c.
  (b) Calculate x when $y = 5$, $m = 4$ and $c = 3$.

11 You are given the formula $v = u + at$.
  (a) Rearrange the formula to give t in terms of v, u and a.
  (b) Work out the value of t when $v = 8$, $u = 20$ and $a = -6$.

12 $A = \dfrac{bh}{2}$
  (a) Rearrange the formula to give b in terms of A and h.
  (b) Calculate b when $A = 9.6$ and $h = 3$.

What you need to know

- A **formula** is an algebraic rule written using numbers, letters and mathematical signs.

You should be able to:

- Write simple algebraic expressions and formulae.
- Substitute positive and negative numbers in expressions and formulae.
- Substitute numbers in simple formulae to solve problems.
- Rearrange simple formulae to make another letter (variable) the subject.

Review Exercise 15

1 What is the value of $2g + 3h$ when $g = 5$ and $h = 2$?

2 Given that $x = 3$ and $y = 4$, find the value of (a) $x + y$, (b) $x - y$, (c) xy.

3 Jean has a Saturday job.
Her pay, in pounds, is worked out using this rule:

Pay = number of hours worked × 7

Last Saturday Jean worked for 8 hours.
How much was she paid?

4 $V = a + bc$. Find the value of V when $a = 5$, $b = 3$ and $c = 4$.

5 What is the value of $5m + 2n$ when $m = 2$ and $n = -3$?

6 $S = pq + r$.　Find the value of S when $p = -3$, $q = 4$ and $r = -2$.

7 $P = 3(m + n)$.　Find the value of P when $m = 0.5$ and $n = 2$.

8 What is the value of $3x^2$ when $x = 6$?

9 What is the value of $t^3 - t$ when $t = 2$?

10 $T = m^2 - 7m$.　Work out the value of T when $m = -5$.

11 (a) Cars can be hired from Andy's Car Hire.

> **ANDY'S CAR HIRE**
> **£23 per day plus a fixed charge of £15**

   (i) How much does it cost to hire a car for 5 days?
   (ii) Mrs Mansi paid £222 altogether to hire a car.
   For how many days did she hire the car?
   (b) Cars can also be hired from Belinda's Car Hire.

> **BELINDA'S CAR HIRE**
> **£25 per day no fixed charge**

   Using T for the total cost in £, and d for the number of days hired, write a formula for T in terms of d.
   (c) Mr Li wants to hire a car for 6 days.
   Which is cheaper, Andy's or Belinda's, and by how much?　AQA

12 To change Pounds (£s) into Euros:

> Multiply the number of Pounds (£s) by 1.52

   (a) Change £250 into Euros.
   (b) Write the statement in the box as an equation connecting e, (Euros) and p, (Pounds).　AQA

13 Segville High School has a disco for Year 7 each year.
A teacher works out how many cans of drink to buy, using this rule:

> two cans for each ticket sold, plus 20 spare cans

   (a) This year, 160 tickets have been sold. How many cans will he buy?
   (b) Using N for the number of cans and T for the number of tickets, write down the teacher's formula for N in terms of T.
   (c) Last year, he bought 300 cans. How many tickets were sold last year?　AQA

14 The cost of hiring a coach is £60 plus £3 for every mile travelled.
   (a) How much will it cost for a journey of 72 miles?
   (b) Write down an expression for the cost in £ of a journey of M miles.
   (c) A journey costs £186.
      (i) Use your answer to part (b) to form an equation using this information.
      (ii) How many miles did the coach travel on this journey?　AQA

15 A formula to estimate the number of rolls of wallpaper, R, for a room is $R = \dfrac{ph}{5}$ where p is the perimeter of the room in metres and h is the height of the room in metres.
The perimeter of Carol's bedroom is 15.5 m and it is 2.25 m high.
How many rolls of wallpaper will she have to buy?　AQA

16 A formula is given as $t = 7p - 50$.　Rearrange the formula to make p the subject.　AQA

17 (a) Use the formula $y = mx + c$ to find the value of y when $m = 3$, $x = -2$ and $c = 9$.
   (b) Rearrange the formula $y = mx + c$ to make x the subject.

CHAPTER 16

Sequences ●●●●●●●●●●●●●

Continuing a sequence

A **sequence** is a list of numbers made according to some rule.
For example:

$$5, \quad 9, \quad 13, \quad 17, \quad 21, \quad \ldots$$

The first term is 5.
To find the next term in the sequence, add 4 to the last term.
The next term in this sequence is $21 + 4 = 25$.
What are the next three terms in the sequence?

> The numbers in a sequence are called **terms**.
> The start number is the **first term**, the next is the second term, and so on.

To continue a sequence:
1. Work out the rule to get from one term to the next.
2. Apply the same rule to find further terms in the sequence.

EXAMPLES

Find the next three terms in each of these sequences.

1 $5, \quad 8, \quad 11, \quad 14, \quad 17, \quad \ldots$

To find the next term in the sequence, add 3 to the last term.
$17 + 3 = 20, \qquad 20 + 3 = 23, \qquad 23 + 3 = 26.$
The next three terms in the sequence are: 20, 23, 26.

2 $2, \quad 4, \quad 8, \quad 16, \quad \ldots$

To find the next term in the sequence, multiply the last term by 2.
$16 \times 2 = 32, \qquad 32 \times 2 = 64, \qquad 64 \times 2 = 128.$
The next three terms in the sequence are: 32, 64, 128.

3 $1, \quad 1, \quad 2, \quad 3, \quad 5, \quad 8, \quad \ldots$

To find the next term in the sequence, add the last two terms.
$5 + 8 = 13, \qquad 8 + 13 = 21, \qquad 13 + 21 = 34.$
The next three terms in the sequence are: 13, 21, 34.
This is a special sequence called the **Fibonacci sequence**.

Exercise 16.1

1 Find the next three terms in these sequences.

(a) $1, \quad 5, \quad 9, \quad 13, \quad \ldots$

(b) $6, \quad 8, \quad 10, \quad 12, \quad \ldots$

(c) $28, \quad 25, \quad 22, \quad 19, \quad \ldots$

(d) $3, \quad 8, \quad 13, \quad 18, \quad 23, \quad \ldots$

(e) $3, \quad 6, \quad 12, \quad 24, \quad \ldots$

(f) $\frac{1}{4}, \quad \frac{1}{2}, \quad \frac{3}{4}, \quad 1, \quad 1\frac{1}{4}, \quad \ldots$

(g) $32, \quad 16, \quad 8, \quad 4, \quad \ldots$

(h) $0.5, \quad 0.6, \quad 0.7, \quad 0.8, \quad \ldots$

(i) $10, \quad 8, \quad 6, \quad 4, \quad \ldots$

(j) $80, \quad 40, \quad 20, \quad 10, \quad \ldots$

(k) $1, \quad 3, \quad 6, \quad 10, \quad 15, \quad \ldots$

(l) $1, \quad 3, \quad 4, \quad 7, \quad 11, \quad 18, \quad \ldots$

2 Find the missing terms from these sequences.

(a) 2, 4, 6, __, 10, 12, __, 16, ... (b) 2, 6, __, 14, 18, __, 26, ...
(c) 1, 2, 4, __, 16, __, 64, ... (d) 28, 22, __, 10, 4, __, ...
(e) 1, 4, 9, __, 25, __, 49, ... (f) 1, 2, 3, 5, __, 13, __, 34, ...
(g) __, 8, 14, __, __, 32, 38, ...

3 Write down the rule, in words, used to get from one term to the next for each sequence.
Then use the rule to find the next two terms.

(a) 2, 9, 16, 23, 30, ... (b) 3, 5, 7, 9, 11, ...
(c) 1, 5, 9, 13, 17, ... (d) 31, 26, 21, 16, ...
(e) 64, 32, 16, 8, 4, ... (f) 1, 3, 9, 27, ...
(g) −2, −4, −6, −8, ... (h) 10, 7, 4, 1, −2, ...

4 A sequence begins 1, 4, 7, 10, ...
(a) What is the 10th number in this sequence?
(b) Explain how you found your answer.

5 A number sequence begins 1, 2, 4, ...
David says that the next number is 8.
Tony says that the next number is 7.
(a) Explain why they could both be correct.
(b) Find the 10th number in David's sequence.
(c) Find the 10th number in Tony's sequence.

6 Here is part of a number sequence: 3, 9, 15, 21, ...
Is the number 50 in this sequence?
Explain your answer.

Using rules

Sometimes you will be given a rule and asked to use it to find the terms of a sequence.

For example:
A sequence begins: 1, 4, 13, ...
The rule for the sequence is:

> Multiply the last number by 3, then add 1

The next term in the sequence is given by:
$$13 \times 3 + 1 = 39 + 1 = 40$$

The following term is given by:
$$40 \times 3 + 1 = 120 + 1 = 121$$

So, the sequence can be extended to: 1, 4, 13, 40, 121, ...

Use the rule to find the next two terms in the sequence.

The **same rule** can be used to make different sequences.

For example:
Another sequence begins: 2, 7, 22, ...
Using the same rule, the next term is given by:
$$22 \times 3 + 1 = 66 + 1 = 67$$

The following term is given by:
$$67 \times 3 + 1 = 201 + 1 = 202$$

So, the sequence can be extended to: 2, 7, 22, 67, 202, ...

Use the rule to find the next two terms in the sequence.

EXAMPLE

This rule is used to find each number in a sequence from the number before it.

> Subtract 3 and then multiply by 4

Starting with 5 we get the following sequence:

5, 8, 20, 68, ...

(a) Write down the next number in the sequence.
(b) Using the same rule, but a different starting number, the second number is 16.
Find the starting number.

(a) $(68 - 3) \times 4 = 65 \times 4 = 260$
Notice that, following the rule, 3 is subtracted first and the result is then multiplied by 4.
The next number in the sequence is 260.
(b) Imagine the first number is x.

Working backwards.

The starting number is 7.

Exercise 16.2

1 Write down the first five terms of these sequences.
(a) First term: 1

Rule: Add 4 to the last term

(b) First term: 1

Rule: Double the last term

(c) First term: 40

Rule: Subtract 5 from the last term

(d) First term: 4

Rule: Double the last term and then subtract 3

(e) First term: 47

Rule: Subtract 1 from the last term and then halve the result

(f) First term: 2   Second term: 6

Rule: Add the last two terms and then halve the result

2 This rule is used to get each number from the number before it:

> Multiply by 2

Use the rule to find the next three numbers when the first number is:
(a) 1,   (b) 3,   (c) −1.

3 This rule is used to get each number from the number before it:

> Add 1 and then double the result

Use the rule to find the next three numbers when the first number is:
(a) 1,   (b) 3,   (c) −3.

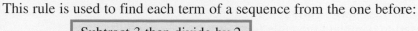

4 This rule is used to find each term of a sequence from the one before:

> Subtract 3 then divide by 2

   (a) The first term is 45.
      (i) What is the second term?
      (ii) What is the **fourth** term?
   (b) Using the same rule, but a different starting number, the second term is 17.
      What is the starting number for the sequence?

5 This rule is used to find each term of a sequence from the one before:

> Add 5 then multiply by 3

   (a) The first term is 7.
      (i) What is the second term?
      (ii) What is the **third** term?
   (b) Using the same rule, but a different starting number, the second term is 45.
      What is the starting number for the sequence?

6 A sequence is formed from this rule:

> Add together the last two terms to find the next term

Part of the sequence is … 5, 9, 14, 23, …
   (a) Write down the next two terms after 23 in the sequence.
   (b) Write down the two terms that come before 5 in the sequence.

7 A sequence begins: 1, -3, …
The sequence is continued using the rule:

> Add the previous two numbers and then multiply by 3

Use the rule to find the next two numbers in the sequence.

8 A sequence begins: 4, 7, 13, 25, …
The next number in the sequence can be found using the rule:

> "Multiply the last term by 2 then subtract 1."

   (a) Write down the next **two** terms in the sequence.
   (b) The 11th term in the sequence is 3073.
      Use this information to find the 10th term in the sequence.

Number sequences

A number sequence which increases (or decreases) by the same amount from one term to the next is called a **linear sequence**.
For example, terms in the sequence:

$$6, \quad 11, \quad 16, \quad 21, \quad …$$

increase by 5 from one term to the next.
We say that the sequence has a **common difference** of 5.

By comparing a sequence with multiples of the counting numbers:

$$1, \quad 2, \quad 3, \quad 4, \quad …$$

we can write a rule to find the nth term of the sequence.

Sequence: 6   11   16   21   …
Multiples of 5: 5   10   15   20   …

To get the nth term add one to the multiples of 5.
So, the nth term is $5n + 1$.

> Compare the sequence with multiples of the common difference.
> In this case the common difference is 5, so, compare the sequence with multiples of 5.

A table can be used to find the nth term of a sequence.

The sequence 2, 8, 14, 20, ... has a common difference of 6.
The common difference is used to complete the table for terms 1, 2, 3, 4 and n.

Term	Term × common difference	Sequence	Difference
1	$1 \times 6 = 6$	2	$2 - 6 = -4$
2	$2 \times 6 = 12$	8	$8 - 12 = -4$
3	$3 \times 6 = 18$	14	$14 - 18 = -4$
4	$4 \times 6 = 24$	20	$20 - 24 = -4$
n	$n \times 6 = 6n$	$6n - 4$	

○○○○○○○○○○

Term:
1 represents the first term,
2 the second term, and so on.
n represents the nth term.

Common difference:
2nd term − 1st term
$= 8 - 2 = 6$.

Differences:
Check that each pair of entries gives the same result.

The nth term of the sequence 2, 8, 14, 20, ... is $6n - 4$.

The rule for the nth term can be used to find the value of any term in the sequence.
To find the fifth term, substitute $n = 5$ into $6n - 4$.
$6 \times 5 - 4 = 30 - 4 = 26$.
The fifth term is 26.

EXAMPLE

(a) Find the nth term in the sequence 31, 28, 25, 22, ...
(b) Find the value of the 10th term in the sequence.

(a) The common difference is -3.

Term	Term × common difference	Sequence	Difference
1	$1 \times (-3) = -3$	31	$31 - (-3) = 34$
2	$2 \times (-3) = -6$	28	$28 - (-6) = 34$
3	$3 \times (-3) = -9$	25	$25 - (-9) = 34$
4	$4 \times (-3) = -12$	22	$22 - (-12) = 34$
n	$n \times (-3) = -3n$	$-3n + 34$	

The nth term is $-3n + 34$.
This can also be written as $34 - 3n$.

(b) The nth term is $34 - 3n$.
Substitute $n = 10$.
$34 - 3 \times 10 = 34 - 30 = 4$
The 10th term is 4.

Exercise 16.3

1 Find the common differences of the following sequences.
(a) 3, 6, 9, 12, ...
(b) 2, 5, 8, 11, ...
(c) 7, 13, 19, 25, ...
(d) 12, 20, 28, 36, ...
(e) 20, 18, 16, 14, ...
(f) 7, 3, −1, −5, ...

2 (a) The multiples of 3 are   3, 6, 9, 12, …
What is the nth multiple of 3?

(b) What is the nth multiple of 8?

(c) What is the nth multiple of 12?

(d) What is the nth even number?

3 A sequence of numbers starts: 4, 7, 10, 13, …

(a) What is the common difference?

(b) Copy and complete this table.

Term	Term × common difference	Sequence	Difference
1	$1 \times \dots =$	4	$4 - \dots =$
2	$2 \times \dots =$	7	$7 - \dots =$
3	$3 \times \dots =$	10	$10 - \dots =$
4	$4 \times \dots =$	13	$13 - \dots =$
n	$n \times \dots = \dots n$	$\dots n + \dots$	

(c) Write down the nth term of the sequence.

(d) What is the value of the 8th term of the sequence?

4 A sequence of numbers starts: 9, 11, 13, 15, …

(a) What is the common difference?

(b) Copy and complete this table for the first four terms of the sequence.

Term	Term × common difference	Sequence	Difference
1	2	9	7
2	4	11	…
		13	

(c) Write down the nth term of the sequence.

(d) What is the value of the 20th term of the sequence?

5 A sequence of numbers starts: 20, 16, 12, 8, …

(a) What is the common difference?

(b) Copy and complete this table for the first four terms of the sequence.

Term	Term × common difference	Sequence	Difference
1	…	20	…
		16	

(c) Write down the nth term of the sequence.

6 Find the nth term of the following sequences.

(a) 1, 4, 7, 10, …

(b) 19, 16, 13, 10, …

(c) 5, 9, 13, 17, …

(d) 4, 8, 12, 16, …

(e) 1, 3, 5, 7, …

(f) 7, 11, 15, 19, …

(g) 6, 4, 2, 0, …

(h) 5, 8, 11, 14, …

(i) 3, 8, 13, 18, …

(j) 40, 35, 30, 25, …

(k) 0, 1, 2, 3, …

(l) -1, 1, 3, 5, …

7 Write down the first three terms of a sequence where the nth term is $n^2 + 3$.
Explain how you found your answer.

16

Sequences . . . Sequences . . . Sequences . . .

173

Activity

These patterns are made using squares.

Pattern 1
3 squares

Pattern 2
5 squares

Pattern 3
7 squares

How many squares are used to make: (a) Pattern 4, (b) Pattern 10, (c) Pattern 100?

The number of squares used to make each pattern forms a **sequence**.

Pattern 4 is made using 9 squares.
You could have answered this: by drawing Pattern 4 or,
by continuing the sequence of numbers 3, 5, 7, …
It is possible to do the same for Pattern 10, though it would involve a lot of work, but it would be unreasonable to use either method for Pattern 100.
Instead we can investigate how each pattern is made.

$2 \times 1 + 1 = 3$ squares $2 \times 2 + 1 = 5$ squares $2 \times 3 + 1 = 7$ squares

Each pattern is made using a **rule**.
The rule can be **described in words**.
To find the number of squares used to make a pattern use the rule:

"Double the pattern number and add 1."

Pattern number	Rule	Number of squares
4	$2 \times 4 + 1$	9
10	$2 \times 10 + 1$	21
100	$2 \times 100 + 1$	201

The same rule can be **written using symbols**.
We can then answer a very important question: How many squares are used to make Pattern n?

Pattern n will have $2 \times n + 1$ squares.
This can be written as $2n + 1$ squares.

Special sequences of numbers

Square numbers

The sequence starts: 1, 4, 9, 16, …
The numbers in this sequence are called **square numbers**.

Triangular numbers

The sequence starts: 1, 3, 6, 10, …
The numbers in this sequence are called **triangular numbers**.

174

1 These patterns are the start of a sequence.

Pattern 1   Pattern 2   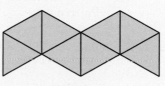 Pattern 3

Draw the next pattern in the sequence.

2 A sequence of patterns is made using equilateral triangles.

 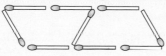 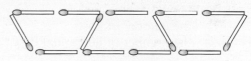

Pattern 1   Pattern 2   Pattern 3   Pattern 4

(a) Draw the next pattern in the sequence.
(b) Copy and complete the table.

Pattern number	1	2	3	4	5	6
Number of equilateral triangles						

(c) Explain why a pattern in this sequence cannot have 27 triangles.
(d) Write an expression, in terms of p, for the number of triangles in Pattern p.
(e) How many triangles are used to make Pattern 12?

3 A sequence of patterns is made using sticks.

    Pattern 1   Pattern 2   Pattern 3

(a) Draw Pattern 4.
(b) How many more sticks are used to make Pattern 5 from Pattern 4?
(c) Copy and complete the table.

Pattern number	1	2	3	4	5
Number of sticks					

(d) Write an expression, in terms of n, for the number of sticks used to make Pattern n.
(e) How many sticks are used to make Pattern 20?

4 A sequence of patterns is made using black and white counters.

    1 black, 3 white   2 black, 6 white   3 black, 9 white

(a) How many white counters are there in a pattern with
     (i) 5 black counters,   (ii) 10 black counters,   (iii) 100 black counters?
(b) How many white counters are there in a pattern with n black counters?

5 These patterns are made using matches.

  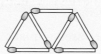

    Pattern 1   Pattern 2   Pattern 3

(a) How many matches are used to make Pattern 5?
(b) Which pattern uses 15 matches?
(c) Find a formula for the number of matches, m, in Pattern p.

6 A sequence of patterns is made using matches.

Pattern 1	Pattern 2	Pattern 3
6 matches	10 matches	14 matches

(a) How many matches are used to make
   (i) Pattern 4,   (ii) Pattern 20?
(b) Which pattern in the sequence uses 30 matches?
(c) Explain why any pattern in the sequence will use an even number of matches.
(d) How many matches are used to make Pattern n?

7 Fences are made by placing fence posts 1 m apart and using 2 horizontal bars between them.

The fence above is 4 m long.
It has 5 posts and 8 horizontal bars.
(a) A fence is 50 m long.
   (i) How many posts does it have?
   (ii) How many horizontal bars does it have?
(b) A fence is x metres long.
   Write down expressions for
   (i) the number of posts,
   (ii) the number of horizontal bars.

8 Linking cubes of side 1 cm are used to make rods.
This rod is made using 4 linking cubes.

The surface area of the rod is 18 square centimetres.
(4 squares on each of the long sides plus one square at each end.)
(a) What is the surface area of a rod made using 5 linking cubes?
(b) What is the surface area of a rod made using 10 linking cubes?
(c) What is the surface area of a rod made using n linking cubes?
(d) How many linking cubes are used to make a rod with a surface area of
   38 square centimetres?

9 A sequence of patterns is made using sticks.

Pattern 1	Pattern 2	Pattern 3	Pattern 4

(a) How many sticks are used to make Pattern 5?
(b) Pattern n uses T sticks.
   Write a formula for T in terms of n.
(c) Use your formula to find the number of sticks used to make Pattern 10.
(d) One pattern uses 77 sticks.
   What is the pattern number?

What you need to know

- A **sequence** is a list of numbers made according to some rule.
  The numbers in a sequence are called **terms**.
- **To continue a sequence:** 1. Work out the rule to get from one term to the next.
  2. Apply the same rule to find further terms in the sequence.
- A number sequence which increases (or decreases) by the same amount from one term to the next is called a **linear sequence**.
  The sequence: 2, 8, 14, 20, 26, … has a **common difference** of 6.
- Special sequences **Square numbers:** 1, 4, 9, 16, 25, …
  Triangular numbers: 1, 3, 6, 10, 15, …
- Patterns of shapes can be drawn to represent a number sequence.
  For example, this pattern represents the sequence 3, 5, 7, …

You should be able to:

- Draw patterns of shapes which represent number sequences.
- Continue a given number sequence.
- Find an expression for the n th term of a sequence.

Review Exercise 16

1 (a) What is the next number in each of these sequences?
   (i) 1, 4, 7, 10, 13, … (ii) 1, 2, 4, 8, 16, … (iii) 10, 6, 2, −2, …
   (b) The rule to continue sequence (i) is: **Add 3 to the last number.**
   Write down the rule to continue each of the other sequences.

2 (a) (i) Find the next two numbers in this sequence: 64, 32, 16, …, …
   (ii) What is the rule for this sequence?

   (b) A rule for another sequence is: **Multiply by 3 and add 1.**
   Using this rule, find the next two numbers in this sequence: 2, 7, 22, …, …

   (c) Here are the rules for a third sequence:
   If a number is even, halve it; If a number is odd, multiply by 3 and add 1.
   Here are the first four terms for a sequence using these rules: 12, 6, 3, 10.
   Write down the next three terms.   AQA

3 Patterns are made of sticks.

   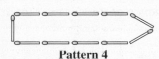

  Pattern 1   Pattern 2   Pattern 3   Pattern 4

   (a) Copy and complete the table for Pattern 4.

Pattern number	1	2	3	4
Number of sticks	5	7	9	

   (b) Sketch Pattern 5.
   (c) Which pattern will have exactly 25 sticks?
   (d) Here is a rule for working out the number of sticks.
       Multiply pattern number by 2 and add 3.
   How many sticks will be in Pattern 100?   AQA

4 What is the next number in each of these sequences?
   (a) 3, 6, 11, 18, 27, … (b) 8, 4, 2, 1, $\frac{1}{2}$, …   AQA

5 (a) A sequence of numbers begins: 1, 7, 13, 19, 25, ..., ...
    (i) Write down the next number in the sequence.
    (ii) Write down the rule for finding the next number in the sequence.
    (iii) What is the 10th number in the sequence?

(b) Another sequence is formed from this rule:

> **Add together the last 2 numbers to find the next number.**

Part of the sequence is: ..., 3, 4, 7, 11, 18, ...
    (i) Write down the number that comes after 18 in this sequence.
    (ii) Write down the number that comes before 3 in this sequence.   AQA

6 The first two terms of a sequence are: 4, 8.

(a) Using the rule: | Add the two previous numbers and divide by two |

write down the third and fourth terms of the sequence.

(b) If the sequence had begun 8, 4 instead of 4, 8 would the third and fourth terms be the same as those in part (a)? Give a reason for your answer.   AQA

7 A sequence of numbers begins: 2, 5, 8, 11, ...
(a) Write down the next number in this sequence.
(b) Work out the 20th number in this sequence.
(c) Find the nth term of this sequence.   AQA

8 (a) Write down the next two terms in the sequence: 3, 5, 9, 15, 23, ...
(b) (i) Write down the next term in the sequence: 3, 5, 9, 17, 33, ..., ...
    (ii) Explain how you got your answer.
(c) Write down the nth term for the sequence: 3, 5, 7, 9, 11, ...   AQA

9 Write down the nth term for each of the following sequences.
(a) 3, 6, 9, 12, ...   (b) 1, 4, 7, 10, ...   AQA

10 A sequence of patterns is formed using equilateral triangles.

Pattern 1   Pattern 2   Pattern 3

(a) How many triangles are in Pattern 10?
(b) Explain why a pattern in this sequence cannot have 40 triangles.
(c) Write an expression, in terms of p, for the number of triangles in Pattern p.   AQA

11 A sequence of patterns is shown.

Pattern 1   Pattern 2   Pattern 3

(a) A pattern in this sequence has 7 black squares.
How many white squares are in the pattern?
(b) How many white squares are used in Pattern 20?
(c) Write an expression, in terms of n, for the number of white squares in the nth pattern of the sequence.   AQA

12 A sequence of numbers begins: 1, 3, 6, 10, 15, 21, ..., ...
(a) Write down the next number in the sequence.
(b) Explain how you worked out your answer.
(c) What type of numbers form this sequence?   AQA

CHAPTER 17

Coordinates and Graphs ● ● ●

Coordinates

Coordinates are used to describe the position of a point.

Two lines are drawn at right angles to each other.
The horizontal line is called the **x axis**. The vertical line is called the **y axis**.
The plural of axis is **axes**.
The two axes cross at the point called the **origin**.

On the diagram, the coordinates of point *A* are (3, 2).
To find point *A*: start at the origin and go right 3 squares then up 2 squares.

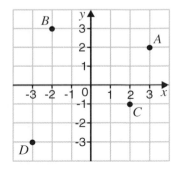

Notation

A is the name, or label, of the point.

The first number is the **x coordinate**.
    If the number is **positive**, go to the **right**.
    If the number is **negative**, go to the **left**.

The second number is the **y coordinate**.
    If the number is **positive**, go **upwards**.
    If the number is **negative**, go **downwards**.

The coordinates of point *B* are (−2, 3).

What are the coordinates of the points C and D?

Exercise **17.1**

1   (a)   Write down the coordinates of points *A* and *B*.
    (b)   Copy the diagram.
         Plot points *C* (4, 2) and *D* (1, 0) on your diagram.

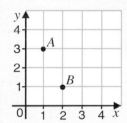

2 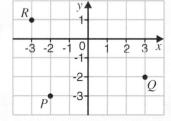   The diagram shows the positions of points *P*, *Q* and *R*.
       Write down the coordinates of these points.

3   Draw *x* and *y* axes from 0 to 5.
    (a)   Plot the points (2, 1), (5, 1) and (5, 4).
    (b)   These are three corners of a square.
         What are the coordinates of the fourth corner of the square?

4   Draw *x* and *y* axes from −5 to 5.
    (a)   Plot the points *S* (3, −2) and *T* (−5, 4).
         Join *ST*.
    (b)   Write down the coordinates of the midpoint of *ST*.

5   Draw *x* and *y* axes from −2 to 4.
    (a)   Plot the points *A* (3, 2), *B* (3, −1) and *C* (−1, −1).
    (b)   Points *A*, *B* and *C* are three corners of a rectangle.
         Point *D* is the fourth corner of the rectangle.
         Plot point *D* on your diagram.
    (c)   What are the coordinates of point *D*?

Linear functions

Look at these coordinates $(0, 1)$, $(1, 2)$, $(2, 3)$, $(3, 4)$.
Can you see any number patterns?

The same coordinates can be shown in a **table**.

x	0	1	2	3
y	1	2	3	4

The diagram shows the coordinates plotted on a **graph**.
The points all lie on a **straight line**.

A **rule** connects the x coordinate with the y coordinate.
All points on the line obey the rule $y = x + 1$.

$y = x + 1$ is an example of a **linear function**.
The graph of a linear function is a **straight line**.

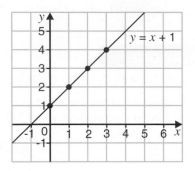

Drawing a graph of a linear function

To draw a linear graph:
- Find at least two corresponding values of x and y.
- Plot the points.
- Join the points with a straight line.

EXAMPLE

(a) Complete the table of values for $y = 2x - 3$.

x	0	1	2	3
y			1	

(b) Draw the graph of the equation $y = 2x - 3$.

(a)

x	0	1	2	3
y	−3	−1	1	3

When $x = 0$, $y = 2 \times 0 - 3 = -3$.
When $x = 1$, $y = 2 \times 1 - 3 = -1$.
When $x = 3$, $y = 2 \times 3 - 3 = 3$.

(b) Plot the points $(0, -3)$, $(1, -1)$, $(2, 1)$, and $(3, 3)$.

The straight line which passes through these points is the graph of the equation $y = 2x - 3$.

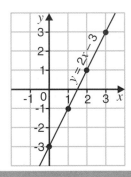

Special graphs

This diagram shows the graphs: $x = 4$ $y = 1$
$x = -2$ $y = -5$

Notice that:
The graph of $x = 4$ is a **vertical** line.
All points on the line have x coordinate 4.

The graph of $y = 1$ is a **horizontal** line.
All points on the line have y coordinate 1.

$x = 0$ is the y axis.
$y = 0$ is the x axis.

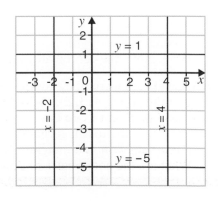

180

EXAMPLE

(a) Draw the graph of the equation $y = 4 - 2x$.

(b) Use your graph to find the value of y when $x = 2.5$.

(a) If values for x are not given in the question you must choose at least two of your own.

When $x = -1$, $y = 4 - 2 \times (-1) = 4 + 2 = 6$.

When $x = 0$, $y = 4 - 2 \times 0 = 4 - 0 = 4$.

When $x = 3$, $y = 4 - 2 \times 3 = 4 - 6 = -2$.

Plot the points $(-1, 6)$, $(0, 4)$ and $(3, -2)$.

The straight line which passes through these points is the graph of the equation $y = 4 - 2x$.

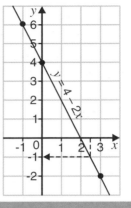

(b) Using the graph:

From 2.5 on the x axis, go down to meet the line $y = 4 - 2x$.

Then go left to meet the y axis at -1.

When $x = 2.5$, $y = -1$.

Exercise 17.2

1 Write down the equations of the labelled lines drawn on these diagrams.

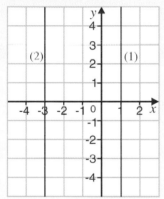 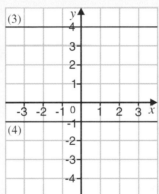 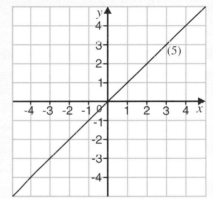

2 Draw x and y axes from -4 to 4.

On your diagram, draw and label the graphs of these equations.

(a) $x = 3$ (b) $y = 2$ (c) $x = -2$ (d) $y = -1$

3 (a) Copy and complete a table of values, like the one shown, for each of these equations.

x	0	1	2	3
y				

(i) $y = x + 2$ (ii) $y = 2x$ (iii) $y = -x$ (iv) $y = 2 - x$

(b) Draw graphs for each of the equations in part (a).

4 Draw tables of values and use them to draw graphs of:

(a) $y = x - 1$ (b) $y = 2x + 1$

Draw and label the x axis from 0 to 4 and the y axis from -2 to 10.

5 (a) Draw the graph of $y = 3x - 2$ for values of x from -1 to 2.

(b) What are the coordinates of the point where the graph crosses the y axis?

6 (a) Copy and complete the table of values for the equation $y = 3 - 2x$.

x	-1	1	3
y			

(b) Draw the graph of $y = 3 - 2x$ for values of x from -1 to 3.
(c) What are the coordinates of the point where the graph crosses the x axis?

7 (a) Copy and complete this table and use it to draw the straight line graph of $y = 4 - x$.

x	-2	-1	0	1	2
y		5			2

Draw and label the x axis from -3 to 3 and the y axis from -1 to 6.
(b) Use your graph to find the value of:
   (i) y when $x = 1.5$,   (ii) y when $x = -0.5$.

8 (a) Draw the graph of $y = 4x + 1$ for values of x from -2 to 2.
(b) Use your graph to find the value of:
   (i) y when $x = -1.5$,   (ii) x when $y = 3$.

9 (a) Draw the graph of $y = 2x - 1$ for values of x from -2 to 3.
(b) Use your graph to find the value of x when $y = 0$.

10 (a) Draw the graph of $y = 5 - 2x$ for values of x from -2 to 3.
(b) Use your graph to find:
   (i) the value of y when $x = 0$,
   (ii) the value of x when $y = 0$,
   (iii) the value of x when $y = 8$.

Gradient and intercept

The gradient of a straight line graph is found by drawing a right-angled triangle.

$$\text{Gradient} = \frac{\text{distance up}}{\text{distance along}}$$

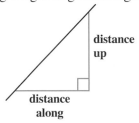

distance up

distance along

The gradient of a line can be positive, zero or negative.

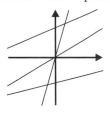

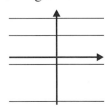

Positive gradients
go "uphill"

Zero gradients
are "flat"

Negative gradients
go "downhill"

Activity

Draw these graphs **on the same diagram**:

$$y = 2x + 2 \qquad y = 2x + 1 \qquad y = 2x \qquad y = 2x - 1$$

Draw and label the x axis from 0 to 3 and the y axis from -1 to 8.
What do they all have in common?
What is different?

The graphs of:
$y = 2x + 2$, $y = 2x + 1$, $y = 2x$, $y = 2x - 1$, go 2 squares up for every 1 square along.
The graphs are all parallel and have a gradient of 2.

Lines that are **parallel** have the same **slope** or **gradient**.

The point where a graph crosses the y axis is called the **y-intercept**.

The y-intercept of the graph $y = 2x + 2$ is 2.
The y-intercept of the graph $y = 2x + 1$ is 1.

What are the y-intercepts of the graphs $y = 2x$ and $y = 2x - 1$?

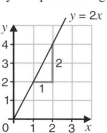

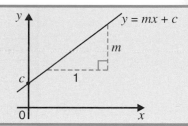

In general, the equation of any straight line can be written in the form

$$y = mx + c$$

where m is the **gradient** of the line and c is the **y-intercept**.

EXAMPLES

1 Write down the gradient and y-intercept for each of the following graphs.
    (a) $y = 3x + 5$ (b) $y = 4x - 1$ (c) $y = 6 - x$

    (a)   Gradient $= 3$, y-intercept $= 5$.
    (b)   Gradient $= 4$, y-intercept $= -1$.
    (c)   Gradient $= -1$, y-intercept $= 6$.

2 Write down the equation of the straight line which has gradient -7 and cuts the y axis at the point $(0, 4)$.

    The general form for the equation of a straight line is $y = mx + c$.
    The gradient, $m = -7$, and the y-intercept, $c = 4$.
    Substitute these values into the general equation.
    The equation of the line is $y = -7x + 4$.
    This can be written as $y = 4 - 7x$.

3 Find the equation of the line shown on this graph.

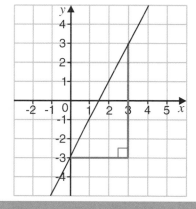

    First, work out the gradient of the line.
    Draw a right-angled triangle.

    Gradient $= \dfrac{\text{distance up}}{\text{distance along}}$

              $= \dfrac{6}{3}$
              $= 2$

    The graph crosses the y axis at the point $(0, -3)$,
    so, the y-intercept is -3.
    The equation of the line is $y = 2x - 3$.

Exercise 17.3

1 (a) Draw these graphs **on the same diagram**:
        (i) $y = x + 2$ (ii) $y = x + 1$ (iii) $y = x$ (iv) $y = x - 1$
        Draw and label the x axis from 0 to 3 and the y axis from -1 to 5.
    (b) What do they all have in common?
        What is different?

Coordinates and Graphs

17

2 (a) Write down the gradient and *y*-intercept of $y = 3x - 1$.
(b) Draw the graph of $y = 3x - 1$ to check your answer.

3 Which of the following graphs are parallel?

| $y = 3x$ | $y = x + 2$ | $y = 2x + 3$ | $y = 3x + 2$ |

4 Copy and complete this table.

Graph	gradient	*y*-intercept
$y = 4x + 3$	4	3
$y = 3x + 5$	3	
$y = 2x - 3$		
$y = 4 - 2x$		4
$y = \frac{1}{2}x + 3$		
$y = 2x$		

5 (a) Write down the equation of the straight line which has gradient 5 and crosses the *y* axis at the point $(0, -4)$.
(b) Write down the equation of the straight line which has gradient $-\frac{1}{2}$ and cuts the *y* axis at the point $(0, 6)$.

6 Match the following equations to their graphs.

(1) $y = x - 6$

(2) $y = 6 - x$

(3) $y = 2x + 1$

(4) $y = 2x - 1$

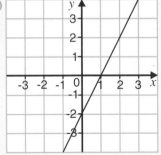

7 Find the equations of the lines shown on the following graphs.

(a)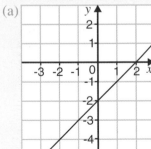

(b)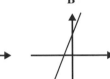

(c)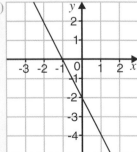

8 (a) Draw *x* and *y* axes from -8 to 8.
(b) Plot the points $A(-2, -6)$ and $B(8, 4)$.
(c) Find the gradient of the line which passes through the points *A* and *B*.
(d) Write down the coordinates of the point where the line crosses the *y* axis.
(e) Find the equation of the line which passes through the points *A* and *B*.

9 (a) Draw *x* and *y* axes from -8 to 8.
(b) Plot the points $P(-2, 3)$ and $Q(3, -7)$.
(c) Find the equation of the line which passes through the points *P* and *Q*.

10 What can you say about the slope of a line if the gradient is (a) 5,   (b) −5,   (c) 0?

11 A line, with a gradient of 3, passes through the origin.
What is the equation of the line?

12 A plumber charges a fixed call-out charge and an hourly rate.
The graph shows the charges made for jobs up to 4 hours.

(a) What is the fixed call-out charge?

(b) What is the hourly rate?

(c) Write down the equation of the line in the
form $y = mx + c$.

(d) Calculate the total charge for a job which
takes 8 hours.

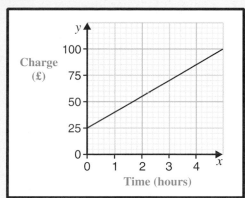

13 The graph shows the taxi fare for journeys up to 3 km.

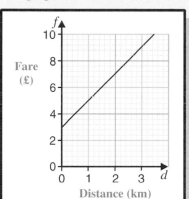

(a) What is the fixed charge?

(b) What is the charge per kilometre?

(c) Write down the equation of the line in the
form $f = md + c$.

(d) Calculate the taxi fare for a journey of 5 km.

14 In an experiment, weights are added to a spring and the length of the spring is measured.
The graph shows the results.

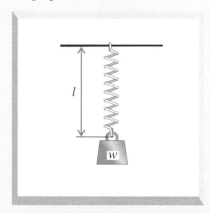

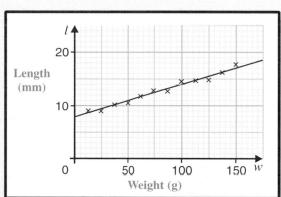

A line of best fit has been drawn.
(a) Estimate the length of the spring when no weight is added.
(b) Calculate the gradient of the line.
(c) Write down the equation of the line in the form $l = mw + c$.
(d) Use your equation to estimate the length of the spring for a weight of 300 g.

Coordinates and Graphs

Drawing graphs of other linear equations

EXAMPLE

Draw the graph of the line given by the equation $x + 2y = 6$.

> By substituting $x = 0$ into the equation we can find the coordinates of the point where the line crosses the y axis.

$x + 2y = 6$
Substitute $x = 0$.
$0 + 2y = 6$
$2y = 6$
$y = 3$

The line crosses the y axis at the point $(0, 3)$.

> By substituting $y = 0$ into the equation we can find the coordinates of the point where the line crosses the x axis.

$x + 2y = 6$
Substitute $y = 0$.
$x + 2 \times 0 = 6$
$x = 6$

The line crosses the x axis at the point $(6, 0)$.

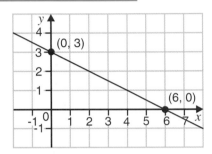

To draw the graph of $x + 2y = 6$:
1. Plot the points $(0, 3)$ and $(6, 0)$.
2. Using a ruler, draw a straight line which passes through the two points.

Exercise 17.4

1 A straight line has equation $x + y = 7$.
   (a) By substituting $x = 0$ find the coordinates of the point where the line crosses the y axis.
   (b) By substituting $y = 0$ find the coordinates of the point where the line crosses the x axis.
   (c) Draw the graph of the line $x + y = 7$.

2 (a) Draw these graphs on the same diagram.
      (i) $x + y = 2$ (ii) $x + y = 3$ (iii) $x + y = 5$
   (b) What do they all have in common?

3 A straight line has equation $3x + y = 6$.
   (a) By substituting $x = 0$ find the coordinates of the point where the line crosses the y axis.
   (b) By substituting $y = 0$ find the coordinates of the point where the line crosses the x axis.
   (c) Draw the graph of the line $3x + y = 6$.

4 Draw the graphs of lines with the following equations.
   (a) $x + 2y = 6$ (b) $2y = 4 - x$ (c) $2y = x + 4$

5 A straight line has equation $3y + 5x = 15$.
   (a) By substituting $x = 0$ find the coordinates of the point where the line crosses the y axis.
   (b) By substituting $y = 0$ find the coordinates of the point where the line crosses the x axis.
   (c) Draw the graph of the line $3y + 5x = 15$.

6 Draw the graphs of lines with the following equations, marking clearly the coordinates of the points where the lines cross the axes.
   (a) $5y + 4x = 20$ (b) $4x - y = 4$ (c) $3y + 2x = 12$

Coordinates and Graphs

EXAMPLE

(a) Complete the tables for $y = x + 1$ and $y = 8 - x$.

x	1	2	3
$y = x + 1$			

x	1	2	3
$y = 8 - x$			

(b) Draw the graphs of $y = x + 1$ and $y = 8 - x$ on the same diagram.
(c) Use your graphs to solve the equation $8 - x = x + 1$.

(a)

x	1	2	3
$y = x + 1$	2	3	4

x	1	2	3
$y = 8 - x$	7	6	5

(b)

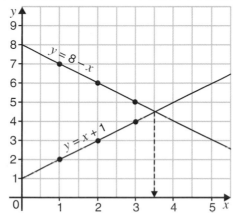

(c) The x value of the point where the two graphs cross gives the solution of the
     equation $8 - x = x + 1$.
     Reading from the graph: $x = 3.5$

Check the graphical solution of the equation by solving $8 - x = x + 1$ algebraically.

Exercise 17.5

1 (a) Copy and complete the tables for $y = x + 2$ and $y = 5 - x$.

x	1	2	3
$y = x + 2$			

x	1	2	3
$y = 5 - x$			

   (b) Draw the graphs of $y = x + 2$ and $y = 5 - x$ on the same diagram.
   (c) Write down the coordinates of the point where the two lines cross.
   (d) Use your graphs to solve $x + 2 = 5 - x$.

2 (a) Draw the graphs of $y = 3x + 1$ and $y = x + 6$ on the same diagram.
   (b) Write down the coordinates of the point where the two lines cross.
   (c) Use your graphs to solve the equation $3x + 1 = x + 6$.

3 (a) Draw the graphs of $y = x$ and $y = 3x - 1$.
   (b) Use your graphs to solve the equation $x = 3x - 1$.

4 By drawing the graphs of $y = 2x$ and $y = 3 - x$, solve the equation $2x = 3 - x$.

5 (a) Draw the graph of $y = 3 + 2x$.
   (b) What graph should be drawn to solve the equation $3 + 2x = 9$?
   (c) Draw the graph and use it to solve the equation $3 + 2x = 9$.

What you need to know

- **Coordinates** (involving positive and negative numbers) are used to describe the position of a point on a graph. For example, $A(-3, 2)$ is the point where the lines $x = -3$ and $y = 2$ cross.

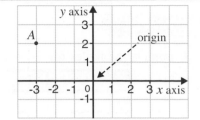

- The x axis is the line $y = 0$. The y axis is the line $x = 0$.
- The graph of a **linear function** is a **straight line**. The general equation of a linear function is $y = mx + c$, where m is the **gradient** of the line and c is the **y-intercept**.
- The **gradient** of a line can be found by drawing a right-angled triangle.

$$\text{Gradient} = \frac{\text{distance up}}{\text{distance along}}$$

Gradient can be positive, zero or negative.

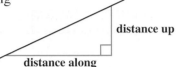

distance up

distance along

You should be able to:

- Substitute values into given functions to generate points.
- Plot graphs of **linear functions**.
- Use graphs of linear functions to solve equations.

Review Exercise 17

1 The diagram shows points P and Q.

(a) What are the coordinates of P?
(b) What are the coordinates of Q?

2

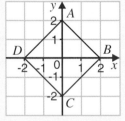

The diagram shows a square $ABCD$.

(a) What are the coordinates of A?
(b) What are the coordinates of D?

3 The diagram shows two sides of a rectangle $KLMN$.

(a) What are the coordinates of M?
(b) Find the coordinates of N.

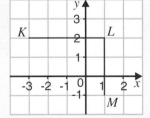

4

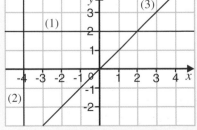

(a) Write down the equations of the lines labelled on this graph.
(b) What are the coordinates of the point where lines (1) and (2) cross?

5 (a) On graph paper, plot the points $P(-2, 3)$ and $Q(4, -1)$.
(b) R is the midpoint of the line PQ. What are the coordinates of R?

188

6 Match the equations to the graphs.

A: $y = 2$ **B:** $x + y = 2$ **C:** $y = x + 2$ **D:** $x = 2$

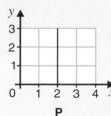

P

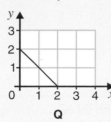

Q

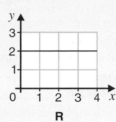

R

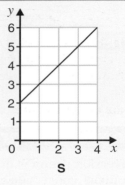

S

7 (a) On the same diagram draw and label the lines: $x = 3$ and $y = 4$.
  (b) Write down the coordinates of the point where the lines cross.

8 (a) Copy and complete the table of values for the equation $y = x + 3$.

x	-3	0	2
y			

  (b) Draw the graph of the line $y = x + 3$.
      Label the x axis from -3 to 2 and the y axis from 0 to 5.

9 (a) Draw the graph of $y = 2x - 1$ for values of x from -1 to 3.
  (b) **Use your graph** to estimate the value of x when $y = 2.5$.   AQA

10 (a) Copy and complete the table of values for the graph of $x + y = 4$.

x	0	1	2	3	4
y	4			1	

  (b) Draw the graph of $x + y = 4$.
  (c) P is a point on the line $x + y = 4$.
      David says, "The x coordinate of P is one greater than the y coordinate of P."
      Write down the coordinates of P.   AQA

11 (a) On the same diagram draw and label the lines: $y = 2x$ and $x + y = 6$
  (b) Write down the coordinates of the point where the lines cross.

12 Draw the graph of $2y + x = 4$ for values of x from -2 to 4.

13 Two firms hire out scaffolding. The charges made by each firm are shown.
      Scaffold Plus: £50 plus £10 per day.   Scaffold Ltd: £100 plus £5 per day.
The graph shows the charges made by Scaffold Ltd for up to 25 days.

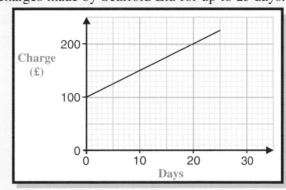

  (a) Copy the graph and draw a line to show the charges made by Scaffold Plus.
  (b) For how many days do both firms make the same charge?
  (c) A builder needs to hire scaffolding for 20 days.
      Which firm would be cheaper and by how much?

Coordinates and Graphs

CHAPTER 18

Using Graphs

Graphs are used in many real-life situations to represent information.

Conversion graphs

A **conversion graph** is used to change one quantity into an equivalent quantity.
For example, conversion graphs can be drawn and used to change:
>   weight – between pounds and kilograms,
>   temperature – between degrees Celsius and degrees Fahrenheit,
>   currency – between pounds, £, and euros, €.

EXAMPLE

Use 16 euros = £10 to draw a conversion graph for pounds and euros.

Use your graph to find: (a)   20 euros in £,   (b)   £4 in euros.

16 euros = £10   Plot the point (16, 10).
 0 euros = £0   Plot the point (0, 0).

The straight line through the points (0, 0) and (16, 10) is the conversion graph for pounds into euros.

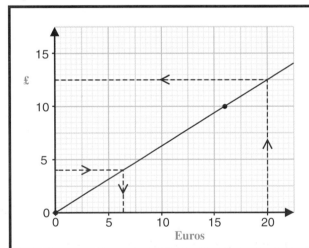

Reading from graph:
(a)   20 euros = £12.50.
(b)   £4 = 6.4 euros.

Exercise 18.1

1   This conversion graph can be used to change measurements from inches into centimetres.

Use the graph to find:
(a)   10 centimetres in inches,
(b)   10 inches in centimetres,
(c)   16 inches in centimetres.

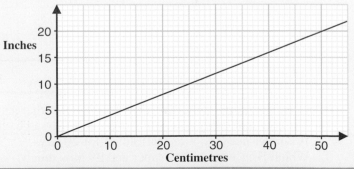

2   Use   £10 = 16 dollars   to draw a conversion graph for £ and dollars.
    Use your graph to find: (a)   £8 in dollars,   (b)   10 dollars in £.

3   Use   5 miles = 8 kilometres   to draw a conversion graph for miles and kilometres.
    Use your graph to find: (a)   3.5 miles in kilometres,   (b)   5 kilometres in miles.

4   Use   10 kilograms = 22 pounds (lb)   to draw a conversion graph for kilograms and pounds.
    Use your graph to find: (a)   4 kilograms in pounds,   (b)   15 pounds in kilograms.

5   Use   32°F = 0°C and 212°F = 100°C   to draw a conversion graph for degrees Fahrenheit
    and degrees Celsius.
    Use your graph to find: (a)   50°F in degrees Celsius,   (b)   75°C in degrees Fahrenheit.

Distance-time graphs

Distance-time graphs are used to illustrate journeys.

Speed is given by the gradient, or slope, of the line.
The faster the speed the steeper the gradient.
Zero gradient (horizontal line) means zero speed (not moving).

> **Calculations** involving speed, distance and time are covered in Chapter 11.

EXAMPLES

1   The graph shows a bus journey.
    (a)   How many times does the bus stop?
    (b)   On which part of the journey
         does the bus travel fastest?

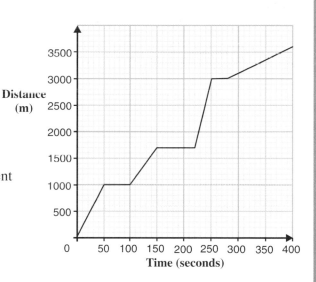

    (a)   At zero speed the distance-time graph
         is horizontal.
         So, the bus stops 3 times.
    (b)   The bus travels fastest when the gradient
         of the distance-time graph is steepest.
         So, the bus travels fastest between the
         second and third stops.

2   The graph represents a train journey
    from Woking.

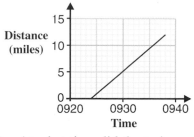

    (a)   At what time did the train
         leave Woking?
    (b)   How far did the train travel?

    (a)   0924
    (b)   12 miles

3   What speed is shown by this distance-time graph?
    Give your answer in metres per second.

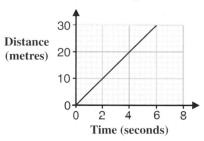

    A distance of 30 metres is travelled in a time of
    6 seconds.

    Using Speed = Distance ÷ Time
         Speed = 30 ÷ 6
              = 5 metres per second

1 The graph represents a bus journey from Poole.

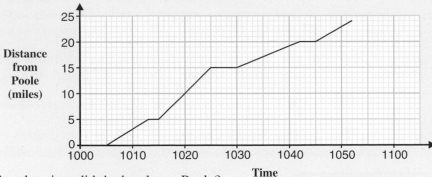

(a) At what time did the bus leave Poole?
(b) How far did the bus travel?
(c) How many times did the bus stop on the journey?

2 The graph represents the journey of a cyclist from Hambone to Boneham.

(a) What time did the cyclist leave Hambone?
(b) The cyclist arrived in Boneham at 1200. How far is Boneham from Hambone?
(c) The cyclist made one stop on his journey.
  (i) At what time did the cyclist stop?
  (ii) How far was the cyclist from Boneham when he stopped?

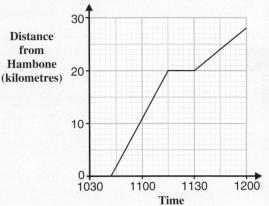

3 The distance-time graph shows the journey of a man from Durham to Leeds and back.

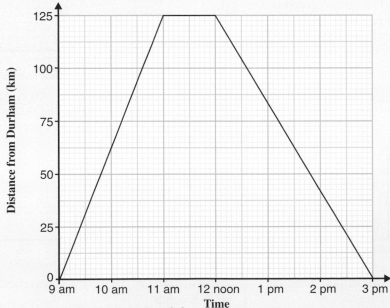

(a) How far is it from Durham to Leeds?
(b) How long did the man stop in Leeds?
(c) Did he travel at a faster speed going to Leeds or on the return journey? Explain your answer.

4

(a) The graph represents the journey of a car. What is the speed of the car in kilometres per hour?

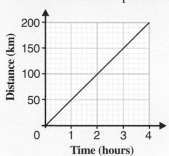

Distance (km) / **Time (hours)**

(b) The graph represents the journey of a train. What is the speed of the train in metres per second?

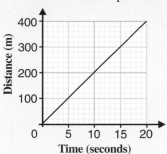

Distance (m) / **Time (seconds)**

(c) The graph represents the speed of a cyclist. What is the speed of the cyclist in miles per hour?

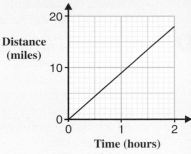

Distance (miles) / **Time (hours)**

5 The distance-time graph shows the journey of a coach from Hove to Southampton.

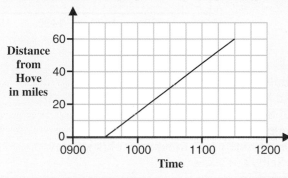

Distance from Hove in miles / **Time**

(a) At what time did the coach leave Hove?

(b) How long did the coach take to travel from Hove to Southampton?

(c) What is the average speed of the coach in miles per hour?

6 Pat cycles from home to the town centre. The graph represents her journey.

Pat takes $\frac{1}{2}$ hour to reach the town centre from her home.

What is her average speed for the journey in kilometres per hour?

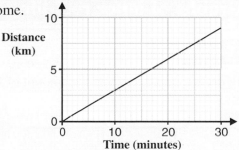

Distance (km) / **Time (minutes)**

7 This graph shows the progress made by a runner during the first 20 km of a marathon race.

Find the average speed of the runner:

(a) during the first 10 km of the race,

(b) during the second 10 km of the race,

(c) during the first 20 km of the race.

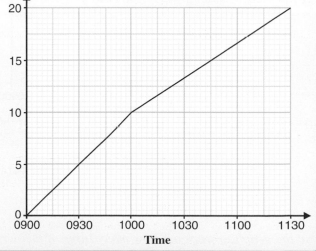

Distance from start (km) / **Time**

Using Graphs . . . Using Graphs . . .

18

193

8 A motorist has to travel to Swansea, a distance of 80 kilometres.
He sets off at 0930 and travels at an average speed of 50 km/h for one hour before stopping.
He stops for 30 minutes and then completes the rest of the journey at an average speed
of 60 km/h.
   (a)   Draw a distance-time graph to represent his journey.
   (b)   At what time did he reach Swansea?

9 The graph represents the journey of a cyclist from Bournemouth to the New Forest.
   (a)   What is the average speed of the cyclist in miles per hour?
   (b)   Another cyclist is travelling from the
         New Forest to Bournemouth at an
         average speed of 12 miles per hour.
         At 1300 the cyclist is 15 miles from
         Bournemouth.
         (i)   Draw a graph to show the journey
               of the cyclist to Bournemouth.
         (ii) At what time does the cyclist
               arrive in Bournemouth?

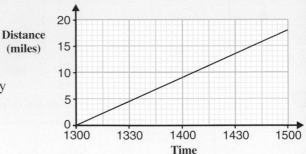

10 Selby is 20 miles from York.
   (a)   Kathryn leaves Selby at 1030 and drives to York.
         She travels at an average speed of 20 miles per hour.
         Draw a distance-time graph to represent her journey.
   (b)   At 1030 Matt leaves York and drives to Selby.
         He travels at an average speed of 30 miles per hour.
         (i)   On the same diagram draw a distance-time graph to represent his journey.
         (ii) At what time does Matt arrive in Selby?

Graphs of other real-life situations

EXAMPLE

Craig drew a graph to show the amount of fuel in the family car as they travelled to their
holiday destination.
He also made some notes:

Part of Graph	Event
A	Leave home.
A to B	Motorway.
B to C	Car breaks down.
C to D	On our way again.
D to E	Stop for lunch.
E to F	Fill tank with fuel.
F to G	Country roads.
G	Arrive, at last!

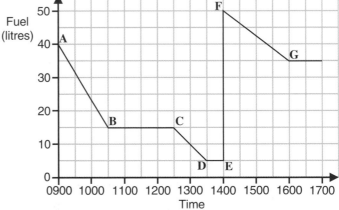

   (a)   How much fuel was in the tank at the start of the journey?
   (b)   At what time did the car break down?
   (c)   How long did the family stop for lunch?
   (d)   How much fuel was put into the tank at the garage?
   (e)   At what time did the journey end?

   (a)   40 litres   (b)   1030   (c) $\frac{1}{2}$ hour   (d)   45 litres   (e)   1600

Use the graph to work out how many litres of fuel were used for the journey.

194

1 Cans of drink can be bought from a vending machine in the school canteen.
The graph shows the number of cans in the machine between 1000 and 1500 one day.

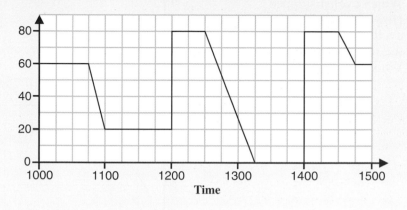

(a) At 1000 the machine is three-quarters full.
How many cans does the machine hold when it is full?

(b) How many drinks were sold between 1045 and 1100?

(c) The machine was filled up twice during the day.
At what times was the machine filled up?

(d) Between what times was the machine empty?

(e) How many cans of drink were sold altogether between 1000 and 1500?

2 Graphs of the average heights and weights for men and women are shown.

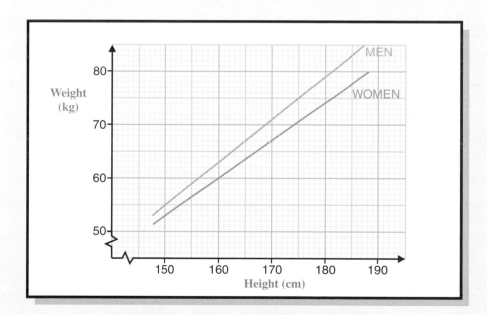

(a) John and his wife are both 170 cm in height.
Use the graphs to estimate the difference in their weights.

(b) Fred and Mary both weigh 75 kg.
Use the graphs to estimate the difference in their heights.

3 The graph illustrates a 10 km cycle race between Afzal and Brian.

(a) Who was cycling faster at the beginning of the race?

(b) Which cyclist stopped?

(c) How far apart were the cyclists 10 minutes after the start of the race?

(d) Who won the race?

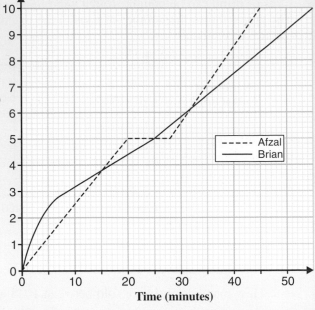

4 The cost of removals includes a fixed amount and a charge per kilometre for the distance moved.
The graph shows the cost, in £, for removals up to a distance of 50 kilometres.

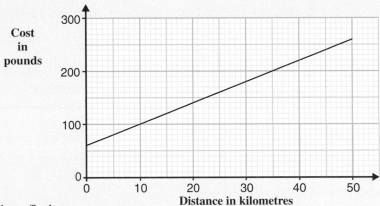

Use the graph to find:

(a) the cost of removals for a distance of 20 kilometres,

(b) the distance moved when the cost of removals is £200,

(c) the fixed amount charged.

5 The diagram shows the distance from the starting position of a swimmer in a race in a swimming pool.

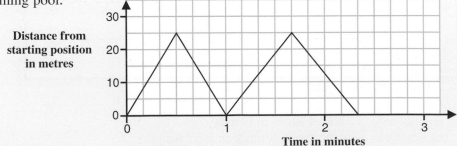

(a) What is the length of the swimming pool?

(b) What is the distance of the race?

(c) How long did the swimmer take to complete the race?

- A graph used to change from one quantity into an equivalent quantity, such as pounds into kilograms, is called a **conversion graph**.
- **Distance-time graphs** are used to illustrate journeys.
  On a distance-time graph: speed can be calculated from the gradient of a line,
  the faster the speed the steeper the gradient,
  zero gradient (horizontal line) means zero speed (not moving).
- Construct and interpret graphs arising from real-life situations.

Review Exercise 18

1   You can use this conversion graph to change measurements from inches to centimetres.

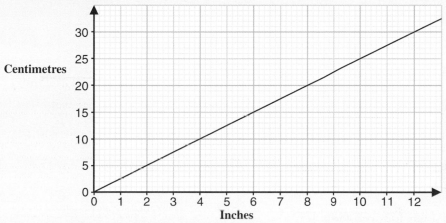

(a)   Change 7 inches to centimetres.
(b)   How many inches are there in 25 centimetres?
(c)   Use information from the graph to change 36 inches to centimetres.

AQA

2   A shop in Dover sells gifts.
The gifts can be bought in either pounds or euros.
The prices are shown for the following gifts in pounds and euros.

(a)   Use this information to draw a conversion
graph for pounds (£) and euros (€).
(b)   A watch costs 20 euros.
How much is the watch in pounds (£)?
(c)   A camera costs £60.
How much is the camera in euros (€)?

£5
8 euros

£15
24 euros

3   A mechanic charges a fixed amount and an hourly rate for call-outs.
The graph shows the charges made for
call-outs up to 2 hours.

(a)   What is the total charge for a call-out
which takes 1 hour to complete?
(b)   How much is the fixed amount?
(c)   How much is the hourly rate?

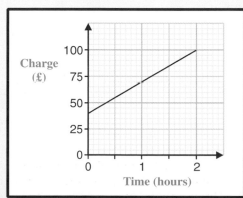

4 Lisa went on a cycling holiday. The graph shows her journey for the first day.

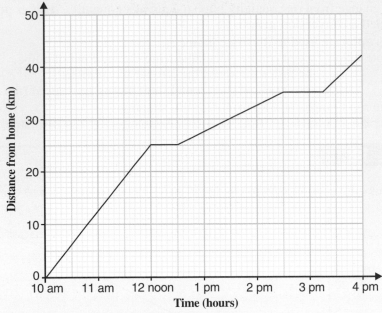

(a) (i) How many times did she stop before 4 pm?
    (ii) For how long did she stop at 2.30 pm?
(b) (i) What was her speed between 10 am and noon?
    (ii) What was her average speed for the whole journey?
    (iii) Between what times was she travelling fastest?

AQA

5 A motorist has to travel 60 miles.
She sets off at 0900 and travels the first 30 miles at an average speed of 40 miles per hour before stopping.
She stops for 15 minutes and then completes her journey, to arrive at 1030.
(a) Draw a distance-time graph to represent her journey.
(b) What is her average speed for the whole journey?

6 The graph represents a coach journey from Bath.

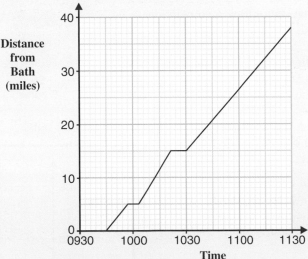

The coach completed its journey at 1130.
(a) At what time did the coach leave Bath?
(b) How far did the coach travel?
(c) How many times did the coach stop on the journey?
(d) What was the average speed of the coach between 1030 and 1130?

CHAPTER 19

Inequalities

Activity

For all children who enter the competition we can say that
Age < 16 years

For anyone riding the Big Dipper we can say that
Height ⩾ 1.2 m

For all items sold in the store we can say that
Cost ⩽ £1

These are examples of inequalities.
Can you think of other situations where inequalities are used?

Inequalities

An **inequality** is a mathematical statement, such as $x > 1$, $a \leqslant 2$ or $-3 \leqslant n < 2$.

In the following, x is an integer.

Sign	Meaning	Example	Possible values of x
<	is less than	$x < 4$	3, 2, 1, 0, −1, −2, −3, …
⩽	is less than or equal to	$x \leqslant 4$	4, 3, 2, 1, 0, −1, −2, −3, …
>	is greater than	$x > 6$	7, 8, 9, 10, …
⩾	is greater than or equal to	$x \geqslant 2$	2, 3, 4, 5, …

An **integer** is a positive or negative whole number or zero.

Explain the difference between the meanings of the signs < and ⩽.
Explain the difference between the meanings of the signs > and ⩾.

Number lines

Inequalities can be shown on a **number line**.

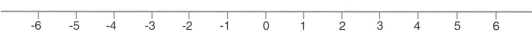

As you move to the right, numbers get bigger.
As you move to the left, numbers get smaller.

The number line below shows the inequality $-2 < x \leqslant 3$.

The circle at 3 is **filled** to show that 3 is **included**.
The circle at −2 is **not filled** to show that −2 is **not included**.

EXAMPLES

1 Draw number lines to show these inequalities:
(a) $x < 1.5$ (b) $x \geqslant -2$ (c) $x \leqslant 4$ and $x > -1$

(a) $x < 1.5$

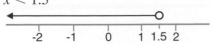

The circle is:
filled if the inequality is **included**,
not filled if the inequality is **not included**.

(b) $x \geqslant -2$

(c) $x \leqslant 4$ and $x > -1$.

x has to satisfy two inequalities.

2 (a) Draw a number line to show the inequality $3 \leqslant x < 8$.
(b) x is an integer.
Write down the values of x which satisfy the inequality.

$3 \leqslant x < 8$
is a shorthand method
of writing
$3 \leqslant x$ **and** $x < 8$.

(a)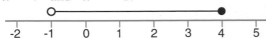

(b) The values of x which satisfy the inequality $3 \leqslant x < 8$ are: 3, 4, 5, 6, 7.

Exercise **19.1**

1 Write down the following mathematical statements and say whether each is true or false.
(a) $4 < 7$ (b) $3 > -3$ (c) $4 \geqslant 4$ (d) $-2 > -1$
(e) $-8 \leqslant -8$ (f) $1.5 \geqslant 2.1$ (g) $3 \times 5 \leqslant 7 \times 2$ (h) $-4 \times (-2) > -4 - 4$

2 Write down an integer which could replace the letter.
(a) $x < 6$ (b) $a \geqslant -2$ (c) $c + 2 < 8$ (d) $2d \leqslant 14$
(e) $f - 3 > 7$ (f) $-2 < h < 0$ (g) $t \leqslant 5$ **and** $t > 4$ (h) $r \geqslant -6$ **and** $r < -1$

3 In this question x is an integer. Write down all the values of x which satisfy these inequalities.
(a) $1 < x < 5$ (b) $-2 < x \leqslant 3$ (c) $-4 \leqslant x \leqslant -1$ (d) $-1 \leqslant x < 3$

4 Write down a mathematical statement, using inequalities, for each of these diagrams.

(a)   (b)

(c)   (d)
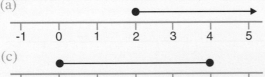   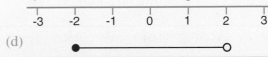

(e)   (f)
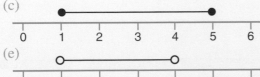   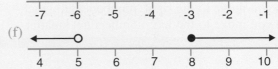

5 Draw number lines to show the following inequalities.
For each part, draw and label a number line from -5 to 5.
(a) $x > 2$ (b) $x \leqslant 1$ (c) $x \geqslant -4$ (d) $x < -2$
(e) $-2 \leqslant x < 3$ (f) $1 < x \leqslant 4$ (g) $-3 < x < 0$ (h) $x < -3$ and $x \geqslant 1$

200

Solving inequalities

Solve means to find the values of x which make the inequality true. The aim is to end up with **one letter** on one side of the inequality and a **number** on the other side of the inequality.

Solving inequalities is similar to solving equations.

EXAMPLE

Solve the inequality $5x - 3 < 27$ and show the solution on a number line.

$5x - 3 < 27$

Add 3 to both sides.

$5x < 30$

Divide both sides by 5.

$x < 6$

The solution is shown on a number line as:

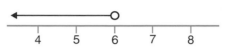

This means that the inequality is **true** for all values of x which are less than 6.

Exercise 19.2

1 Solve each of the following inequalities and show the solution on a number line.

(a) $3n > 6$ (b) $2x < -4$ (c) $a + 1 < 5$ (d) $a - 3 < 1$

(e) $2d - 5 \leqslant 1$ (f) $t + 2 < -1$ (g) $5 + 2g > 1$ (h) $4 + 3y \geqslant 4$

2 Solve the following inequalities.
Show your working clearly.

(a) $a + 3 < 7$ (b) $5 + x \geqslant 3$ (c) $y + 2 < -1$ (d) $3c > 15$

(e) $2d < -6$ (f) $b - 3 \geqslant -2$ (g) $-2 + b \leqslant -1$ (h) $2c + 5 \leqslant 11$

(i) $3d - 4 > 8$ (j) $4 + 3f < -2$ (k) $8g - 1 \leqslant 3$ (l) $5h < h + 8$

(m) $3x < x - 6$ (n) $6j \geqslant 2j + 10$ (o) $7k > 3k - 16$ (p) $6m - 7 \leqslant m$

Double inequalities

EXAMPLES

1 Find the values of x such that $-3 < x - 2 \leqslant 1$ and show the solution on a number line.

$-3 < x - 2 \leqslant 1$

Add 2 to each part of the inequality.

$-1 < x \leqslant 3$

The solution is shown on a number line as:

2 Find the integer values of n for which $-1 \leqslant 2n + 3 < 7$.

$-1 \leqslant 2n + 3 < 7$

Subtract 3 from each part.

$-4 \leqslant 2n < 4$

Divide each part by 2.

$-2 \leqslant n < 2$

Integer values which satisfy the inequality $-1 \leqslant 2n + 3 < 7$ are: $-2, -1, 0, 1$.

1 Solve each of the following inequalities and show the solution on a number line.

(a) $5 < x + 4 \leqslant 9$ (b) $-3 \leqslant x - 2 < 7$ (c) $2 < 9 + x \leqslant 13$

2 Find the values of x such that:

(a) $2 < 2x \leqslant 6$ (b) $-6 \leqslant 3x < 12$ (c) $-1 \leqslant 2x + 3 \leqslant 1$

(d) $5 < 2x - 1 < 8$ (e) $-2 \leqslant 3x - 1 \leqslant 11$ (f) $12 < 5x + 2 \leqslant 27$

3 Find the integer values of n for which:

(a) $3 < n - 2 < 7$ (b) $-2 < n + 1 \leqslant 5$ (c) $-2 < 2n \leqslant 4$

(d) $5 \leqslant 2n - 3 < 13$ (e) $0 < 2n - 8 < 3$ (f) $5 < 4n + 1 \leqslant 13$

What you need to know

● Inequalities can be described using words or numbers and symbols.

Sign	Meaning
$<$	is less than
$\leqslant$	is less than or equal to

Sign	Meaning
$>$	is greater than
$\geqslant$	is greater than or equal to

● Inequalities can be shown on a **number line**.

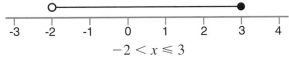

$-2 < x \leqslant 3$

The circle is:

     filled if the inequality is **included** (i.e. $\leqslant$ or $\geqslant$),

     not filled if the inequality is **not included** (i.e. $<$ or $>$).

● **Solve** means find the values of x which make the inequality true.

Review Exercise **19**

1 Draw number lines to show each of these inequalities.

(a) $x \leqslant -1$ (b) $x > 3$ (c) $4 < x \leqslant 9$ (d) $x < -2$ **and** $x > 5$

2 Solve each of these inequalities and show the solution on a number line.

(a) $2x < 6$ (b) $3x \geqslant -15$ (c) $x + 1 \geqslant 5$ (d) $7x - 3 \leqslant 18$

3 Solve the inequality $5x + 1 \geqslant 11$.

4 Solve the inequality $3x + 5 < 2$.

5 List the values of x, where x is an integer number, such that $-3 \leqslant x < 5$.   AQA

6 List the values of n, where n is an integer, such that $3 \leqslant n + 4 < 6$.   AQA

7 Solve these inequalities and show the solution on a number line.

(a) $x - 3 < 1$ (b) $-2 < 2x \leqslant 4$ (c) $-1 \leqslant 3x - 4 < 5$ (d) $-1 < 2x + 5 < 3$

8 Find the integer values of n such that:

(a) $-4 < 2n \leqslant 8$ (b) $-3 \leqslant 3n + 6 < 12$ (c) $-4 \leqslant 5n + 6 \leqslant 1$

9 (a) List all the possible values of x, where x is an integer, such that $-4 \leqslant x < 2$.

     (b) Solve $4x - 5 < -3$.   AQA

10 Solve the inequality $-1 \leqslant 3x + 2 < 5$.   AQA

Quadratic functions

Look at these coordinates: $(-3, 9)$, $(-2, 4)$, $(-1, 1)$, $(0, 0)$, $(1, 1)$, $(2, 4)$, $(3, 9)$.
Can you see any number patterns?

The diagram shows the coordinates plotted on a **graph**.
The points all lie on a **smooth curve**.

A **rule** connects the x coordinate with the y coordinate.
All points on the line obey the rule $y = x^2$.

$y = x^2$ is an example of a **quadratic function**.

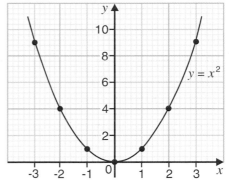

The graph of a **quadratic function** is always a smooth curve and is called a **parabola**.
The general equation of a quadratic function is $y = ax^2 + bx + c$, where a cannot be equal to zero.

The graph of a quadratic function is symmetrical and has a **maximum** or a **minimum** value.

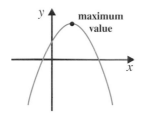 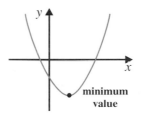

Drawing a graph of a quadratic function

To draw a quadratic graph:
● Make a table of values connecting x and y.
● Plot the points.
● Join the points with a smooth curve.

EXAMPLE

Draw the graph of $y = x^2 - 3$ for values of x from -3 to 3.

First make a table of values for $y = x^2 - 3$.

x	-3	-2	-1	0	1	2	3
y	6	1	-2	-3	-2	1	6

Plot these points.
The curve which passes through these points
is the graph of the equation $y = x^2 - 3$.

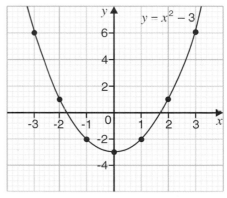

● Quadratic graphs are always symmetrical.

● Join plotted points using smooth curves and
not a series of straight lines.

1　(a)　Copy and complete this table of values for $y = x^2 - 1$.

x	-3	-2	-1	0	1	2	3
y	8		0			3	

　(b)　Draw the graph of $y = x^2 - 1$.
　　　Label the x axis from -3 to 3 and the y axis from -2 to 10.

2　(a)　Copy and complete this table of values for $y = x^2 - 2$.

x	-3	-2	-1	0	1	2	3
y		2			-1		7

　(b)　Draw the graph of $y = x^2 - 2$.
　　　Label the x axis from -3 to 3 and the y axis from -3 to 8.
　(c)　Use your graph to find the value of y when $x = -1.5$.
　(d)　Write down the coordinates of the points where the graph of $y = x^2 - 2$
　　　crosses the x axis.

3　(a)　Draw the graph of $y = x^2 - 4$ for values of x from -3 to 3.
　(b)　Use your graph to find the value of y when $x = -1.5$.

4　Draw the graphs of $y = x^2$, $y = x^2 + 2$ and $y = x^2 - 2$ on the same diagram.
Draw the x axis from -2 to 2 and the y axis from -2 to 6.
What do you notice about the graphs?

5　(a)　Draw the graph of $y = x^2 + 1$ for values of x from -2 to 4.
　(b)　Use your graph to find the value of y when $x = 2.5$.
　(c)　Use your graph to find the values of x when $y = 4$.
　(d)　Write down the coordinates of the point at which the graph has a minimum value.

6　(a)　Copy and complete this table of values for $y = 6 - x^2$.

x	-3	-2	-1	0	1	2	3
y		2		6			-3

　(b)　Draw the graph of $y = 6 - x^2$ for values of x from -3 to 3.
　(c)　Write down the coordinates of the points where the graph of $y = 6 - x^2$
　　　crosses the x axis.
　(d)　Find the coordinates of the point at which the graph has a maximum value.

7　(a)　Copy and complete this table of values for $y = x^2 + x$.

x	0	1	2	3	4	5	6
y	0			12			42

　(b)　Draw the graph of $y = x^2 + x$ for values of x from 0 to 6.
　(c)　Use your graph to find a value of x when $y = 25$.

8　Draw the graph of $y = 2x^2$ for values of x from -2 to 2.

9　(a)　Copy and complete this table of values for $y = x^2 - x + 2$.

x	-2	-1	0	1	2	3
y	8		2		4	

　(b)　Draw the graph of $y = x^2 - x + 2$ for values of x from -2 to 3.

10　Draw the graph of $y = x^2 + 3x - 2$ for values of x from -5 to 2.

Using graphs to solve quadratic equations

The diagram shows the graph of $y = x^2 - 4$.

The values of x where the graphs of quadratic functions cross (or touch) the x axis give the **solutions to quadratic equations**.

At the point where the graph $y = x^2 - 4$ crosses the x axis the value of $y = 0$.

$$x^2 - 4 = 0$$

The solutions of this quadratic equation can be read from the graph: $x = -2$ and $x = 2$

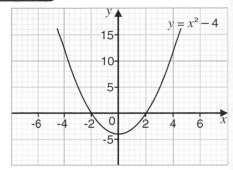

EXAMPLE

(a) Draw the graph of
$y = x^2 + 2x - 3$ for values of x from -4 to 2.
(b) Use your graph to find the solutions of the equation
$x^2 + 2x - 3 = 0$.

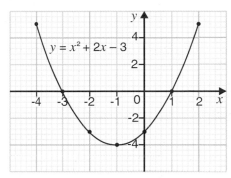

(a) First make a table of values for $y = x^2 + 2x - 3$.

x	-4	-3	-2	-1	0	1	2
y	5	0	-3	-4	-3	0	5

Plot these points.
The curve which passes through these points is the graph of the equation $y = x^2 + 2x - 3$.

(b) To solve the equation $x^2 + 2x - 3 = 0$ read the values of x where the graph of $y = x^2 + 2x - 3$ crosses the x axis.
Reading from the graph: $x = -3$ and $x = 1$

Exercise 20.2

1 The diagram shows the graph of $y = 6x - x^2$.
Use the graph to solve the equation
$6x - x^2 = 0$.

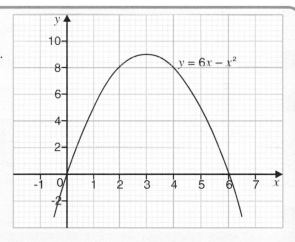

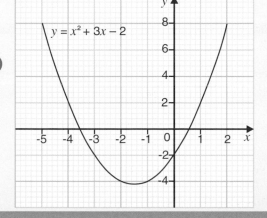

2 The graph of $y = x^2 + 3x - 2$ is shown.
Use the graph to solve the equation
$$x^2 + 3x - 2 = 0.$$
Give your answers correct to 1 d.p.

3 (a) Copy and complete this table of values for $y = x^2 - 8$.

x	-4	-3	-2	-1	0	1	2	3	4
y	8				-8		-4		

(b) Draw the graph of $y = x^2 - 8$.

(c) Use your graph to solve the equation $x^2 - 8 = 0$.

4 (a) Draw the graph of $y = x^2 + x$ for values of x from -3 to 2.

(b) Use your graph to solve the equation $x^2 + x = 0$.

(c) Find the coordinates of the point at which the graph has a minimum value.

5 (a) Copy and complete this table of values for $y = x^2 - 2x + 1$.

x	-2	-1	0	1	2	3
y						

(b) Draw the graph of $y = x^2 - 2x + 1$.

(c) Use your graph to solve the equation $x^2 - 2x + 1 = 0$.

6 (a) Draw the graph of $y = x^2 + 2x - 3$ for values of x from -5 to 3.

(b) Use your graph to solve the equation $x^2 + 2x - 3 = 0$.

7 (a) Draw the graph of $y = 10 - x^2$ for values of x from -4 to 4.

(b) Use your graph to solve the equation $10 - x^2 = 0$.

(c) Find the coordinates of the point at which the graph has a maximum value.

8 (a) Draw the graph of $y = 2x^2 - 5$ for values of x from -2 to 2.

(b) Use your graph to solve the equation $2x^2 - 5 = 0$.

9 (a) Copy and complete this table of values for $y = 15 - 2x^2$.

x	-3	-2	-1	0	1	2	3
y							

(b) Draw the graph of $y = 15 - 2x^2$.

(c) Use your graph to solve the equation $15 - 2x^2 = 0$.

10 Draw suitable graphs to solve the following equations.

(a) $x^2 - 10 = 0$ (b) $5 - x^2 = 0$ (c) $y = x^2 - 3x + 2$ (d) $12 - 2x^2 = 0$

11 (a) Copy and complete this table of values for $y = 2x^2 - 4x - 2$.

x	-1	0	1	2	3
y		-2			4

(b) Draw the graph of $y = 2x^2 - 4x - 2$.

(c) Hence, solve the equation $2x^2 - 4x - 2 = 0$.

12 By drawing a suitable graph solve the equation $x^2 + 4x - 5 = 0$.

What you need to know

- The graph of a **quadratic function** is a **smooth curve**.
- The general equation for a **quadratic function** is
  $y = ax^2 + bx + c$,   where a cannot be zero.
  The graph of a quadratic function is symmetrical
  and has a **maximum** or **minimum** value.
- **You should be able to:**   **substitute** values into given functions to generate points,
  plot graphs of **quadratic functions**,
  use graphs of quadratic functions to solve equations.

1 The graph of $y = x^2 - 4x$ is shown.

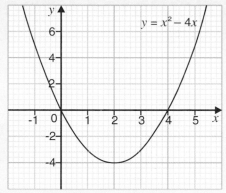

(a) Use the graph to solve the equation $x^2 - 4x = 0$.
(b) Find the coordinates of the point at which the graph has a minimum value.

2 (a) Copy and complete the table of values for $y = x^2 - 5$.

x	-3	-2	-1	0	1	2	3
y		-1		-5			4

(b) Draw the graph of $y = x^2 - 5$.
(c) Write down the values of x where the line $y = -1$ crosses your graph.

3 (a) Draw the graph of $y = 8 - x^2$ for values of x from -3 to 3.
(b) Find the coordinates of the point at which the graph has a maximum value.
(c) Use your graph to find the values of x when $y = 5$.

4 (a) Copy and complete this table of values for $y = x^2 - x - 1$.

x	-2	-1	0	1	2	3
y	5			-1		

(b) Draw the graph of $y = x^2 - x - 1$.
(c) Use your graph to solve the equation $x^2 - x - 1 = 0$.

5 (a) Copy and complete the table of values for $y = x^2 - 2x - 2$.

x	-2	-1	0	1	2	3	4
y	6		-2			1	

(b) Draw the graph of $y = x^2 - 2x - 2$.
(c) Use your graph to solve the equation $x^2 - 2x - 2 = 0$.

AQA

6 (a) Draw the graph of $y = x^2 + 3x - 5$ for values of x from -5 to 2.
(b) Use your graph to solve the equation $x^2 + 3x - 5 = 0$.

7 (a) Complete the table of values for $y = 2x^2 - 4x - 1$.

x	-2	-1	0	1	2	3
y	15		-1		-1	5

(b) Draw the graph of $y = 2x^2 - 4x - 1$ for values of x from -2 to 3.
(c) An approximate solution of the equation $2x^2 - 4x - 1 = 0$ is $x = 2.2$.
   (i) Explain how you can find this from the graph.
   (ii) Use your graph to write down another solution of this equation.

AQA

8 (a) Draw the graph of $y = 2x^2 - x - 1$ for values of x from -1 to 2.
(b) Use your graph to solve the equation $2x^2 - x - 1 = 0$.

Quadratic Graphs . . . Quadratic Graphs . . . Quadratic Graphs . . .

Do not use a calculator for this exercise.

1 (a) Find the missing numbers in these sequences.
        (i) 11, 18, 25, 32, …   (ii) …, 3, 7, 11, 15
    (b) Draw the next pattern in this sequence.

2 Work out the missing values in these calculations.
    (a)   3 → +5 → ×2 → .....   (b)   ..... → +5 → ×2 → 30

3 This rule is used to find the time, in minutes, it takes to cook a chicken.

$$20 \times \text{weight in pounds} + 20$$

A chicken weighs 5 pounds. How long will it take to cook?

4 Emma made these shapes with matchsticks.

Shape 1   **Shape 2**   **Shape 3**
6 matchsticks   **11 matchsticks**   **16 matchsticks**

    (a) Draw shape 4 for Emma.
    (b) (i) Copy and complete this table.

Shape number	1	2	3	4	5
Number of matchsticks	6	11	16		

        (ii) What pattern do you notice in the "number of matchsticks" row?
        (iii) How many matchsticks are needed to make shape 9?
             Explain how you can work it out without doing any drawings.   AQA

5 (a) Crayons cost 12p each.
        Write an expression in pence for the cost of t crayons.
    (b) Find the value of $\frac{2a + 3b}{5}$ when $a = 7$ and $b = -3$.

6 (a) Simplify $a + 2a + 3a$.
    (b) A circle is divided into 3 parts, as shown.
        Work out the value of angle a.   AQA

7 (a) On graph paper, plot the points $A(1, 3)$ and $B(-3, -1)$.
    (b) M is the midpoint of the line segment AB. What are the coordinates of M?

8 Solve.   (a) $7x = 35$ (b) $3 = x + 5$ (c) $5x + 3 = 18$

9 Tara buys x rulers at 25 pence each and y biros at 70 pence each.
    Write an expression for the total cost of the rulers and biros.   AQA

10 Trent thinks of a number.   He multiplies it by 3 and then adds 2.   His answer is 23.
    What number did Trent think of?

11 Solve.   (a) $\frac{x}{2} = 6$ (b) $3x - 2 = 28$

⑫ (a) A sequence begins 3, 4, 6, 10, … The rule for continuing the sequence is:

$$\boxed{\textbf{Double the last number and subtract 2.}}$$

What is the next number in this sequence?
(b) A different sequence begins $-2,$ $-4,$ $-6,$ $-8,$ …
What is the next number in this sequence?

AQA

⑬ A building supplier hires out cement mixers. He calculates the hire charge using this formula.

$$\boxed{\textbf{Five pounds per day plus a fixed charge of seven pounds.}}$$

(a) How much would it cost to hire the cement mixer for 3 days?
(b) A builder paid £52 altogether to hire a cement mixer. For how many days did he hire it?

AQA

⑭ (a) Simplify $3m + 4m + 5m$.
(b) Solve the equations. (i) $y + 3 = 11$ (ii) $4a = 32$

⑮ (a) Pens cost x pence each. Write an expression for the cost of 3 pens.
(b) A ruler costs 7 pence more than a pen. Write an expression for the cost of a ruler.

⑯ (a) Simplify $3m + 5n + m - 3n + 2m$.
(b) Solve. (i) $\frac{t}{6} = 2.5$ (ii) $3p - 2 = 7$

⑰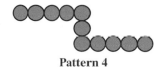

Pattern 1   **Pattern 2**   **Pattern 3**   **Pattern 4**

(a) Copy and complete the table.

Pattern number	1	2	3	4	5	6
Number of circles	5	7	9	11		

(b) How many circles are in Pattern 50? Explain how you worked out your answer.

⑱ If $x = 5$ and $y = -7,$ find the value of: (a) $4x + 3y$ (b) $\frac{x - y}{4}$

AQA

⑲ A rule is used to produce lists of numbers. The rule is shown below.

$$\boxed{\textbf{Add 3 to the last number and then double the result.}}$$

(a) A list starts: 1, 8, 22, ….   What is the next number in the list?
(b) A different list starts with the number -1.   What is the second number in this list?
(c) The second number in another list is 16.   What is the first number in this list?

AQA

⑳ The conversion graph can be used for changing between miles and kilometres.

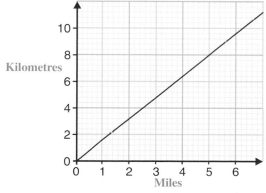

(a) Use the graph to change (i) 4 kilometres to miles,   (ii) 4 miles to kilometres.
(b) Explain how you can use the graph to change 100 miles to kilometres.

㉑ Solve the equations (a) $4x + 7 = 3,$ (b) $3y - 11 = 9 - y.$

AQA

22 (a) (i) Complete this table of values for $y = 2x + 3$.

x	-2	-1	0	1	2	3
y	-1		3			

　　(ii) Draw the graph of $y = 2x + 3$.

　(b) On the same grid, draw the graph of $y = 1$.

　(c) Write down the coordinates of the point where the lines cross.

23 Apples cost t pence per kilogram.

　(a) What is the cost of 5 kg of apples?

　(b) Pears cost 10 pence per kilogram less than apples.
　　What is the cost of one kilogram of pears?

　(c) Bananas cost twice as much per kilogram as apples.
　　What is the cost of one kilogram of bananas?

24 (a) What is the value of $2xy$ when $x = 3$ and $y = 4$?

　(b) Work out the value of $ab + c$ when $a = 3$, $b = -2$ and $c = 5$.

　(c) What is the value of $3p - q$ when $p = -1$ and $q = 2$?

25 (a) Simplify $6x + 7 - 2x + 4$.

　(b) Using the formula $a = 5b - \frac{c}{4}$, find the value of a when $b = 12$ and $c = 24$. AQA

26 Solve (a) $8x + 4 = 20$, (b) $3x - 2 = 10$.

27 (a) Simplify (i) $g + g + g$, (ii) $g \times g \times g$.

　(b) Solve the equations (i) $2g = 9$, (ii) $2g + 1 = 9$, (iii) $2(g + 1) = 9$.

28 (a) Work out the value of $2p^2$ when $p = 3$.

　(b) Simplify $2(x - 3) + 5$.

　(c) Solve $3y - 7 = 5$.

29 Mrs Crawley drove from her home to the supermarket, did her shopping and then returned home. The distance-time graph shows her journey.

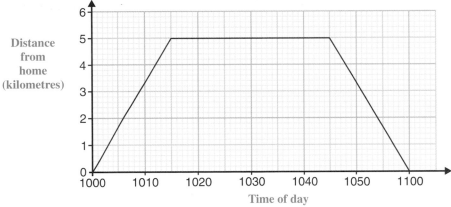

　(a) At what time did Mrs Crawley leave home?

　(b) How far is the supermarket from Mrs Crawley's home?

　(c) How many minutes did Mrs Crawley spend at the supermarket?

　(d) Work out Mrs Crawley's average speed on her journey home?
　　Give your answer in kilometres per hour.

30 (a) Complete this table of values for $x + y = 5$.

x	-1	0	1	2	3	4	5
y		5					

　(b) Draw the graph of $x + y = 5$ for values of x from -1 to 5.

31 (a) Using the values $u = 4$, $v = -3$ and $w = 5$, work out: (i) $u^2 + v^2$ (ii) $\dfrac{uv}{2w}$

    (b) Simplify. $m \times n \times 3$

    (c) Multiply out the brackets. (i) $7(3x + 2y)$ (ii) $a(a - 3)$

32 Solve these equations. (a) $4x - 7 = 5$ (b) $2(y + 5) = 28$ (c) $7z + 2 = 9 - 3z$

AQA

33 (a) Buns cost x pence each. How much will 2 buns cost?

    (b) A doughnut costs 5 pence more than a bun. How much will 3 doughnuts cost?

    (c) The cost of buying 2 buns and 3 doughnuts is 95 pence.
By forming an equation find the cost of a bun.

34 (a) Draw the graph of $y + 2x = 5$ for values of x from -1 to 4.

    (b) Use your graph to find the value of x when $y = -2$.

35 Factorise (a) $4a + 2b$,   (b) $3t - t^2$.

36 (a) Multiply out and simplify where possible. (i) $x(x - 5)$ (ii) $5(3p + 2) + 5p$

    (b) Factorise. (i) $6n + 9$ (ii) $2m^2 + m$

    (c) Solve. (i) $\dfrac{x}{4} = 25$ (ii) $3x + 5 = x - 2$

37 (a) Solve the inequality $3x - 5 < 7$.

    (b) An equality is shown on the number line.

    Write down the inequality.

38 Simplify. (a) $c \times c \times c \times c$ (b) $d^3 \times d^2$ (c) $\dfrac{e^8}{e}$ AQA

39 (a) Complete this table of values for $y = x^2$ 3.

x	-3	-2	-1	0	1	2	3
y	6	1		-3	-2	1	6

    (b) Draw the graph of $y = x^2 - 3$ for values of x from -3 to 3.

    (c) Write down the values of x at the points where the line $y = 2$ crosses your graph.   AQA

40 (a) Use the formula $y = mx + c$ to find the value of y
when $m = -4$, $x = -3$ and $c = -5$.

    (b) Rearrange the formula $y = mx + c$ to make x the subject.

41

$x + 2$
x ▭

The area, y, of this rectangle is given by $y = x^2 + 2x$.

    (a) Copy and complete this table of values for y.

x	0	1	2	3	4	5
y	0		8	15		35

    (b) Draw the graph of $y = x^2 + 2x$.

    (c) Use your curve to find the value of x if the area of the rectangle is $20\,\text{cm}^2$.

42 (a) Solve the equation $4(x + 3) = 22 + x$

    (b) (i) Write down the integer values of n for which $2 < 5n \leqslant 12$.

        (ii) Solve the inequality $5x + 3 \geqslant 4$.

43 Solve the inequality $6x - 3 < 15$.
Show the solution on a number line.

Algebra Calculator Paper

You may use a calculator for this exercise.

1 Copy the diagram.

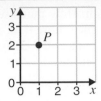

   (a) What are the coordinates of *P*?
   (b) Plot the point *Q* (3, 1).

2 (a) Here are the first five odd numbers: 1, 3, 5, 7, 9.
      (i) Write down the tenth odd number.
      (ii) What is the twentieth odd number?
   (b) Here are the first four terms of a sequence: 21,   20,   17,   12,   ...,   ...
      Write down the next two terms in the sequence.   AQA

3 (a) A pattern of numbers is shown.   15,   19,   23,   27,   ...
      (i) What is the next number in the pattern?
      (ii) Explain how you found your answer.
   (b) A second pattern of numbers uses the rule:

   > **Take four from the previous number.**

   Continue this pattern by writing down the **next two** numbers in the pattern.
   $$19,   15,   11,   7,   ...,   ...$$
   (c) A third pattern of numbers uses the rule:

   > **Add five to the previous number.**

   What number comes before the number 33 in the pattern?   AQA

4 Edwina does a part-time job. Her pay is calculated using this rule:

   > **Pay = Hourly rate × Number of hours.**

   Calculate her pay when she works 7 hours at an hourly rate of £6.

5 Using the input-output diagram, copy and complete the following table.

Input	9	...
Output	...	24

                                                                             AQA

6 (a) Solve.   (i) $7x = 42$ (ii) $x - 3 = 7$
   (b) Simplify. $7g - g + 5g$
   (c) Work out the value of $5m - 3n$ when $m = 2.2$ and $n = 3.5$.

7 Here are the first five terms of a sequence: 50   49   47   44   40   ...
   Write down the next term.
   Explain how you worked out the answer.

8 The formula for the perimeter, *P* cm,
   of this shape is $P = 5a + 3b$.
   Find *P* when $a = 3.2$ and $b = 3.8$.

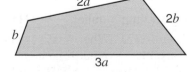

9 (a) (i) Write down the next **two** numbers of the sequence: 29,   26,   23,   20,   17,   ...
      (ii) Explain how you worked out these answers.
   (b) A different sequence of numbers is: 1,   3,   9,   27,   81,   ...
      (i) Write down the next **two** numbers in this sequence.
      (ii) Explain how you worked out these answers.
      (iii) One number in this sequence is 19 683.
         Work out the previous number in the sequence.   AQA

10 If $a = 3$, $b = 4$ and $c = \frac{1}{2}$, work out the value of:

(a) $2a + 3b$,   (b) $a - b + 3c$.   AQA

11 A sequence of patterns is made using diamonds.

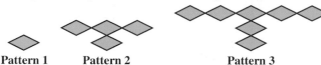

Pattern 1   Pattern 2   Pattern 3   Pattern 4

(a) Copy and complete the table.

Pattern number	1	2	3	4	5	6
Number of diamonds	1	4	7			

(b) How many more diamonds are needed for Pattern 7 than Pattern 6?

(c) How many diamonds are needed for Pattern 100? Explain how you worked out your answer.

12 n represents any whole number.

(a) What type of whole number is $2n$?

(b) Which of the statements below describes the number $3n + 1$? Explain your answer.

       always even   always odd   could be even or odd   AQA

13 There is a rule in the box below.

> **To change temperatures from Centigrade to Fahrenheit, "Double the Centigrade and add 30"**

(a) Use this rule to change 20° Centigrade to Fahrenheit.

(b) C is the temperature in °Centigrade. F is the temperature in °Fahrenheit. Write the rule as an equation using C and F.

(c) Use your equation to calculate C when $F = 100$.   AQA

14 (a) Simplify   (i) $3a + a$,   (ii) $3a - 4b + 2a + 3b$,   (iii) $3 \times a \times a$.

(b) Solve the equation $13 - x = 7$.

(c) $d = 3a - 1$.   Find the value of d when $a = 10$.

15 The conversion graph is used to compare £ with euros.

(a) How many euros are equivalent to £5?

(b) Whilst on holiday in Spain, Joan bought a camera for 120 euros. What was the price of the camera in £s?

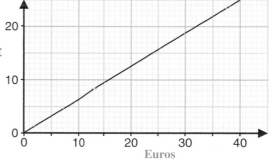

16 Rowena and Hugh are following these instructions:

Choose any starting number.	→	Halve it.	→	Halve the result.	→	Final answer.

(a) Rowena chooses 32. What final answer will she get?

(b) Hugh chooses 30. What final answer will he get?

(c) Rowena's final answer is a whole number, but Hugh's is not. Find two more numbers that will give final answers which are whole numbers.

(d) What do you notice about the starting numbers that give whole number answers?   AQA

17 (a) Pencils cost 18p each. How much does Janet pay for x pencils?

(b) Graham goes to a different shop. At this shop pencils cost n pence each. Rulers cost m pence each. Use the letters n and m to write down the total cost of 3 pencils and 2 rulers.   AQA

18 Match the equations to the graphs.

A: $x = 3$ **B**: $y = 3$ **C**: $y = x + 3$ **D**: $y + x = 3$

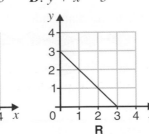

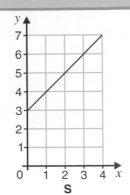

P   Q   R   S

19 The graph shows the journey made by a cyclist from Guildford to Brighton.

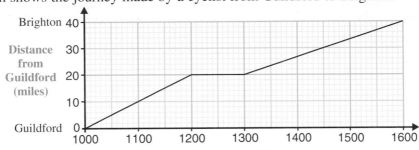

(a) (i) How far is Brighton from Guildford?
    (ii) Describe what happened between 1200 and 1300 hours.
    (iii) Find the average speed of the cyclist between 1000 and 1200.
(b) The cyclist later cycled back to Guildford at an average speed of 16 miles per hour.
    How long did it take the cyclist to get back to Guildford?

20 (a) Simplify. $3d - 5e + 4d + e$
   (b) Multiply out. $4(3x + 7)$
   (c) Solve. $4(3x + 7) = 28$ AQA

21 (a) On the same diagram, draw and label the lines $y = x$ and $y = 6 - x$.
   (b) What are the coordinates of the point where the two lines cross?

22 Solve the equations.   (a) $5x - 3 = 7$ (b) $5x + 5 = 7 + x$ AQA

23

> **Think of a number.   Multiply it by 2.   Now subtract 7.**

(a) Ken starts with 3. What is his answer?
(b) Lubna starts with x. What is her answer?   AQA

24 (a) Solve the equation $5x - 7 = 3x + 5$.
   (b) Expand and simplify $2x + 3(x - 4)$.   AQA

25 (a) Sarah thinks of a number. She doubles it and then subtracts 3.
    (i) What was her number when the answer is 27?
    (ii) What was her number when the answer is -1?
   (b) Solve the equation $2(2x - 3) = 6$.   AQA

26 (a) Write, as simply as possible, an expression
    for the total length of these rods.
   (b) The total length of the rods is 23 cm.
    By forming an equation find the value of a.

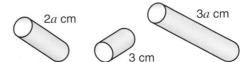

27 The nth term of a sequence is $3n - 1$.
   (a) Write down the first and second terms of the sequence.
   (b) Which term in the sequence is equal to 32?
   (c) Explain why 85 is not a term in this sequence.   AQA

28 The diagram shows a sketch of the line $2y + x = 10$.
Find the coordinates of points G and H.

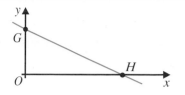

29 Factorise (a) $6a + 3$, (b) $t - t^2$.

30 (a) Write down the next term in the sequence: 2, 6, 10, 14, …
(b) Write an expression, in terms of n, for the nth term of the sequence.

31 (a) Solve the equation $3(x + 2) = 4 - x$.
(b) Factorise $m^2 - 7m$.
(c) List the values of n, where n is an integer, such that $1 \leqslant n - 5 < 4$. AQA

32 Using trial and improvement copy and continue the table to find a solution to the
equation $x^3 + x = 20$.

x	$x^3 + x$	Comment
2	10	Too low

Give your answer correct to 1 decimal place. AQA

33 (a) Complete this table of values for the graph $y = x^2 - 7$.

x	-3	-2	-1	0	1	2	3
y	2			-7		-3	2

(b) Draw the graph of $y = x^2 - 7$.
(c) Use your graph to find the values of x for which $x^2 - 7 = 0$.

34 Write down an expression for the nth term of these sequences.
(a) 3, 6, 9, 12, … (b) 4, 7, 10, 13, …

35 (a) Expand and simplify. $2(4x + 3) - 5x$

(b) Solve the following equations. (i) $\frac{x}{3} = 9$ (ii) $6x + 7 = x + 3$

(c) Simplify. (i) $m^2 \times m^3$ (ii) $\frac{n^6}{n^3}$

36 Use trial and improvement to solve the equation $x^3 + 3x = 80$.
Show all your trials. Give your answer correct to one decimal place.

37 (a) Write down the integer values of n for which $-4 \leqslant 4n < 8$.
(b) Solve the inequality $3x + 7 \geqslant 22$.

38 (a) Factorise. (i) $8y + 4$ (ii) $x^3 - 5x$

(b) Simplify. $x^2 y^3 \times \frac{x^3}{y}$

(c) Rearrange the formula $y = 5x + 10$ to make x the subject.

(d) Solve. (i) $\frac{t}{3} = 7$ (ii) $5(x - 3) = 3x + 1$

39 (a) Copy and complete the table for the equation $y = x^2 - 2x + 2$.

x	-1	0	1	2	3
y					

(b) Draw the graph of $y = x^2 - 2x + 2$ for $-1 \leqslant x \leqslant 3$.
(c) Use your graph to find the values of x when $y = 3$. AQA

40 The equation $x^3 - x = 400$ has a solution between 7 and 8.
Use a trial and improvement method to find this solution.
Show all your trials. Give your answer correct to one decimal place.

Angles ●●●●●●●●●●●●●●●

The diagram shows a stopwatch with a second hand.
Every minute the second hand will make one complete turn.
An **angle** is a measure of turn.
Angles are measured in **degrees**.
In one minute the second hand will turn through an angle of 360°.

Types and names of angles

Acute angle	Right angle	Obtuse angle	Reflex angle
			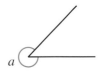
An angle less than 90° is called an **acute angle**.	A quarter-turn is called a **right angle**. A right angle is 90°.	An angle between 90° and 180° is called an **obtuse angle**.	An angle greater than 180° is called a **reflex angle**.
$0° < a < 90°$	$a = 90°$	$90° < a < 180°$	$180° < a < 360°$

Exercise 21.1

1 Which of these angles is obtuse? Explain why.

18°	118°	180°	298°	318°

2 Through what angle will the second hand of a clock turn in:

(a) half a minute, (b) quarter of a minute, (c) three-quarters of a minute,

(d) 15 seconds, (e) 20 seconds, (f) 1 second,

(g) 7 seconds, (h) 2 minutes, (i) $1\frac{1}{2}$ minutes?

3 This clock shows 4.30.
(a) What size is the acute angle between the hands of the clock?
(b) What is the size of the reflex angle between the hands?

4 Through what angle will the hour hand of the clock turn between:
(a) 10.00 am and 11.30 am,
(b) 10.00 am and 10.00 pm?

5 Say whether each of the marked angles is acute, obtuse, reflex or a right angle.

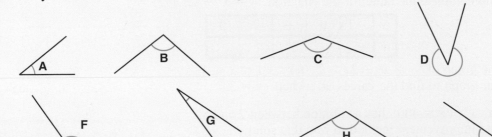

Measuring angles

To measure an angle accurately we need to use a **protractor**.

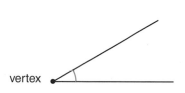

Some protractors have two scales.
Look at the type of angle
(acute/obtuse) you are measuring and
use the correct scale.

To measure an angle, the protractor is placed so that its centre point is on the corner (vertex) of the angle, with the base along one of the arms of the angle, as shown.

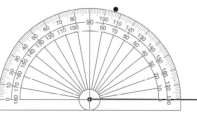

vertex

This angle measures 30°.

How can you measure the size of a reflex angle?

Drawing angles

Draw an angle of 74°.

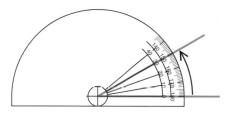

Draw a line.
Mark the vertex of the angle.

Position the protractor as if you were measuring an angle.
Mark a dot at 74°.

Draw a line from the vertex through the dot.

Exercise **21.2**

1   Use a protractor to measure these angles.

(a)   (b)   (c)

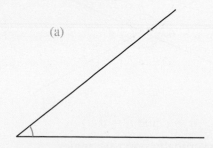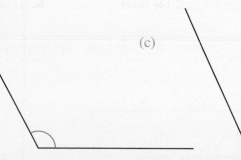

Angles . . . . Angles . . . . Angles . . . .

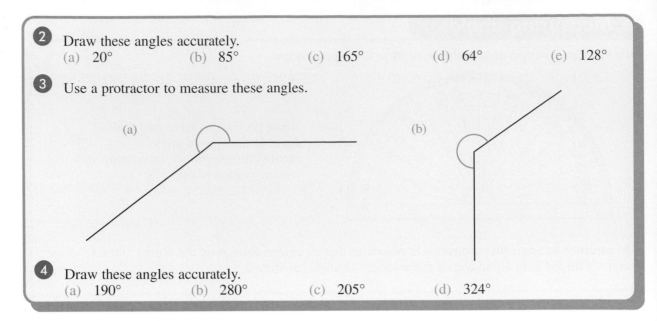

2 Draw these angles accurately.
   (a) 20°   (b) 85°   (c) 165°   (d) 64°   (e) 128°

3 Use a protractor to measure these angles.

(a)   (b)

4 Draw these angles accurately.
   (a) 190°   (b) 280°   (c) 205°   (d) 324°

Angle properties

Complementary angles

When two angles add up to 90°, the angles are called **complementary**.

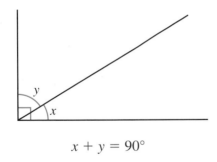

$$x + y = 90°$$

x and y are complementary angles.

Supplementary angles

Angles which can be placed together on a straight line add up to 180°.
When two angles add up to 180°, the angles are called **supplementary**.

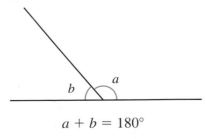

$$a + b = 180°$$

a and b are supplementary angles.

Angles at a point

When angles meet at a point, the sum of all the angles is 360°.

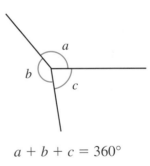

$$a + b + c = 360°$$

Vertically opposite angles

When two lines cross each other the angles between the lines make two pairs of equal angles.

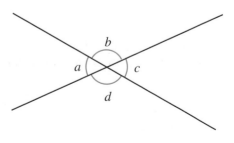

$$a = c \text{ and } b = d$$

a and c are vertically opposite angles.
b and d are vertically opposite angles.

EXAMPLES

1 Work out the size of the angle marked *a*.

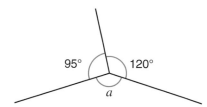

Angles at a point add up to 360°.
$$a + 95° + 120° = 360°$$
$$a = 360° - 95° - 120°$$
$$a = 145°$$

2 Work out the size of the angles marked with letters.

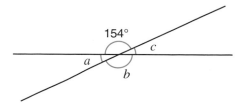

$b = 154°$ (vertically opposite angles)

$a + 154° = 180°$ (supplementary angles)
$$a = 180° - 154°$$
$$a = 26°$$

$c = 26°$

$a = 26°, b = 154°, c = 26°$

Exercise 21.3   The diagrams in this exercise have **not** been drawn accurately.

1 These angles are complementary.
Work out the size of angle *p* in each diagram.

(a)

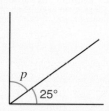

(b)

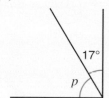

(c)

2 These angles are supplementary.
Work out the size of angle *q* in each diagram.

(a)

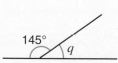

(b)

(c)

3 *PQ* and *RS* are straight lines.
Work out the size of angle *x* in each diagram.

(a)

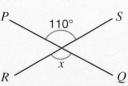

(b)

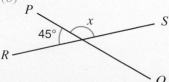

(c)

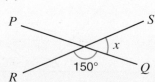

4 Work out the size of angle *y* in each diagram.

(a)

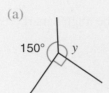
150° *y*

(b)

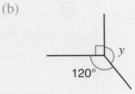
120° *y*

(c)

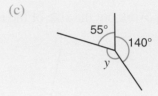
55° 140° *y*

5 Work out the size of the angles marked with letters.
Give a reason for each answer.

(a)

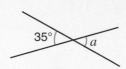
35° *a*

(b)

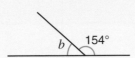
b 154°

(c)

c 135° 120°

6 Work out the size of the angles marked with letters.

(a)

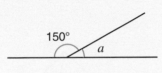
150° *a*

(b)

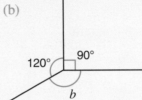
120° 90° *b*

(c)

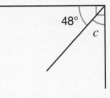
48° *c*

(d)

d 40° 30°

(e)

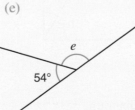
e 54°

(f)

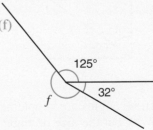
125° 32° *f*

(g)

g 47° *h*

(h)

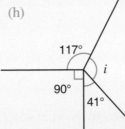
117° *i* 90° 41°

(i)

53° 90° *j* *l* *k*

(j)

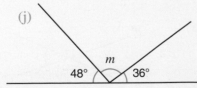
m 48° 36°

(k)

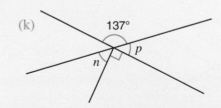
137° *p* *n*

220

7 Work out the value of *x* in each diagram.

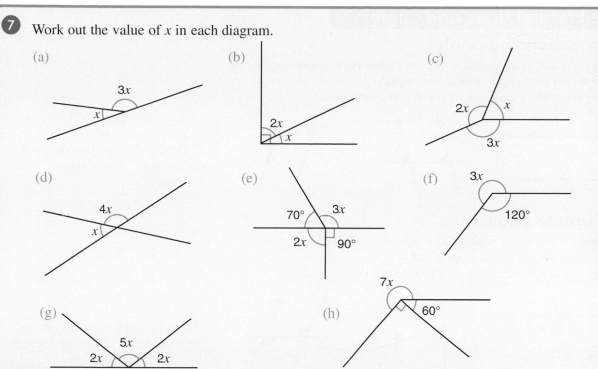

(a)

3x
x

(b)

2x
x

(c)

2x   x
3x

(d)

4x
x

(e)

70°   3x
2x   90°

(f)

3x
120°

(g)

5x
2x   2x

(h)

7x
60°

Lines

A straight line joining two points is called a **line segment**.

Perpendicular lines

Lines which meet at right angles are **perpendicular** to each other.

Parallel lines

Parallel lines are lines which never meet.
Which of the following pairs of lines are parallel?

(a) (b) (c) (d)

The pairs of lines in (a) and (c) are parallel.

Activity

The diagram shows two parallel lines crossed by another straight line called a **transversal**.

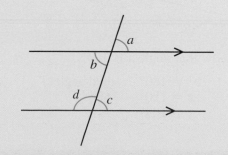

Arrowheads are used to show that lines are parallel.

a
b
d *c*

Measure the marked angles.
What do you notice?

Parallel lines and angles

Corresponding angles

Angles *a* and *c* are equal. They are called **corresponding** angles.
Corresponding angles are always equal.
Here are some examples of corresponding angles.

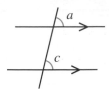 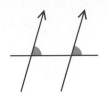 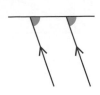

Corresponding angles are always on the same side of the transversal.

Alternate angles

Angles *b* and *c* are equal. They are called **alternate** angles.
Alternate angles are always equal.
Here are some examples of alternate angles.

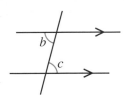 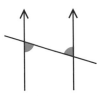

Alternate angles are always on opposite sides of the transversal.

Allied angles

Angles *b* and *d* add up to 180°. They are called **allied** angles.
Allied angles are supplementary, they always add up to 180°.
Here are some examples of allied angles.

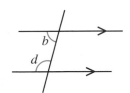 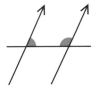

Allied angles are always between parallels on the same side of the transversal.

$b + d = 180°$

EXAMPLES

1 Work out the size of the angles marked with letters. Give a reason for each answer.

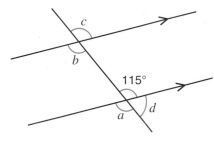

$a = 115°$ (vertically opposite angles)

$b = 115°$ (alternate angles)

$c = 115°$ (corresponding angles)

$d + 115° = 180°$ (supplementary angles)
$$d = 180° - 115°$$
$$d = 65°$$

2 Show that angle *x* is 104°.

$x + 76° = 180°$ (allied angles)
$$x = 180° - 76°$$
$$x = 104°$$

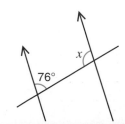

Exercise **21.4**   The diagrams in this exercise have **not** been drawn accurately.

㉑

Angles . . . Angles . . . Angles . . .

1   Work out the size of the angles marked with letters.
Give a reason for each answer.

(a)

(b)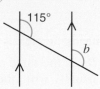

(c)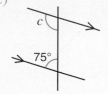

(d)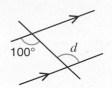

2   Work out the size of the angles marked with letters.

(a)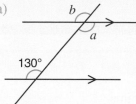

(b)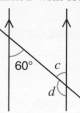

(c)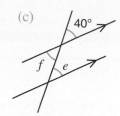

(d)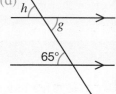

3   Calculate the size of the angles marked with letters.

(a)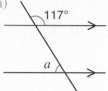

(b)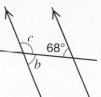

(c)

(d)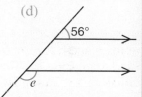

(e)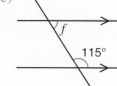

(f)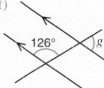

(g)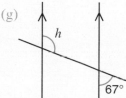

(h)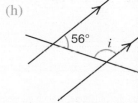

4   Calculate the size of the angles marked with letters.

(a)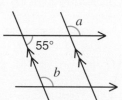

(b)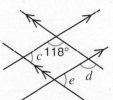

(c)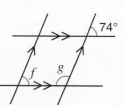

(d)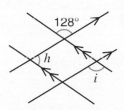

5   Calculate the size of the angles marked with letters.

(a)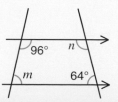

(b)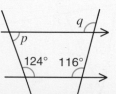

(c)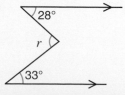

(d)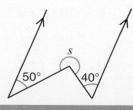

Naming angles

Up to now we have used small letters to name angles. This is not always convenient. Another method is to use three capital letters.

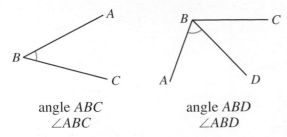

angle *ABC*
∠*ABC*

angle *ABD*
∠*ABD*

∠ means 'angle'.

∠*CBA* is the same as ∠*ABC*.
We usually write the letters either side of the vertex (shown by the middle letter) in alphabetical order.

Notice that the middle letter is where the angle is made.

Exercise 21.5

1 Use three letters to name the marked angles in each of these diagrams.

(a)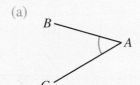

(b)

(c)

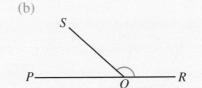 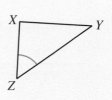

2 Use three letters to name the angles marked with small letters in this diagram.

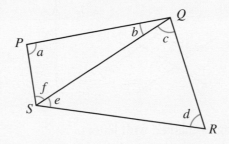

3 Use your protractor to measure accurately the size of these angles.

(a) ∠*ABH*

(b) ∠*HGF*

(c) ∠*BCD*

(d) ∠*AJE*

(e) ∠*GFJ*

(f) reflex ∠*GFJ*

(g) reflex ∠*BHG*

(h) reflex ∠*DEJ*

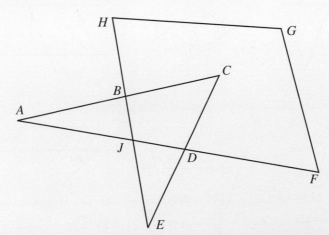

4 These diagrams have not been drawn accurately.
  (i)   Work out the size of the required angles.
  (ii) Give a reason for each of your answers.

(a) *PQ* is a straight line.
Find ∠*QOR*.

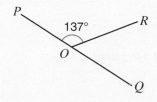

(b) *AB* and *CD* are straight lines.
Find ∠*AOD*.

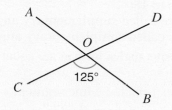

(c) *PQ* and *RT* are parallel.
Find ∠*XOQ*.

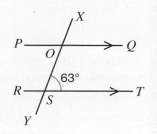

5 These diagrams are not drawn accurately.
Work out the size of the required angles.

(a)

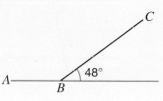

Find ∠*ABC*.

(b)

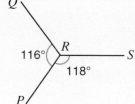

Find ∠*QRS*.

(c)

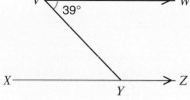

Find ∠*ZYV*.

(d)

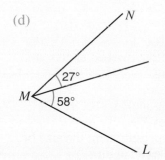

Find ∠*LMN*.

(e)

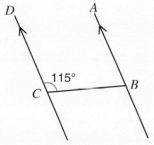

Find ∠*ABC*.

(f)

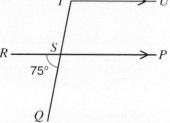

Find ∠*QSP* and ∠*STU*.

6 These diagrams are not drawn accurately.
Work out the size of the required angles.

(a)

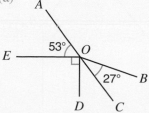

Find ∠*AOB* and ∠*COD*.

(b)

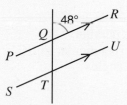

Find ∠*QTU* and ∠*QTS*.

(c)

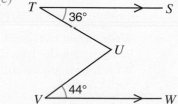

Find reflex angle *TUV*.

Angles . . . . Angles . . . . Angles . . . . Angles . . . .

- An angle of 90° is called a **right angle**.
  An angle less than 90° is called an **acute angle**.
  An angle between 90° and 180° is called an **obtuse angle**.
  An angle greater than 180° is called a **reflex angle**.
- The sum of the angles at a point is 360°.
- Angles on a straight line add up to 180°.
  Angles which add up to 180° are called **supplementary angles**.
  Angles which add up to 90° are called **complementary angles**.
- When two lines cross, the opposite angles formed are equal and are called
  vertically opposite angles.
- A straight line joining two points is called a **line segment**.
- Lines which meet at right angles are **perpendicular** to each other.
- Lines which never meet and are always the same distance apart are **parallel**.
- When two parallel lines are crossed by a transversal the following pairs of angles are formed.

Corresponding angles   **Alternate angles**   **Allied angles**

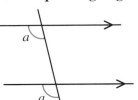   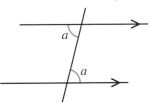   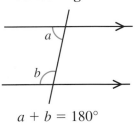

$a + b = 180°$

- You should be able to use a protractor to measure and draw angles accurately.

Review Exercise 21

1 The diagram shows a four-sided shape.

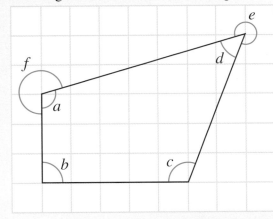

(a) Which of the marked angles are:
    (i) acute,
    (ii) obtuse,
    (iii) right-angled,
    (iv) reflex?
(b) Find by measurement the size of all the marked angles.

2 The diagram shows a shape, *ABCDE*.

(a) Write down two lines that are parallel.
(b) Write down the line that is perpendicular to *BC*.
(c) Which **two** angles are obtuse angles?

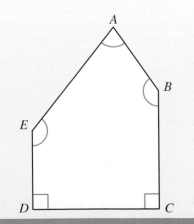

AQA

226

3 (a) What is the size of angle a?
Give a reason for your answer.

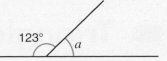

123° a

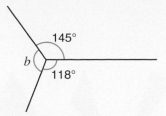

145°
b
118°

(b) What is the size of angle b?
Give a reason for your answer.

x 5x

(c) Show that angle x is 30°.

4 Copy the diagram onto squared paper.

(a) (i) Draw a line through P which is parallel to AB.
(ii) Draw a line through Q which is perpendicular to AB.

(b) In the diagram below, ABC is a straight line.

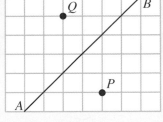

Q B P A

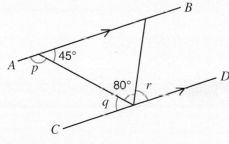

95°
$x°$ 28°
A B C

(i) Work out the value of x.
(ii) Which of the following describes an angle of 95°?

acute   half-turn   obtuse   reflex   right-angle

AQA

5 AB is parallel to DC.
(a) Work out the size of angle x.
Give a reason for your answer.
(b) Work out the size of angle y.
Give a reason for your answer.

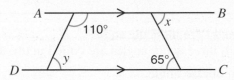

A B
110° x
y 65°
D C

AQA

6 In the diagram the line AB is parallel to the line CD.

(a) Work out the size of angle p.
(b) Work out the size of angle q.
(c) Work out the size of angle r.

B
45°
A p
80° r D
q
C

7 (a) What is the size of angle AOB?
(b) Work out the size of angle AOD.

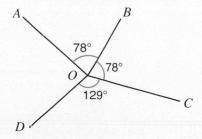

A B
78°
78°
O
129°
C
D

8

63° b
c
71°
a

Find the size of:
(a) angle a,
(b) angle b,
(c) angle c.

AQA

A **triangle** is a shape made by three straight lines.

The smallest number of straight lines needed to make a shape is 3. Can you explain why?

Types of triangle

Measure the angles in each of these triangles.
What do you notice?

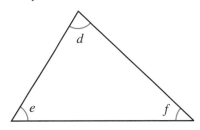 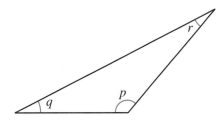 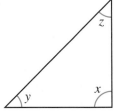

Angles *d*, *e* and *f* are all acute angles.
Triangles with three acute angles are called **acute-angled** triangles.

Angle *p* is an obtuse angle.
Triangles with an obtuse angle are called **obtuse-angled** triangles.

Angle *x* is a right angle.
Triangles with a right angle are called **right-angled** triangles.

Add up the three angles *d*, *e* and *f* in the triangle above.
Do the same for the other two triangles.
What do you notice?

The sum of the angles in a triangle

The sum of the three angles in a triangle is 180°.

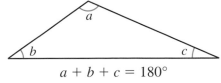

$$a + b + c = 180°$$

This result can easily be proved.

Draw a line which is parallel to one side of the triangle, as shown.

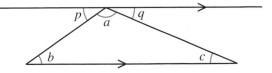

$p + a + q = 180°$ (supplementary angles)
$p = b$ (alternate angles)
$q = c$ (alternate angles)

Substitute $p = b$ and $q = c$ into $p + a + q = 180°$.
So, $b + a + c = 180°$, which can be written as $a + b + c = 180°$.

EXAMPLE

Without measuring, work out the size of the angle marked a.

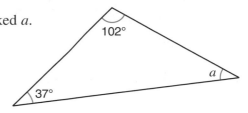

The sum of the angles in a triangle is 180°.

$$a + 102° + 37° = 180°$$
$$a + 139° = 180°$$
$$a = 180° - 139°$$
$$a = 41°$$

Exercise 22.1

1. Is it possible to draw triangles with the following types of angles?
   Give a reason for each of your answers.
   (a) three acute angles,
   (b) one obtuse angle and two acute angles,
   (c) two obtuse angles and one acute angle,
   (d) three obtuse angles,
   (e) one right angle and two acute angles,
   (f) two right angles and one acute angle.

2. Is it possible to draw a triangle with these angles?
   If a triangle can be drawn, what type of triangle is it?
   Give a reason for each of your answers.
   (a) 95°, 78°, 7°   (b) 48°, 62°, 90°   (c) 48°, 62°, 70°
   (d) 90°, 38°, 52°   (e) 130°, 35°, 15°   (f) 27°, 100°, 63°

3. Without measuring, work out the size of the third angle in each of these triangles.

   (a)   (b)   (c)   (d)

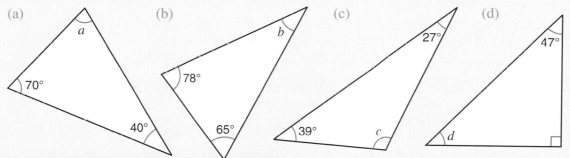

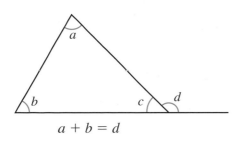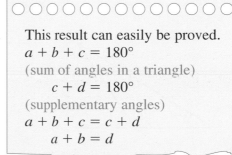

Exterior angle of a triangle

When one side of a triangle is extended, as shown, the angle formed is called an **exterior angle**.

This result can easily be proved.
$a + b + c = 180°$
(sum of angles in a triangle)
$c + d = 180°$
(supplementary angles)
$a + b + c = c + d$
$a + b = d$

$$a + b = d$$

In any triangle the exterior angle is always equal to the sum of the two opposite interior angles.
Check this by measuring the angles a, b and d in the diagram.

Find the size of the angles marked a and b.

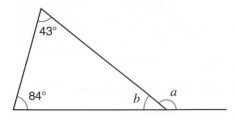

Short but fully accurate:
In geometry we often abbreviate words and use symbols to provide the reader with full details using the minimum amount of writing.

Δ is short for triangle.
ext. $\angle$ of a Δ means exterior angle of a triangle.
supp. $\angle$'s means supplementary angles.

$a = 84° + 43°$ (ext. $\angle$ of a Δ)
$a = 127°$

$b + 127° = 180°$ (supp. $\angle$'s)
$\quad\quad b = 180° - 127°$
$\quad\quad b = 53°$

Exercise 22.2

The diagrams in this exercise have **not** been drawn accurately.

1 The diagram shows a triangle with one side extended.
Explain why angle x is 35°.

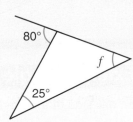

2 Work out the size of the marked angles.

(a)

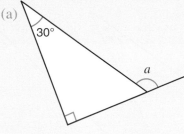

(b)

(c)

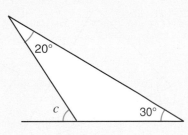

(d)

(e)

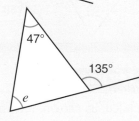

(f)

3 Work out the size of the marked angles.

(a)

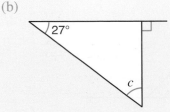

(b)

(c)

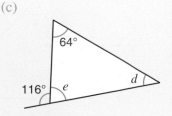

4 A triangle has angles of 27°, 85° and 68°.
One side of the triangle is extended to form an exterior angle.
Explain why this exterior angle must be an obtuse angle.

Naming parts of a triangle

Triangles are named by labelling each vertex with a capital letter.
Triangle ABC can be written as ΔABC.

Triangle ABC is formed by the sides AB, BC and AC.
Triangles and lines are often named in alphabetical order.
ΔABC is the same as ΔBCA.

The angles of a triangle are also described in terms of the vertices.
For example, the angle marked on the diagram is angle ACB or $\angle ACB$.
The middle letter is the vertex where the angle is made.

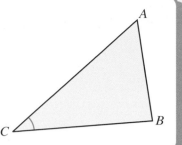

Special triangles

We have already seen that triangles can be described in terms of their angles but they can also be described in terms of their sides.

Activity

Measure the lengths of the sides of these triangles.
What do you notice?

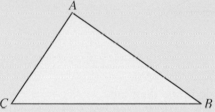
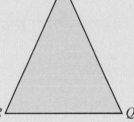
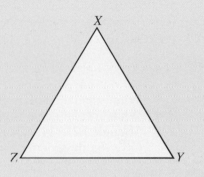

Now measure the size of the angles of triangles PQR and XYZ.
What do you notice?

Triangle ABC has sides of different lengths.
A triangle with sides of different lengths is called **scalene**.

Triangle PQR has two equal sides. $PQ = PR$.
A triangle with two equal sides is called **isosceles**.

Triangle XYZ has three equal sides. $XY = YZ = XZ$.
A triangle with three equal sides is called **equilateral**.

In triangle PQR, angle PQR = angle PRQ.
An **isosceles** triangle has two equal sides and two equal angles.

In triangle XYZ, all the angles are equal to $60°$.
An **equilateral** triangle has three equal sides and three equal angles.

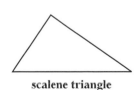
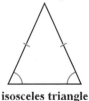
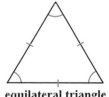

scalene triangle	isosceles triangle	equilateral triangle
Sides have different lengths. Angles are all different.	Two equal sides. Two angles equal.	Three equal sides. All angles are 60°.

A sketch is used when an accurate drawing is not required.
Dashes across lines show sides that are equal in length.
Equal angles are marked using arcs.

1 (a) Name three different triangles in the diagram.

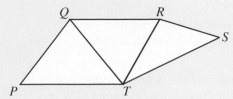

(b) Draw a sketch of the diagram.
Mark on your diagram: angle *TPQ* with the letter *a*,
angle *QRT* with the letter *b*,
angle *STR* with the letter *c*.

2 In the diagram, *AE* = *BE* = *BD* = *DE* and *CDE* is a straight line.

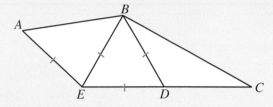

(a) What special name is given to Δ*ABE*?
(b) What special name is given to Δ*BDE*?
(c) Triangle *BDC* is scalene.
Give the three-letter name of another scalene triangle in the diagram.

3 (a) On squared paper, draw triangles with the following coordinates:
(i) (1, 1), (6, 1), (3, 5),
(ii) (1, 1), (5, 1), (1, 4),
(iii) (1, 1), (5, 1), (3, 4),
(iv) (1, 1), (6, 1), (9, 5).
(b) Which of the following words could be used to describe each of the triangles you have drawn?

Acute-angled, Obtuse-angled or Right-angled.

Scalene, Equilateral or Isosceles.

4 On squared paper, draw an isosceles triangle with coordinates:
A (3, 3), *B* (9, 3) and *C* (6, 10).
Which two sides are equal?
Which two angles are equal?

5 Triangle *PQR* is isosceles with angle *RPQ* = angle *QRP*.
P is the point (3, 5) and *R* is the point (9, 5).
Give the coordinates of the two possible positions of *Q* so that angle *PQR* is a right angle.

6 These triangles have not been drawn accurately.
Work out the size of angle *a* in each triangle.

(a)

(b)

(c)

(d)

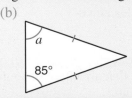
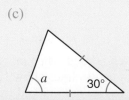
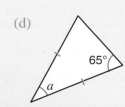

7 The following diagrams have not been drawn accurately.
Work out the size of the angles marked with letters.

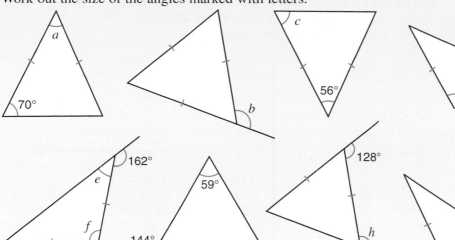

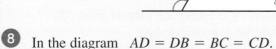

8 In the diagram $AD = DB = BC = CD$.

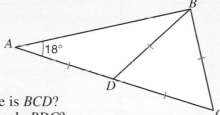

(a) What type of triangle is BCD?
(b) What is the size of angle BDC?
(c) Work out the size of angle ABC.

9 In the diagram $AB = BD = DA$ and $BC = CD$. CD is extended to E.

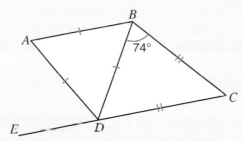

(a) What type of triangle is BCD?
(b) What is the size of angle BDC?
(c) Work out the size of angle ADE.

10 These diagrams have not been drawn accurately.
Work out the size of the required angles.

(a)

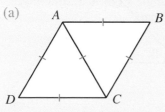

Find $\angle BCD$.

(b)

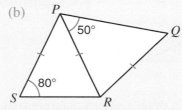

Find $\angle PRQ$ and $\angle QRS$.

(c)

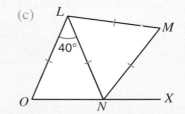

ONX is a straight line.
Find $\angle MNX$.

Your ruler, compasses and protractor can be used to draw triangles accurately.
Drawings can be made from written information or sketch diagrams.
Follow the instructions below to accurately draw two triangles.

Sketch diagram

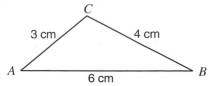

Information given:
Lengths of three sides of the triangle.

Step 1
Start by drawing the longest side, *AB*.
Draw a line 6 cm long.

Step 2
Set your compasses to a radius of 4 cm.
Draw an arc from *B*.

Step 3
Set your compasses to a radius of 3 cm.
Draw an arc from *A* to intersect (cross) the
arc drawn in step 2.
Label the point *C*.

Step 4
Draw the sides *AC* and *BC*.
Add labels.

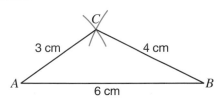

Sketch diagram

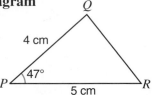

Information given:
Lengths of two sides of the triangle and the
size of the angle between the two sides.

Step 1
Start by drawing the longest side, *PR*.
Draw a line 5 cm long.

Step 2
∠*QPR* = 47° (acute angle)
Use your protractor to measure 47°.

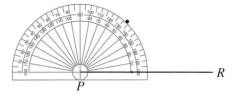

Step 3
Using the dot as a guide, draw a line,
4 cm long, from *P*.
Label point *Q*.

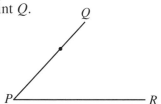

Step 4
Draw the line *QR* to complete the triangle.
Add labels.

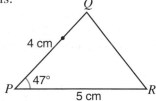

If you are given written information draw a sketch diagram first.

For example, information for triangle *ABC* could be given as:
    Draw accurately triangle *ABC* with sides   *AB* = 6 cm,   *BC* = 4 cm   and   *AC* = 3 cm.

Activity

Write instructions which someone could follow to draw the following triangles accurately.

(a)

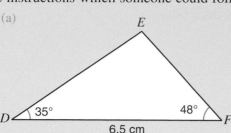

(b)

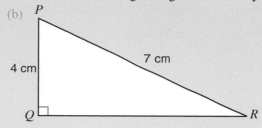

Exercise 22.4

1 Use a ruler and compasses to draw accurately triangles with the following sides.
   (a) 4 cm, 5 cm, 6 cm.
   (b) 3.5 cm, 4.5 cm, 5 cm.
   (c) $AB = 4.8$ cm, $BC = 3.6$ cm, $AC = 6.2$ cm.
   (d) $PQ = 6$ cm, $QR = 6.5$ cm, $PR = 2.5$ cm.

2 Use a ruler and compasses to construct an equilateral triangle of side 5 cm.

3 Draw these triangles accurately using the information given.

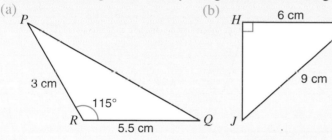

4 Use a ruler and protractor to draw the following triangles.
   (a) $AB = 4$ cm, $BC = 4$ cm, $\angle ABC = 40°$.
   (b) $PQ = 3.5$ cm, $PR = 5$ cm, $\angle QPR = 100°$.
   (c) $XY = YZ = ZX = 4$ cm.
   (d) $FG = 5$ cm, $FH = 5$ cm, $\angle FGH = 40°$.

5 A sketch of triangle PQR is shown.

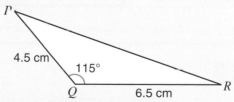

   (a) Make an accurate drawing of triangle PQR.
   (b) Measure and write down the length of PR.
   (c) Measure and write down the size of angle QPR.

6 A sketch of triangle ABC is shown.

   (a) Make an accurate drawing of this triangle.
   (b) What is the length of CB?
   (c) What is the size of angle ABC?

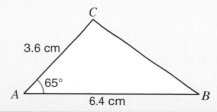

Perimeter of a triangle

The **perimeter** is the distance round the outside of a shape.
The perimeter of a triangle is the sum of the lengths of its three sides.

Measure the sides of this triangle.
What is the perimeter?

You should find:
$AB = 4\,cm$, $BC = 5\,cm$ and $AC = 6\,cm$.
Perimeter $= 4 + 5 + 6 = 15\,cm$.

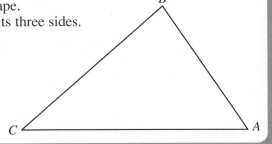

Finding the area of a triangle

Area is the amount of surface covered by a shape.
The standard unit for measuring area is the **square centimetre, cm²**.

Activity

Finding areas by counting squares
The diagram shows three triangles which have been drawn on centimetre-squared paper.

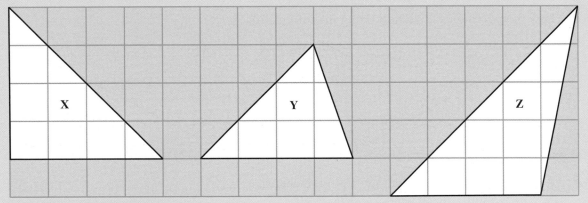

The area of each square on the grid is $1\,cm^2$.
Triangle X covers a total of 8 squares.
The area of triangle X is $8\,cm^2$.

1. What is the area of triangle Y?
2. What is the area of triangle Z?

> **Can you find a rule?**
> Does your rule work for all triangles?
> Try to explain why your rule works.

Is there a quicker way to find the areas of triangles without having to count squares?

Area of a triangle

The area of a triangle is given by: Area $= \frac{1}{2} \times$ base $\times$ perpendicular height.

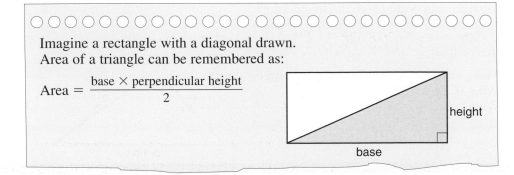

Imagine a rectangle with a diagonal drawn.
Area of a triangle can be remembered as:

$$Area = \frac{base \times perpendicular\ height}{2}$$

In these triangles b is the base and h is the **perpendicular height**.

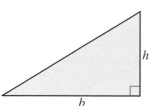 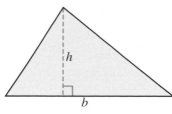 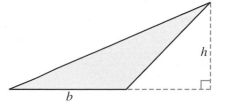

The area, A, can be found using the formula: $A = \frac{1}{2} \times b \times h$

EXAMPLES

1 Calculate the area of this triangle.

$A = \frac{1}{2} \times b \times h$
$\quad = \frac{1}{2} \times 12 \times 7$
$\quad = 42\,\text{cm}^2$

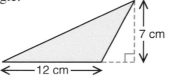

7 cm

12 cm

2 This triangle has area $36\,\text{cm}^2$. Find the height of the triangle.

$A = \frac{1}{2} \times b \times h$
$36 = \frac{1}{2} \times 16 \times h$
$36 = 8h$
$h = \frac{36}{8}$
$h = 4.5\,\text{cm}$

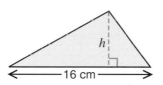

h

16 cm

Exercise 22.5

Do not use a calculator for questions 1 to 6.

1 The two shorter sides of a right-angled triangle are 3 cm and 4 cm.
Draw the triangle accurately on centimetre-squared paper.
Find, by measurement, the perimeter of the triangle.

2 Work out the lengths of the perimeters of these triangles.

(a)
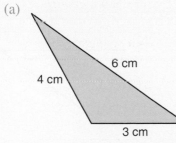
6 cm
4 cm
3 cm

(b)
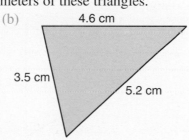
4.6 cm
3.5 cm
5.2 cm

(c)
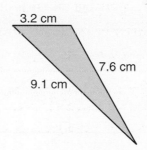
3.2 cm
7.6 cm
9.1 cm

3 Which of the triangles PQR, QRS or RST has the largest perimeter?

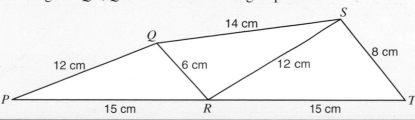

14 cm
Q
S
12 cm
6 cm
12 cm
8 cm
P
15 cm
R
15 cm
T

4 These triangles each have a perimeter of length 20 cm.
Work out the lengths of the marked sides.

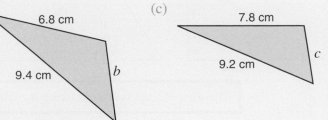

(a) a 5 cm   8 cm

(b) 6.8 cm   9.4 cm b

(c) 7.8 cm   9.2 cm c

5 These triangles have been drawn on 1 cm squared paper.
Find the area of each triangle.

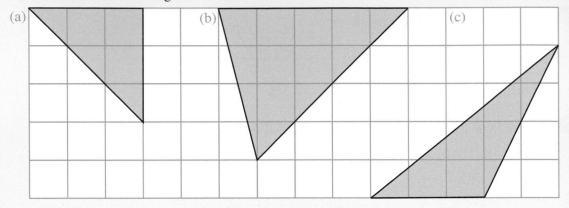

(a)   (b)   (c)

6 Calculate the areas of these triangles.

(a)

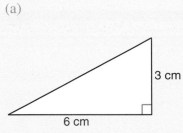

3 cm
6 cm

(b)

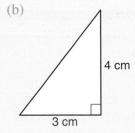

4 cm
3 cm

(c)

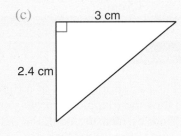

3 cm
2.4 cm

7 Work out the areas of these triangles.

(a)

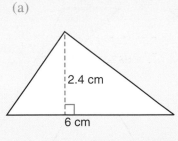

2.4 cm
6 cm

(b)

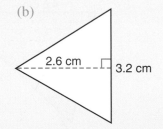

2.6 cm   3.2 cm

(c)

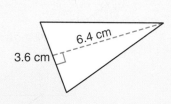

6.4 cm
3.6 cm

8 Find the areas of the shaded triangles.

(a)

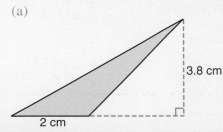

3.8 cm
2 cm

(b)

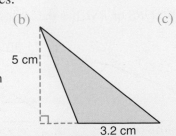

5 cm
3.2 cm

(c)

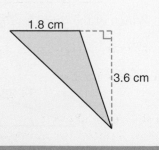

1.8 cm
3.6 cm

9 These triangles each have an area of 24 cm². Calculate the height of each triangle.

(a)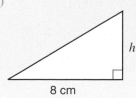
8 cm

(b)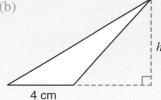
4 cm

(c)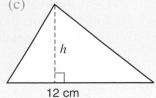
12 cm

10 These triangles each have an area of 32 cm². Calculate the lengths of the marked sides.

(a)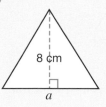
8 cm
a

(b)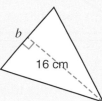
b
16 cm

(c)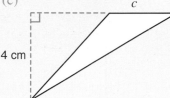
c
4 cm

11 This triangle has a perimeter of 45 cm. Calculate the area of the triangle.

19.5 cm
18 cm

12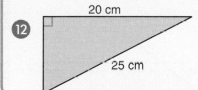
20 cm
25 cm

This triangle has an area of 150 cm². Calculate the perimeter of the triangle.

What you need to know

- Triangles can be: **acute-angled**, **obtuse-angled**, **right-angled**.
- The sum of the angles in a triangle is 180°.
  $a + b + c = 180°$
- The exterior angle is equal to the sum of the two opposite interior angles.
  $a + b = d$

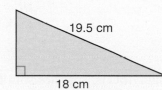

- Types of triangle:

  Scalene triangle   **Isosceles triangle**   **Equilateral triangle**

  Sides have different lengths.   Two equal sides.   Three equal sides.
  Angles are all different.   Two equal angles.   Three equal angles, 60°.

- Perimeter of a triangle is the sum of its three sides.
- Area of a triangle = $\dfrac{\text{base} \times \text{perpendicular height}}{2}$

  $A = \frac{1}{2} \times b \times h$

You should be able to:
- Draw triangles accurately using ruler, compasses, protractor.

Triangles . . . Triangles . . . Triangles . . .

1 (a) On one centimetre squared paper, plot the points $P(2, 1)$, $Q(4, 5)$, $R(6, 1)$.
Join the points to form triangle PQR.
(b) (i) What special name is given to triangle PQR?
(ii) What is the area of the triangle?
(c) On the same diagram draw another triangle PRS, which has the same area as triangle PQR.

2 Calculate the area of this triangle.

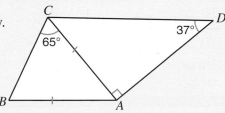

11 cm

24 cm

AQA

3

A

y

B

64°

x

C

D

ABC is an isosceles triangle. BCD is a straight line.
(a) Work out the size of angle x.
(b) $AB = AC$.
Work out the size of angle y.

AQA

4 In the diagram, triangle ABC is isosceles with $BA = AC$, and triangle ACD is right-angled with angle $CAD = 90°$.
The diagram has not been drawn accurately.
(a) Angle $ADC = 37°$.
Work out the size of angle DCA.
(b) Angle $ACB = 65°$.
Work out the size of angle BAC.
Give a reason for your answer.

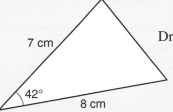

C

65°

D

37°

B

A

5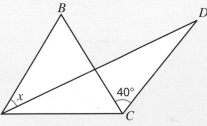

7 cm

42°

8 cm

Draw an accurate full size copy of this triangle.

AQA

6 The diagram shows a sketch of triangle PQR.
(a) Make an accurate drawing of the triangle.
(b) By measuring the height of your triangle calculate the area of triangle PQR.

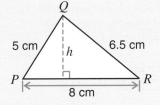

Q

5 cm

h

6.5 cm

P

8 cm

R

7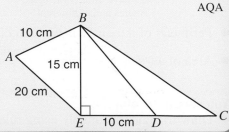

B

D

x

40°

A

C

ABC is an equilateral triangle.
ACD is an isosceles triangle. $\angle BCD = 40°$.
Work out the size of angle x,
giving a reason for your answer.

AQA

8 The diagram shows three triangles, BAE, BED and BDC.
(a) Calculate the perimeter of triangle BAE.
(b) Calculate the area of triangle BED.
(c) The areas of triangles BED and BDC are equal.
Calculate the length of DC.

A

10 cm

B

15 cm

20 cm

E

10 cm

D

C

CHAPTER 23

Symmetry and Congruence

Lines of symmetry

These shapes are **symmetrical**.

When each shape is folded along the dashed line one side will fit exactly over the other side.
The dashed line is called a **line of symmetry**.

Some shapes have more than one line of symmetry.

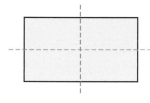 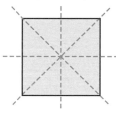  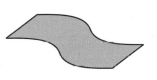

Rectangle	Square	Circle	Shape with no lines of symmetry.
2 lines of symmetry.	4 lines of symmetry.	Infinite number of lines of symmetry. Each diameter is a line of symmetry.	

Rotational symmetry

Is this shape symmetrical?

The shape does not have line symmetry.

Try placing a copy of the shape over the original and rotating it about the centre of the circle.

After 180° (a half-turn) the shape fits into its own outline.
The shape has **rotational symmetry**.
The point about which the shape is rotated is called the **centre of rotation**.
The **order of rotational symmetry** is 2. When rotating the shape through 360° it fits into its own outline twice (once after a half-turn and again after a full-turn).
A shape is only described as having rotational symmetry if the order of rotational symmetry is 2 or more.

A shape can have both line symmetry and rotational symmetry.

EXAMPLES

Order of rotational symmetry 5.   Order of rotational symmetry 4.   Order of rotational symmetry 1.
4 lines of symmetry.   The shape is **not** described as having rotational symmetry.

1 These shapes have **line symmetry**.
Copy each shape and draw the line of symmetry.

(a) 　　(b) 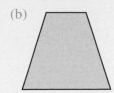　　(c) 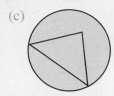　　(d)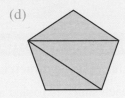

2 The following diagrams show half a shape.
The dashed line is the line of symmetry for the complete shape.
Copy the diagrams and complete each shape.

(a) 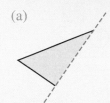　　(b) 　　(c) 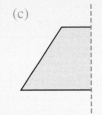　　(d)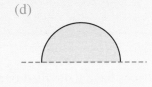

3 These shapes have been drawn accurately.
How many lines of symmetry has each shape?

(a) 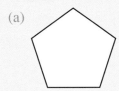　　(b) 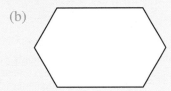　　(c)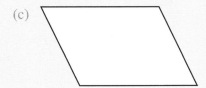

4 How many lines of symmetry has each of these letters?

A C E H K

5 What is the order of rotational symmetry for each of these shapes?

(a) 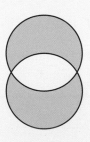　　(b) 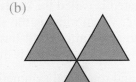　　(c)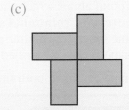

(d) 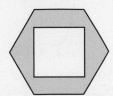　　(e) 　　(f)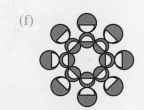

6 Look at these triangles.

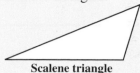 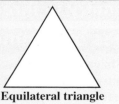

Scalene triangle **Isosceles triangle** **Equilateral triangle**

What is the order of rotational symmetry of each triangle?

7 Copy each of these diagrams.

A   B 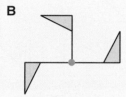 C

(a) Add **one** more flag to each of your diagrams so that the final diagrams have rotational symmetry.

(b) What is the order of rotational symmetry for each of your diagrams?

8 Look at these letters of the alphabet.

J M N O P X Y Z

(a) Which two letters have only line symmetry?
(b) Which two letters have only rotational symmetry?
(c) Which letters have rotational symmetry of order 2?
(d) Which letters have neither rotational nor line symmetry?

9 Make a copy of this shape.

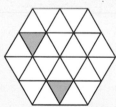

(a) How many lines of symmetry does the shape have?
(b) (i) Colour one more square so that your shape has rotational symmetry of order 2.
    (ii) Mark the centre of rotational symmetry on your shape.

10 Make a copy of this shape.

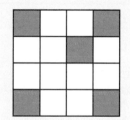

(a) How many lines of symmetry does the shape have?
(b) (i) Colour one more triangle so that your shape has rotational symmetry of order 3.
    (ii) How many lines of symmetry does your shape have?

11 For each shape state:
(i) the number of lines of symmetry, (ii) the order of rotational symmetry.

(a) (b) (c) (d) (e)

Symmetry in three-dimensions

Planes of symmetry

So far we have looked at two-dimensional (flat) shapes.
Two-dimensional shapes can have line symmetry.
Three-dimensional objects can have **plane symmetry**.
A **plane of symmetry** slices through an object so that one half is the mirror image of the other half.
A cuboid has three planes of symmetry as shown.

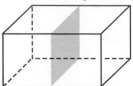 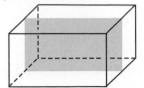 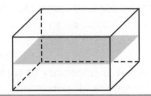

Axes of symmetry

A wall is built using cuboids.

In how many different ways can the next cuboid be placed in position?

If the cuboid can be placed in more than one way, it must have rotational symmetry about one or more **axes**.

A cuboid has three axes of symmetry.
The diagram shows one **axis of symmetry**.
The order of rotational symmetry about this axis is two.

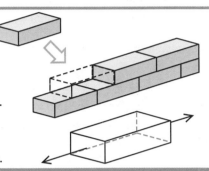

Exercise 23.2

1 How many planes of symmetry has a cube?

2 State the order of rotational symmetry about the axis shown in each of the following.
  (a) Cube   (b) Square-based pyramid   (c) Cylinder   (d) Cone

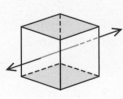 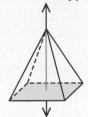 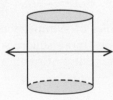 

3

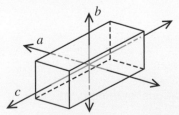

Each end of this cuboid is a square.
The axes of symmetry are labelled *a*, *b* and *c*.
What is the order of rotational symmetry about:
  (a) axis *a*,
  (b) axis *b*,
  (c) axis *c*?

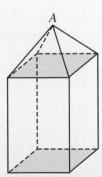

4 The diagram shows a cuboid, with a square base.
On top of the cuboid is a square-based pyramid with vertex *A* above the centre of the top of the cuboid.
  (a) How many planes of symmetry has the figure?
  (b) How many axes of symmetry has the figure?
      Give the order of rotational symmetry about each axis.

5 The diagram shows a triangular prism.
The ends of the prism are equilateral triangles.

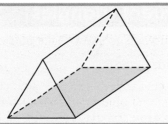

(a) How many axes of symmetry has the prism?
(b) How many planes of symmetry has the prism?

Congruent shapes

When two shapes are the same shape and size they are said to be **congruent**.
A copy of one shape would fit exactly over the second shape.
Sometimes it is necessary to turn the copy over to get an exact fit.
These shapes are all congruent.

Exercise 23.3

1 Look at the shapes below. List five **pairs** of congruent shapes.

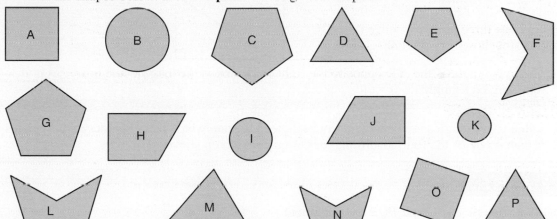

2 Which of these shapes are congruent to each other?

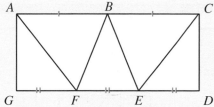

3 The diagram shows a rectangle that has been divided into five triangles.

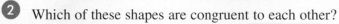

(a) Which triangle is congruent to triangle *AFG*?
(b) Which quadrilateral is congruent to quadrilateral *ABEF*?

4 Triangle *ABC* has been divided into
four smaller triangles as shown.
Name two pairs of congruent triangles.

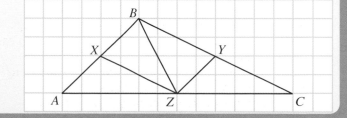

There are four ways to show that a pair of triangles are congruent.

1 Three sides. SSS

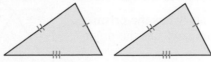

2 Two sides and the included angle. SAS

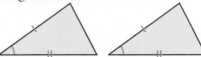

The included angle is the angle between the two sides.

3 Two angles and a corresponding side. ASA

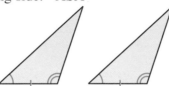

This can be written as AAS if the corresponding side is not between the angles.

4 Right angle, hypotenuse and one side. RHS

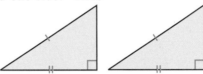

The hypotenuse is the side opposite the right angle and is the longest side in a right-angled triangle.

To show that two triangles are congruent you will need to state which pairs of sides and/or angles are equal, to match one of the four conditions for congruency given above.

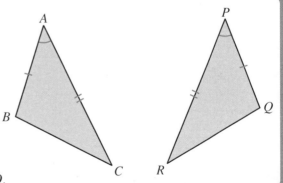

EXAMPLE

Show that triangles *ABC* and *PQR* are congruent.

AB = *PQ*   (equal lengths, given)
AC = *PR*   (equal lengths, given)
∠*BAC* = ∠*QPR*   (equal angles, given)

So, triangles *ABC* and *PQR* are congruent.
Reason: SAS   (Two sides and the included angle.)

Since the triangles are congruent we also know that
BC = *QR*,   ∠*ABC* = ∠*PQR*   and   ∠*ACB* = ∠*PRQ*.

Exercise 23.4

The triangles in this exercise have **not** been drawn accurately.

1 Which two of these triangles are congruent to each other?
Give a reason for your answer.

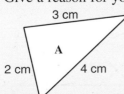

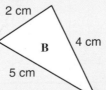

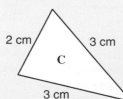

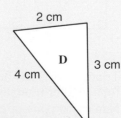

2 Which two of these triangles are congruent to each other?
Give a reason for your answer.

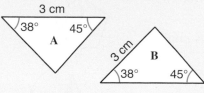
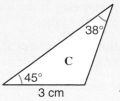
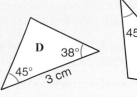

3 State whether each pair of triangles is congruent or not.
Where triangles are congruent give the reason.

(a)
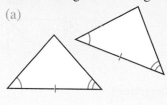

(b)
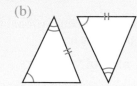

(c)
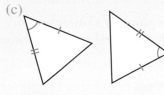

(d)
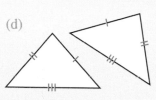

(e)
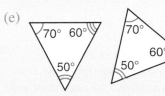

(f)
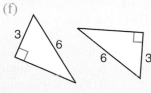

4 For each of the following, is it possible to draw a congruent triangle without taking any other measurements from the original triangle?
If a triangle can be drawn give the reason for congruence which applies.

(a)
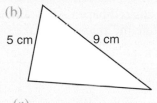

(b)
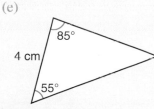

(c)
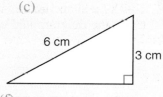

(d)
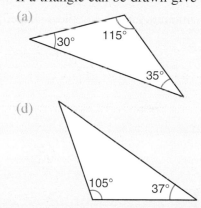

(e)
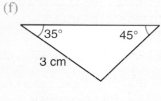

(f)

5 Look at the following triangles.
Equal sides and equal angles have been marked.
Using only the information given identify pairs of congruent triangles.
Give a reason for each of your answers.

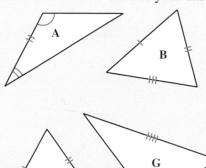

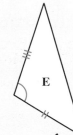
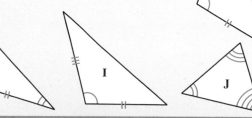

Symmetry and Congruence

What you need to know

- A two-dimensional shape has **line symmetry** if the line divides the shape so that one side fits exactly over the other.

- A two-dimensional shape has **rotational symmetry** if it fits into a copy of its outline as it is rotated through 360°.

- A shape is only described as having rotational symmetry if the order of rotational symmetry is 2 or more.

- The number of times a shape fits into its outline in a single turn is the **order of rotational symmetry**.

- A **plane of symmetry** slices through a three-dimensional object so that one half is the mirror image of the other half.

- Three-dimensional objects can have **axes of symmetry**.

- When two shapes are the same shape and size they are said to be **congruent**.

- There are four ways to show that a pair of triangles are congruent:
  - SSS   Three equal sides.
  - SAS   Two sides and the included angle.
  - ASA   Two angles and a corresponding side.
  - RHS   Right angle, hypotenuse and one other side.

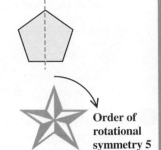

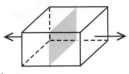

Order of rotational symmetry 5

Review Exercise 23

1 Half of a shape is drawn on squared paper. *AB* is a line of symmetry for the complete shape. Copy the diagram and complete the shape.

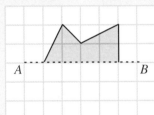

2 Which of these letters have:
  (a) line symmetry,
  (b) rotational symmetry of order 2?

AQA

3 (a) This diagram shows a six-sided shape. Copy the diagram and draw the line of symmetry.

(b) Three of these shapes are used in a design for a badge.

     (i)   How many lines of symmetry does this badge have?
     (ii)   What is its order of rotational symmetry?

AQA

4 All these patterns have five sides.

 A   B   C   D

  (a) Which pattern has more than one line of symmetry?
  (b) Which pattern does not have rotational symmetry?

AQA

5 The diagram shows a five-pointed star.
  (a) Write down the order of rotational symmetry of the star.
  (b) (i) Draw a shape that has rotational symmetry of order 3.
       (ii) On your drawing mark the centre of rotational symmetry with an X.

AQA

6 Copy the diagram and shade two more squares so that the final pattern has line symmetry **and** rotational symmetry.

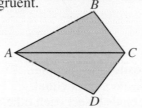

AQA

7 Triangle *ABC* is reflected in the line *AC*. The reflection is another triangle, *ADC*.
Triangles *ABC* and *ADC* are congruent.
Explain what this means.

AQA

8 Some of these shapes are congruent to each other. Which shapes are congruent?

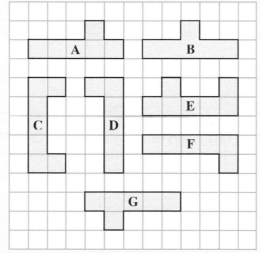

AQA

9 These triangles are congruent.
What is the size of:
  (a) *x*,
  (b) *y*?

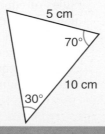 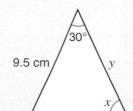

Not drawn accurately

AQA

Symmetry and Congruence

Quadrilaterals

A **quadrilateral** is a shape made by four straight lines.

Special quadrilaterals

Parallelogram

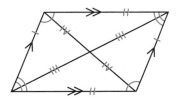

Opposite sides equal and parallel.
Opposite angles equal.
Diagonals bisect each other.

Rhombus

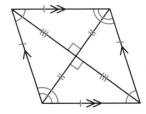

Four equal sides.
Opposite sides parallel.
Opposite angles equal.
Diagonals bisect each other at 90°.

Trapezium

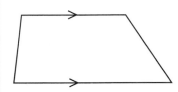

One pair of parallel sides.

Rectangle

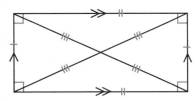

Opposite sides equal and parallel.
Angles of 90°.
Diagonals bisect each other.

Kite

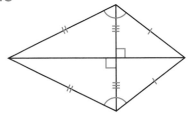

Two pairs of adjacent sides equal.
One pair of opposite angles equal.
One diagonal bisects the other at 90°.

Isosceles trapezium

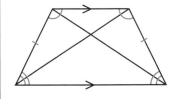

One pair of parallel sides.
Non-parallel sides equal.
Two pairs of equal angles.
Diagonals equal.

Square

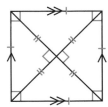

Four equal sides.
Opposite sides parallel.
Angles of 90°.
Diagonals bisect each other at 90°.

Remember:
Sides of equal length are marked with the same number of **dashes**.
Lines which are parallel are marked with the same number of **arrowheads**.
Angles of equal size are marked with the same number of **arcs**.

Sum of the angles of a quadrilateral

The sum of the four angles of a quadrilateral is 360°.

Measure the angles of this quadrilateral.
Do the angles add up to 360°?

You may not always get 360°.
Can you explain why?

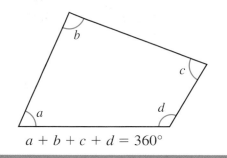

$a + b + c + d = 360°$

This result can easily be proved.

Split the quadrilateral into two triangles A and B, as shown.

In triangle A:
$\quad p + q + r = 180°$ (sum of angles in a triangle)
In triangle B:
$\quad w + x + y = 180°$ (sum of angles in a triangle)

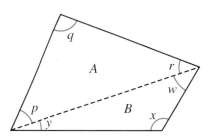

The sum of the angles of the quadrilateral is given by:
$\quad p + q + r + w + x + y$
$\quad = 180° + 180°$
$\quad = 360°$

EXAMPLE

PQRS is a parallelogram.
Work out the size of the angle marked x.

The opposite angles of a parallelogram are equal.

$55° + 55° + x + x = 360°$
$\qquad 110° + 2x = 360°$
$\qquad\qquad 2x = 360° - 110°$
$\qquad\qquad 2x = 250°$
$\qquad\qquad\ x = 125°$

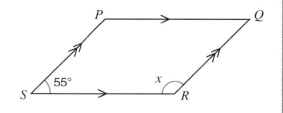

Exercise 24.1   Use squared paper to answer questions 1 to 8.

1   (a)   Draw quadrilaterals with the following coordinates.
        (i) $A(3, 1)$, $B(1, 3)$, $C(2, 6)$, $D(6, 2)$
        (ii) $E(1, 0)$, $F(6, 2)$, $G(8, 9)$, $H(3, 7)$
        (iii) $J(3, 0)$, $K(0, 4)$, $L(3, 8)$, $M(6, 4)$
        (iv) $P(1, 1)$, $Q(2, 4)$, $R(4, 4)$, $S(4, 2)$
        (v) $W(3, 1)$, $X(1, 3)$, $Y(3, 5)$, $Z(5, 3)$
   (b)   What special name is given to each of these quadrilaterals?

2   *JKLM* is a square. *J* is the point $(1, 1)$, $K(4, 1)$, $L(4, 4)$.
     Find the coordinates of *M*.

3   *PQRS* is a rectangle. *P* is the point $(1, 3)$, $Q(4, 6)$, $R(6, 4)$.
     Find the coordinates of *S*.

4   *ABCD* is a rhombus. *A* is the point $(3, 0)$, $B(0, 4)$ and $D(8, 0)$.
     Find the coordinates of *C*.

5 *WXYZ* is a parallelogram. *W* is the point (1, 0), *X* (4, 1), *Z* (3, 3).
Find the coordinates of *Y*.

6 *OABC* is a kite. *O* is the point (0, 0), *B* (5, 5), *C* (3, 1).
Find the coordinates of *A*.

7 *KLMN* is an isosceles trapezium with *K* at (1, 1), *M* (4, 3) and *N* (5, 1).
Find the coordinates of *L*.

8 *STUV* is a square with *S* at (1, 3) and *U* at (5, 3).
Find the coordinates of *T* and *V*.

9 Work out the size of angle *a* in each of these quadrilaterals.

(a)   (b) 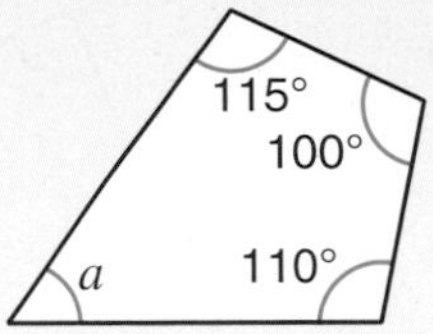   (c) 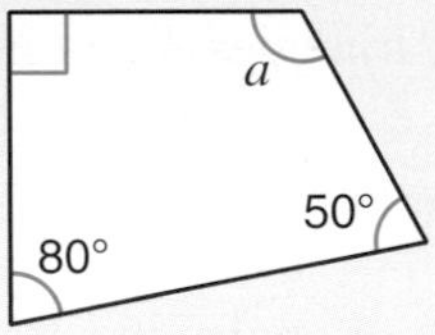   (d)

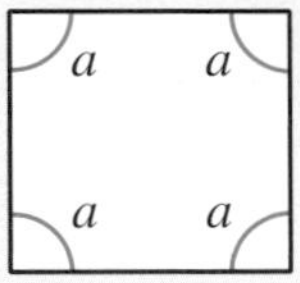

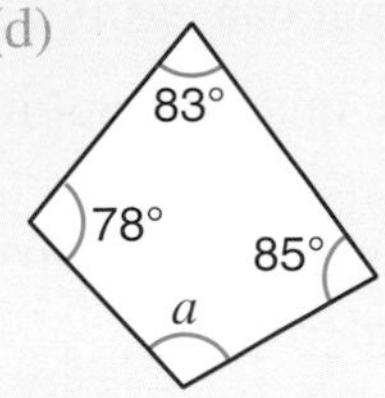

10 Work out the size of the angles marked with letters in each of these rectangles.

(a)   (b) 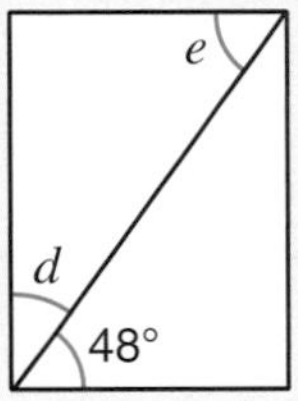   (c) 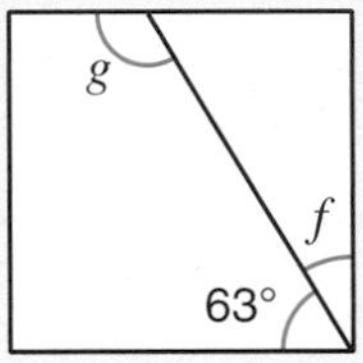   (d)

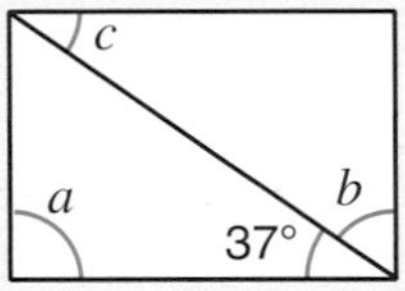

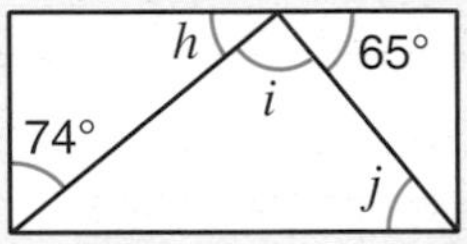

11 Work out the size of the angles marked with letters in each of these parallelograms.

(a)   (b) 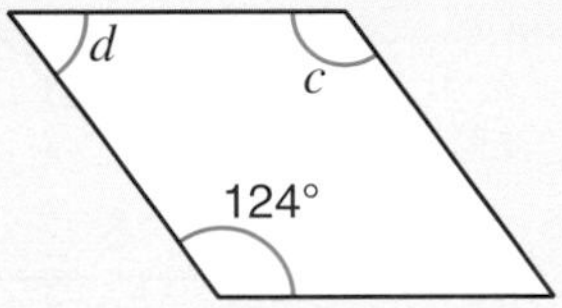   (c) 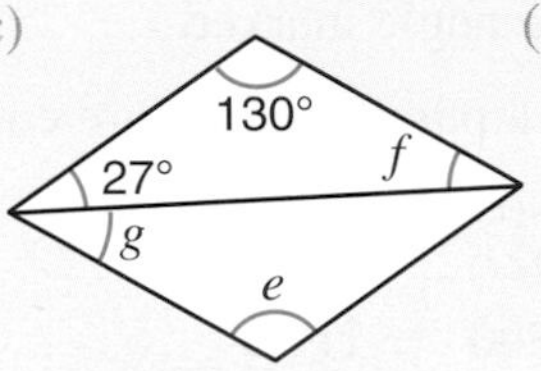   (d)

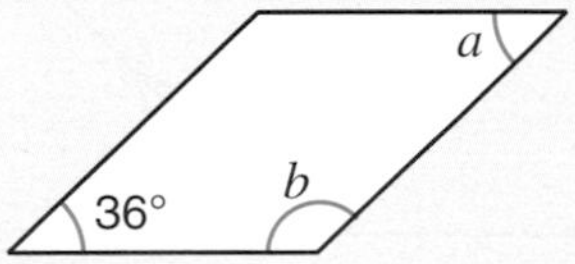

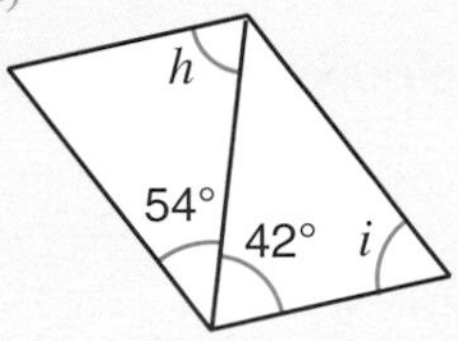

12 Work out the size of the angles marked with letters in each of these kites.

(a)   (b) 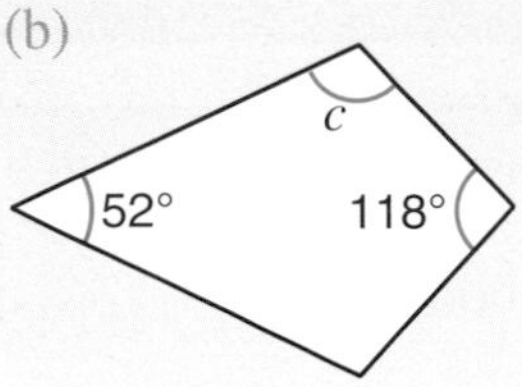   (c) 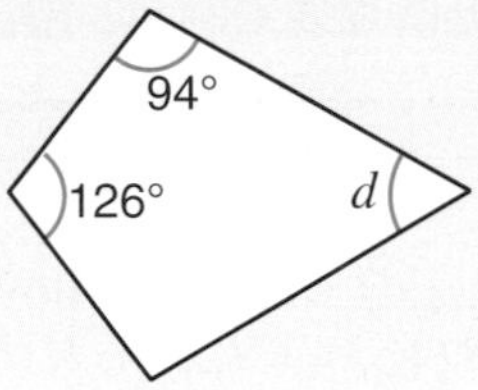   (d)

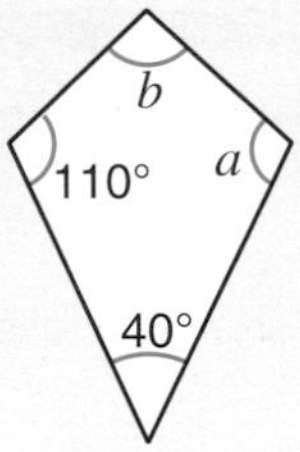

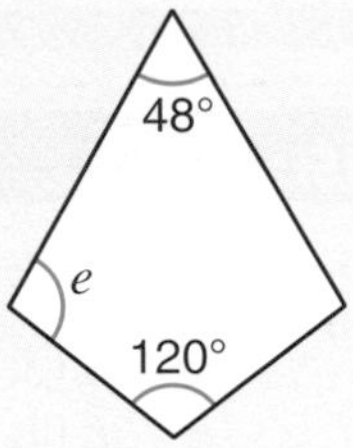

13 The diagram shows a trapezium.
Find the size of angle *a* and angle *b*.

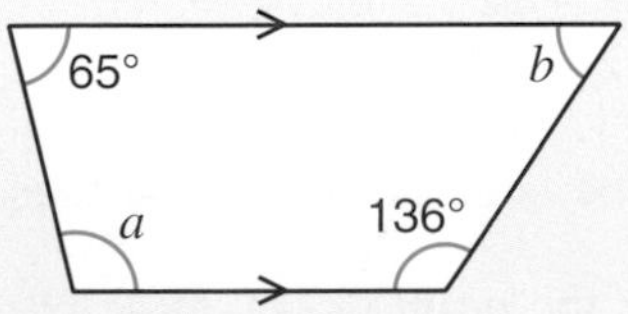

14 *WXYZ* is an isosceles trapezium.
Work out the size of angle *WXY* and angle *XYZ*.

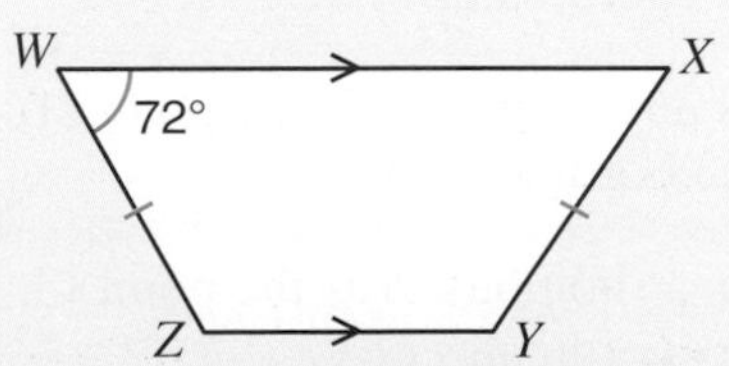

15 The following diagrams have not been drawn accurately.
Work out the size of the angles marked with letters.

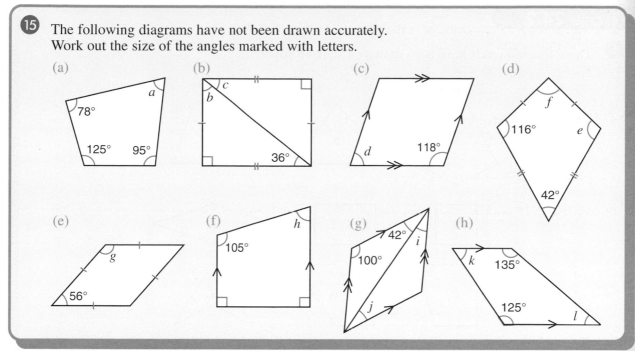

(a)   (b)   (c)   (d)

(e)   (f)   (g)   (h)

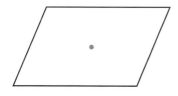
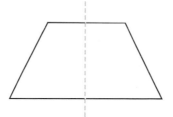

Symmetry of quadrilaterals

Remember:
A two-dimensional shape has line symmetry if the line divides the shape so that one side fits exactly over the other.
A two-dimensional shape has rotational symmetry if it fits into a copy of its own outline as it is rotated through 360°.

Parallelogram

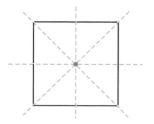

0 lines of symmetry.
Order of rotational symmetry 2.

Isosceles trapezium

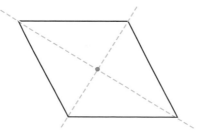

1 line of symmetry.

Rectangle

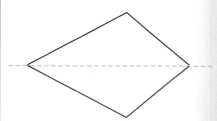

2 lines of symmetry.
Order of rotational symmetry 2.

Square

4 lines of symmetry.
Order of rotational symmetry 4.

Rhombus

2 lines of symmetry.
Order of rotational symmetry 2.

Kite

1 line of symmetry.

1 These quadrilaterals have been drawn on squared paper.

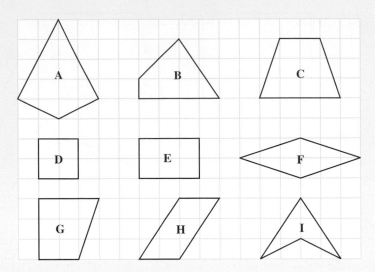

Copy and complete the table for each shape.

Shape	A	B	C	D	E	F	G	H	I
Number of lines of symmetry									
Order of rotational symmetry									

2 How many lines of symmetry has each of these quadrilaterals?

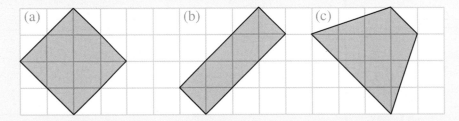

3 What is the order of rotational symmetry for each of these quadrilaterals?

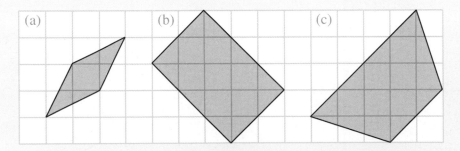

4 (a) Draw a rhombus of side 4 cm, with angles of 80° and 100°.
    (b) Mark on your diagram any lines of symmetry.
    (c) What order of rotational symmetry has the rhombus?

Perimeters of rectangles and squares

The **perimeter** is the distance round the outside of a shape.
The perimeter of a rectangle (or square) is the sum of the lengths of its four sides.

Measure the sides of this rectangle.
What is the perimeter of the rectangle?

You should find:

$AB = 3\,cm$, $BC = 4\,cm$,
$CD = 3\,cm$, $DA = 4\,cm$.
Perimeter $= 3 + 4 + 3 + 4$
        $= 14\,cm$

Area

Area is the amount of surface covered by a shape.
The standard unit for measuring area is the square centimetre, cm².
Small areas are measured using square millimetres, mm².
Large areas are measured using square metres, m², or square kilometres, km².

Activity

Finding areas by counting squares
The diagram shows two rectangles and a square drawn on centimetre-squared paper.

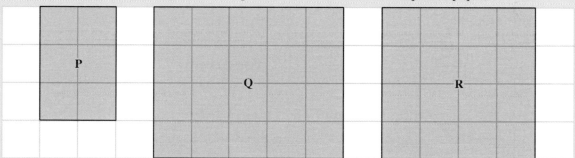

Rectangle **P** covers 6 squares.   The area of each square is 1 cm².   The area of rectangle **P** is 6 cm².

What is the area of rectangle **Q**?

What is the area of square **R**?

Can you find a rule to find the areas of rectangles and squares without having to count squares?

Look at the following diagram.
It shows a rectangle and a parallelogram drawn on centimetre-squared paper.

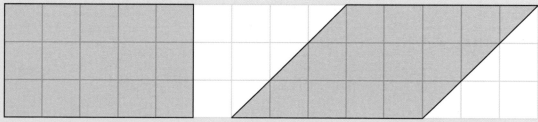

Find the area of the rectangle by counting squares.
Find the area of the parallelogram by counting squares.
What do you notice?

Can you find a rule to find the area of a parallelogram without having to count squares?

Quadrilaterals . . . . Quadrilaterals . . . .

Activity

A trapezium has been drawn on centimetre-squared paper.

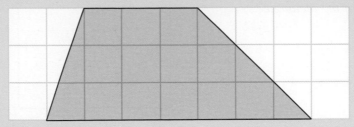

Find the area of the trapezium by counting squares.
Can you find a rule to find the area of the trapezium without having to count squares?

Area formulae

Rectangle

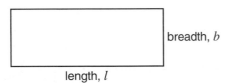

breadth, b

length, l

Area = length × breadth
  $A = lb$

Parallelogram

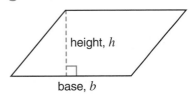

height, h

base, b

Area = base × height
  $A = bh$

Square

length, l

Area = length × breadth
In a square, length = breadth
Area = (length)²
  $A = l^2$

Trapezium

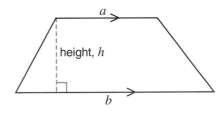

a

height, h

b

Area = half the sum of the parallel sides × perpendicular height
  $A = \frac{1}{2}(a + b)h$

The **base** is the side of the shape from which the height is measured.
The base does not have to be at the bottom of the shape.
The height of a shape, measured at right angles to the base, is called the **perpendicular height**.

EXAMPLE

 Find the perimeter and area of this rectangle.

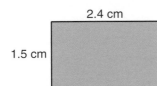

2.4 cm

1.5 cm

Perimeter = 1.5 + 2.4 + 1.5 + 2.4
           = 7.8 cm

Area = length × breadth
      = 2.4 × 1.5
      = 3.6 cm²

EXAMPLES

2 Find the area of this trapezium.

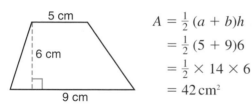

5 cm

6 cm

9 cm

$A = \frac{1}{2}(a + b)h$

$= \frac{1}{2}(5 + 9)6$

$= \frac{1}{2} \times 14 \times 6$

$= 42\,cm^2$

3 The area of a rectangular room is $17.5\,m^2$.
The room is 5 m long.
Find the width of the room.

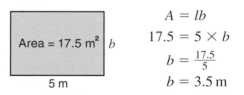

Area = 17.5 m² b

5 m

$A = lb$

$17.5 = 5 \times b$

$b = \frac{17.5}{5}$

$b = 3.5\,m$

Exercise 24.3   You should be able to do questions 1 to 4 without using a calculator.

1 These rectangles have been drawn on 1 cm squared paper.
Find (i)   the perimeter and
      (ii) the area of each rectangle.

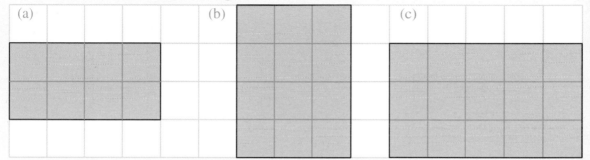

(a)

(b)

(c)

2 These squares have been drawn on 1 cm squared paper.
Find (i)   the perimeter and
      (ii) the area of each square.

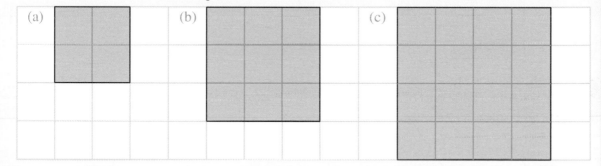

(a)

(b)

(c)

3 Four rectangles are shown.

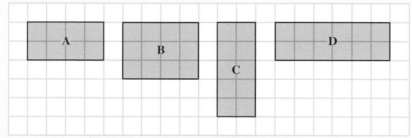

A

B

C

D

(a) Which of these rectangles have the same perimeter?
(b) Which of these rectangles have the same area?

4 These shapes have been drawn on 1 cm squared paper.
Find the area of each shape.

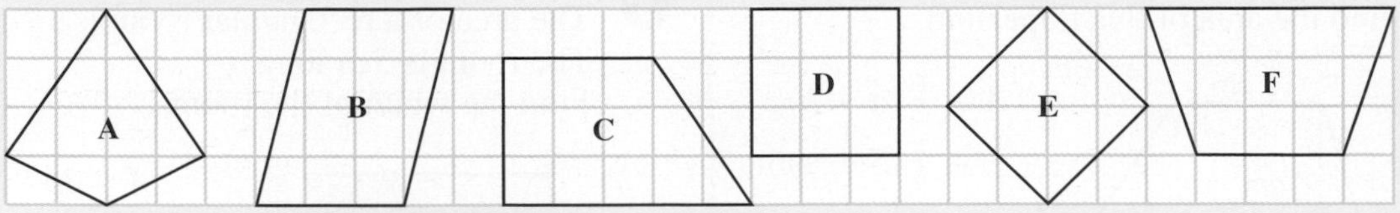

5 Calculate the perimeters of these rectangles and squares.

(a)

2.5 cm
1.5 cm

(b)

3.1 cm
0.9 cm

(c)

2.8 cm
2.8 cm

(d)

1.4 cm
1.4 cm

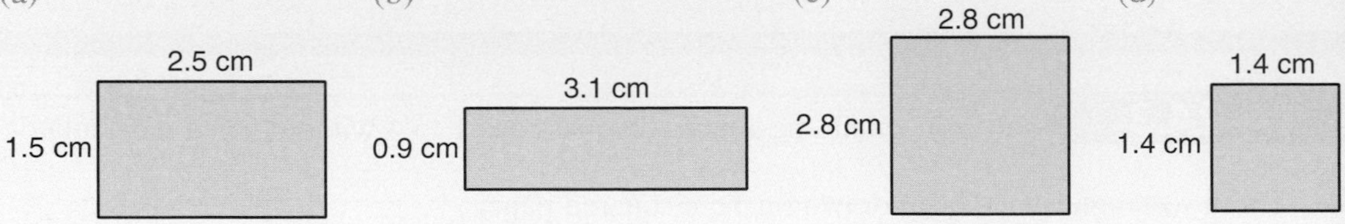

6 Calculate the areas of these rectangles.

(a)

4 cm
1.5 cm

(b)

3 cm
1.8 cm

(c)

2.5 cm
4.6 cm

(d)

3.6 cm
2.2 cm

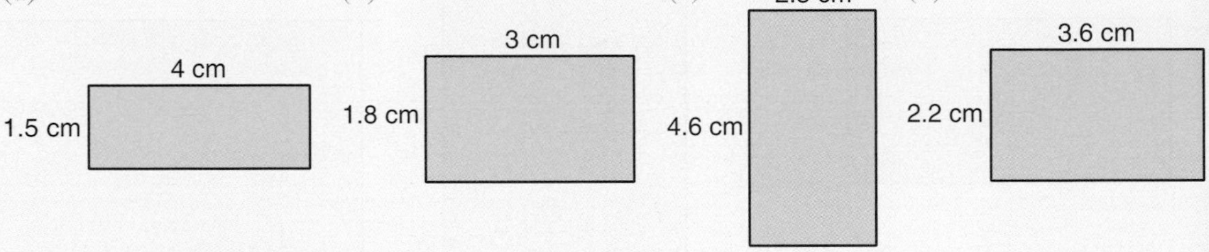

7 Calculate the areas of these squares.

(a)

7 cm

(b)

2.4 cm

(c)

4.3 cm

(d)

1.8 cm

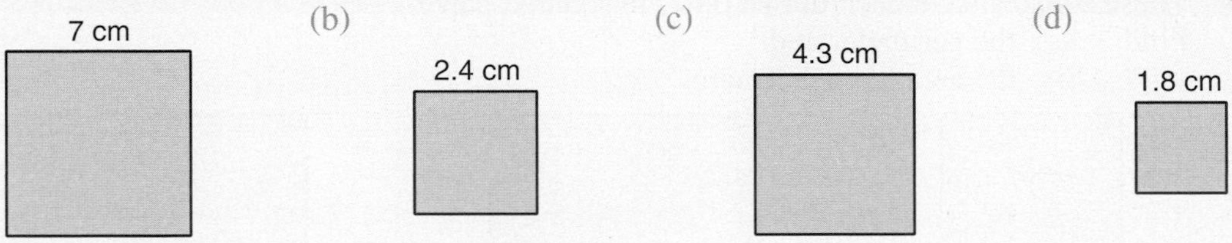

8 Calculate the areas of these parallelograms.

(a)

5 cm
6 cm

(b)

2.5 cm
4 cm

(c)

3 cm
4.5 cm

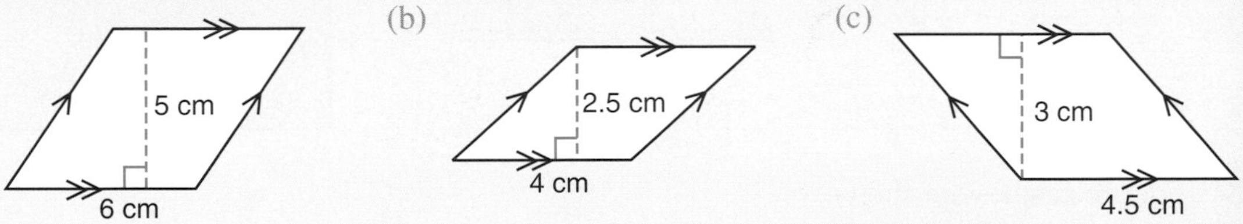

9 Calculate the areas of these trapeziums.

(a)

3 cm
4 cm
7 cm

(b)

2 cm
1.5 cm
8 cm

(c)

1.4 cm
3.6 cm 2.3 cm

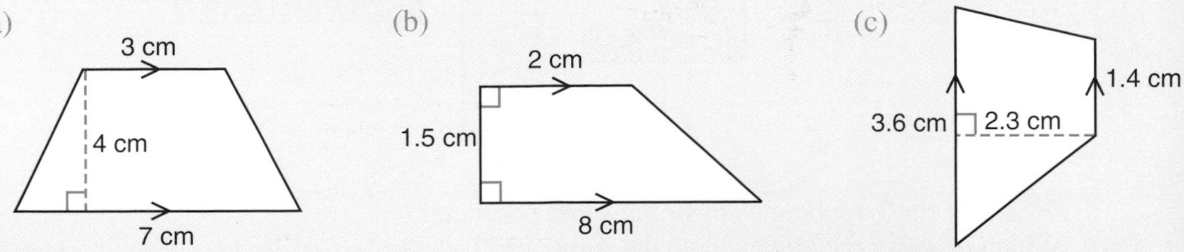

10 These rectangles each have an area of 24 cm².
Find the breadth, b, of each rectangle.

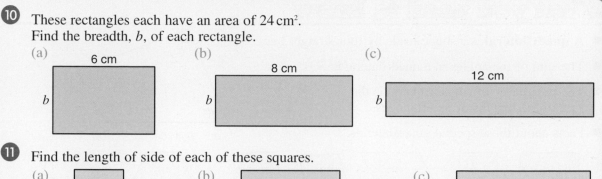

(a) 6 cm b

(b) 8 cm b

(c) 12 cm b

11 Find the length of side of each of these squares.

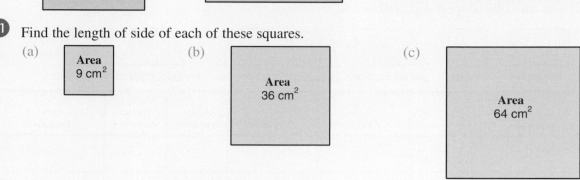

(a) Area 9 cm²

(b) Area 36 cm²

(c) Area 64 cm²

12 A carpet measuring 4 m by 4 m is placed on a rectangular floor measuring 5 m by 6 m.
What area of floor is not carpeted?

13 The diagram shows a picture in a rectangular frame.

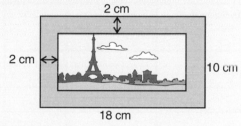

2 cm

2 cm

10 cm

18 cm

The outer dimensions of the frame are 18 cm by 10 cm.
The frame is 2 cm wide.
What is the area of the picture?

14 A rectangle has an area of 36 cm².
The length of the rectangle is 9 cm.
What is the breadth?

15 The diagram shows a square drawn inside a rectangle.
Calculate the shaded area.

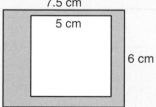

7.5 cm

5 cm

6 cm

16 The parallelogram has the same area as the square.
Calculate the height of the parallelogram.

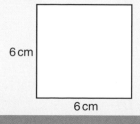

6 cm

6 cm

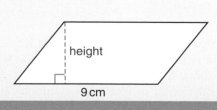

height

9 cm

- A **quadrilateral** is a shape made by four straight lines.

- The sum of the angles in a quadrilateral is 360°.

- The **perimeter** of a quadrilateral is the sum of the lengths of its four sides.

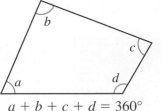

- Facts about these special quadrilaterals:

$$a + b + c + d = 360°$$

rectangle   square   parallelogram   rhombus   trapezium   isosceles trapezium   kite

Quadrilateral	Sides	Angles	Diagonals	Line symmetry	Order of rotational symmetry	Area formula
Rectangle	Opposite sides equal and parallel	All 90°	Bisect each other	2	2	$A = \text{length} \times \text{breadth}$ $A = lb$
Square	4 equal sides, opposite sides parallel	All 90°	Bisect each other at 90°	4	4	$A = (\text{length})^2$ $A = l^2$
Parallelogram	Opposite sides equal and parallel	Opposite angles equal	Bisect each other	0	2	$A = \text{base} \times \text{height}$ $A = bh$
Rhombus	4 equal sides, opposite sides parallel	Opposite angles equal	Bisect each other at 90°	2	2	$A = \text{base} \times \text{height}$ $A = bh$
Trapezium	1 pair of parallel sides					$A = \frac{1}{2}(a+b)h$
Isosceles trapezium	1 pair of parallel sides, non-parallel sides equal	2 pairs of equal angles	Equal in length	1	1*	$A = \frac{1}{2}(a+b)h$
Kite	2 pairs of adjacent sides equal	1 pair of opposite angles equal	One bisects the other at 90°	1	1*	

*A shape is only described as having rotational symmetry if the order of rotational symmetry is 2 or more.

Review Exercise 24

1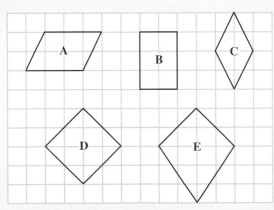

(a) Which of these shapes is a square?
(b) What special name is given to shape **E**?
(c) How many lines of symmetry has shape **B**?
(d) Which shape has no lines of symmetry?
(e) What is the order of rotational symmetry of shape **D**?

2 These shapes have been drawn on 1 cm squared paper.

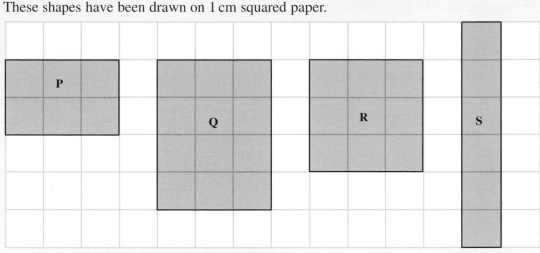

(a) (i) What is the perimeter of shape **R**?
   (ii) Which two shapes have the same perimeter?
(b) (i) What is the area of shape **Q**?
   (ii) Which two shapes have the same area?

3 Point A is shown on the diagram.
Copy the diagram.

(a) Plot the point $B(2, 4)$ and the point $C(-2, -2)$.
(b) (i) Mark the point D so that $ABCD$ is a rectangle.
   (ii) Write down the coordinates of D.

AQA

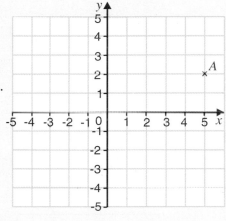

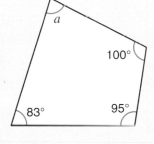

4 The diagram shows a quadrilateral.
Work out the size of angle a.

5 Jonathan is describing a quadrilateral.

Two sides are parallel but have different lengths. The other two sides are not parallel.

(a) Sketch the quadrilateral.
(b) Write down the special name of this quadrilateral.

AQA

6 $PQRS$ is a parallelogram with vertices at
$P(1, 1)$, $R(7, 8)$ and $S(5, 3)$.
A sketch of the parallelogram is shown.
Write down the coordinates of Q.

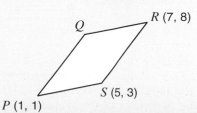

AQA

7 The diagram shows a kite *PQRS*.
   (a) How many lines of symmetry has *PQRS*?

   Angle *QRS* = 93° and angle *PQR* = 118°.
   (b) Work out the size of angle *PSR*.

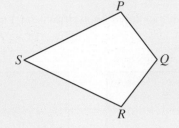

AQA

8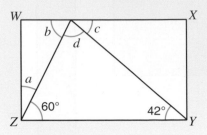

   WXYZ is a rectangle.
   Calculate angles *a*, *b*, *c* and *d*.

9 *ABCD* is a parallelogram.
   Angle *DAC* = 87° and angle *ACD* = 44°.
   Calculate angle *ABC*, giving reasons for each step of your working.

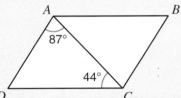

AQA

10 A square has a perimeter of 20 cm.
   Calculate the area of the square.

11 The area of a rectangle is 63 cm².
   The height of the rectangle is 9 cm.
   What is the width of the rectangle?

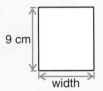

AQA

12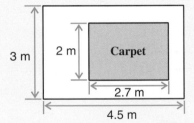

   The plan of a bedroom floor is shown.
   (a) What is the area of the bedroom floor?
   (b) What area of floor is **not** covered by the carpet?

AQA

13 *ABCD* is a trapezium.
   (a) Find the area of *ABCD*.
   (b) Find the area of triangle *BCD*.

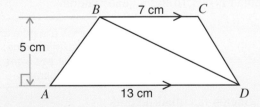

AQA

14 The length of a rectangle is twice its width.
   The area of the rectangle is 50 cm².

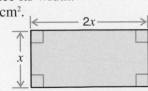

   (a) Show that the length of the rectangle is 10 cm.
   (b) Work out the perimeter of the rectangle.

AQA

CHAPTER 25

Polygons

A **polygon** is a shape made by straight lines.
A three-sided polygon is a **triangle**.

A four-sided polygon is called a **quadrilateral**.

A polygon is a many-sided shape.
Look at these polygons.

Pentagon
5 sides

Hexagon
6 sides

Heptagon
7 sides

Octagon
8 sides

Interior and exterior angles of a polygon

Angles formed by sides inside a polygon are called **interior angles**.

When a side of a polygon is extended, as shown, the angle formed is called an **exterior angle**.

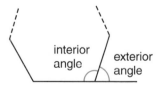

interior
angle

exterior
angle

At each vertex of the polygon: interior angle + exterior angle = 180°

Sum of the interior angles of a polygon

The diagram shows polygons with the diagonals from one vertex drawn.

P   Q   R   S

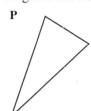 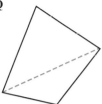 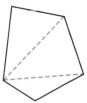 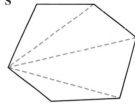

The diagonals divide the polygons into triangles.

Shape	Number of sides	Number of triangles	Sum of interior angles
P	3	1	1 × 180° = 180°
Q	4	2	2 × 180° = 360°
R	5	3	3 × 180° = 540°
S	6	4	4 × 180° = 720°

In general, for any n-sided polygon, the sum of the interior angles is $(n - 2) \times 180°$.

Sum of the exterior angles of a polygon

The sum of the exterior angles of **any** polygon is 360°.

$$a + b + c + d + e = 360°$$

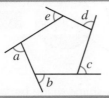

EXAMPLES

1 Find the size of angle x.

The sum of the exterior angles is 360°.
$x + 100° + 45° + 150° = 360°$
$x + 295° = 360°$
$\quad x = 360° - 295°$
$\quad x = 65°$

2 Find the sum of the interior angles of a pentagon.

To find the sum of the interior angles of a pentagon substitute $n = 5$ into $(n - 2) \times 180°$.
$(5 - 2) \times 180°$
$= 3 \times 180°$
$= 540°$

3 Find the size of the angles marked a and b.

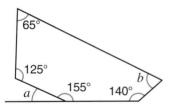

$155° + a = 180°$
(int. angle + ext. angle = 180°)
$a = 180° - 155°$
$a = 25°$

The sum of the interior angles of a pentagon is 540°.
$b + 140° + 155° + 125° + 65° = 540°$
$b + 485° = 540°$
$\quad b = 540° - 485°$
$\quad b = 55°$

Exercise 25.1   The diagrams in this exercise have **not** been drawn accurately.

1 What special name is given to each of these polygons?

(a)

(b)

(c)

(d)

2 In the diagram, ABC is a straight line.

(a) Explain why angle $x = 50°$.
(b) Show that angle $y = 60°$.

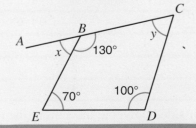

3 Work out the size of the angles marked with letters.

(a)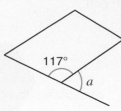

(b)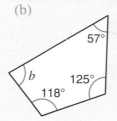

(c)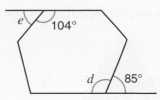

4 Work out the size of the angles marked with letters.

(a)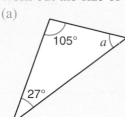

(b)

(c)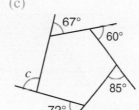

(d)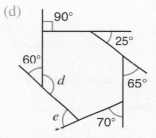

5 Work out the size of the angles marked with letters.

(a)

(b)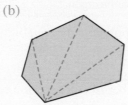

(c)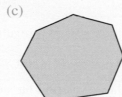

(d)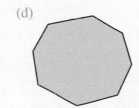

6 Work out the sum of the interior angles of these polygons.

(a) (b) (c) (d)

7 Work out the size of the angles marked with letters.

(a)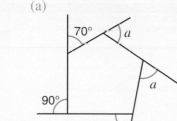

(b)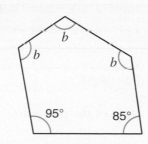

(c)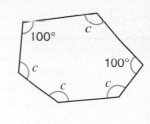

8 Work out the size of the angles marked with letters.

(a)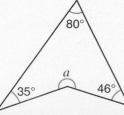

(b)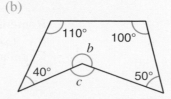

(c)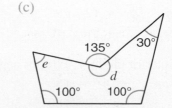

Regular polygons

A polygon with all sides equal and all angles equal is called a **regular polygon**.

These are the first four regular polygons.

Regular triangle

Regular quadrilateral

Regular pentagon

Regular hexagon

A regular triangle is usually called an **equilateral triangle**.
A regular quadrilateral is usually called a **square**.

Exterior angles of regular polygons

Measure the exterior angles of these regular polygons.
What do you find?

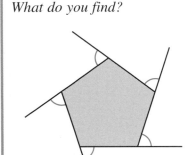
Regular pentagon

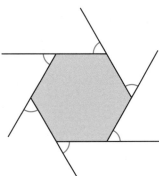
Regular hexagon

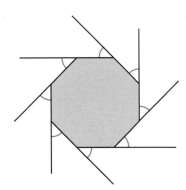
Regular octagon

You should find the exterior angles of a regular polygon are equal.

In general, for any regular n-sided polygon: exterior angle $= \dfrac{360°}{n}$

By rearranging the formula we can find the number of sides, n, of a regular polygon when we know the exterior angle.

$$n = \dfrac{360°}{\text{exterior angle}}$$

EXAMPLE

A regular polygon has an exterior angle of 30°.
(a) How many sides has the polygon?
(b) What is the size of an interior angle of the polygon?

(a) $n = \dfrac{360°}{\text{exterior angle}}$

$n = \dfrac{360°}{30°} = 12$

The polygon has 12 sides.

(b) interior angle + exterior angle = 180°

int. $\angle + 30° = 180°$

int. $\angle = 180° - 30°$

interior angle = 150°

Remember:
It is a good idea to write down the formula you are using.

266

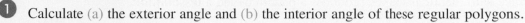

1   Calculate (a) the exterior angle and (b) the interior angle of these regular polygons.

(i)   (ii)      (iii)   (iv)

2   A regular polygon has an exterior angle of 18°. How many sides has the polygon?

3   Calculate the number of sides of regular polygons with an exterior angle of:
(a)   9°   (b)   24°   (c)   40°   (d)   60°

4   A regular polygon has an interior angle of 135°. How many sides has the polygon?

5   Calculate the number of sides of regular polygons with an interior angle of:
(a)   108°   (b)   162°   (c)   171°   (d)   90°

6   (a)   Calculate the size of an exterior angle of a regular pentagon.
(b)   What is the size of an interior angle of a regular pentagon?
(c)   What is the sum of the interior angles of a pentagon?

7   The following diagrams are drawn using regular polygons.
Work out the values of the marked angles.

(a)   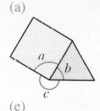   (b)   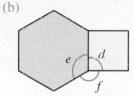   (c)   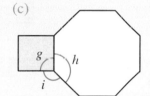   (d)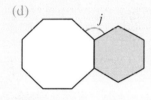

(e)   (f)   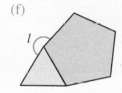   (g)   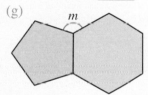   (h)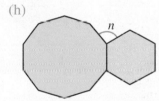

8   The following diagrams are drawn using regular polygons.
Work out the values of the marked angles.

(a)   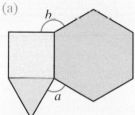   (b)   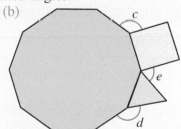   (c)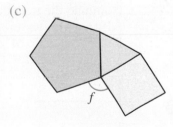

9   The following diagrams are drawn using regular polygons.
Work out the values of the marked angles.

(a)   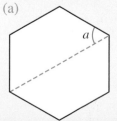   (b)   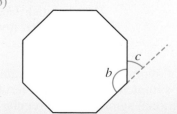   (c)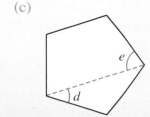

Tessellations

Covering a surface with identical shapes produces a pattern called a **tessellation**.

To tessellate the shape must not overlap and there must be no gaps.

Regular tessellations

This pattern shows a tessellation of regular hexagons.

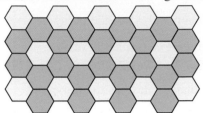

This pattern is called a **regular tessellation** because it is made by using a single regular polygon.

Exercise 25.3

1 Draw diagrams to show tessellations of these shapes.

(a)
(b)
(c)

2 Copy these regular tessellations. Continue the tessellation by drawing four more shapes.

(a)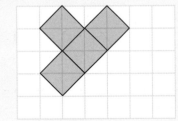
(b)

3 (a) The diagram shows part of a tessellation.
Copy the diagram.
Continue the tessellation by drawing four more triangles.
(b) All triangles tessellate.
Draw a triangle of your own, make copies,
and show that it will tessellate.

4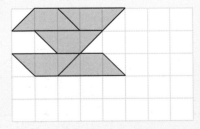

(a) The diagram shows part of a tessellation.
Copy the diagram.
Continue the tessellation by drawing
four more quadrilaterals.
(b) All quadrilaterals tessellate.
Draw a quadrilateral of your own, make copies,
and show that it will tessellate.

5 (a) Draw a regular hexagon.
(b) (i) Draw all the lines of symmetry on your diagram.
(ii) How many lines of symmetry has a regular hexagon?
(c) What is the order of rotational symmetry of a regular hexagon?

6 The diagram shows three regular polygons.

(a)

(b)

(c)

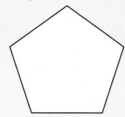  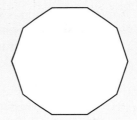

For each regular polygon find:
(i) the number of lines of symmetry,
(ii) the order of rotational symmetry.

What you need to know

- A **polygon** is a many-sided shape made by straight lines.
- A polygon with all sides equal and all angles equal is called a **regular polygon**.
- Shapes you need to know:
  A 3-sided polygon is called a **triangle**. A 5-sided polygon is called a **pentagon**.
  A 4-sided polygon is called a **quadrilateral**. A 6-sided polygon is called a **hexagon**.
- The sum of the exterior angles of any polygon is 360°.
- At each vertex of a polygon: interior angle + exterior angle = 180°
- The sum of the interior angles of an n-sided polygon is given by:
  $(n - 2) \times 180°$
- For a regular n-sided polygon: exterior angle $= \dfrac{360°}{n}$
- A shape will **tessellate** if it covers a surface without overlapping and leaves no gaps.
- Equilateral triangles, squares and hexagons can be used to make **regular tessellations**.

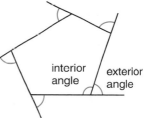

interior angle   exterior angle

 IDEAS FOR INVESTIGATION

Inscribed regular polygons

Inscribed regular polygons can be constructed by equal divisions of a circle.
To draw an inscribed regular polygon follow these steps.

Step 1 Find the exterior angle of the polygon.
$$\text{Exterior angle} = \frac{360°}{\text{number of sides}}$$

Step 2 Draw a circle.
Divide the circle into equal sectors, where the sector angles are equal to the exterior angle of the polygon.

Step 3 Join the divisions on the circumference of the circle to form the polygon.

Example Draw an inscribed regular hexagon.

Step 1

A hexagon has 6 sides.

Exterior angle $= \frac{360°}{6} = 60°$

Step 2

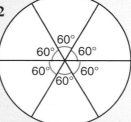

Step 3

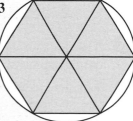

(a) Draw an inscribed equilateral triangle.
(b) Draw an inscribed square.
(c) Draw other inscribed regular polygons.
Which regular polygons are difficult to draw accurately? Explain why.

1 A regular polygon is shown.
   (a) What special name is given to this type of polygon?
   (b) Copy the diagram and draw all the lines of symmetry.
   (c) What is the order of rotational symmetry of this polygon?   AQA

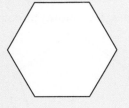

2 (a) Explain why equilateral triangles tessellate.
   (b) On dotty paper draw at least five regular hexagons in the form of a tessellation.
   (c) This tessellation is formed from regular octagons and squares.

   Calculate the size of angle $x°$.
   Show all your working.

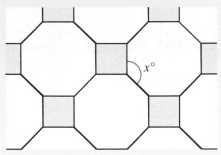

                                                                              AQA

3 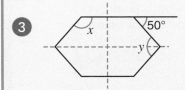   The diagram shows a hexagon.
       The hexagon has two lines of symmetry, as shown.
       Work out   (a) the value of x,   (b) the value of y.   AQA

4 The diagram shows a pentagon with just one line of symmetry.
   (a) Explain why angle x is 90°.
   (b) Calculate the size of angle y.
   (c) Write down the size of angle z.

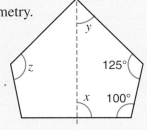

                                                                              AQA

5 (a) This diagram shows a shape with 5 sides.

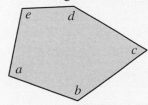

        (i) What is the mathematical name given to a shape with 5 sides?
        (ii) Explain why the shape in the diagram is **not** regular.

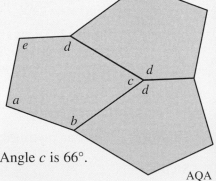

   (b) Three of the shapes are fitted together.
        Copy the diagram.
        (i) The angles a, b, c, d and e have been marked
             on one of the shapes.
             Mark the positions of angles a, b, c and e
             on the other two shapes.
        (ii) The three angles c, d, and d fit together exactly. Angle c is 66°.
             Calculate the size of angle d.   AQA

6 The diagram shows a shape which consists of two regular polygons.
   Work out the size of angle x.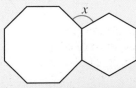

                                                                              AQA

7 Copy each of the following diagrams onto squared paper and draw six more shapes to form a tessellation.

(a)

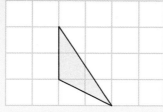

(b)

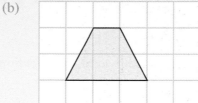

8 Work out the size of angle *x*.

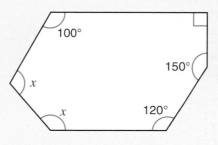

9

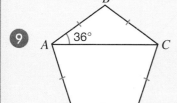

ABCDE is a regular pentagon.
Given that $\angle BAC = 36°$, explain why *AC* is parallel to *ED*.

10 The diagram shows a regular octagon.

(a) Work out the size of angle *x*, giving a reason for your answer.
(b) What name is given to the quadrilateral *ABCD*?

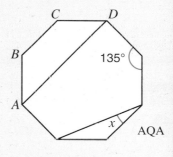

AQA

11

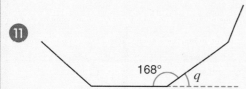

Part of a regular polygon is shown.
(a) What is the size of angle *q*?
(b) How many sides has the polygon?

12 *ABCDEFGHIJ* is a regular decagon.

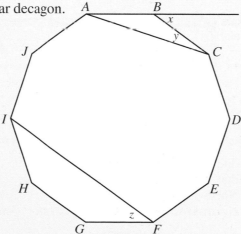

(a) (i) Calculate the size of the angle marked *x*.
   (ii) Calculate the size of the angle marked *y*.
(b) (i) What can you say about the lines *GH* and *FI*?
   (ii) Calculate the size of the angle marked *z*.

AQA

Polygons . . . Polygons . . . Polygons . . .

CHAPTER 26

Direction and Distance

Journeys are often described in terms of **direction** and **distance**.

When planning journeys we often use **maps**.

To interpret maps we need to understand:

angles in order to describe **direction**,

scales in order to find **distances**.

Compass points and **three-figure bearings** are used to describe direction.

Compass points

The diagram shows the points of the compass.

The angle between North and East is 90°.

The angle between North and North-East is 45°.

Do you know the names of any other compass points?

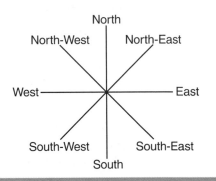

Exercise 26.1

1 A map of a cycle track is shown.

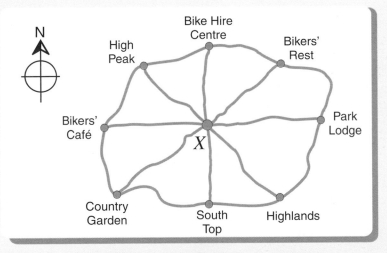

(a) (i) Which place is due North of South Top?
(ii) Which place is due West of Park Lodge?
(iii) Which place is North-East of Country Garden?
(iv) Which place is North-West of Highlands?

(b) John is at the position marked *X* on the map.
(i) Which place is South-East of John?
(ii) In which direction does he need to cycle to reach Bikers' Rest?
(iii) He cycles South-West.
Which place will he reach?

2 The diagram shows a road junction.

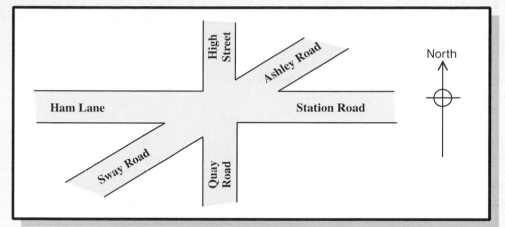

(a) A car travels from the junction along Quay Road.
In which direction is it travelling?

(b) A coach travels from the junction along Station Road.
In which direction is it travelling?

(c) A taxi drives along Station Road towards the junction.
In which direction is it travelling?

(d) Sway Road is directly opposite Ashley Road.
When traffic goes from the junction along Ashley Road it is travelling North-East.
In which direction is traffic travelling when it goes from the junction along Sway Road?

3 (a) What is the angle between North and North-West?
(b) What is the angle between South and North-West?
(c) What is the angle between South-West and South-East?
(d) What is the angle between North-West and South-West?
(e) What is the angle between South-East and North-West?

4 (a) Lyn is facing North.
She turns through an angle of 180°.
In which direction is she now facing?

(b) Tony is facing West.
He turns through an angle of 90° clockwise.
In which direction is he now facing?

5 (a) Claire is facing South.
In which direction will she face after turning clockwise through an angle of 135°?

(b) Kevin turned anticlockwise through an angle of 270°.
He is now facing South-East.
In which direction was he facing?

6 Copy and complete this table.
The first line has been done for you.

Start facing	Amount of turn	Finish facing
South	135° clockwise	North-West
North-East	90° clockwise	
West	135° anticlockwise	
	270° clockwise	East
	45° anticlockwise	West

Direction and Distance

Bearings are used to describe the direction in which you must travel to get from one place to another.
A bearing is an angle measured from the North line in a clockwise direction.
The angle, which can be from 0° to 360°, is written as a three-figure number.
Bearings which are less than 100° include noughts to make up the three figures, e.g. 005°, 087°.

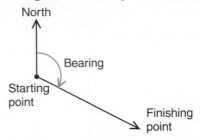

To show the direction given by a bearing

Example
The bearing of C from D is 153°.
Draw a diagram to show this information.

The bearing of C **from** D tells you that D is the starting point.
- Draw a North line. Mark and label point D on the North line.
- Using your protractor, centred on point D, mark an angle of 153° measured in a clockwise direction from the North line.
- Draw a line from D through the marked point.
  An arrow is drawn on the line to show the direction in which you must travel to get to C.

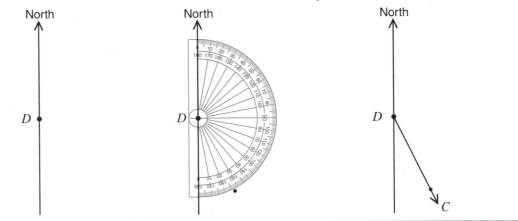

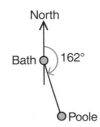

This diagram shows the positions of Bath and Poole.

The bearing of Poole from Bath is 162°.

If you are at Bath, facing North, and turn through 162° in a clockwise direction you will be facing in the direction of Poole.

Back bearings

The return bearing of Bath from Poole is called a **back bearing**.
Back bearings can be found by using parallel lines and alternate angles.

The bearing of Poole from Bath is 162°.

$a = 162°$ (alternate angles)

Required angle $= 180° + 162° = 342°$.
The bearing of Bath from Poole (the back bearing) is 342°.

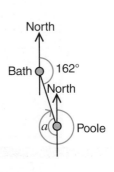

Exercise 26.2

1 The diagram shows the position of points *P* and *Q*.

   (a) Write down the three-figure bearing of *Q* from *P*.
   (b) A point *R* is due South of *Q*.
       Write down the three-figure bearing of *R* from *Q*.
   (c) A point *S* is due East of *P*.
       Write down the three-figure bearing of *S* from *P*.

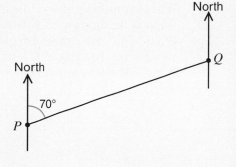

Questions 2 to 7.
Diagrams have been drawn accurately.
Use your protractor to measure angles.

2 Measure and write down the three-figure bearings of *A* from *B* in each of the following.

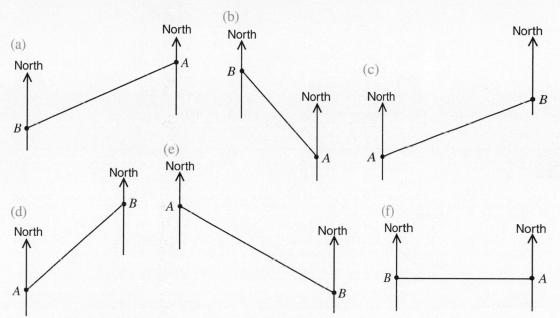

3 (a) Draw sketches to show the following information.
   (i)   The bearing of *F* from *E* is 050°.
   (ii) The bearing of *C* from *H* is 125°.
   (iii) The bearing of *K* from *Q* is 195°.
   (iv) The bearing of *L* from *B* is 260°.
   (v)   The bearing of *A* from *J* is 305°.
   (vi) The bearing of *X* from *T* is 175°.
   (b) Use your sketches to give the back bearings for each of the directions in part (a).

4 Copy the diagram.

   R is on a bearing of 100° from *P*.
   R is on a bearing of 060° from *Q*.
   Mark the position of *R* on your diagram.

5 The diagram shows the positions of three towns.

   (a) What is the bearing of *B* from *A*?
   (b) What is the bearing of *C* from *A*?
   (c) What is the bearing of *A* from *C*?
   (d) What is the bearing of *C* from *B*?
   (e) What is the bearing of *B* from *C*?

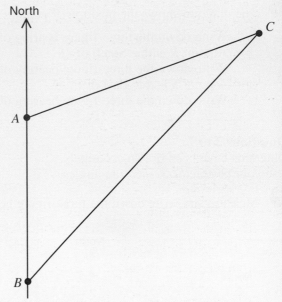

6 The diagram shows the positions of three oil rigs at *A*, *B* and *C*.

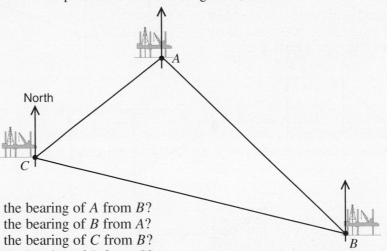

   (a) What is the bearing of *A* from *B*?
   (b) What is the bearing of *B* from *A*?
   (c) What is the bearing of *C* from *B*?
   (d) What is the bearing of *B* from *C*?
   (e) What is the bearing of *C* from *A*?
   (f) What is the bearing of *A* from *C*?

7 Copy the diagram.

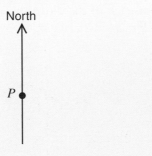

Q is on a bearing of 070° from *P*.
Q is on a bearing of 320° from *R*.
Mark the position of *Q* on your diagram.

Scale drawing

Maps and plans are scaled down representations of real-life situations.
The **scale** used in drawing a map or plan determines the amount of detail that can be shown.

The distances between different points on a map are all drawn to the same scale.
There are two ways to describe a scale.

1 A scale of 1 cm to 10 km means that a distance of 1 cm on the map represents an actual distance of 10 km.

2 A scale of 1 : 10 000 means that all distances measured on the map have to be multiplied by 10 000 to find the real distance.

EXAMPLES

1 A road is 3.7 cm long on a map.
The scale given on the map is '1 cm represents 10 km'.
What is the actual length of the road?

1 cm represents 10 km.
Scale up, so multiply.
3.7 cm represents 3.7 × 10 km = 37 km
The road is 37 km long.

2 A plan of a field is to be drawn using a scale of 1 : 500.
Two trees in the field are 350 metres apart.
How far apart will they be on the plan?

Scale down, so divide.
Distance on plan − 350 m ÷ 500
Change 350 m to centimetres.
            = 35 000 cm ÷ 500 = 70 cm
The trees will be 70 cm apart on the plan.

Exercise 26.3

1 A forest walk measures 8.4 cm on a map.
The scale given on the map is "1 cm represents 2 km".
What is the actual length of the walk in kilometres?

2 A motor-racing circuit is 9.6 km in length.
A plan of the circuit has been drawn to a scale of 1 cm to 3 km.
What is the length of the circuit on the plan?

3 On a piste map a ski run measures 3.8 cm.
The map has been drawn to a scale of 1 cm to 500 m.
What is the actual length of the ski run in metres?

4 Here is a map of an island.
  (a) Use your protractor to find:
      (i) the bearing of Q from P,
      (ii) the bearing of P from Q.

  (b) (i) Measure the distance between P and Q on the map.
      (ii) What is the actual distance between P and Q?

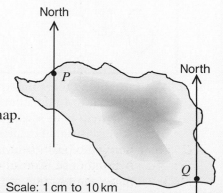

Scale: 1 cm to 10 km

5 A scale drawing of part of a golf course is shown.
The diagram has been drawn to a scale of 1 cm to 20 m.

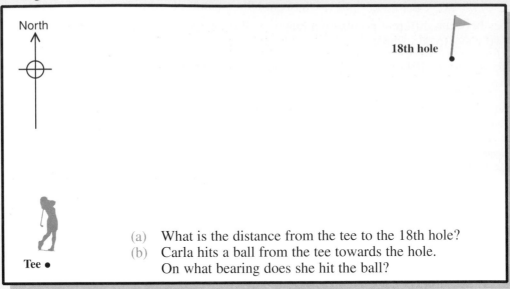

North

18th hole

(a) What is the distance from the tee to the 18th hole?
(b) Carla hits a ball from the tee towards the hole.
   On what bearing does she hit the ball?

Tee •

6 The diagram shows a group of islands. The map has been drawn to a scale of 1 cm to 5 km.

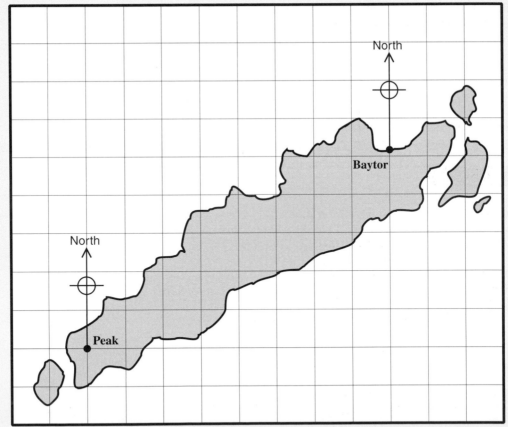

North

Baytor

North

Peak

(a) A straight road joins Baytor to Peak.
   Use the map to find the length of this road in kilometres.
(b) Copy the map onto squared paper.
   (i) A ship is on a bearing of 200° from Baytor and due East of Peak.
      Show the position of the ship on your map.
   (ii) A lighthouse is on a bearing of 035° from Peak and 253° from Baytor.
      Show the position of the lighthouse on your map.

7 The plan of a house is drawn using a scale of 1 : 100.
The lounge is 4.6 m in length.
What is the length of the lounge on the plan?

8 The diagram shows the plan of a building plot.
The plan has been drawn to a scale of 1 : 1000.

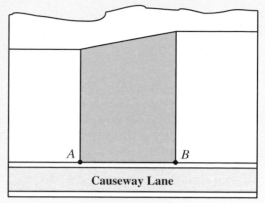

AB is the width of the plot.
(a) Measure *AB*.
(b) What is the width of the plot in metres?

9 The scale of a map is 1 : 200.
(a) On the map a house is 3.5 cm long.
     How long is the actual house?
(b) A field is 60 m wide.
     How wide is the field on the map?

10 The diagram shows the flight path of a plane between two airports.
The diagram has been drawn to a scale of 1 : 250 000.

Use the diagram to find:
(a) the actual distance between the airports, in kilometres,
(b) the bearing of *B* from *A*,
(c) the bearing of *A* from *B*.

11 The diagram shows the plan of a cross-country course.
Runners have to go round markers at *A*, *B* and *C*.
(a) What is the bearing of *B* from *A*?
(b) What is the bearing of *A* from *C*?

The plan has been drawn to a scale of 1 : 20 000.
(c) What is the distance from *A* to *C* in metres?

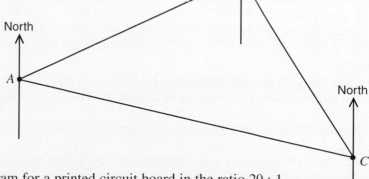

12 Claire draws a diagram for a printed circuit board in the ratio 20 : 1.
On Claire's diagram the distance between two components is 35 mm.
What is the actual distance between the components?

13 The sketch shows the positions of Ayton, Boulder, Carey and Dole.

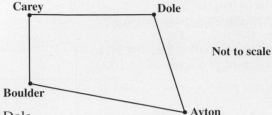

Not to scale

Carey is 12 km due West of Dole.
Carey is 8 km due North of Boulder.
Ayton is on a bearing 100° from Boulder and 160° from Dole.
By using a scale of 1 cm to 2 km, find by scale drawing the distance of Ayton from Carey.

14 A boat leaves port and sails on a bearing of 144° for 4 km.
It then changes course and sails due East for 5 km to reach an island.
Find by scale drawing:
  (a) the distance of the island from the port,
  (b) the bearing of the island from the port,
  (c) the bearing on which the boat must sail to return directly to the port.

15 A yacht sails on a bearing of 040° for 5000 m and then a further 3000 m on a bearing of 120°.
Find by scale drawing:
  (a) the distance of the yacht from its starting position,
  (b) the bearing on which it must sail to return directly to its starting position.

16 An aircraft leaves an airport, at A, and flies on a bearing of 035° for 50 km and then on a bearing of 280° for a further 40 km before landing at an airport, at B.
Find by scale drawing:
  (a) the distance between the airports,
  (b) the bearing of B from A,
  (c) the bearing of A from B.

What you need to know

- Compass points

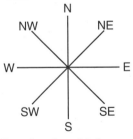

- **Bearings** are used to describe the direction in which you must travel to get from one place to another. A bearing is an angle measured from the North line in a clockwise direction.
- A bearing can be any angle from 0° to 360° and is written as a three-figure number.
- To find a bearing:
      measure angle a to find the bearing of Y from X,
      measure angle b to find the bearing of X from Y.

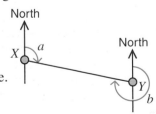

- **Scales**
  The distances between points on a map are all drawn to the same scale.
  There are two ways to describe a scale.
  1. A scale of 1 cm to 10 km means that a distance of 1 cm on the map represents an actual distance of 10 km.
  2. A scale of 1 : 10 000 means that all distances measured on the map have to be multiplied by 10 000 to find the real distance.

3-dimensional coordinates

One coordinate identifies a point on a line.
Two coordinates identify a point on a plane.
Three coordinates identify a point in space.

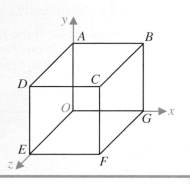

The diagram shows a cuboid drawn in 3-dimensions.

Using the axes x, y and z shown:

Point A is given as $(0, 2, 0)$.

Point B is given as $(2, 2, 0)$.

Point C is given as $(2, 2, 3)$.

Give the 3-dimensional coordinates of points D, E, F, and G.

Review Exercise 26

1 (a) The church is due North of the bridge.
Copy and complete the sentence:

> The bridge is due .... of the church.

(b) The castle is South-West of the school.
Copy and complete the sentence:

> The school is .... of the castle.

AQA

2 The diagram shows a roundabout.

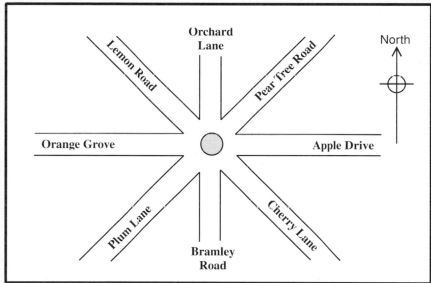

(a) Tim is at the centre of the roundabout.
He is facing due North. Which road is he looking at?
(b) Simon is walking along Apple Drive towards the roundabout.
In which direction is he walking?
(c) In which direction is traffic travelling when it goes from the roundabout along Pear Tree Road?
(d) Helen leaves the roundabout and cycles along Lemon Road.
In which direction is she cycling?

3 A bus route measures 7.3 cm on a map.
The scale given on the map is "1 cm represents 2 km".
What is the actual length of the route in kilometres?

4 The diagram shows the positions of three villages.
The diagram has not been drawn accurately.

(a) Work out the size of angle x.

(b) Give the three-figure bearing of Aire from Tees.

(c) Work out the three-figure bearing of:
  (i) Swale from Tees,
  (ii) Tees from Aire.

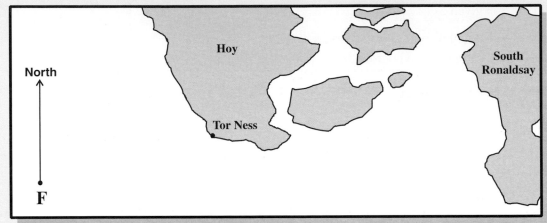

5 A map of Pentland Firth is shown. A ferry is at the point marked **F**.

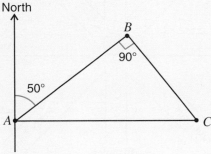

(a) Write down the bearing of Tor Ness from the ferry.
(b) The scale of the map is 1 cm = 25 km.
Calculate the distance of the ferry from Tor Ness.

AQA

6 The diagram shows the plan of a sailboard race.
The sailboards have to go round buoys at A, B and C.
Buoy B is on a bearing of 050° from buoy A.
Angle ABC is 90°.

(a) What is the bearing of A from B?
(b) What is the bearing of C from B?

The plan has been drawn to a scale of 1 : 20 000.
(c) (i) Measure AB.
  (ii) What is the distance from A to B in metres?

7 Axford is 70 km from Moxley on a bearing of 065°.
Parley is 55 km from Moxley on a bearing of 125°.
(a) By using a scale of 1 cm to 10 km, draw an accurate diagram to show the positions of Axford, Parley and Moxley.
(b) What is the bearing of Moxley from Axford?
(c) By taking measurements from your diagram work out:
  (i) the distance of Parley from Axford,
  (ii) the bearing of Parley from Axford.

A **circle** is the shape drawn by keeping a pencil the same distance from a fixed point on a piece of paper. Compasses can be used to draw circles accurately.

It is important that you understand the meaning of the following words:

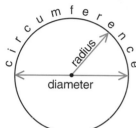

Circumference – special name used for the perimeter of a circle.

Radius – distance from the centre of the circle to any point on the circumference. The plural of radius is **radii**.

Diameter – distance right across the circle, passing through the centre point. Notice that the diameter is twice as long as the radius.

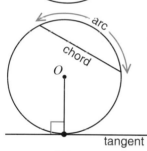

Chord – a line joining two points on the circumference. The longest chord of a circle is the diameter.

Tangent – a line which touches the circumference of a circle at one point only. A tangent is perpendicular to the radius at the point of contact.

Arc – part of the circumference of a circle.

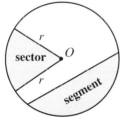

Segment – a chord divides a circle into two segments.

Sector – two radii divide a circle into two sectors.

Activity

Draw a circle with radius 2 cm.
Use thread or the edge of a strip of paper to measure the circumference of your circle.
Draw circles with radii 3 cm, 4 cm and so on.
Measure the circumference of each circle and write your results in a table.

Radius (cm)	2	3	4	5	6	7	8
Diameter (cm)							
Circumference (cm)							

What do you notice?

The Greek letter π

The circumference of any circle is just a bit bigger than three times the diameter of the circle.
The Greek letter π is used to represent this number.

We use an approximate value for π, such as 3, $3\frac{1}{7}$, 3.14, or the π key on a calculator, depending on the accuracy we require.

Circumference of a circle

The diagram shows a circle with radius r and diameter d.

The **circumference** of a circle can be found using the formulae:

$$C = \pi \times d$$

$$\text{or} \quad C = 2 \times \pi \times r$$

Remember: $d = 2 \times r$

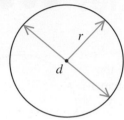

We sometimes use letters in place of words.

C is short for circumference.
r is short for radius.
d is short for diameter.

These formulae can be rearranged to find the radius, or diameter, when given the circumference.

For $C = 2\pi r$ For $C = \pi d$

$$r = \frac{C}{2\pi} \qquad\qquad d = \frac{C}{\pi}$$

EXAMPLES

1   Find the circumference of a circle with diameter 80 cm. Take π to be 3.14.
Give your answer to the nearest centimetre.

$C = \pi \times d$
   $= 3.14 \times 80$ cm
   $= 251.2$ cm
Circumference is 251 cm to the nearest centimetre.

2   A circle has circumference 37.2 cm.
Find the radius of the circle, giving your answer to the nearest millimetre. Take $\pi = 3.14$.

$C = 2 \times \pi \times r$
$37.2 = 2 \times 3.14 \times r$
$37.2 = 6.28 \times r$
   $r = \frac{37.2}{6.28}$
   $r = 5.923\ldots$
   $r = 5.9$ cm, to the nearest millimetre.

Exercise 27.1   Do not use a calculator for questions 1 to 3.

1   Estimate the circumference of these circles.
Use the approximate rule: Circumference = 3 × diameter.

(a)   (b)   (c)

4 cm

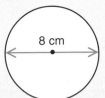

8 cm

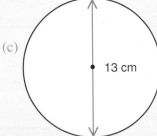

13 cm

2   Use the approximate rule to estimate the circumference of these circles.

Remember: diameter = 2 × radius

(a)   (b)   (c)

2.5 cm

5 cm

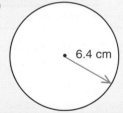

6.4 cm

3 The diagram shows the actual size of a 1p coin and a 2p coin.

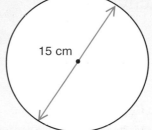

(a) Find, by measurement, the diameter of each coin.
(b) Use the approximate rule:
Circumference = 3 × diameter to estimate the circumference of each coin.

In questions 4 to 16, take π to be 3.14 or use the π key on your calculator.

4 Calculate the circumference of these circles.
Use the formula $C = \pi \times d$.

(a)

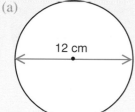

12 cm

(b)

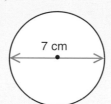

7 cm

(c)

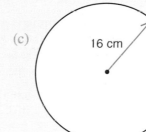

15 cm

5 Calculate the circumference of these circles.
Use the formula $C = 2 \times \pi \times r$.

(a)

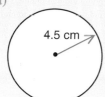

4.5 cm

(b)

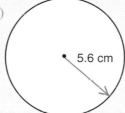

5.6 cm

(c)

16 cm

6 A circle has a diameter of 9 cm. Calculate the circumference of the circle.
Give your answer correct to the nearest whole number.

7 A circular biscuit tin has a diameter of 24 cm. What is the circumference of the tin?

8 A dinner plate has a radius of 13 cm. Calculate the circumference of the plate.
Give your answer to an appropriate degree of accuracy.

9 A circle has a radius of 6.5 cm. Calculate the circumference of the circle.
Give your answer correct to one decimal place.

10 Stan marks the centre circle of a football pitch.
The circle has a radius of 9.15 m.
What is the circumference of the circle?

11 The radius of a tractor wheel is 0.8 m.
Calculate the circumference of the wheel.

12 Two cyclists go once round a circular track.
Eddy cycles on the inside of the track which has a radius of 20 m.
Reg cycles on the outside of the track which has a radius of 25 m.
How much further does Reg cycle?

13 A circular mug has a circumference of 24 cm.
Find the radius of the mug, giving your answer to the nearest millimetre.

14 The circumference of a copper pipe is 94 mm.
Find, to the nearest millimetre, the diameter of the pipe.

15 The circumference of a bicycle wheel is 190 cm.
Find the diameter of the wheel,
giving your answer to the nearest centimetre.

16 The circumference of the London Eye is approximately 420 metres.
What is the radius? Give your answer correct to the nearest metre.

Area of a circle

Activity

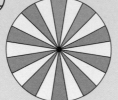 

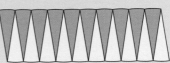
Cut out the 20 sectors.
Arrange them like this.

Draw a circle.
Divide it into 20 equal **sectors**.
Colour the sectors using two colours.

The circumference of a circle is given by $2 \times \pi \times r$.
Half of the circumference is $\pi \times r$.
So, the length of the rectangle is $\pi \times r$.
The width of the rectangle is the same as the radius of the circle, r.

Using area of a rectangle = length × breadth
   area of a circle = $\pi \times r \times r$
   area of a circle = $\pi \times r^2$

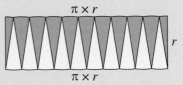
π × r

π × r

r

π × r

Take the end sector and cut it in half.
Place one piece at each end of the pattern

Area of a circle

The area of a circle can be found using the formula: $A = \pi \times r^2$

This formula can be rearranged to find the
radius when given the area of the circle.

For $A = \pi r^2$

$$r^2 = \frac{A}{\pi}$$

$$r = \sqrt{\frac{A}{\pi}}$$

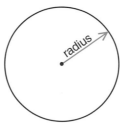
radius

EXAMPLES

1 Estimate the area of a circle with radius of 6 cm. Take π to be 3.

$A = \pi \times r \times r$
   $= 3 \times 6 \times 6$
   $= 108$ cm^2

Area is approximately 108 cm^2.

2 Calculate the area of a circle with diameter 9 cm. Use the π key on your calculator.
Give your answer correct to the nearest whole number.

$A = \pi \times r \times r$
   $= \pi \times 4.5 \times 4.5$
   $= 63.617\ldots$ cm^2

Remember: $r = \frac{d}{2}$

Area is 64 cm^2, to the nearest whole number.

EXAMPLE

3 The top of a tin of cat food has an area of 78.5 cm².
What is the radius of the tin? Take $\pi = 3.14$.

$A = \pi \times r^2$
Substitute values for A and π.
$\quad 78.5 = 3.14 \times r^2$
Solve this equation to find r.
Divide both sides of the equation by 3.14.
$\quad \frac{78.5}{3.14} = r^2$
$\quad r^2 = 25$
Take the square root of both sides.
$\quad r = 5$
The radius of the tin is 5 cm.

Exercise 27.2 Do not use a calculator for questions 1 and 2.

1 Estimate the areas of these circles. Use the approximate rule: Area = 3 × (radius)².
(a) 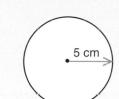 5 cm
(b)  7 cm
(c) 9 cm

2 Estimate the areas of these circles.
Take π to be 3.

Remember: Radius = $\frac{\text{diameter}}{2}$

(a) 6 cm
(b) 10 cm
(c) 16 cm

In questions 3 to 13, take π to be 3.14 or use the π key on your calculator.

3 Calculate the areas of these circles. Give your answers to the nearest whole number.
(a) 4 cm
(b) 6.5 cm
(c) 12 cm

4 Calculate the areas of these circles. Give your answers correct to one decimal place.
(a) 6.4 cm
(b) 7.6 cm
(c) 26 cm

5 The base of a paddling pool is a circle with radius 84 cm.
Find the area of the base.

6 The lid on a tin of paint is a circle of radius 72 mm.
Calculate the area of the lid.

7 A dinner plate has a diameter of 25 cm.
Find the area of the plate.
Give your answer correct to the nearest whole number.

8 A circular table has a diameter of 1.2 m.
Calculate the area of the table.
Give your answer correct to one decimal place.

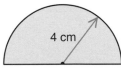

9 A circular rug has a radius of 0.5 m.
Calculate the area of the rug.
Give your answer correct to two decimal places.

10 A mug has a diameter of 8 cm.
Calculate the area of the base of the mug.
Give your answer to an appropriate degree of accuracy.

11 A circle has an area of 50 cm².
Calculate the radius of the circle.

12 A circular flower bed has an area of 40 m².
Calculate the diameter of the flower bed.
Give your answer to a suitable degree of accuracy.

13 The face of a circular disc has an area of 5.3 cm².
Calculate the radius of the disc.
Give your answer to the nearest millimetre.

Mixed questions involving circumferences and areas of circles

Some questions will involve finding the area and some the circumference of a circle.
Remember: Choose the correct formula for area or circumference.
You need to think about whether to use the radius or the diameter.

EXAMPLES

1 Gina's bicycle wheel has a radius of 24 cm.
How many complete rotations of the wheel are needed to cycle 500 cm?
Take $\pi = 3.14$.

Find the circumference of the wheel.
$C = 2 \times \pi \times r$
$\quad = 2 \times 3.14 \times 24 \text{ cm}$
$\quad = 150.72 \text{ cm}$

Number of rotations
$\quad = 500 \text{ cm} \div \text{circumference}$
$\quad = 500 \text{ cm} \div 150.72 \text{ cm}$
$\quad = 3.317\ldots$

So, 4 complete rotations of the wheel are needed.

2 Find the area of a semi-circle with radius 4 cm.
Take $\pi = 3.14$.
Give your answer to the nearest whole number.

4 cm

Begin by finding the area of a circle with radius 4 cm.

$A = \pi \times r^2$
$\quad = 3.14 \times 4 \times 4 \text{ cm}^2$
$\quad = 50.24 \text{ cm}^2$

Area of semi-circle
$\quad = \frac{1}{2} \times \text{area of circle}$
$\quad = \frac{1}{2} \times 50.24 \text{ cm}^2$
$\quad = 25.12 \text{ cm}^2$

Area of semi-circle is 25 cm², to the nearest whole number.

In this exercise take π to be 3.14 or use the π key on your calculator.

1 A tea plate has a radius of 9 cm.
(a) What is the circumference of the plate?
(b) What is the area of the plate?

2 The top of a tin of cat food is a circle of diameter 8.4 cm.
(a) Calculate the circumference of the tin.
Give your answer correct to the nearest whole number.
(b) Calculate the area of the top of the tin.
Give your answer correct to one decimal place.

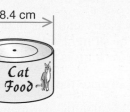

3 The front wheel on Nick's tricycle has a diameter of 18 cm.
(a) Calculate the circumference of the front wheel.
(b) How far does Nick have to cycle for the front wheel to make 20 complete turns?

4 The letter O is cut from card.
The inside radius of the letter is 3 cm.
The outside radius of the letter is 4 cm.
Calculate the area of the letter.

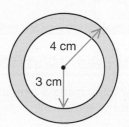

5 Which has the greater area:
a circle with radius 4 cm, or a semi-circle with diameter 11 cm?
You must show all your working.

6 The radius of a circular plate is 15 cm.
(a) What is its area?
(b) What is its circumference?
Give your answers in terms of π.

7 A circle is drawn inside a square, as shown.
The square has sides of length 10 cm.
Calculate the area of the shaded region.

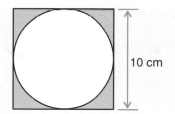

8 Thirty students join hands to form a circle.
The diameter of the circle is 8.5 m.
(a) Find the circumference of the circle.
Give your answer to the nearest metre.
(b) What area is enclosed by the circle?
Give your answer to an appropriate degree of accuracy.

9 Find the perimeter of a semi-circle with radius 6 cm.
Give your answer to the nearest whole number.

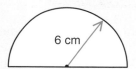

10 A wheel on Adrian's bicycle has a diameter of 66 cm.
(a) What is the circumference of the wheel?
(b) Adrian cycles a distance of 1000 cm.
How many complete rotations does the wheel make?

11 A cotton reel is a cylinder with radius 1.3 cm.
200 cm of cotton is wrapped round the reel.
How many times does it wrap round?

12 A circular flower bed has diameter 12 m.
(a) How much edging is needed to go right round the bed?
(b) The gardener needs one bag of fertiliser for each 7 m².
    How many bags of fertiliser are needed for this bed?

13 A pastry cutter is in the shape of a semi-circle.
The straight side of the semi-circle is 12 cm long.
How long is the curved side?

14 Calculate the area of a semi-circle with a diameter of 6.8 cm.

15 Calculate the perimeter of a semi-circle with a radius of 7.5 cm.

16 A circle has a circumference of 64 cm.
Calculate the area of the circle.

17 A circle has an area of 128 cm².
Calculate the circumference of the circle.

What you need to know

- A **circle** is the shape drawn by keeping a pencil the same distance from a fixed point on a piece of paper.
- Words associated with circles:

    Circumference – perimeter of a circle.

    Radius – distance from the centre of the circle to any point on the circumference. The plural of radius is **radii**.

    Diameter – distance right across the circle, passing through the centre point.

    Chord – a line joining two points on the circumference.

    Tangent – a line which touches the circumference of a circle at one point only. A tangent is perpendicular to the radius at the point of contact.

    Arc – part of the circumference of a circle.

    Segment – a chord divides a circle into two segments.

    Sector – two radii divide a circle into two sectors.

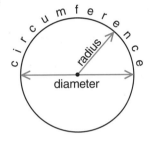

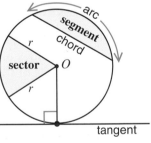

- Diameter = 2 × radius
- The **circumference** of a circle is given by: $C = \pi \times d$ or $C = 2 \times \pi \times r$
- The **area** of a circle is given by: $A = \pi \times r^2$

Review Exercise 27

Take π to be 3.14 or use the π key on your calculator.

1 A cardboard party plate has a diameter of 22 cm.
(a) Calculate the circumference of the plate.
(b) (i) What is the radius of the plate?
    (ii) Calculate the area of the plate correct to the nearest whole number.

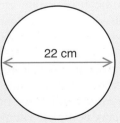

22 cm

AQA

2 A circle has a radius of 15 cm.
Calculate the area of the circle.

AQA

3 A circular flower bed has an area of 84 m².
Explain why the radius of the flower bed must be bigger than 5 m.

AQA

4 (a) Calculate the area of a circle of radius 1.2 cm.
   (b) Calculate the circumference of a circle of diameter 3.5 cm.
   AQA

5 A circular pond has a radius of 2.2 m.
   (a) Calculate the circumference of the pond.
   (b) Calculate the area of the pond.
   AQA

6  Andre is rolling a hoop along the ground.
   The hoop has a diameter of 90 cm.
   What is the minimum number of complete turns the hoop must make to cover a distance of 5 m?
   AQA

7 The diagram shows a cardboard ring.
   The inner radius is 9 cm.
   The outer radius is 11 cm.
   Calculate the area of the ring.

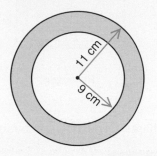

8 Suki rides to school on her bike.
   The distance is 2500 m.
   The radius of a wheel on Suki's bike is 0.3 m.
   Calculate the number of times that the wheel turns during Suki's journey.
   AQA

9 This millstone is a circle of radius 24 inches.
   A square of side length 12 inches is cut out of the centre of the circle, as shown.

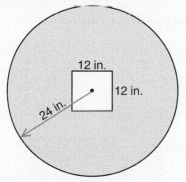

   (a) Calculate the circumference of the millstone.
   (b) Calculate the shaded area.
   AQA

10 A circular table top has radius 1.3 m.
   (a) Calculate the circumference of the top of the table.
   (b) The top of the table is varnished. 1 litre of varnish covers 5 m².
        Find, correct to one decimal place, the number of litres of varnish needed to cover the top of the table?

11 (a) Calculate the area of a circle of radius 4 cm.
        Give your answers in terms of π.
   (b) Calculate the perimeter of a semi-circle, radius 10 cm.

12 The Smith family has a circular dining table.
   The circumference of the dining table is 320 centimetres.
   (a) Calculate its radius.

   The family uses circular table mats each of radius 11 centimetres.
   (b) Work out the area covered by one of these mats.
   AQA

Areas and Volumes ● ● ● ●

Areas of shapes

In earlier chapters you found the areas of various shapes by counting squares and using formulae. Here is a reminder of some of those shapes.

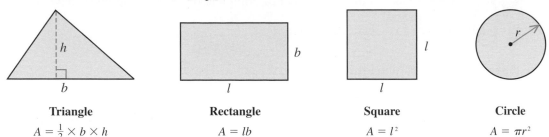

Triangle
$A = \frac{1}{2} \times b \times h$

Rectangle
$A = lb$

Square
$A = l^2$

Circle
$A = \pi r^2$

We are now going to look at shapes formed by combining rectangles, triangles, etc.

Compound shapes

Shapes formed by joining different shapes together are called **compound shapes**.
To find the area of a compound shape we must first divide the shape up into rectangles, triangles, circles, etc, and then find the area of each part.
Shapes can be divided in different ways, but they should all give the same answer.

EXAMPLE

1 Calculate the area of this shape.

This shape can be divided in different ways, as shown.

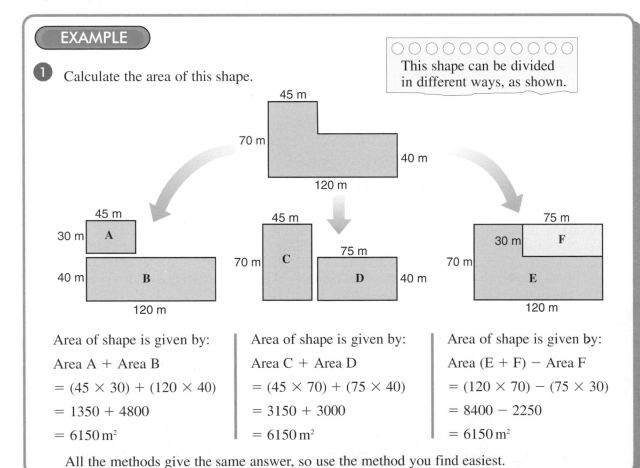

Area of shape is given by:

Area A + Area B

$= (45 \times 30) + (120 \times 40)$

$= 1350 + 4800$

$= 6150 \, \text{m}^2$

Area of shape is given by:

Area C + Area D

$= (45 \times 70) + (75 \times 40)$

$= 3150 + 3000$

$= 6150 \, \text{m}^2$

Area of shape is given by:

Area (E + F) − Area F

$= (120 \times 70) - (75 \times 30)$

$= 8400 - 2250$

$= 6150 \, \text{m}^2$

All the methods give the same answer, so use the method you find easiest.

EXAMPLE

❷ Find the area of this shape.

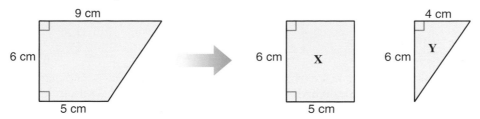

Area of shape is given by: Area X + Area Y

$$= (6 \times 5) + \left(\tfrac{1}{2} \times 6 \times 4\right)$$

$$= 30 + 12$$

$$= 42 \, cm^2$$

Exercise 28.1

Do not use a calculator for questions 1 to 3.

❶ These letters have been drawn on centimetre-squared paper.

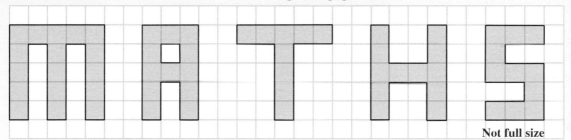

Not full size

(a)   Find the area of each letter.
(b)   Which letter has the largest area?
(c)   Which letter has the smallest area?
(d)   Which two letters have the same area?

❷ A square has sides of length 5 cm.
A square with sides of 2 cm is cut from the corner of the larger square.

(a)   What is the area of the larger square?
(b)   What is the area of the smaller square?
(c)   Find the shaded area in the diagram.

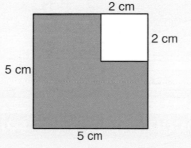

❸ Find the areas of these shapes which are made up of rectangles and squares.

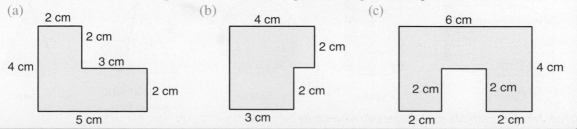

4 Find the areas of these shapes which are made up of rectangles and right-angled triangles.

(a)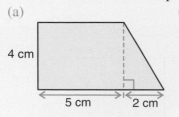

4 cm

5 cm | 2 cm

(b)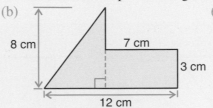

8 cm

7 cm

3 cm

12 cm

(c)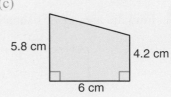

5.8 cm

4.2 cm

6 cm

5 Find the areas of these shapes which are made up of rectangles and semi-circles.

(a)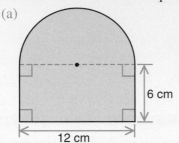

6 cm

12 cm

(b)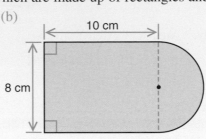

10 cm

8 cm

(c)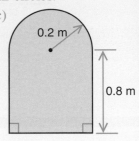

0.2 m

0.8 m

6 The diagram shows a car park. Calculate the area of the car park.

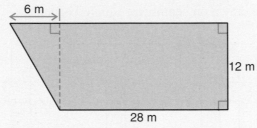

6 m

12 m

28 m

7 Find the areas of the shaded shapes.

(a)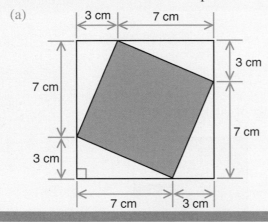

3 cm | 7 cm

3 cm

7 cm

7 cm

3 cm

7 cm | 3 cm

(b)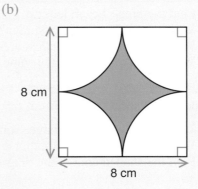

8 cm

8 cm

3-dimensional shapes (or solids)

These are all examples of 3-dimensional shapes.

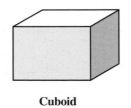 

Cuboid **Cylinder** **Sphere** **Pyramid with square base** **Cone**

What other 3-dimensional shapes do you know?

Nets

3-dimensional shapes can be made using **nets**.

This is the net of a cube.   The net can be folded to make a cube.

2-dimensional drawings of 3-dimensional shapes

Isometric drawings are used to draw 3-dimensional shapes.
Here are two isometric drawings of a cube of side 2 cm.

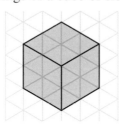 

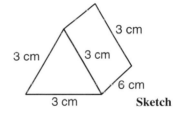

EXAMPLE

This prism is 6 cm long.
The ends are equilateral triangles with sides of 3 cm.
Draw an accurate net of the prism.

Sketch

Step 1: Draw the rectangular faces of the prism.
Each rectangle is 6 cm long and 3 cm wide.

Step 2: The ends of the prism are equilateral triangles.
The length of each side of the triangles is 3 cm.
Use your compasses to construct the
equilateral triangles.

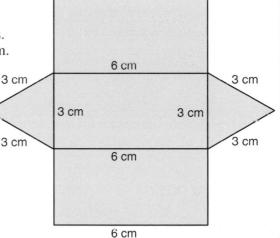

Naming parts of a solid shape

Each flat surface is called a **face**.
Two faces meet at an **edge**.
Edges of a shape meet at a corner, or point, called a **vertex**.
The plural of vertex is **vertices**.

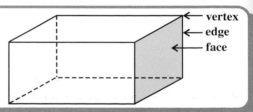

Areas and Volumes

1 This is a cuboid.

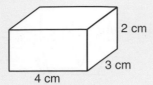

2 cm
3 cm
4 cm

Sarah has started to draw a net of the cuboid on squared paper.

(a) Copy the diagram.
(b) Complete the net of the cuboid.

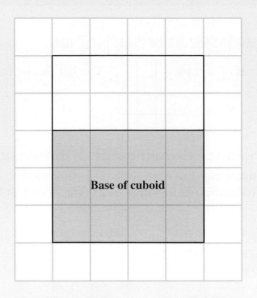

Base of cuboid

2 A cube has edges of length 2 cm.
Use squared paper to draw an accurate net of the cube.

3 (a) The diagram shows part of a net of a cube.   (b)

In how many different ways can you complete the net?
Draw each of your nets.

Explain why the diagram above is **not** the net of a cube.

4 Use squared paper to draw an accurate net of this cuboid.

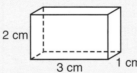

2 cm
3 cm
1 cm

5 Draw an accurate net for each of these 3-dimensional shapes.

(a)

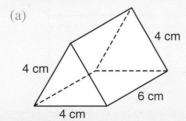

4 cm
4 cm
6 cm
4 cm

(b)

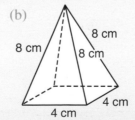

8 cm
8 cm
8 cm
4 cm
4 cm

(c)

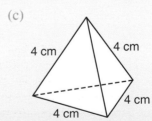

4 cm
4 cm
4 cm
4 cm
4 cm

6 The diagram shows a cuboid measuring 2 cm × 1 cm × 1 cm drawn on isometric paper.
Draw a net of the cuboid.

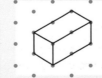

7 Draw these 3-dimensional shapes on isometric paper.
(a) A cube of side 3 cm.
(b) A 3 cm by 2 cm by 1 cm cuboid.
(c) A 3 cm by 4 cm by 5 cm cuboid.

8 There are 8 different 3-dimensional shapes which can be made using 4 linking cubes of side 1 cm.
One of them is shown.

(a) Make all the 3-dimensional shapes using four linking cubes.
(b) Draw the 3-dimensional shapes on isometric paper.

9 Look at these diagrams of 3-dimensional shapes.
Dotted lines are used to show the edges which cannot be seen when you look at the shape from one side.

(a)   (b)   (c)   (d)

  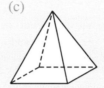 

Copy and complete this table.

	Name of shape	Number of faces	Number of vertices	Number of edges
(a)				
(b)				
(c)				
(d)				

10 The diagram shows a pyramid. A model of the pyramid is to be made using straws.
The straws are each 10 cm long.

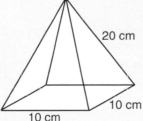

20 cm

10 cm

10 cm

(a) How many edges does the pyramid have?
(b) How many vertices does the pyramid have?
(c) How many straws are needed to make the pyramid?
(d) What is the total length of the edges of the pyramid?

Plans and Elevations

When an architect designs a building he has to draw diagrams to show what the building will look like from different directions.
These diagrams are called **plans and elevations**.

The view of a building looking from above is called the **plan**.
The views of a building from the front or sides are called **elevations**.

To show all the information about a 3-dimensional shape we often need to draw several diagrams.

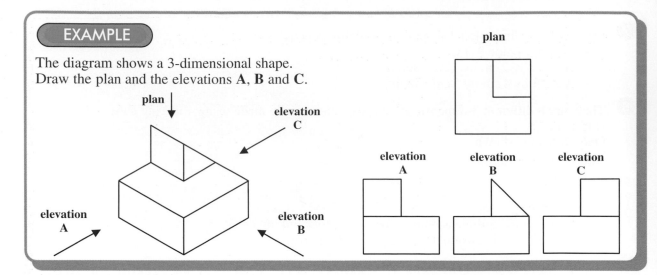

EXAMPLE

The diagram shows a 3-dimensional shape.
Draw the plan and the elevations **A**, **B** and **C**.

plan

elevation C

elevation A

elevation B

plan

elevation A

elevation B

elevation C

Exercise 28.3

1 Draw a sketch to show the plan view of each of these 3-dimensional shapes.

(a) **a staircase**

(b) **a pyramid**

(c) **a cup**

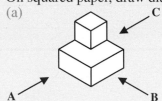

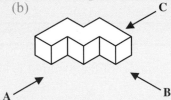

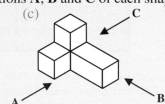

2 Each of these 3-dimensional shapes has been made using 5 linking cubes of side 1 cm.
On squared paper, draw diagrams to show the plan and the elevations **A**, **B** and **C** of each shape.

(a)   C   (b)   C   (c)   C

A   B   A   B   A   B

3 The diagram shows a plastic cylinder of height 3 cm and radius 2 cm with a hole of radius 1 cm drilled through the centre.
Draw the plan and a side elevation of the cylinder.

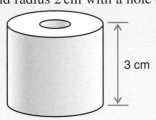

3 cm

4 The diagram shows an open box containing 3 balls of radius 2 cm.

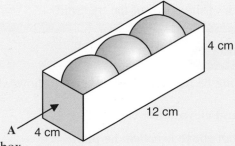

4 cm

12 cm

A   4 cm

(a) Draw a plan of the box.
(b) Draw an elevation of the box from the direction marked **A**.

5 The plans and elevations of two 3-D shapes made from linking cubes of side 1 cm are shown. Draw both of these 3-D shapes on isometric paper.

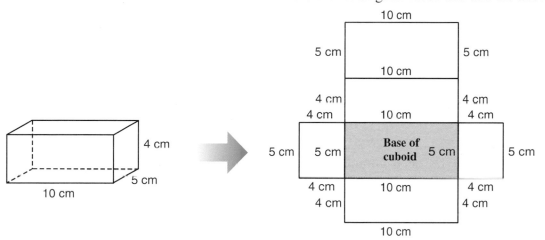

(a) Plan | Left - side elevation | Front elevation | Right - side elevation

(b) Plan | Left - side elevation | Front elevation | Right - side elevation

Surface area of a cuboid

Opposite faces of a cuboid are the same shape and size.

To find the surface area of a cuboid find the areas of the six rectangular faces and add the answers.

The surface area of a cuboid can also be found by finding the area of its net.

Volume

Volume is the amount of space occupied by a three-dimensional shape.

This **cube** is 1 cm long, 1 cm wide and 1 cm high.
It has a volume of **1 cubic centimetre**.
The volume of this cube can be written as 1 cm³.

Small volumes can be measured using cubic millimetres (mm³).
Large volumes can be measured using cubic metres (m³).

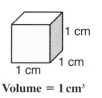

Volume = 1 cm³

Volume of a cuboid

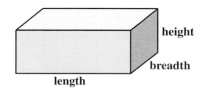

Volume of a cube:
A cube is a special cuboid in which the length, breadth and height all have the same measurement.
Volume = length × length × length
$V = l^3$

The formula for the volume of a cuboid is: Volume = length × breadth × height
This formula can be written using letters as: $V = lbh$

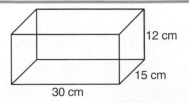

EXAMPLE

A cuboid measures 30 cm by 15 cm by 12 cm.

(a) Find the surface area of the cuboid.

(b) Find the volume of the cuboid.

(a)

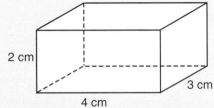

Top and bottom faces.
Each 30 cm × 15 cm.

Two side faces.
Each 15 cm × 12 cm.

Front and back faces.
Each 30 cm × 12 cm.

Surface area = $(2 \times 30 \times 15) + (2 \times 15 \times 12) + (2 \times 30 \times 12)$
$= 900 + 360 + 720$
$= 1980 \text{ cm}^2$

(b) Volume = length × breadth × height
$= 30 \text{ cm} \times 15 \text{ cm} \times 12 \text{ cm}$
$= 5400 \text{ cm}^3$

Exercise 28.4

Do not use a calculator for questions 1 to 5.

1 This is a cuboid.

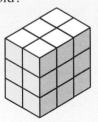

(a) Draw a net of the cuboid on one-centimetre squared paper.

(b) Calculate the area of the net.

(c) What is the surface area of the cuboid?

2 These cuboids are made using one-centimetre cubes.
What is the volume of each cuboid?

(a)   (b)   (c)

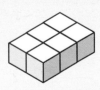

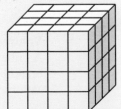

3 Large cubes are made from small cubes of edge 1 cm.

(iv)

(iii)

(ii)

(i)

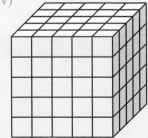

(a) How many small cubes are in each of the large cubes?

(b) What is the surface area of each large cube?

4 Calculate the volumes and surface areas of these cubes and cuboids.

(a)
3 cm
3 cm
3 cm

(b)
2 cm
3 cm
5 cm

(c)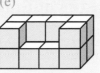
5 cm
4 cm
7 cm

5 Shapes are made using one-centimetre cubes. Find the volume and surface area of each shape.

(a)　　(b)　　(c)　　(d)　　(e)　　(f)

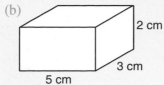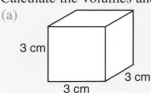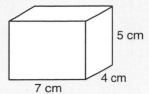

6 Calculate the volumes and surface areas of these cuboids.
Where necessary give your answer to an appropriate degree of accuracy.
(a)　3 cm by 5 cm by 10 cm.
(b)　2.4 cm by 3.6 cm by 6 cm.
(c)　18 cm by 24 cm by 45 cm.
(d)　3.2 cm by 4.8 cm by 6.3 cm.
(e)　5.8 cm by 10.6 cm by 14.9 cm.

7 A cuboid has a volume of 76.8 cm³.
The length of the cuboid is 3.2 cm. The breadth of the cuboid is 2.4 cm.
What is the height of the cuboid?

8 A cuboid has a square base of side 3.6 cm. The volume of the cuboid is 58.32 cm³.
Calculate the height of the cuboid.

9 A cuboid has a volume of 2250 cm³.
The length of the cuboid is 25 cm. The height of the cuboid is 12 cm.
Calculate the surface area of the cuboid.

10 The surface area of this cuboid is 294 cm².
Calculate the volume of a cube with the same surface area.

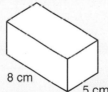
8 cm
5 cm

Prisms

These shapes are all **prisms**.

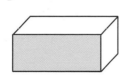

What do these 3-dimensional shapes have in common?
Draw a different 3-dimensional shape which is a prism.

Explain why these shapes are not prisms.

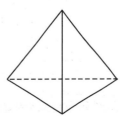

Volume of a prism

The formula for the volume of a prism is:

  Volume = area of cross-section × length.

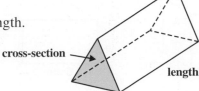

cross-section ← | length

Volume of a cylinder

A **cylinder** is a prism.

The **volume of a cylinder** can be written as:

  Volume = area of cross-section × height

  $V = \pi r^2 h$

Notice that length has been replaced by height.

EXAMPLES

Find the volumes of these prisms.

(a)

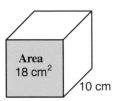

Area 18 cm²
10 cm

(b)

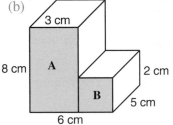

3 cm
8 cm | A | 2 cm
B | 5 cm
6 cm

(c)

5 cm
6 cm

(a)	(b)	(c)
Volume = area of cross-section × length = 18 × 10 = 180 cm³	Area A = 8 × 3 = 24 cm² Area B = 3 × 2 = 6 cm² Total area = 30 cm² Volume = 30 × 5 = 150 cm³	$V = \pi r^2 h$ = $\pi \times 5^2 \times 6$ = 471.238… = 471 cm³, correct to 3 s.f.

Exercise 28.5

1 Find the volumes of these prisms.

(a)

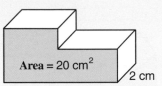

Area = 20 cm²
2 cm

(b)

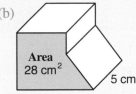

Area 28 cm²
5 cm

(c)

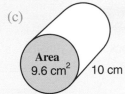

Area 9.6 cm²
10 cm

2 Calculate the shaded areas and the volumes of these prisms.

(a)

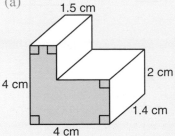

1.5 cm
2 cm
4 cm
1.4 cm
4 cm

(b)

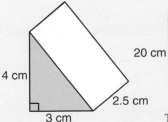

4 cm
2.5 cm
3 cm

(c)

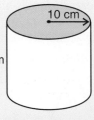

10 cm
20 cm
Take π to be 3.14

(d)

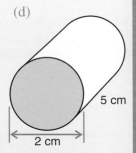

5 cm
2 cm
Take π to be 3.14

3 Find the volumes of these prisms.
Where necessary take π to be 3.14 or use the π key on your calculator.

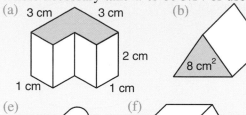

(a) 3 cm 3 cm 2 cm 1 cm 1 cm

(b) 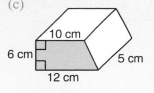 8 cm² 7 cm

(c) 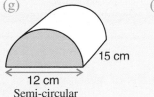 10 cm 6 cm 5 cm 12 cm

(d) 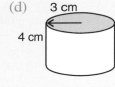 3 cm 4 cm

(e) 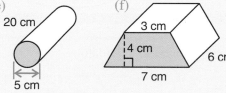 20 cm 5 cm

(f) 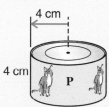 3 cm 4 cm 6 cm 7 cm

(g) 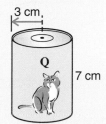 15 cm 12 cm
Semi-circular
cross-section

(h) 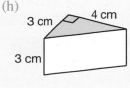 3 cm 4 cm 3 cm

4 Which tin holds more cat food?
Show your working.

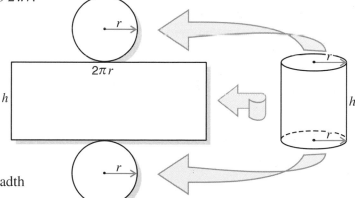

4 cm 4 cm **P**

3 cm **Q** 7 cm

5 Sylvia says, "A cylinder with a radius of 5 cm and a height of 10 cm has the same volume as a cylinder with a radius of 10 cm and a height of 5 cm."
Is she right? Explain your answer.

Surface area of a cylinder

The top and bottom of a cylinder are circles.
The curved surface of a cylinder is a rectangle.

The rectangle has the same height, h, as the cylinder.
The length of the rectangle must be just long enough to "wrap around" the circle.
The lid of the cylinder has radius r and circumference $2\pi r$.
So, the length of the rectangle is also $2\pi r$.

r

$2\pi r$

h

r

h

r

Area of lid = πr^2
Area of base = πr^2
Area of lid and base = $2\pi r^2$
Area of rectangle = length $\times$ breadth
= $2\pi r \times h$
= $2\pi rh$

If a cylinder has radius, r, and height, h,
then the formula for the surface area is:

Surface area = $2\pi r^2 + 2\pi rh$

Area of the top and bottom   Area of the rectangle

The formula for the surface area
is sometimes given as:
Surface area = $2\pi r(r + h)$

Areas and Volumes

28

303

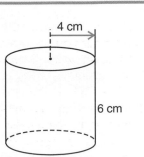

Find the surface area of a cylinder with radius 4 cm and height 6 cm.
Use the π key on your calculator.

$$\begin{aligned}
\text{Area} &= 2\pi rh + 2\pi r^2 \\
&= 2 \times \pi \times 4 \times 6 + 2 \times \pi \times 4^2 \\
&= 150.796 \ldots + 100.530 \ldots \\
&= 251.327 \ldots \\
&= 251.3 \text{ cm}^2, \text{ correct to 1 d.p.}
\end{aligned}$$

Exercise 28.6

Take π to be 3.14 or use the π key on your calculator.

1 Find the surface areas of these cylinders.

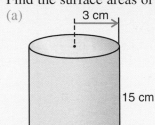
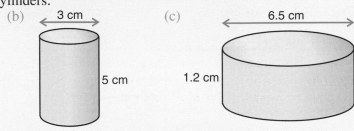

(a) 3 cm, 15 cm

(b) 3 cm, 5 cm

(c) 6.5 cm, 1.2 cm

2 Show that the curved surface area of this can is approximately 75 cm².

3 cm, 4 cm, BEANZ

3 50 cm

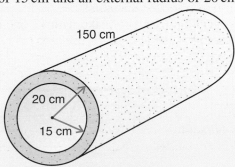

40 cm

A bucket is in the shape of a cylinder.
(a) Calculate the area of the bottom of the bucket.
(b) Calculate the curved surface area of the bucket.

4 A cylinder has a radius of 3.6 cm. The length of the cylinder is 8.5 cm.
Calculate the total surface area of the cylinder.
Give your answer to an appropriate degree of accuracy.

5 A concrete pipe is 150 cm long.
It has an internal radius of 15 cm and an external radius of 20 cm.

150 cm

20 cm

15 cm

Calculate, giving your answers to 3 significant figures,
(a) the area of the curved surface inside of the pipe,
(b) the curved surface area of the outside of the pipe.

6 A cylinder is 15 cm high. The curved surface area of the cylinder is 377 cm².
Calculate the radius of the cylinder.

What you need to know

- Shapes formed by joining different shapes together are called **compound shapes**.
  To find the area of a compound shape we must first divide the shape up
  into rectangles, triangles, circles, etc, and find the area of each part.
  Add the answers to find the total area.

  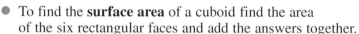

- **Faces**, **vertices** (corners) and **edges**.
  For example, a cube has 6 faces, 8 vertices and 12 edges.

- A **net** can be used to make a solid shape.

-   **Isometric paper** is used to make 2-dimensional drawings
  of 3-dimensional shapes.

- **Plans and Elevations.**
  The view of a 3-dimensional shape looking from above is called a **plan**.
  The view of a 3-dimensional shape from the front or sides is called an **elevation**.

- **Volume** is the amount of space occupied by a 3-dimensional shape.

- The formula for the volume of a **cuboid** is:
  Volume = length × breadth × height
  $$V = l \times b \times h$$

- To find the **surface area** of a cuboid find the area
  of the six rectangular faces and add the answers together.

- Volume of a **cube** is: Volume = (length)³
  $$V = l^3$$

- If you make a cut at right angles to the length of a **prism**
  you will always get the same cross-section.

- Volume of a prism = area of cross-section × length

-  A **cylinder** is a prism.
  Volume of a cylinder is: $V = \pi \times r^2 \times h$
  Surface area of a cylinder is: Surface area = $2\pi r^2 + 2\pi rh$

IDEAS FOR INVESTIGATION

(a) This cuboid is made using 12 one-centimetre cubes.

Find the surface area of the cuboid.
What other cuboids can you make with 12 one-centimetre cubes?
Which cuboid has the largest surface area?
Which cuboid has the smallest surface area?

(b) Investigate the surface areas of cuboids made with 24 one-centimetre cubes.
What do you notice?
Investigate further.

Review Exercise 28   Do not use a calculator for questions 1 to 8.

1 Solids **A**, **B** and **C** are shown.

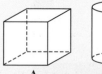

A   B   C

Write down the mathematical name for each of the solids.

AQA

2 This shape is a triangular prism.
How many faces, edges and vertices has the triangular prism?

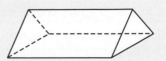

3 Which of these shapes is the net of a cube?

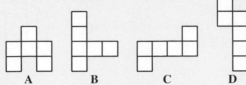

A   B   C   D

4 The diagram shows an open box.

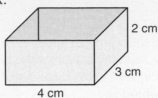

2 cm

3 cm

4 cm

(a) Draw an accurate net of the open box on one-centimetre squared paper.
(b) Calculate the total outside surface area of the open box.

AQA

5 Jane uses one centimetre cubes to make a model. Two views of her model are shown.
(a) How many cubes has Jane used?

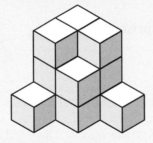 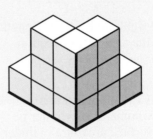

Jane places her model inside a box.
The box is in the shape of a cube with sides 3 cm.
(b) Calculate the volume of space left in the box, stating your units.

AQA

6 This 3-dimensional shape has been made using linking cubes of side 1 cm.

On squared paper, draw diagrams to show:
(a) the plan of the shape,
(b) the elevation from **X**.

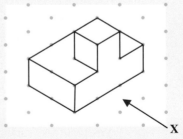

X

7 A length of tape, 1 cm wide, is used to make a letter G.
Find the area of the letter.

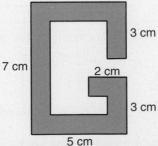

3 cm

7 cm

2 cm

3 cm

5 cm

8 Mary has some cubes of side 1 centimetre.
She makes this shape with her cubes.
(a) What is the volume of this shape?

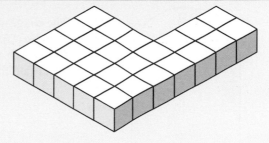

(b) 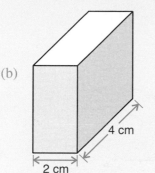 24 of Mary's cubes just fill this box.
What is the height of the box?

4 cm
2 cm

(c) Copy the diagram onto isometric paper and
complete a full size drawing of the box.

4 cm

(d) 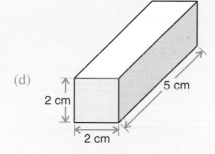 Here is another box.
Will 24 of Mary's cubes fit into it?
Give a reason for your answer.

2 cm
5 cm
2 cm

AQA

9 The diagram shows the shape of a building plot.
Calculate the area of the plot.

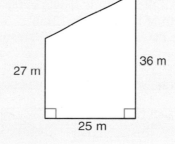

27 m
36 m
25 m

10 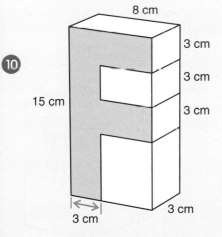 A block of wood measures 15 cm by 8 cm by 3 cm.
A letter F is cut out of the block of wood, as shown.

8 cm
3 cm
3 cm
15 cm
3 cm
3 cm
3 cm

(a) Calculate the shaded area.
(b) Calculate the volume of the letter F.

11 This block of jelly measures 2 cm by 6 cm by 4.5 cm.

(a) What is the volume of this block of jelly?
(b) A shopkeeper buys a box of these jellies.
The box measures 6 cm by 12 cm by 9 cm.
How many blocks of jelly fill the box?

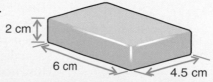

2 cm
6 cm
4.5 cm

AQA

12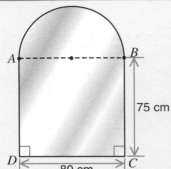

The diagram shows a window.
The arc AB is a semi-circle.
$BC = AD = 75$ cm, $DC = 80$ cm.
Calculate the area of the window.

13 The diagram shows a prism.
The cross section of the prism is a trapezium.
Calculate the volume of the prism.

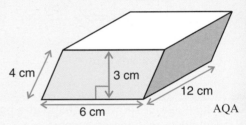

14 The diagram shows a cuboid which is just big enough to hold six tennis balls.

Each tennis ball has a diameter of 6.8 cm.
Calculate the volume of the cuboid.

15 Ice cream is sold in a box that is the shape of a prism.
The ends are parallelograms.
The size of the prism is shown in the diagram.
The length of the prism is 12 cm.
Calculate the volume of the ice cream in the box.

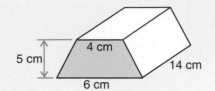

16 The diagram shows a block of wood. The block is a cuboid measuring 8 cm by 13 cm by 16 cm.

A cylindrical hole of radius 5 cm is drilled through the block of wood.
Find the volume of wood remaining.

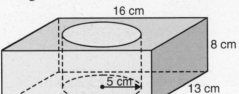

17 Lentil soup is sold in cylindrical tins.
Each tin has a base radius of 3.8 cm and a height of 12.6 cm.
Calculate the surface area of a tin.
Give your answer to a suitable degree of accuracy.

18

"Bradley's Soup" is canned by a small family business.
Each morning they make 200 litres of soup.
This is put in cylindrical tins, each of which is 8.4 cm high and has a diameter of 7.0 cm.
How many of these tins can be filled from the 200 litres of soup?

CHAPTER 29

Loci and Constructions

Following rules

Three students are given rules to follow.

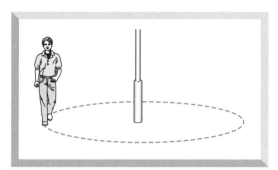

John
Walk so that you are always 2 metres from the lamp post.

His path is a circle, radius 2 metres.

Hanif
Walk along a straight road.
You must keep 30 cm from the edge of the road and stay on the pavement.

His path is a straight line.

Sarah
Start from the corner of the lawn.
Walk across the lawn so that you are always the same distance from two sides.

Her path is a straight line.
The line cuts the angle in two.

Locus

The path of a point which moves according to a rule is called a **locus**.
If we talk about more than one locus we call them **loci**.

EXAMPLES   Draw sketches to show the loci of John, Hanif and Sarah.

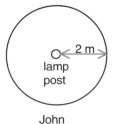

2 m
lamp post

John

road

30 cm   pavement

Hanif

45°
45°

Sarah

1. Adam goes down this slide. Make a sketch of the slide as viewed from the side and show the locus of Adam's head as he goes down the slide.

2. The diagram shows part of a rectangular lawn.
   Starting from the wall, Sally walks across the lawn so that she is always the same distance from both hedges. Draw a sketch to show the locus of Sally's path.

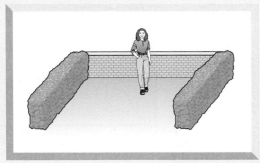

3. *PQRS* is a square of side 8 cm.
   A point *X* is inside the square.
   X is less than 8 cm from *P*.
   X is nearer to *PQ* than to *SR*.
   Make a sketch showing where *X* could be.

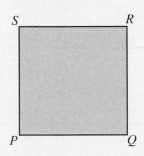

4.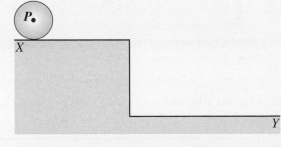

   A ball is rolled down a step.
   Copy the diagram, and sketch the locus of *P*, the centre of the ball, as it rolls from *X* to *Y*.

5. A wire is stretched between two posts.
   A ring slides along the wire and a dog is attached to the ring by a rope.
   Make a sketch to show where the dog can go.

6. A point *P* is 1 cm from this shape.
   Copy the diagram, and draw an accurate locus of **all** the positions of *P*.

Accurate constructions

Sometimes it is necessary to construct loci accurately.
You are expected to use only a ruler and compasses.
Here are the methods for two constructions.

To draw the perpendicular bisector of a line

This means to draw a line at right angles to a given line dividing it into two equal parts.

1 Draw line *AB*.

2 Open your compasses to just over half the distance *AB*. Mark two arcs which cross at *C* and *D*.

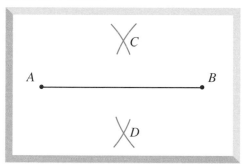

3 Draw a line which passes through the points *C* and *D*.

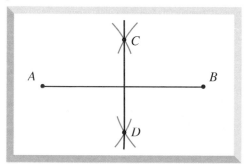

This line is the locus of a point which is the same distance from *A* and *B*.
Points on the line *CD* are **equidistant** (the same distance) from points *A* and *B*.
The line *CD* is at right angles to *AB*.
CD is sometimes called the **perpendicular bisector** of *AB*.

To draw the bisector of an angle

This means to draw a line which divides an angle into two equal parts.

1 Draw the angle *A*.
Use your compasses, centre *A*, to mark points *B* and *C* which are the same distance from *A*.

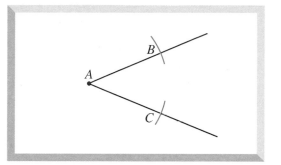

2 Use points *B* and *C* to draw equal arcs which cross at *D*.

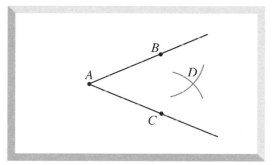

3 Draw a line which passes through the points *A* and *D*.

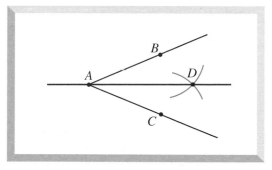

This line is the locus of a point which is the same distance from *AB* and *AC*.
Points on the line *AD* are **equidistant** from the lines through *AB* and *AC*.
The line *AD* cuts angle *BAC* in half.
AD is sometimes called the **bisector** of angle *BAC*.

1 Mark two points, *A* and *B*, 10 cm apart. Construct the perpendicular bisector of *AB*.

2 Use a protractor to draw an angle of 60°. Construct the bisector of the angle.
Check that both angles are 30°.

3 Draw a triangle in the middle of a new page.
Construct the perpendicular bisectors of all three sides. They should meet at a point, *Y*.

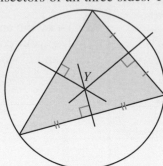

Put the point of your compasses on *Y* and draw the circle which goes through all three vertices of the triangle. This construction is sometimes called the **circumscribed circle of a triangle**.

4 Draw another triangle on a new page. Bisect each angle of the triangle.
The bisectors should meet at a point, *X*.

Put the point of your compasses on point *X* and draw
the circle which just touches each side of the triangle.
This construction is sometimes called the
inscribed circle of a triangle.

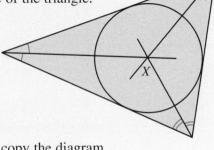

5 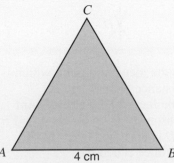 Using a circle of radius 4 cm, copy the diagram.
Draw the perpendicular bisectors of the chords *WX* and *YZ*.
What do you notice about the perpendicular bisectors?

> The perpendicular bisector of a chord always
> passes through the centre of a circle.

6 Two trees are 6 metres apart.
Alan walks so that he is always an equal distance from each tree.
Draw a scale diagram to show his path.

7 *ABC* is an equilateral triangle with sides 4 cm.

A point *X* is inside the triangle.
It is nearer to *AB* than to *BC*.
It is less than 3 cm from *A*.
It is less than 2 cm from *BC*.
Shade the region in which *X* could lie.

8 Draw a rectangle *ABCD* with   *AB* = 6 cm   and   *AD* = 4 cm.
(a) Mark, with a thin line, the locus of a point which is 1 cm from *AB*.
(b) Mark, with a dotted line, the locus of a point which is the same distance from *A* and *B*.
(c) Mark, with a dashed line, the locus of a point which is 3 cm from *A*.

9 Draw a right-angled triangle with sides of 6 cm, 8 cm and 10 cm.

A point X is in the triangle.
It is 4 cm from B.
It is the same distance from A and B.

Mark accurately, the position of X.

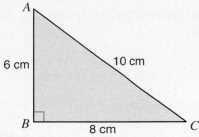

10 Copy the diagram and draw the locus of a point which is the same distance from PQ and RS.

11 Triangle ABC is isosceles with $AB = BC = 7$ cm   and $AC = 6$ cm.
  (a) Construct triangle ABC.
  (b) Point X is equidistant from A, B and C. Mark accurately the position of X.

12 (a) The diagram shows the sketch of a field.
      Make a scale drawing of the field using 1 cm to represent 100 m.

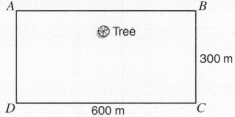

  (b) A tree is 400 m from corner D and 350 m from corner C.
      Mark the position of the tree on your drawing.
  (c) John walks across the field from corner D, keeping the same distance from AD and CD.
      Show his path on your diagram.
  (d) Does John walk within 100 m of the tree in crossing the field?

13 Part of a coastline is shown.

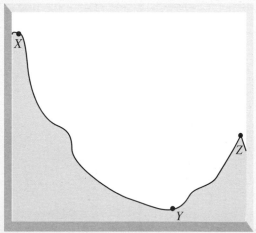

A boat is: (i) equidistant from X and Z,   and   (ii) equidistant from XY and YZ.
Copy the diagram and mark the position of the boat.

To draw the perpendicular from a point to a line

This means to draw a line at right angles to a given line, from a point that **is not on the line**.

1 Open your compasses so that from point *A* you can mark two arcs on the line *PQ*.

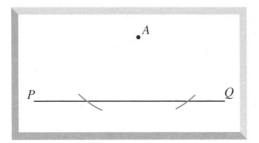

2 Use points *B* and *C* to draw equal arcs which cross at *D*.

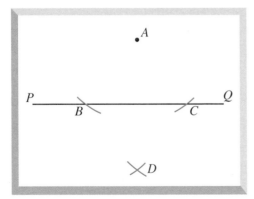

3 Draw a line from *A* to *D*.

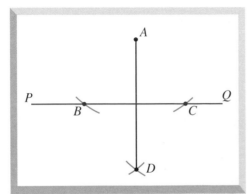

The line *AD* is perpendicular (at right angles) to the line *PQ*.

To draw the perpendicular from a point on a line

This means to draw a line at right angles to a given line, from a point that **is on the line**.

Keep your compasses at the same setting whilst doing this construction.

1 From *A*, draw an arc which cuts the line *PQ*.

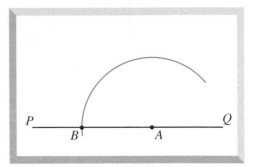

2 From *B*, draw an arc to cut the first arc at *C*. Then from *C*, draw an arc to cut the first arc at *D*.

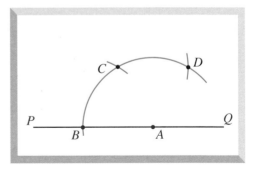

3 From *C* and *D*, draw arcs to meet at *E*. Draw the line *AE*.

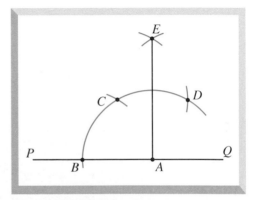

The line *AE* is perpendicular (at right angles) to the line *PQ*.

1 Draw a line *PQ*, 8 cm long.
Mark a point *A*, about 5 cm above the line.
Draw the line which passes through *A* and is perpendicular to line *PQ*.

2 Draw a line *PQ*, 8 cm long.
Mark a point *A*, somewhere on the line.

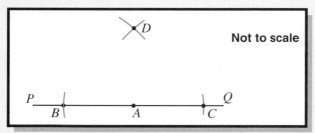

Not to scale

(a) Using your compasses, mark points *B* and *C*, which are 3 cm from *A* on the line *PQ*.
(b) Set your compasses to 5 cm.
Draw arcs from *B* and *C* which intersect at *D*.
(c) Draw the line *AD*.

This is another method of constructing a perpendicular from a given point on a line.

3 Using ruler and compasses only, make an accurate drawing of the triangle shown in this sketch.

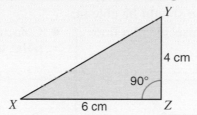

4 (a) Make an accurate drawing of this triangle.

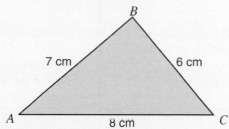

(b) The **altitude** of the triangle is a line perpendicular to a side which passes through the opposite corner of the triangle.
Draw the altitude of the triangle *ABC* which passes through point *B*.

5 (a) Copy the diagram.

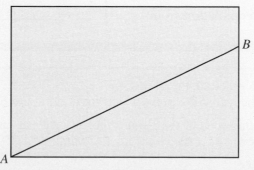

(b) Mark all points inside the rectangle that are less than 2 cm from the line *AB*.

Loci and Constructions

6 Draw a line *PQ*, 5 cm long.
Using compasses, draw an arc centre *P* to cut *PQ* at *X*.
With your compasses at the same setting draw another arc, centre *X*, to cut the first arc at *Y*.
Draw a line through *PY*.
Measure angle *XPY*. What do you find?

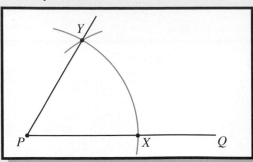

Show how you can use this construction to draw angles of 30° and 120°.

What you need to know

- The path of a point which moves according to a rule is called a **locus**.
- The word **loci** is used when we talk about more than one locus.

Using a ruler and compasses you should be able to:

- Construct the **perpendicular bisector of a line**.

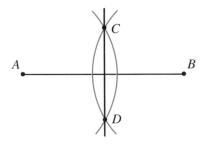

Points on the line *CD* are **equidistant** from the points *A* and *B*.

- Construct the **bisector of an angle**.

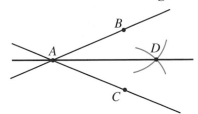

Points on the line *AD* are **equidistant** from the lines *AB* and *AC*.

- Construct the **perpendicular from a point to a line**.

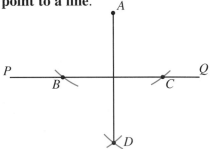

- Construct the **perpendicular from a point on a line**.

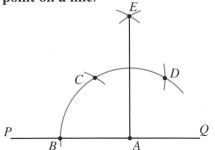

A coin is rolled along a line.
Sketch the locus of a point which starts off at the bottom.

What is the locus of the point if the coin rolls around another coin, or if the coins are not round, or … ?

Investigate.

1 The diagram shows part of a building site.
The scale is 1 cm to 1 m.
People are not allowed to walk anywhere within 2 m of the site.

Building Site

Copy the diagram.
Draw accurately the edge of the region where people may **not** walk.

AQA

2 (a) Use ruler and compasses only to construct an equilateral triangle of side 4 cm.
(b) A point P is 1 cm from the edge of the triangle.
Draw an accurate locus of **all** the possible positions of P.

3 $PQRS$ shows a sketch of a park.
(a) Use ruler and compasses only to construct a plan of the park using a scale of 1 cm to represent 100 m.

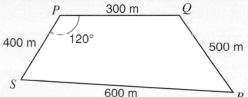

P   300 m   *Q*
400 m   120°   500 m
S   600 m   *R*

A fountain is:
  (i) equidistant from P and Q,
  (ii) equidistant from PS and SR.
(b) Draw the locus for (i) and (ii) on your plan and, hence, find the position of the fountain.
Label it with the letter F.
(c) Find the distance, in metres, of the fountain from R.

4 Ceri and Diane want to find how far away a tower, T, is on the other side of a river.
To do this they mark a base line, AB, 100 metres long, as shown on the diagram.
Next they measure the angles at the ends A and B between the base line and the lines of sight of the tower. These angles are 30° and 60°.

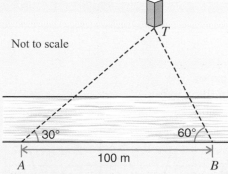

Not to scale

30°   60°
A   100 m   *B*

(a) Use ruler and compasses only to make a scale drawing of the situation.
Use a scale of 1 cm to represent 10 m.
Show clearly all your construction lines.
(b) Find the shortest distance of the tower, T, from the base line AB.

AQA

CHAPTER 30

Transformations

The movement of a shape from one position to another is called a **transformation**.
The change in position of the shape can be described in terms of a **reflection**, a **rotation** or a **translation**.
In a later chapter you will meet another transformation, called an **enlargement**.

Reflection

Look at this diagram.
It shows a **reflection** of a shape in the line PQ.
The line PQ is sometimes called a **mirror line**.
Place a mirror on the line PQ and look at the reflection.
You should see that the image of the shape is the
same distance from the mirror as the original.

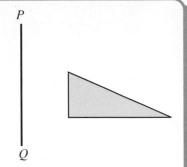

In this diagram the shape $WXYZ$ has been reflected in the line AB to $W_1X_1Y_1Z_1$.

If you join the points W and W_1:
>   the distance from W to the mirror line is the same as the distance from the mirror line to W_1,
>   the line WW_1 is at right angles to the mirror line.

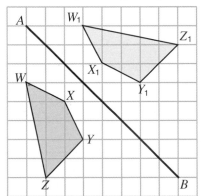

Notation:

$W_1X_1Y_1Z_1$ is the **image** of $WXYZ$.

We can also say that $WXYZ$ is **mapped** onto $W_1X_1Y_1Z_1$.

When a shape is reflected it stays the same shape and size but it is turned over.

For a reflection we need: a mirror line.

EXAMPLE

Copy the shape P onto squared paper.
Draw the reflection of shape P in the y axis.

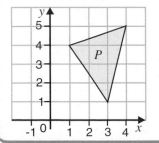 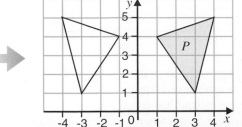

Notice that:

$(1, 4) \rightarrow (-1, 4)$

$(4, 5) \rightarrow (-4, 5)$

$(3, 1) \rightarrow (-3, 1)$

Can you see a pattern?

Exercise **30.1**

1 Copy each of the following shapes and draw the reflection of the shape in the line *AB*.

(a)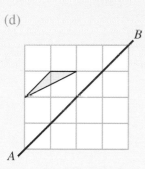

(b)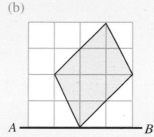

(c)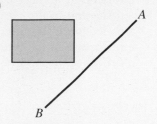

2 Copy each of the following shapes onto squared paper and draw the image of the shape after reflection in the line *AB*.

(a)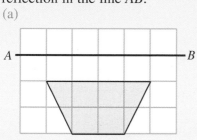

(b)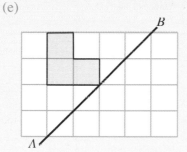

(c)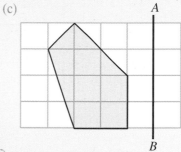

(d)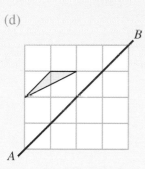

(e)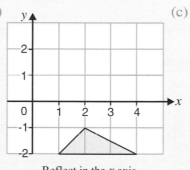

(f)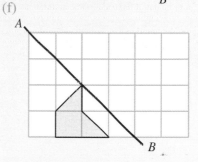

3 Copy each of the following diagrams onto squared paper and draw the reflection of each shape in the line given.

(a)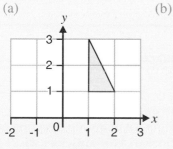

Reflect in the *y* axis.

(b)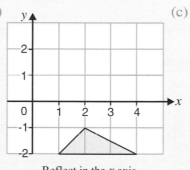

Reflect in the *x* axis.

(c)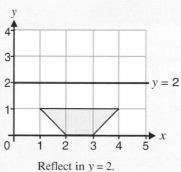

Reflect in *y* = 2.

(d)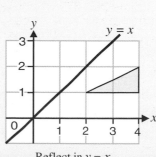

Reflect in *x* = 3.

(e)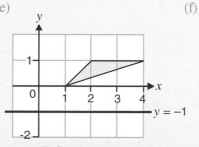

Reflect in *y* = −1.

(f)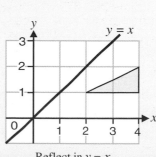

Reflect in *y* = *x*.

4 Copy the diagram onto squared paper.
Draw the image of the shape after:
(a) a reflection in the x axis,
(b) a reflection in the y axis,
(c) a reflection in the line $x = 3$,
(d) a reflection in the line $x = -1$,
(e) a reflection in the line $y = x$.

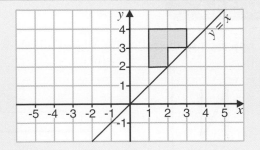

5

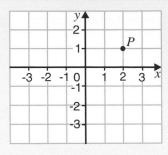

In the diagram, P is the point $(2, 1)$.
Find the coordinates of the image of P under a reflection in:
(a) the x axis,
(b) the y axis,
(c) the line $x = 1$,
(d) the line $y = -1$,
(e) the line $y = x$.

6 The diagram shows a quadrilateral $ABCD$.
Give the coordinates of B after:
(a) a reflection in the x axis,
(b) a reflection in the y axis,
(c) a reflection in the line $x = 4$,
(d) a reflection in the line $x = -1$,
(e) a reflection in the line $y = x$.

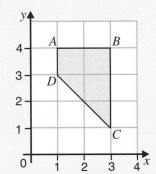

Rotation

Look at this diagram.
It shows the **rotation** of a shape P through $\frac{1}{4}$ turn anticlockwise about centre X.

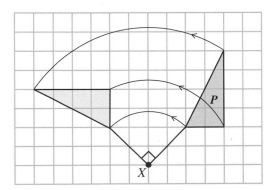

When describing rotations the direction of turn can be **clockwise** or **anticlockwise**.

Remember:

clockwise anticlockwise

All points on shape P are turned through the same angle about the same point.
This point is called the **centre of rotation**.

When a shape is rotated it stays the same shape and size but its **position** on the page changes.

For a rotation we need: a centre of rotation, an amount of turn, a direction of turn.

EXAMPLE

Copy triangle ABC onto squared paper.
Draw the image of triangle ABC after it has been rotated through 90° clockwise about the point $P(1, 1)$.
Label the image $A_1 B_1 C_1$.

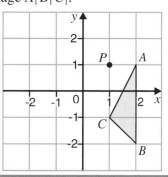 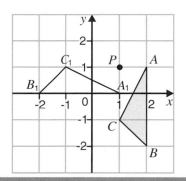

Exercise 30.2

1 Copy each of these shapes onto squared paper.
Draw the new position of each shape after the rotation given.

(a)
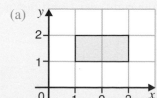

$\frac{1}{4}$ turn clockwise about centre X.

(b)
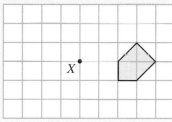

$\frac{1}{2}$ turn clockwise about centre X.

(c)
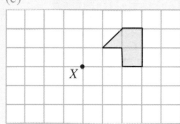

$\frac{1}{4}$ turn anticlockwise about centre X.

(d)
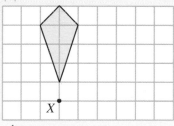

$\frac{1}{2}$ turn about centre X.

(e)
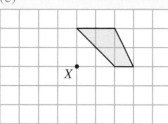

$\frac{3}{4}$ turn clockwise about centre X.

(f)
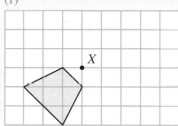

$\frac{3}{4}$ turn anticlockwise about centre X.

2 Copy each of the following shapes onto squared paper.
Draw the new position of the shape after it has been rotated through 90° clockwise about the origin (0, 0).

(a)
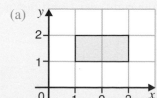

(b)
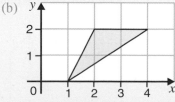

(c)
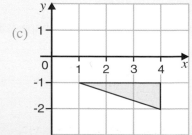

3 Copy each of the following shapes onto squared paper and then draw the new position of the shape after it has been rotated through 180°, about the point *X*.

(a)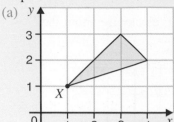

(b)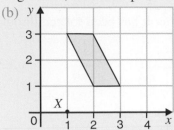

(c)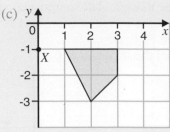

4 The diagram shows a quadrilateral *ABCD*.

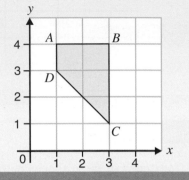

Give the coordinates of *B* after:
(a) a rotation through 90°, clockwise about (0, 0),
(b) a rotation through 90°, anticlockwise about (0, 0),
(c) a rotation through 180°, about (0, 0),
(d) a rotation through 90°, clockwise about (3, 1),
(e) a rotation through 90°, anticlockwise about (3, 1),
(f) a rotation through 180°, about (3, 1).

Translation

Look at this diagram.
It shows a **translation** of a shape *P*.

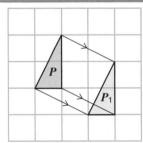

P is mapped onto P_1.
All points on the shape *P* are moved the same distance in the same direction without turning.
A translation can be given:
- in terms of a **distance** and a **direction**, e.g. 2 units to the right and 1 unit down.
- with a vector, e.g. $\binom{2}{-1}$.

When a **vector** is used to describe a translation:
the top number describes the **horizontal** part of the movement:
$$+ = \text{to the right}, \qquad - = \text{to the left}$$
the bottom number describes the **vertical** part of the movement:
$$+ = \text{upwards}, \qquad - = \text{downwards}$$
When a shape is translated it stays the same shape and size and has the same orientation.

EXAMPLE

Draw the new position of shape *P* after it has been translated 3 units to the right and 1 unit down.

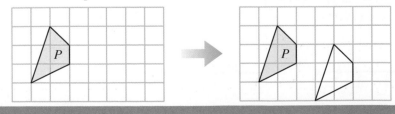

Copy triangle P onto squared paper. The translation $\begin{pmatrix} -3 \\ 2 \end{pmatrix}$ maps P onto P_1. Draw and label P_1.

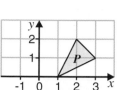
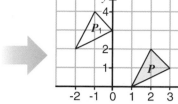

Vector notation:

$\begin{pmatrix} -3 \\ 2 \end{pmatrix}$ means move triangle P
3 units to the left and 2 units up.

Exercise 30.3

1 Copy the shape onto squared paper.
Draw the new position of the shape after
each of the following translations:
(a) 2 units to the right and 3 units up,
(b) 1 unit to the right and 2 units down,
(c) 3 units to the left and 2 units up,
(d) 1 unit to the left and 3 units down.

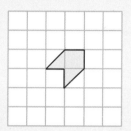

2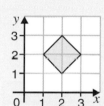

Copy the shape onto squared paper.
Draw the new position of the shape after each of the following
translations:
(a) 3 units to the right and 2 units up,
(b) 2 units to the right and 3 units down,
(c) 2 units to the left and 3 units up,
(d) 2 units to the left and 3 units down.

3 The diagram shows a quadrilateral $ABCD$.
Give the coordinates of B after the shape
has been translated with vector:

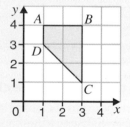

(a) $\begin{pmatrix} 2 \\ 1 \end{pmatrix}$ (b) $\begin{pmatrix} -2 \\ 2 \end{pmatrix}$ (c) $\begin{pmatrix} 1 \\ 3 \end{pmatrix}$ (d) $\begin{pmatrix} -2 \\ -3 \end{pmatrix}$

4 The translation $\begin{pmatrix} 2 \\ -1 \end{pmatrix}$ maps $S(5, 3)$ onto T.
What are the coordinates of T?

5 Write down the translation which maps:
(a) $X(1, 1)$ onto $P(3, 2)$,
(b) $X(1, 1)$ onto $Q(2, -1)$,
(c) $X(1, 1)$ onto $R(-2, 2)$,
(d) $X(1, 1)$ onto $S(-2, -1)$.

6 The diagram shows quadrilateral S.
Copy S onto squared paper.

(a) The translation $\begin{pmatrix} 3 \\ 2 \end{pmatrix}$ maps S onto T.
Draw and label T.
(b) Write down the translation which maps T onto S.

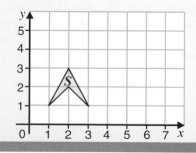

Describing transformations

Look at each of these diagrams.
In each case a shape has been moved to a new position by a **single transformation**.

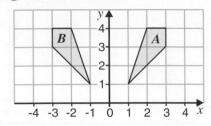

The transformation which takes A onto B is described as:
a **reflection** in the y axis.

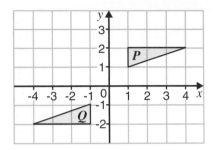

The transformation which takes P onto Q is described as:
a **rotation** of 180° about the origin.

The transformation which takes S onto T is described as:
a **translation** 3 units to the right and 2 units down,

or a translation with vector $\begin{pmatrix} 3 \\ -2 \end{pmatrix}$.

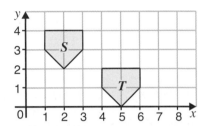

Transformation	Image same shape and size?	Details needed to describe the transformation
Reflection	Yes	Mirror line, sometimes given as an equation.
Rotation	Yes	Centre of rotation, amount of turn, direction of turn.
Translation	Yes	Horizontal movement and vertical movement. Vector: top number = horizontal movement, bottom number = vertical movement.

To find a line of reflection

1 Join each point to its image point.
2 Put a mark halfway along each line.
3 Use a ruler to join the marks.

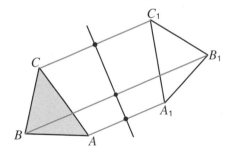

To find the centre and angle of rotation

1 Join each point to its image point.
2 Put a mark halfway along each line.
3 Use a set-square to draw a line at right angles to each line.
    The point where the lines cross is the centre of rotation, R.
4 Join one point and its image to the centre of rotation.
5 The angle of rotation is given by the size of the angle ARA_1.

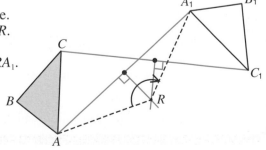

1 Which of these transformations takes *X* onto *Y* in each diagram?

reflection **rotation** **translation**

(a)

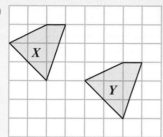

(b)

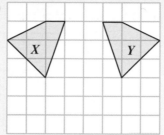

(c)

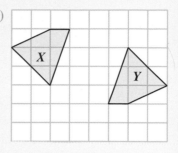

2 Describe fully the single transformation which takes *P* onto *Q* in each diagram.

(a)

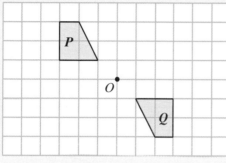

(b)

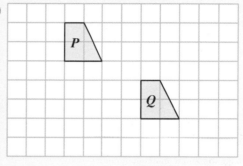

3 Describe fully the single transformation which takes:

(a) **L**₁ onto **L**₂,

(b) **L**₁ onto **L**₃,

(c) **L**₁ onto **L**₄,

(d) **L**₁ onto **L**₅,

(e) **L**₁ onto **L**₆.

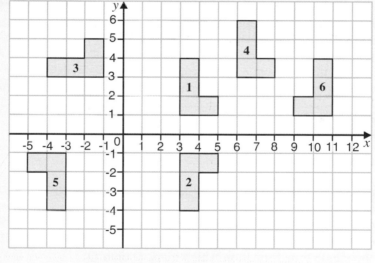

4 Describe fully the single transformation which takes:

(a) *P* onto *Q*,

(b) *P* onto *R*,

(c) *P* onto *S*,

(d) *R* onto *Q*.

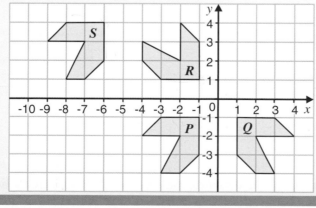

5 Describe fully the single transformation which maps

   (a) T onto U,

   (b) T onto V,

   (c) T onto W.

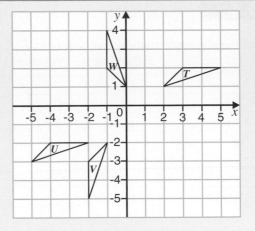

6 Describe fully the single transformation which maps

   (a) A onto B,

   (b) A onto C,

   (c) A onto D.

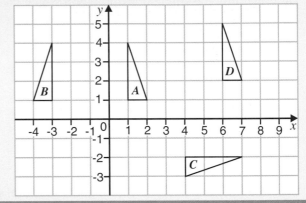

Combinations of transformations

So far we have looked at **single** transformations only.
There is no reason why the image of a transformation cannot be transformed.
The result of applying more than one transformation is called a **combined transformation**.

EXAMPLE

Copy triangle Q onto squared paper.

(a) Q is mapped onto Q_1 by a rotation through $90°$, clockwise about $(0, 0)$.
Draw and label Q_1.

(b) Q_1 is mapped onto Q_2 by a reflection in the line $y = 0$.
Draw and label Q_2.

(c) Describe the single transformation which maps Q onto Q_2.

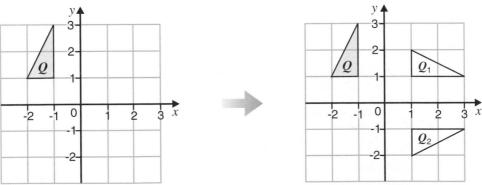

(c)   The single transformation which maps Q onto Q_2 is a reflection in the line $y = x$.

1 The diagram shows a quadrilateral labelled A. Copy the diagram onto squared paper.

(a) A is mapped onto A_1 by a reflection in the line $x = 0$. Draw and label A_1.

(b) A_1 is mapped onto A_2 by a reflection in the line $x = 4$. Draw and label A_2.

(c) Describe fully the single transformation which maps A onto A_2.

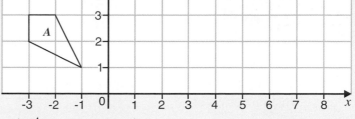

2 The diagram shows a triangle labelled P. Copy the diagram onto squared paper.

(a) P is mapped onto P_1 by a reflection in the line $y = x$. Draw and label P_1.

(b) P_1 is mapped onto P_2 by a reflection in the line $x = 5$. Draw and label P_2.

(c) Describe fully the single transformation which maps P onto P_2.

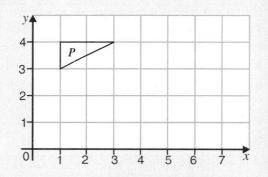

3 The diagram shows a triangle labelled T. Copy the diagram.

(a) Rotate T through 90° clockwise about (0, 0) to T_1. Draw and label T_1.

(b) Reflect T_1 in the line $y = 0$ to T_2. Draw and label T_2.

(c) Reflect T_2 in the line $x = 0$ to T_3. Draw and label T_3.

(d) Describe fully the single transformation which maps T onto T_3.

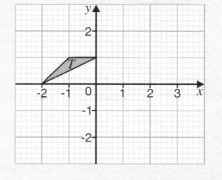

4 The diagram shows a quadrilateral labelled Q. Copy the diagram.

(a) Q is mapped onto Q_1 by a rotation through 90°, anticlockwise about (0, 0). Draw and label Q_1.

(b) Q_1 is mapped onto Q_2 by a rotation through 90°, anticlockwise about (2, 0). Draw and label Q_2.

(c) Describe fully the single transformation which maps Q onto Q_2.

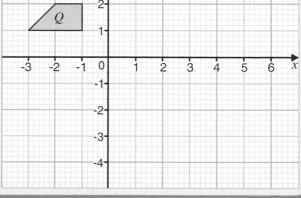

5 The diagram shows a shape labelled S.
Copy the diagram.

(a) The translation $\begin{pmatrix} 4 \\ 2 \end{pmatrix}$ maps S onto S_1.
Draw and label S_1.

(b) The translation $\begin{pmatrix} -8 \\ 1 \end{pmatrix}$ maps S_1 onto S_2.
Draw and label S_2.

(c) Describe fully the single transformation which maps S onto S_2.

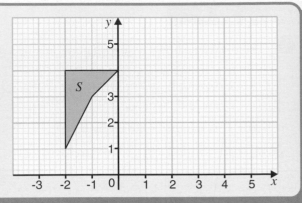

What you need to know

- The movement of a shape from one position to another is called a **transformation**.
- **Single transformations** can be described in terms of a reflection, a rotation or a translation.
- **Reflection**: The image of the shape is the same distance from the mirror line as the original.
- **Rotation**: All points are turned through the same angle about the same point, called a centre of rotation.
- **Translation**: All points are moved the same distance in the same direction without turning.
- How to fully describe a transformation.

Transformation	Image same shape and size?	Details needed to describe the transformation
Reflection	Yes	Mirror line, sometimes given as an equation.
Rotation	Yes	Centre of rotation, amount of turn, direction of turn.
Translation	Yes	Horizontal movement and vertical movement. Vector: top number = horizontal movement, bottom number = vertical movement.

Review Exercise 30

1 A wallpaper pattern is designed using rectangles, as shown.
Using only one of the words reflection, rotation or translation, describe a transformation which would
(a) take A onto B, (b) take A onto C.

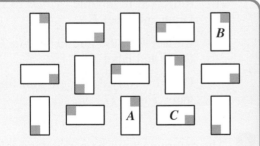

2

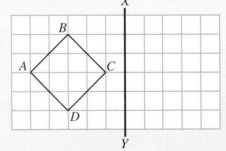

Copy the diagram.
(a) Draw the reflection of $ABCD$ in the mirror line XY.
(b) $ABCD$ has rotational symmetry.
Mark with a cross its centre of rotation. AQA

3 Copy the diagram.
Shape R is reflected in the line $x = 2$.
Draw the new position of R on your diagram.

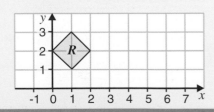

4 Copy the diagram.

(a) Reflect shape *P* in the line *AB*.
Label the new position *Q*.

(b) Rotate shape *P* through $\frac{1}{4}$ turn clockwise,
about centre *O*.
Label the new position *R*.

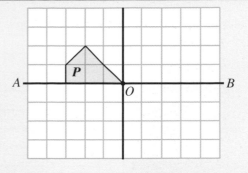

5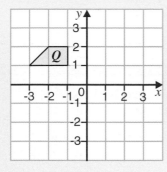

Copy the diagram.
Shape *Q* is rotated 90° anticlockwise about centre (0, 0).
Draw the new position of *Q* on your diagram.

6 Copy the diagram.
Shape *P* is reflected in the line $y = -1$.
Draw the new position of *P* on your diagram.

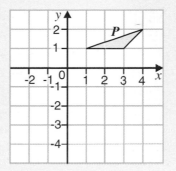

7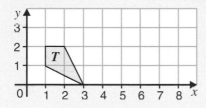

Copy the diagram.
Shape *T* is translated 3 units to the right and 2 units up.

(a) Draw the new position of *T* on your diagram.

(b) Describe fully the single transformation that takes
T back to its original position.

8 (a) Give the letter of the finishing position:

(i) after the shaded shape is reflected in
the *x* axis,

(ii) after the shaded shape is rotated
$\frac{1}{2}$ turn about (0, 0),

(iii) after the shaded shape is translated
3 units right and 4 units down.

(b) Describe fully the single transformation
which will map shape **G** onto shape **H**.

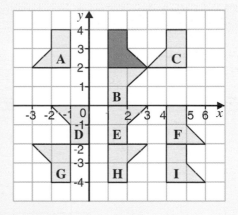

AQA

9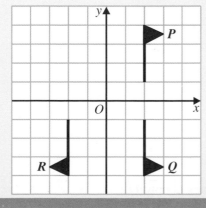

(a) Describe fully the transformation that
maps flag *P* onto flag *Q*.

(b) Describe fully the single transformation that
maps flag *P* onto flag *R*.

AQA

10 Describe fully the single transformation which takes:

(a) T onto U,

(b) V onto W,

(c) W onto X.

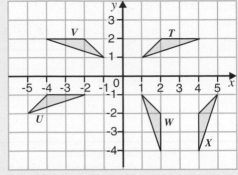

11 (a) Describe the single transformation that will transform shape C to shape B.

(b) Describe the single transformation that will transform shape A to shape C.

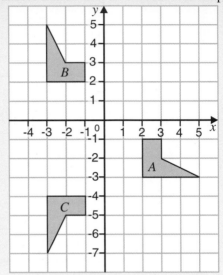

AQA

12 The diagram shows a trapezium labelled Q. Copy the diagram onto squared paper.

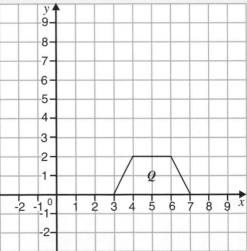

(a) Q is mapped onto Q_1 by a reflection in the x axis.
Draw and label Q_1.

(b) Q_1 is mapped onto Q_2 by a translation with vector $\binom{2}{4}$.
Draw and label Q_2.

(c) Q_2 is mapped onto Q_3 by a reflection in the line $y = x$.
Draw and label Q_3.

(d) Describe fully the single transformation which maps Q onto Q_3.

Enlargements and Similar Figures

Enlargement

This diagram shows another transformation, called an **enlargement**.

When a shape is enlarged:
  angles remain unchanged,
  all **lengths** are multiplied by a **scale factor**.

For example:
  Shape **B** is an enlargement of Shape **A**.
  The scale factor is 2.

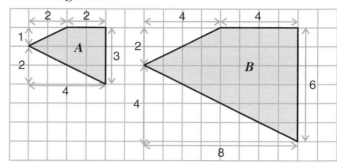

$$\text{Scale factor} = \frac{\text{new length}}{\text{original length}}$$

This can be rearranged to give: new length = original length × scale factor

Exercise 31.1

1   Copy each diagram onto squared paper and draw an enlargement with the given scale factor.

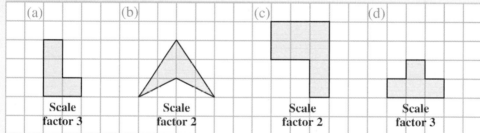

(a)   (b)   (c)   (d)

Scale   Scale   Scale   Scale
factor 3 factor 2 factor 2 factor 3

2   Shape A is enlarged to make shape B. What is the scale factor of the enlargement?

(a)   (b)

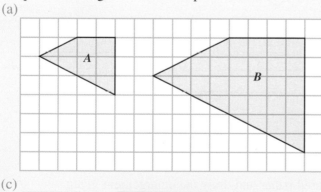

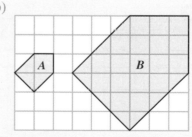

(c)   (d)

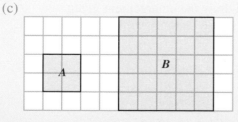

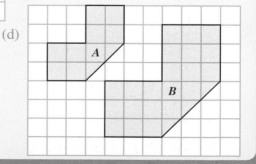

Using a centre of enlargement

A slide projector makes an enlargement of a picture.
The light bulb is the **centre of enlargement.**

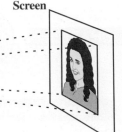

To enlarge a shape using a centre of enlargement:
Draw a line from the centre of enlargement, *P*, to one corner, *A*.
Extend this line to *A′* so that the length of *PA′* = the scale factor × the length of *PA*.
Do the same for other corners of the shape.
Join up the corners to make the enlarged shape. Label the diagram.

EXAMPLES

1 Draw an enlargement of triangle *ABC*, centre *P* (0, 1) and scale factor 3.

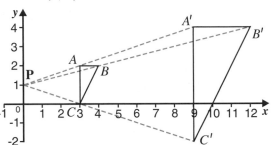

2 Use centre *P* (3, 2) and a scale factor of 2 to enlarge triangle *ABC*.

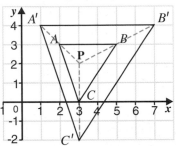

Exercise 31.2

1 Copy the following shapes and draw the enlargement, with scale factor 2, centre *O*.

(a)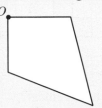

(b)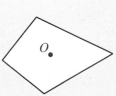

(c)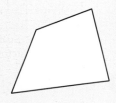

2 Copy the following shapes onto squared paper and draw the enlargement given.

(a) Scale factor 2, centre *X*.   (b) Scale factor 3, centre *X*.

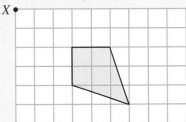

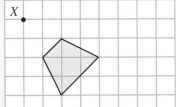

3 Copy the following shapes onto squared paper and draw the enlargement given.

(a) Scale factor 2, centre (0, 0).   (b) Scale factor 3, centre (0, 0).

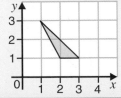

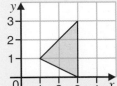

4 Copy the shape onto squared paper.

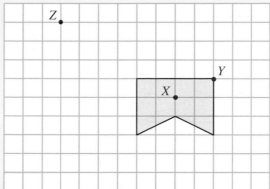

(a) Enlarge the shape with scale factor 2, centre *X*.
(b) Enlarge the shape with scale factor 2, centre *Y*.
(c) Enlarge the shape with scale factor 2, centre *Z*.
(d) What do you notice about the positions of the centres of enlargement and the positions of the enlarged shapes?

5 The diagram shows a quadrilateral *ABCD*.
Give the coordinates of *B* after an enlargement:

(a) scale factor 2, centre (0, 0),
(b) scale factor 3, centre (0, 0),
(c) scale factor 2, centre (0, 2),
(d) scale factor 3, centre *C* (3, 1),
(e) scale factor 2, centre *D* (1, 3).

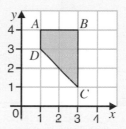

6 Copy each diagram onto squared paper and enlarge it using the centre and scale factor given. You will need longer axes than those shown below.

(a)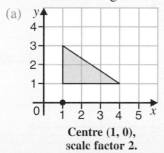
Centre (1, 0), scale factor 2.

(b)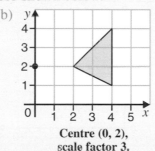
Centre (0, 2), scale factor 3.

(c)
Centre (3, 2), scale factor 2.

To find the centre and scale factor of an enlargement:

- Join pairs of corresponding points.
- Extend the lines until they meet.
  This point is the centre of enlargement.
- Measure a pair of corresponding lengths.

$$\text{Scale factor} = \frac{\text{new length}}{\text{original length}}$$

For example, $X'Y'Z'$ is an enlargement of XYZ.
The centre of enlargement is the point O.

$$\text{Scale factor} = \frac{X'Y'}{XY} = \frac{2.0}{0.8} = 2.5$$

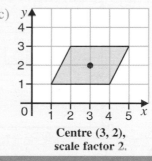

The transformation which maps *X* onto *Y* is: an enlargement, scale factor 2, centre (0, 0).

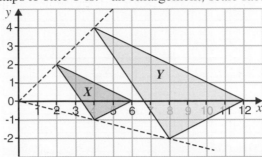

Exercise **31.3**

1 Triangle *ABC* is enlarged, as shown. Write down the scale factor and centre of the enlargement.

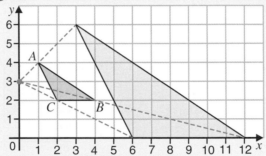

2 In each diagram, *A′B′C′(D′)* is an enlargement of *ABC(D)*.
Find the scale factor and centre of each enlargement.

(a)

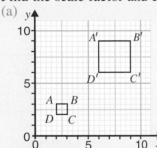

(b)

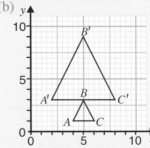

(c)

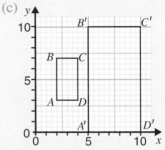

3 Describe fully the single transformation that maps *P* onto *Q*.

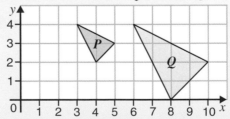

4 Describe the single transformation which maps *ABCD* onto *PQRS*.

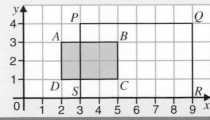

Using a scale factor which is a fraction

When the scale factor is a value between 0 and 1, such as 0.5 or $\frac{1}{3}$, the new shape is smaller than the original shape. Even though the shape gets smaller it is still called an enlargement.

EXAMPLE

Draw an enlargement of this shape with centre (0, 1) and scale factor $\frac{1}{3}$.

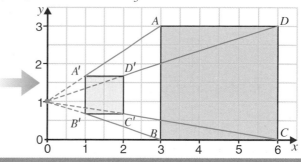

Exercise 31.4

1 Copy the following shapes onto squared paper and draw the enlargement given.

(a) Scale factor $\frac{1}{2}$, centre (1, 2).

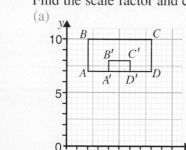

(b) Scale factor $\frac{1}{3}$, centre (0, 0).

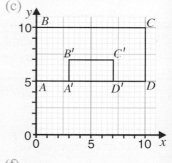

2 In each diagram, $A'B'C'(D')$ is an enlargement of $ABC(D)$.
Find the scale factor and centre of each enlargement.

(a)

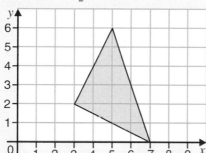

(b)

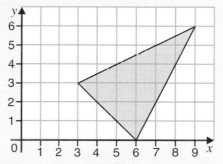

(c)

(d)

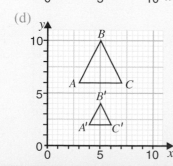

(e)

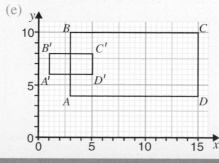

(f)

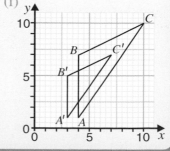

3 Describe fully the single transformation which maps *ABC* onto *XYZ*.

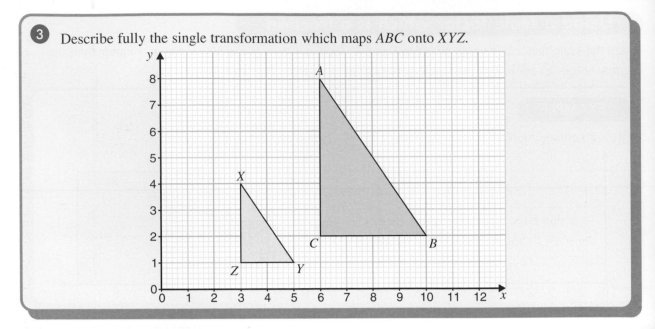

Similar figures

When one figure is an enlargement of another, the two figures are **similar**.

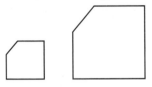

Sometimes one of the figures is rotated or reflected.
For example: Figures **C** and **E** are enlargements of figure **A**.   Figures **A**, **C** and **E** are similar.

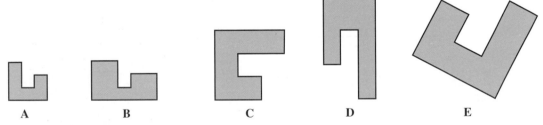

A   B   C   D   E

When two figures are **similar**:
  their **shapes** are the same, their **angles** are the same,
  corresponding **lengths** are in the same ratio, this ratio is the **scale factor** of the enlargement.

Activity

Figures *X* and *Y* are similar.
Y is an enlargement of *X*.

The ratio (or scale factor) is given by

$$\frac{\text{new length}}{\text{original length}}$$

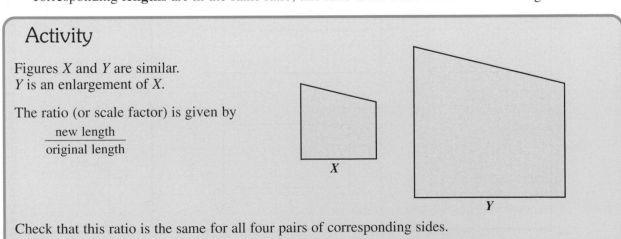

Check that this ratio is the same for all four pairs of corresponding sides.
Check that the angles are the same in the two figures.

EXAMPLES

1 A photo has width 6 cm and height 9 cm.
An enlargement is made, which has width 8 cm.
Calculate the height of the enlargement.

9 cm

h

6 cm

8 cm

Scale factor $= \frac{8}{6}$

$h = 9 \times \frac{8}{6}$

$h = 12$ cm

2 These two figures are similar.
Calculate the lengths of x and y.
Write down the size of the angle marked a.

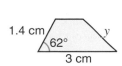 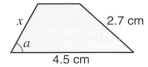

1.4 cm   62° y 3 cm

x 2.7 cm a 4.5 cm

The scale factor $= \frac{4.5}{3} = 1.5$

Lengths in the large figure are given by: length in small figure $\times$ scale factor

$x = 1.4 \times 1.5$

$x = 2.1$ cm

Lengths in the small figure are given by: length in large figure $\div$ scale factor

$y = 2.7 \div 1.5$

$y = 1.8$ cm

The angles in similar figures are the same, so, $a = 62°$.

Exercise **31.5**

1 The shapes in this question have been drawn accurately.
(a) Explain why these two shapes are not similar to each other.

(b) Which two of these shapes are similar to each other?

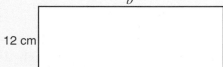

P   Q   R

2 Which of the following must be similar to each other?
(a) Two circles. (b) Two kites. (c) Two parallelograms. (d) Two squares. (e) Two rectangles.

3 These rectangles are all similar. The diagrams have not been drawn accurately.
Work out the lengths of the sides marked a and b.

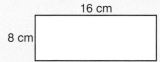

b

16 cm

8 cm   8 cm

a

12 cm

4 These two kites are similar.
(a) What is the scale factor of their lengths?
(b) Find the length of the side marked x.
(c) What is the size of angle a?

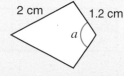 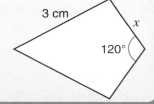

2 cm   1.2 cm a

3 cm x 120°

Enlargements and Similar Figures

5 (a) Explain why triangles *ABC* and *PQR* are similar.
 (b) Calculate the length of *AB*.

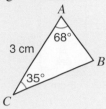
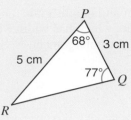

6 A shape has width 8 cm and length 24 cm.
It is enlarged to give a new shape with width 10 cm.
Calculate the length of the new shape.

7 In each part, the two figures are similar. Lengths are in centimetres.
Calculate the lengths and angles marked with letters.

(a)

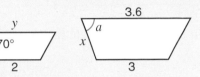

(b)

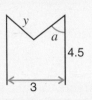

(c)

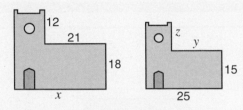

8 These two tubes are similar.

The width of the small size is 2.4 cm and
the height of the small size is 10 cm.
The width of the large size is 3.6 cm.
Calculate the height of the large size.

9 A motor car is 4.2 m long and 1.4 m high.
A scale model of the car is 8.4 cm long.
What is the height of the model?

10 The smallest angle in triangle *T* is 18°.
Triangle *T* is enlarged by a scale factor of 2.
How big is the smallest angle in the enlarged triangle?

11 A castle has height 30 m.
The height of the castle wall is 6 m.
A scale model of the castle has height 25 cm.
Calculate the height of the castle wall in the scale model.

12 The dimensions of three sizes of paper are given.

Length (cm)	24	30	y
Width (cm)	x	20	32

All the sizes are similar.
Calculate the values of x and y.

What you need to know

- When a shape is **enlarged**: all **lengths** are multiplied by a **scale factor**,
                                  angles remain unchanged.

  Scale factor = $\dfrac{\text{new length}}{\text{original length}}$ New length = scale factor × original length

  The size of the original shape is:
      increased by using a scale factor greater than 1,
      reduced by using a scale factor which is a fraction, i.e. between 0 and 1.
- When two figures are **similar**:
      their **shapes** are the same, their **angles** are the same,
      corresponding **lengths** are in the same ratio, this ratio is the **scale factor** of the enlargement.
- All circles are similar to each other.
- All squares are similar to each other.

IDEAS FOR INVESTIGATION

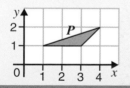

Enlargement and its effects on lengths, areas and volumes.

1 Some cubes have side 2 cm.
They are built together to make a larger cube with side 6 cm.
This represents an enlargement with scale factor 3.

Copy and complete the table.

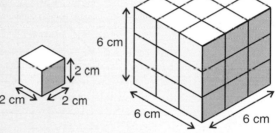

	Length of side (cm)	Area of face (cm²)	Volume of cube (cm³)
Small cube	2		8
Large cube	6	36	
Scale factor	$\dfrac{6}{2} = 3$	$\dfrac{36}{} =$	$\dfrac{}{8} =$

What do you notice about the three scale factors?

2 A cuboid is enlarged with scale factor 2, as shown.

What effect has the enlargement on length, area and volume?

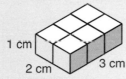

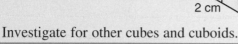
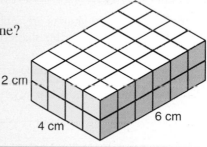

Investigate for other cubes and cuboids.

Review Exercise 31

1 The diagram shows the position of X and a pentagon.
Copy the diagram onto squared paper.
Draw an enlargement of the pentagon, scale factor 2, centre X.

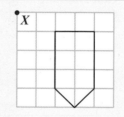

AQA

2 Copy the diagram.
Draw an enlargement of shape P, scale factor 3, centre $(0, 0)$.

3 Describe fully the single transformation which takes *A* onto *B*.

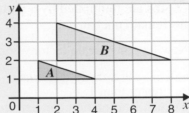

4 The diagram shows triangles *R* and *T*.
Describe fully the single transformation which maps
(a) *R* onto *T*,
(b) *T* onto *R*.

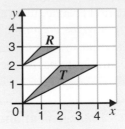

5 Copy the diagram onto squared paper and enlarge the triangle with scale factor $\frac{1}{3}$, centre *P*.

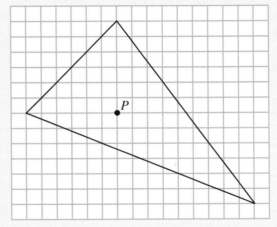

AQA

6 Two similar solid shapes are made.
The height of the smaller shape is 7 cm.
The width of the smaller shape is 6 cm.
The width of the larger shape is 9.6 cm.

Calculate the height of the larger shape.

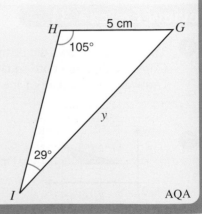

AQA

7 All of these triangles are similar.
(a) Calculate the length *x*.
(b) Calculate the length *y*.
(c) What is the value of angle *z*?

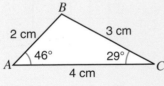
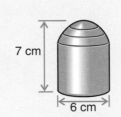

AQA

CHAPTER 32

Pythagoras' Theorem

The longest side in a right-angled triangle is called the **hypotenuse**.

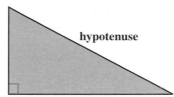

hypotenuse

In any right-angled triangle it can be proved that:

"The square on the hypotenuse is equal to the sum of the squares on the other two sides."

This is known as the **Theorem of Pythagoras**, or **Pythagoras' Theorem**.

Checking the Theorem of Pythagoras

Look at this triangle.

Notice that: the side opposite angle A is labelled a,
the side opposite angle B is labelled b,
the side opposite angle C is labelled c.

ABC is a right-angled triangle because $\angle BAC = 90°$.
$a = 5\,\text{cm}$, so, $a^2 = 25\,\text{cm}^2$.
$b = 4\,\text{cm}$, so, $b^2 = 16\,\text{cm}^2$.
$c = 3\,\text{cm}$, so, $c^2 = 9\,\text{cm}^2$.

$a^2 = b^2 + c^2$

Activity

Use a ruler and a pair of compasses to draw the following triangles accurately.

(a)

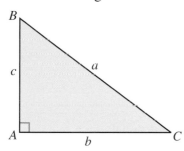

(b)

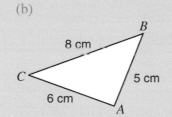

(c)

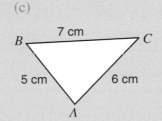

(d)

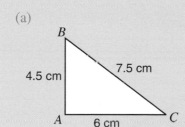

(e)

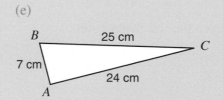

(f)

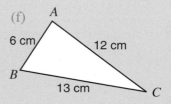

For each triangle: Measure angle BAC.
Is angle $BAC = 90°$?
Does $a^2 = b^2 + c^2$?
Explain your answers.

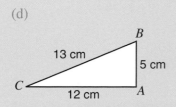

When we know the lengths of two sides of a right-angled triangle, we can use the Theorem of Pythagoras to find the length of the third side.

Finding the hypotenuse

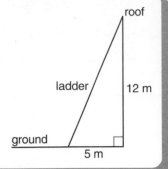

The roof of a house is 12 m above the ground.
What length of ladder is needed to reach the roof, if the foot of the ladder has to be placed 5 m away from the wall of the house?

Using Pythagoras' Theorem.   Take the square root of both sides.
$$l^2 = 5^2 + 12^2$$
$$l^2 = 25 + 144$$ $$l = \sqrt{169}$$
$$l^2 = 169$$ $$l = 13 \text{ m}$$

The ladder needs to be 13 m long.

Exercise 32.1

1 These triangles are right-angled. Calculate the length of the hypotenuse.

(a)   (b)   (c)

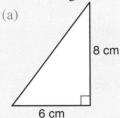
8 cm
6 cm

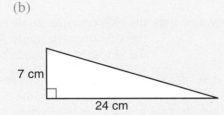
7 cm
24 cm

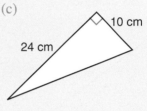
10 cm
24 cm

2 These triangles are right-angled. Calculate the length of side a to one decimal place.

(a)   (b)   (c)

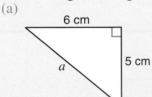
6 cm
5 cm
a

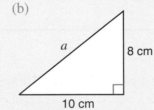
a
8 cm
10 cm

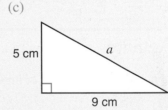
5 cm
a
9 cm

3 AB and CD are line segments, drawn on a centimetre-squared grid.
Calculate the exact length of
(a) AB,   (b) CD.

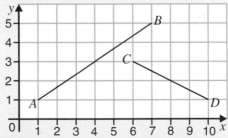

4 Calculate the distance between the following points.
(a) $A(2, 0)$ and $B(6, 3)$.   (b) $C(6, 3)$ and $D(0, 10)$.
(c) $E(2, 2)$ and $F(-3, -10)$.   (d) $G(-2, -2)$ and $H(-6, 5)$.
(e) $I(3, -1)$ and $J(-3, -5)$.

5 The coordinates of the vertices of a parallelogram are $P(1, 1)$, $Q(3, 5)$, $R(x, y)$ and $S(7, 3)$.
(a)   Find the coordinates of R.
(b) X is the midpoint of PQ. Find the coordinates of X.
(c) Y is the midpoint of PS. Find the coordinates of Y.
(d)   Calculate the distance XY.

Finding one of the shorter sides

To find one of the shorter sides we can rearrange the Theorem of Pythagoras.

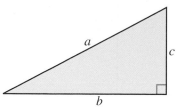

To find b we use:
$b^2 = a^2 - c^2$

To find c we use:
$c^2 = a^2 - b^2$

○○○○○○○○○○○○○○○○○○○○○
To find the length of a shorter side of a right-angled triangle:
Subtract the square of the known short side from the square on the hypotenuse.
Take the square root of the result.

EXAMPLE

A wire used to keep a radio aerial steady is 9 metres long.
The wire is fixed to the ground 4.6 metres from the base of the aerial.
Find the height of the aerial, giving your answer correct to one decimal place.

Using Pythagoras' Theorem. $9^2 = h^2 + 4.6^2$

Rearranging this we get: $h^2 = 9^2 - 4.6^2$
$h^2 = 81 - 21.16$
$h^2 = 59.84$

Take the square root of both sides. $h = \sqrt{59.84}$
$h = 7.735\ldots$
$h = 7.7$ m, correct to 1 d.p.

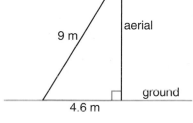

The height of the aerial is 7.7 m, correct to 1 d.p.

Exercise 32.2

1 Work out the length of side b.

(a)

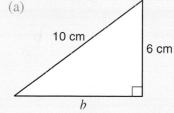

(b)

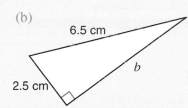

(c)

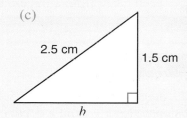

2 Work out the length of side c, correct to one decimal place.

(a)

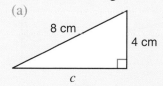

(b)

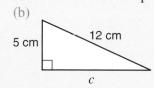

(c)

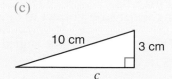

3 Two boats A and B are 360 m apart.
Boat A is 120 m due east of a buoy.
Boat B is due north of the buoy.
How far is boat B from the buoy?

Pythagoras' Theorem

4 The diagram shows a right-angled triangle, *ABC*, and a square, *ACDE*.

$AB = 2.5$ cm and $BC = 6.5$ cm.
Calculate the area of the square *ACDE*.

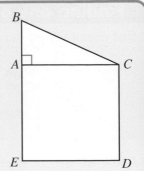

5

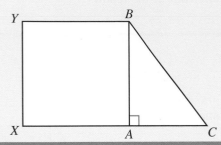

The diagram shows a right-angled triangle, *ABC*, and a square, *XYBA*.

$BC = 6$ cm.
The square *XYBA* has an area of 23.04 cm².
Calculate the length of *AC*.

Problems involving the use of Pythagoras' Theorem

Questions leading to the use of Pythagoras' Theorem often involve:
Understanding the problem.
What information is given?
What are you required to find?
Drawing diagrams.
In some questions a diagram is not given.
Drawing a diagram may help you to understand the problem.
Selecting a suitable right-angled triangle.
In more complex problems you will have to select a right-angled triangle which can be used to answer the question. It is a good idea to draw this triangle on its own, especially if it has been taken from a three-dimensional drawing.

EXAMPLE

The diagram shows the side view of a swimming pool.
It slopes steadily from a depth of 1 m to 3.6 m.
The pool is 20 m long.
Find the length of the sloping bottom of the pool,
giving the answer correct to three significant figures.

$\triangle CDE$ is a suitable right-angled triangle.
$CD = 3.6 - 1 = 2.6$ m

Using Pythagoras' Theorem in $\triangle CDE$.
$DE^2 = CD^2 + CE^2$
$DE^2 = 2.6^2 + 20^2$
$DE^2 = 6.76 + 400$
$DE^2 = 406.76$
$DE = \sqrt{406.76}$ m
$DE = 20.1682\ldots$ m

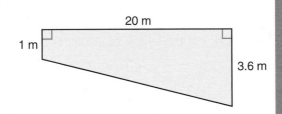

The length of the sloping bottom of the pool is 20.2 m, correct to 3 sig. figs.

1 In each of the following, work out the length of the side marked x.

(a)

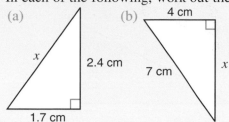

x
2.4 cm
1.7 cm

(b)

4 cm

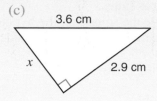

7 cm
x

(c)

3.6 cm

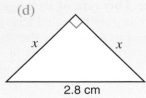

x
2.9 cm

(d)

x   x

2.8 cm

2 A rectangle is 8 cm wide and 15 cm long.
Work out the length of its diagonals.

3 The length of a rectangle is 24 cm. The diagonals of the rectangle are 26 cm.
Work out the width of the rectangle.

4 A square has sides of length 6 cm.
Work out the length of its diagonals.

5 The diagonals of a square are 15 cm.
Work out the length of its sides.

6 The height of an isosceles triangle is 12 cm. The base of the triangle is 18 cm.
Work out the length of the equal sides.

7 An equilateral triangle has sides of length 8 cm.
Work out the height of the triangle.

8 The diagram shows the side view of a car ramp.
The ramp is 110 cm long and 25 cm high.
The top part of the ramp is 40 cm long.
Calculate the length of the sloping part of the ramp.

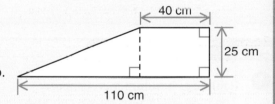

40 cm
25 cm
110 cm

9

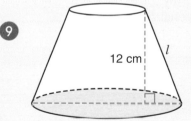

12 cm
l

The top of a lampshade has a diameter of 10 cm.
The bottom of the lampshade has a diameter of 20 cm.
The height of the lampshade is 12 cm.
Calculate the length, l, of the sloping sides.

10 The top of a bucket has a diameter of 30 cm.
The bottom of the bucket has a diameter of 16 cm.
The sloping sides are 25 cm long.
How deep is the bucket?

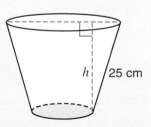

h   25 cm

11

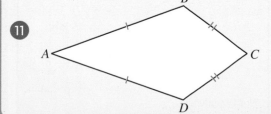

B
A   C
D

ABCD is a kite.
AB = 8.5 cm, *BC* = 5.4 cm and *BD* = 7.6 cm.
(a) Calculate the length of *AC*.
(b) Calculate the area of the kite.

What you need to know

- The longest side in a right-angled triangle is called the **hypotenuse**.
- The **Theorem of Pythagoras** states:
  "In any right-angled triangle the square on the hypotenuse is equal to the sum of the squares on the other two sides."
  $$a^2 = b^2 + c^2$$
  Rearranging gives:
  $$b^2 = a^2 - c^2$$
  $$c^2 = a^2 - b^2$$

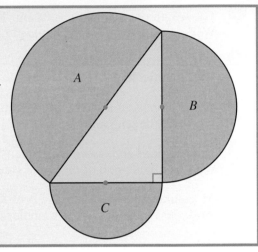

- When we know the lengths of two sides of a right-angled triangle, we can use the Theorem of Pythagoras to find the length of the third side.

Investigate the relationship between the areas of the semi-circles A, B and C.

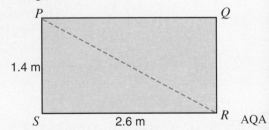

Review Exercise 32

1 Pauline is building a greenhouse.
The base, $PQRS$, of the greenhouse should be a rectangle measuring 2.6 m by 1.4 m.
To check the base is rectangular Pauline has to measure the diagonal PR.

(a) Calculate the length of PR when the base is rectangular.

(b) When building the greenhouse Pauline finds angle $PSR > 90°$. She measures PR. Which of the following statements is **true**?
- X: PR is greater than it should be.
- Y: PR is less than it should be.
- Z: PR is the right length.

AQA

2 The diagram shows the position of a ferry sailing between Folkestone and Calais.
The ferry is at X.
The position of the ferry from Calais is given as:

North of Calais 15 km,
West of Calais 24 km.

Calculate the distance of the ferry from Calais.
Give your answer correct to **one** decimal place.

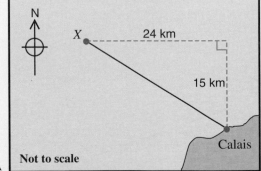

AQA

3 The coordinates of the points *A* and *B* are (6, 8) and (1, 1).
Work out the length of *AB*.

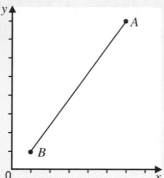

4 The sketch shows triangle *ABC*.
$AB = 40$ cm, $AC = 41$ cm and $CB = 9$ cm.

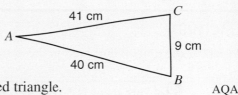

By calculation, show that triangle *ABC* is a right-angled triangle.

AQA

5 John is standing 200 m due west of a power station and 300 m due north of a pylon.
Calculate the distance of the power station from the pylon.

6 The diagram shows the position of a camp, a supply hut and a heliport.

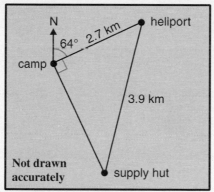

(a) What is the bearing of the supply hut from the camp?
(b) Calculate the distance from the camp to the supply hut.
Give your answer correct to 2 significant figures.

AQA

7 A helicopter flies from its base on a bearing of 045° for 20 km before landing.
How far east of its base is the helicopter when it lands?

8 Helena is playing golf.

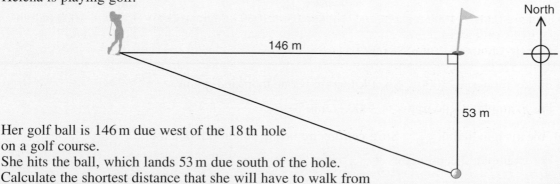

Her golf ball is 146 m due west of the 18th hole
on a golf course.
She hits the ball, which lands 53 m due south of the hole.
Calculate the shortest distance that she will have to walk from
where she hit the ball to where it landed.

AQA

CHAPTER 33

Understanding and Using Measures

Units of measurement

Different units can be used to measure the same quantity.

For example:

> The same **length** can be measured using centimetres, kilometres, inches, miles, …
>
> The same **mass** can be measured using grams, kilograms, pounds, ounces, …
>
> The same **capacity** can be measured using litres, millilitres, gallons, pints, …

There are two sorts of units in common use − **metric** units and **imperial** units.

Activity

Which of the units mentioned in these statements are metric and which are imperial?

This sheet is 8 feet by 4 feet and 6 millimetres thick.

I did about 300 miles last week so I'll need about 8 gallons to fill the tank. How many litres is that?

An Olympic champion runs 100 metres at an average speed of nearly 25 miles per hour.

I weigh 8 stone 2 pounds or 52 kilograms.

Metric units

The common metric units used to measure length, mass (weight) and capacity (volume) are shown below.

Length	Mass	Capacity and volume
1 kilometre (km) = 1000 metres (m)	1 tonne (t) = 1000 kilograms (kg)	1 litre = 1000 millilitres (ml)
1 m = 100 centimetres (cm)	1 kg = 1000 grams (g)	1 cm³ = 1 ml
1 cm = 10 millimetres (mm)		

Kilo means thousand, 1000. So, a **kilo**gram is one thousand grams.

> For example: 3 kilograms = 3000 grams.

Centi means hundredth, $\frac{1}{100}$. So, a **centi**metre is one hundredth of a metre.

> For example: 2 centimetres = $\frac{2}{100}$ metre.

Milli means thousandth, $\frac{1}{1000}$. So, a **milli**litre is one thousandth of a litre.

> For example: 5 millilitres = $\frac{5}{1000}$ litre.

Changing from one metric unit to another

Changing from one metric unit to another involves multiplying, or dividing, by a power of 10 (10, 100 or 1000). Multiplying and dividing by powers of 10 was covered in Chapter 1.

EXAMPLES

1

| 1 centimetre (cm) = 10 millimetres (mm) |

(a) Change 6.3 cm into millimetres.

To change centimetres into millimetres, multiply by 10.
$6.3 \times 10 = 63$
$6.3 \text{ cm} = 63 \text{ mm}$

(b) Change 364 mm into centimetres.

To change millimetres into centimetres, divide by 10.
$364 \div 10 = 36.4$
$364 \text{ mm} = 36.4 \text{ cm}$

2

| 1 metre (m) = 100 centimetres (cm) |

(a) Change 4.2 m into centimetres.

To change metres into centimetres, multiply by 100.
$4.2 \times 100 = 420$
$4.2 \text{ m} = 420 \text{ cm}$

(b) Change 850 cm into metres.

To change centimetres into metres, divide by 100.
$850 \div 100 = 8.5$
$850 \text{ cm} = 8.5 \text{ m}$

3

| 1 kilogram (kg) = 1000 grams (g) |

(a) Change 19.4 kg into grams.

To change kilograms into grams, multiply by 1000.
$19.4 \times 1000 = 19\,400$
$19.4 \text{ kg} = 19\,400 \text{ g}$

(b) Change 245 g into kilograms.

To change grams into kilograms, divide by 1000.
$245 \div 1000 = 0.245$
$245 \text{ g} = 0.245 \text{ kg}$

Changing units - areas and volumes

$$1 \text{ m} = 100 \text{ cm}$$
$$1 \text{ m}^2 = 100 \times 100 \text{ cm}^2 = 10\,000 \text{ cm}^2$$
$$1 \text{ m}^3 = 100 \times 100 \times 100 \text{ cm}^3 = 1\,000\,000 \text{ cm}^3$$

How many mm equal 1 cm?
How many mm² equal 1 cm²?
How many mm³ equal 1 cm³?

Exercise 33.1
Do not use a calculator.

1 Change each of the following lengths into millimetres.
  (a) 6 cm   (b) 32 cm   (c) 632 cm   (d) 8.6 cm   (e) 0.8 cm   (f) 0.08 cm

2 Change each of the following lengths into centimetres.
  (a) 90 mm   (b) 210 mm   (c) 3500 mm   (d) 73.5 mm   (e) 2 mm   (f) 3.5 mm

3 Change each of the following lengths into metres.
  (a) 200 cm   (b) 320 cm   (c) 4550 cm   (d) 66 cm   (e) 8 cm   (f) 9.8 cm

4 Change each of the following lengths into centimetres.
  (a) 6 m   (b) 56 m   (c) 7.6 m   (d) 23.5 m   (e) 0.9 m   (f) 0.07 m

5 Change each of the following lengths into kilometres.
  (a) 4000 m   (b) 35 000 m   (c) 6500 m   (d) 455 m   (e) 75 m   (f) 7 m

6 Change each of the following lengths into metres.
(a) 6 km   (b) 32 km   (c) 650 km   (d) 3.31 km   (e) 0.35 km   (f) 0.085 km

7 Change each of the following areas into square centimetres (cm^2).
(a) $2 m^2$ (b) $10 m^2$ (c) $0.5 m^2$

8 Change each of the following volumes into cubic centimetres (cm^3).
(a) $3 m^3$ (b) $20 m^3$ (c) $0.4 m^3$

9 Change each of the following masses into grams.
(a) 2 kg   (b) 45 kg   (c) 7.5 kg   (d) 42.5 kg   (e) 0.6 kg   (f) 0.025 kg

10 Change each of the following masses into kilograms.
(a) 3000 g   (b) 32 000 g   (c) 9300 g   (d) 220 g   (e) 83 g   (f) 6 g

11 Copy and complete each of the following.
(a) 320 000 ml = ... l
(b) 0.32 t = ... kg = ... g
(c) 3200 g = ... kg = ... t
(d) 320 mm = ... cm = ... m
(e) 32 000 cm = ... m = ... km
(f) 3.2 km = ... m = ... cm

12 Find the number of kilograms in:
(a) 6 t   (b) 8000 g   (c) 800 g   (d) 0.65 t

13 Find the number of metres in:
(a) 4 km   (b) 8000 mm   (c) 8.6 cm   (d) 0.04 km

14 Find the number of millilitres in:
(a) $2 l$ (b) $\frac{1}{2} l$ (c) $0.85 l$ (d) $0.03 l$

15 Which two lengths are the same?   2000 m   20 km   200 m   2 km   0.02 km

16 Which two weights are the same?   8 g   8 kg   8000 g   0.8 kg   80 kg

17 Which length is longest?   0.5 km   50 m   5000 mm   500 cm

18 Which weight is heaviest?   0.3 t   3000 g   3 kg   30 kg

19 How many:
(a) metres are there in 3123 mm,
(b) centimetres are there in 4.5 m,
(c) metres are there in 3.24 km,
(d) grams are there in 1 tonne,
(e) litres are there in 400 ml?

20 Which area is larger? $0.5 m^2$ or $500 cm^2$ Give a reason for your answer.

21 Which volume is larger? $0.08 m^3$ or $800 000 cm^3$ Give a reason for your answer.

22 A can of coke contains 330 ml.
How many litres of coke are there in 6 cans?

23 One lap of a running track is 400 m.
How many laps are run in an 8 km race?

24 Twenty children at a party share equally 1 kg of fruit pastilles.
How many grams of pastilles does each child receive?

25 A recipe for a dozen biscuits uses 240 g of flour. James has 1.2 kg of flour.
How many biscuits can he make?

26 Ben takes two 5 ml doses of medicine four times a day.
Ben stops taking the medicine after 5 days.
Originally, there was $\frac{1}{4}$ of a litre of medicine.
How much medicine is left?

Estimating length, mass and capacity

It is a useful skill to be able to estimate length, mass and capacity. These facts might help you.

Length

Most adults are between 1.5 m and 1.8 m tall. The door to your classroom is about 2 m high.
Find some more facts which will help you to estimate length and distance.

Mass

A biro weighs about 5 g. A standard bag of sugar weighs 1 kg.
Find some more facts which will help you to estimate weight or mass.

Capacity

A teaspoon holds about 5 ml. A can of pop holds about 330 ml.
Find some more facts which will help you to estimate volume and capacity.

EXAMPLES

1 The diagram shows a house.
The door is approximately 2 m high.
Estimate the height of the house.

The height of the house is roughly the same as
the height of four doors.
So, the height of the house is about $4 \times 2 = 8$ m.

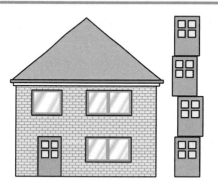

2 Complete each of the following statements by choosing one of the quantities given.
(a) A cup holds about ...... of tea.
(b) The height of a car is about ......
(c) The weight of this book is about ......
(d) The weight of an elephant is about ......
(e) The distance from London to Newcastle is about ......

| 8 g | 200 ml | 7 *l* | 8 kg | 15 cm | 700 kg | 1.5 m | 4 km | 2 kg | 400 km | 0.07 *l* | 15 mm |

(a) 200 ml   (b) 1.5 m   (c) 2 kg   (d) 700 kg   (e) 400 km

Choosing an appropriate unit

EXAMPLES

1 The Great Wall of China is the longest man-made structure in the world.
It is the only man-made structure that can be seen from space.
What unit should be used for its length?

For very long lengths use the **kilometre**.
The Great Wall of China is actually about 2350 km long.

2 The smallest known mammal is the Kitti's hog-nosed bat.
It is not much bigger than a pea.
(a) What unit should be used for its mass?
(b) What unit should be used for its length?

(a) For very small masses use the **gram**.
The Kitti's hog-nosed bat actually weighs about 1.5 g.
(b) For very small lengths use the **millimetre**.
The length of the Kitti's hog-nosed bat is about 10 mm.

1 Estimate, in metres, centimetres or millimetres, the following lengths in your classroom.
- (a) The length of the room.
- (b) The height of the room.
- (c) The height of your desk.
- (d) The thickness of a watch strap.
- (e) The width of the door.
- (f) The diameter of your pen (or pencil).
- (g) The length of your exercise book.
- (h) The length of a pencil sharpener.

2 Estimate, in grams or kilograms, the weights of the following.
- (a) A pea
- (b) A chair
- (c) A cat
- (d) A calculator
- (e) A pencil
- (f) A dinner plate
- (g) A car
- (h) Your desk

3 Estimate, in litres or millilitres, the following.
- (a) The volume of milk you would add to a cup of tea.
- (b) The volume of milk you would pour on a bowl of breakfast cereal.
- (c) The volume of water you drink in a day.

4 Which of the following is the best estimate for the mass of a banana?
1 kg   5 g   250 g   30 g   3 kg   750 g

5 Which of the following is the best estimate for the diameter of a football?
2 m   50 mm   30 cm   1.5 m   0.6 m   800 mm

6 Which of the following would be the best estimate for the capacity of a mug?
15 m   1200 ml   2 l 0.5 l 200 ml   800 ml

7 Give a sensible estimate using an appropriate unit for the following measures:
- (a) the length of a matchstick,
- (b) the length of a football pitch,
- (c) the weight of a 30 cm ruler,
- (d) the weight of a double decker bus,
- (e) the volume of drink in a glass.

8 Give the most appropriate metric unit that you would use to measure the following.
- (a) The distance from London to York.
- (b) The distance across a road.
- (c) The length of your foot.
- (d) The length of your little finger nail.
- (e) The weight of a bag of potatoes.
- (f) The weight of an egg.
- (g) The capacity of a bucket.
- (h) The capacity of a medicine bottle.

9 The diagram, which is drawn to scale, shows a man standing next to a tree.
Using an appropriate metric unit estimate the height of the tree.
State the degree of accuracy that you have used in making your estimate.

10

> "My teacher's height is about 1.7 mm."
> **This statement is incorrect.**

It can be corrected by changing the unit: "My teacher's height is about 1.7 m."
It can also be corrected by changing the quantity: "My teacher's height is about 1700 mm."

Each of these statements is also incorrect.
"Tyrannosaurus, a large meat-eating dinosaur, is estimated to have been about 12 cm long."
"The tallest mammal is the giraffe which grows up to about 5.9 mm tall."
"My car used 5 ml of petrol on a journey of 35 miles."
"The area of the school hall is about 500 mm²."

Correct each statement:
- (a) by changing the unit,
- (b) by changing the quantity.

Imperial units

The following imperial units of measurement are in everyday use.

Length	Mass	Capacity and volume
1 foot = 12 inches	1 pound = 16 ounces	1 gallon = 8 pints
1 yard = 3 feet	14 pounds = 1 stone	

EXAMPLES

1 Jane is 5 feet 3 inches tall.
How many inches is this?

There are 12 inches in 1 foot.
5 feet = 5 × 12 = 60 inches.
60 + 3 = 63
Jane is 63 inches tall.

2 Jane weighs 89 pounds.
Give her weight in stones and pounds.

There are 14 pounds in 1 stone.
6 × 14 = 84
So, 6 stones is the same as 84 pounds.
89 − 84 = 5
Jane weighs 6 stones 5 pounds.

Metric and imperial conversions

In order to convert to and from metric and imperial units you need to know these facts.

Length	Mass	Capacity and volume
5 miles is about 8 km	1 kg is about 2.2 pounds	1 litre is about 1.75 pints
1 inch is about 2.5 cm		1 gallon is about 4.5 litres
1 foot is about 30 cm		

EXAMPLES

1 Convert 40 cm to inches.

1 inch is about 2.5 cm.
40 cm is about 40 ÷ 2.5 inches.
40 cm is about 16 inches.

2 How many pints are there in a
4 litre carton of milk?

1 litre is about 1.75 pints.
4 litres is about 4 × 1.75 pints.
4 litres is about 7 pints.

3 Change 5 kg to pounds.

1 kg is about 2.2 pounds.
5 kg is about 5 × 2.2 pounds.
5 kg is about 11 pounds.

4 Tim is 6 feet 2 inches tall.
Estimate Tim's height in centimetres.

6 feet 2 inches = 6 × 12 + 2 = 74 inches.
74 inches is about 74 × 2.5 cm.
6 feet 2 inches is about 185 cm.

5 The capacity of a car's petrol tank is 12 gallons.
How much does the petrol tank hold in litres?

1 gallon is about 4.5 litres.
12 gallons is about 12 × 4.5 litres.
12 gallons is about 54 litres.

6 How far is 32 km in miles?

There are 32 ÷ 8 = 4 lots of 8 km in 32 km.
So, there must be 4 lots of 5 miles in 32 km.
4 × 5 = 20.
There are 20 miles in 32 km.

Understanding and Using Measures

1 Change these lengths into centimetres. (a) 2 inches (b) 2 feet

2 Change these lengths into inches. (a) 2 m (b) 20 cm

3 Wendy is 162 cm tall. What is Wendy's height in feet and inches?

4 Change these lengths into kilometres. (a) 5 miles (b) 45 miles

5 Change these lengths into miles. (a) 8 km (b) 40 km

6 Poole is about 260 miles from Ormskirk. What is this distance in kilometres?

7 Change these weights into pounds. (a) 25 kg (b) 1 t

8 Change these weights into kilograms. (a) 100 pounds (b) 6 stones 11 pounds

9 Georgina weighs 48 kg. What is Georgina's weight in stones and pounds?

10 How many litres are there in: (a) 5 pints, (b) 2 gallons 3 pints?

11 Convert each quantity to the units given.
(a) 15 kg to pounds. (b) 20 litres to pints. (c) 5 metres to inches.
(d) 6 inches to millimetres. (e) 50 cm to inches. (f) 6 feet to centimetres.

12 A box contains 200 balls. Each ball weighs 50 g.
Estimate the total weight of the balls in pounds.

13 Kelvin is 5 feet 8 inches tall.
Estimate Kelvin's height in centimetres.

14 Paddy weighs 9 stones 12 pounds.
Estimate Paddy's weight in kilograms.

15 Estimate the number of:
(a) metres in 2000 feet, (b) kilometres in 3 miles, (c) feet in 150 centimetres,
(d) pounds in 1250 grams, (e) litres in 10 gallons.

16 A sheet of card measures 12 inches by 20 inches.
What is the area of the card in square centimetres?

17 The height of a pile of magazines is 20 inches. Each magazine is 5 mm thick.
How many magazines are in the pile?

18 Lauren says 10 kg of potatoes weighs the same as 20 lb of sugar.
Is she correct? Show all your working.

19 Alfie cycles 6 miles. Jacob cycles 10 kilometres.
Alfie claims that he has cycled further than Jacob.
Is he correct? Show all your working.

20 Nick weighs 10 stones 6 pounds. Last year he weighed 10 kg more.
How much did Nick weigh last year?

21 Lubna has 250 g of butter. She uses 4 oz of butter to make a cake.
What weight of butter is left?

22 A stair carpet is 85 cm wide and 4.5 m in length.
Calculate the area of the carpet in square feet.

23 Convert the following speeds to kilometres per hour.
(a) 30 miles per hour. (b) 50 miles per hour. (c) 30 metres per second.

24 Convert the following speeds to miles per hour.
   (a) 60 km per hour.   (b) 40 metres per second.

25 Convert a speed of 60 miles per hour to metres per second.
Give your answer to a suitable degree of accuracy.

26 (a)   A car does 40 miles to the gallon. How many kilometres does it do per litre?
     (b)   A car does 9.6 kilometres to the litre. How many miles does it do per gallon?

27 30 g of grass seed is needed to sow 1 m² of lawn.
What weight of grass seed is needed to sow a rectangular lawn measuring 40 foot by 30 foot?

28 Concrete is sold by the cubic metre.
A path, 31 feet long, 5 feet wide and 1 foot 6 inches deep, is to be made of concrete.
How many cubic metres of concrete are needed?

Reading scales

EXAMPLES

1 (a) Use an appropriate metric unit to measure accurately the length of each of lines **A**, **B** and **C**.

    A ————————————————
    B ——————————————————————
    C ——————

   (b) What is the total length of lines **A**, **B** and **C**?
   (c) What is the difference in length between lines **B** and **C**?

   (a) Line **A** is 7 cm (or 70 mm) long.
         Line **B** is 12.5 cm (or 125 mm) long.
         Line **C** is 4.6 cm (or 46 mm) long.
   (b) 7 + 12.5 + 4.6 = 24.1 cm (or 241 mm).
   (c) 12.5 − 4.6 = 7.9 cm (or 79 mm).

2 Chandni, Jill and Susan measured their weights.
The diagram shows the readings on the scale.
What are each of their weights?

Chandni weighs 46 kg.   Jill weighs 55 kg.   Susan weighs 51 kg.

Exercise **33.4**

1 Measure the lengths of these lines.

    X ————————————————————
    Y ————————

   (a) What is the length of each line in centimetres?
   (b) What is the length of each line in millimetres?

2 Read each of the following scales at pointers **A**, **B** and **C**.
(a)   (b)   (c)

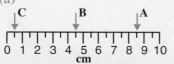
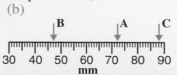
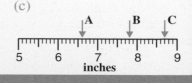

Understanding and Using Measures

3 (a) Read each of the following scales at pointers **A**, **B** and **C**.

(i)

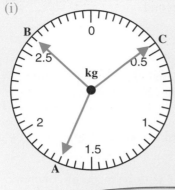

(ii)

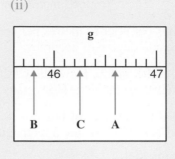

(iii)

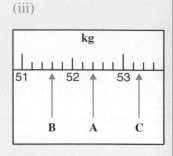

(iv)

(v)

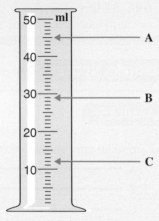

(b) For each of the above scales work out the difference between the highest and lowest readings.

4 What is the temperature shown by pointers **A**, **B** and **C**?

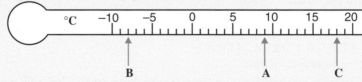

5

The diagram shows the petrol gauge on a car.
The car's petrol tank holds 12 gallons when full.

(a) How many gallons are in the petrol tank?
(b) How many litres are in the petrol tank?

6 This diagram shows a speedometer on a car.
What is the speed when the pointer is at **A**, **B** and **C**?

(a) Give your answers in miles per hour.
(b) Give your answers in kilometres per hour.

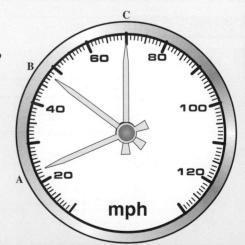

Accuracy in measurement

No measurement is ever exact and measurements given to the nearest whole unit may be inaccurate by up to one half of a unit in either direction.

For example: Harry weighs 54 kg, correct to the nearest kilogram.
Whole unit = 1 kg.   Possible inaccuracy = 1 kg ÷ 2 = 0.5 kg.
So, Harry's actual weight is any weight from 53.5 kg to 54.5 kg.
This can be written as the inequality: 53.5 kg ≤ Harry's weight < 54.5 kg.

EXAMPLE

Harry is 1.63 metres tall, to the nearest centimetre.
What is Harry's minimum possible height?
Whole unit = 1 cm.   Possible inaccuracy = 1 cm ÷ 2 = 0.5 cm = 0.005 m.
Harry's minimum height = 1.63 − 0.005 = 1.625 m.

Exercise 33.5

1   The length of a field is 264 metres, correct to the nearest metre.
What is the minimum possible length of the field?

2   The weight of a necklace is 32 grams, to the nearest gram.
Copy and complete the inequality   …… ≤ weight of necklace < ……

3   A post is 1.6 m in height, correct to the nearest tenth of a metre.
What are the minimum and maximum possible heights of the post?

4   Jayne runs 4.8 km, correct to the nearest hundred metres.
What is the minimum distance Jayne has run?

5   A concrete block weighs 650 grams, correct to the nearest 10 grams.
What are the limits between which the weight of the block lies?

6   What is the minimum time for a race timed at 12.63 seconds, measured to the nearest one hundredth of a second?

Dimensions and formulae

Formulae can be used to calculate perimeters, areas and volumes of various shapes.
By analysing the **dimensions** involved it is possible to decide whether a given formula represents a perimeter, an area or a volume.

Length (L)   has **dimension 1.**
Length (L) × Length (L) = **Area** (L²)   has **dimension 2.**
Length (L) × Length (L) × Length (L) = **Volume** (L³) has **dimension 3.**

The size of this square based cuboid depends on:
x, the length of the side of the square base, y, the height of the cuboid.

The total **length** of the edges of the cuboid is given by the formula: $E = 8x + 4y$
This formula involves: Numbers: 8 and 4
Lengths (L): x and y
The formula has **dimension 1**.

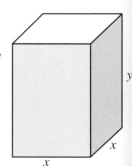

The total **surface area** of the cuboid is given by the formula: $S = 2x^2 + 4xy$
This formula involves: Numbers: 2 and 4
Areas (L²): $x \times x$ and $x \times y$
The formula has **dimension 2**.

The **volume** of the cuboid is given by the formula: $V = x^2y$
This formula involves: Volume (L³): $x \times x \times y$
This formula has **dimension 3**.

Understanding and Using Measures

EXAMPLE

In each of these expressions the letters a, b and c represent lengths.
Use dimensions to check whether the expressions could represent a perimeter, an area or a volume.

(a) $2a + 3b + 4c$ (b) $3a^2 + 2b(a + c)$
(c) $2a^2b + abc$ (d) $3a + 2ab + c^3$

> **Note:**
> When checking formulae and expressions, numbers can be ignored because they have no dimension.
> $\equiv$ means 'is equivalent to'.

(a) $2a + 3b + 4c$
Write this using dimensions.
$L + L + L \equiv 3L \equiv L$
$2a + 3b + 4c$ has dimension 1 and could represent a perimeter.

(b) $3a^2 + 2b(a + c)$
Write this using dimensions.
$L^2 + L(L + L)$
$\equiv L^2 + L(2L)$
$\equiv L^2 + 2L^2$
$\equiv 3L^2$
$\equiv L^2$
$3a^2 + 2b(a + c)$ has dimension 2 and could represent an area.

(c) $2a^2b + abc$
Write this using dimensions.
$L^2 \times L + L \times L \times L$
$\equiv L^3 + L^3$
$\equiv 2L^3$
$\equiv L^3$
$2a^2b + abc$ has dimension 3 and could represent a volume.

(d) $3a + 2ab + c^3$
Write this using dimensions.
$L + L \times L + L^3$
$\equiv L + L^2 + L^3$
The dimensions are **inconsistent**.
$3a + 2ab + c^3$ does not represent a perimeter, an area or a volume.

Exercise 33.6

1 p, q, r and x, y, z represent lengths.
For each formula state whether it represents a length, an area or a volume.

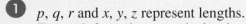
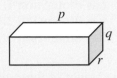
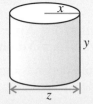

(a) pq (b) $2\pi x$ (c) $p + q + r$ (d) πz
(e) pqr (f) $2(pq + qr + pr)$ (g) $\pi x^2 y$ (h) $2\pi x(x + y)$

2 In each of the expressions below, x, y and z represent lengths.
By using dimensions decide whether each expression could represent a perimeter, an area, a volume or none of these. Explain your answer in each case.

(a) $x + y + z$ (b) $xy + xz$ (c) xyz (d) $x^2(y^2 + z^2)$

(e) $x(y + z)$ (f) $\dfrac{x^2}{y}$ (g) $\dfrac{xz}{y}$ (h) $x + y^2 + z^3$

(i) $xy(y + z)$ (j) $x^3 + x^2(y + z)$ (k) $xy(y^2 + z)$ (l) $x(y + z) + z^2$

3 The diagram shows a discus. x and y are the lengths shown on the diagram.
These expressions could represent certain quantities relating to the discus.

$$\pi(x^2 + y^2) \qquad \pi x^2 y^2 \qquad \pi xy \qquad 2\pi(x + y)$$

(a) Which of them could be an expression for:
  (i) the longest possible distance around the discus,
  (ii) the surface area of the discus?
(b) Use dimensions to explain your answers to part (a).

4 p, q, r and s represent the lengths of the edges of this triangular prism.
Match the formulas to the measurements.

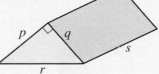

Formulas $\frac{1}{2}pqs \qquad 2\left(p + q + r + \frac{3s}{2}\right) \qquad s(p + q + r) + pq$

Measurements Edge length   Surface area   Volume

5 These arrows are similar.

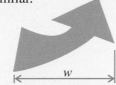

w represents the width of any arrow.
k and c are numbers.
H represents the height of the arrow and A its area.

Which of the following statements could be correct and which **must** be wrong?
(a) $H = kw$ (b) $H = ckw$ (c) $H = kw + c$
(d) $A = cw$ (e) $A = kw^2$ (f) $A = kw^3$
Give a reason for each of your answers and where you think the formula **must** be wrong suggest what it might be for.

What you need to know

- The common units – both **metric** and **imperial** – used to measure **length**, **mass** and **capacity**.
- How to convert from one unit to another. This includes knowing the connection between one metric unit and another and the approximate equivalents between metric and imperial units.

Metric Units	**Imperial Units**	**Conversions**
Length	**Length**	**Length**
1 kilometre (km) = 1000 metres (m)	1 foot = 12 inches	5 miles is about 8 km
1 m = 100 centimetres (cm)	1 yard = 3 feet	1 inch is about 2.5 cm
1 cm = 10 millimetres (mm)		1 foot is about 30 cm
	Mass	
Mass	1 pound = 16 ounces	**Mass**
1 tonne (t) = 1000 kilograms (kg)	14 pounds = 1 stone	1 kg is about 2.2 pounds
1 kg = 1000 grams (g)		
	Capacity and volume	**Capacity and volume**
Capacity and volume	1 gallon = 8 pints	1 litre is about 1.75 pints
1 litre = 1000 millilitres (ml)		1 gallon is about 4.5 litres
$1 \, cm^3 = 1 \, ml$		

- How to change between units of area. For example $1 \, m^2 = 10\,000 \, cm^2$.
- How to change between units of volume. For example $1 \, m^3 = 1\,000\,000 \, cm^3$.
- How to estimate length, mass and capacity using appropriate units.
- How to read scales accurately.

You should be able to:
- Recognise limitations on the accuracy of measurements. A measurement given to the nearest whole unit may be inaccurate by up to one half of a unit in either direction.
- Analyse the **dimensions** of a formula to decide whether a given formula represents a **length** (dimension 1), an **area** (dimension 2) or a **volume** (dimension 3).

Do not use a calculator for questions 1 to 5.

1 The diagram shows the amount of milk in a measuring jug.

(a) How many millilitres of milk are in the jug?
(b) A different jug holds 1500 ml.
How many litres is this?

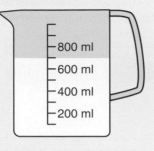

AQA

2 Sid's height is measured in metres.

(a) Write down Sid's height in metres.
(b) Toby's height is 1.86 metres.
How many centimetres taller is Toby than Sid?

AQA

3 How many magazines, each 0.6 cm thick, will fit on a bookcase shelf which is exactly 1.2 m wide?

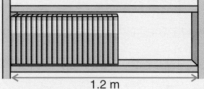

1.2 m

4 One glass of lemonade contains 300 ml.
How many glasses of lemonade can be poured from a jug which contains 2.4 litres?

5 From the list:

| 300 km | 300 m | 300 cm | 300 mm |

choose the best one to complete each of the following sentences.
(a) A school ruler is …… long.
(b) The distance from London to Manchester is ……
(c) A school field is …… long.

AQA

6 Use the following conversions to change 2 tonnes into pounds (lb).

> 1 tonne = 1000 kg
> 1 kg = 2.205 pounds (lb)

AQA

7 (a) How many centimetres are there in 39 millimetres?
(b) The distance from London to Edinburgh is 400 miles.
How many kilometres is this?

AQA

8 Ben is 5 feet 10 inches tall and weighs 72 kg.
Sam is 165 cm tall and weighs 11 stone 7 pounds.
Who is taller? Who is heavier?

9 The diagram shows the petrol gauge on a car.
The car's petrol tank holds 60 litres when full.

(a) Estimate how many litres are in the petrol tank.
(b) Estimate how many gallons are in the petrol tank.

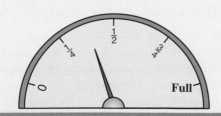

10 One fluid ounce = 28.4 millilitres
   (a) Convert 8 fluid ounces to millilitres.
   (b) Use your answer in (a) to explain why 8 fluid ounces is less than $\frac{1}{4}$ litre.   AQA

11 Change   1.5 m² to cm².

12 Dilip has a space in his living room which is $2\frac{1}{2}$ feet wide.
He has a bookcase which is 80 cm wide.
Will the bookcase fit into the space? You **must** show your working.   AQA

13 A rectangular strip of card measures 0.8 metres by 5 cm.

0.8 m

5 cm

Calculate the area of the card in:
   (a) square metres,   (b) square centimetres.

14 One litre of water weighs 1 kg. One litre of water is approximately 0.22 of a gallon.
   (a) Estimate the weight of one gallon of water in pounds.
   (b) Calculate how many litres are equivalent to one gallon.
        Give your answer correct to one decimal place.   AQA

15 Susan is driving through a French town.
The speed limit in the French town is 50 kilometres per hour.
Her speed is 35 miles per hour.
Is she breaking the speed limit? You must show all your working.   AQA

16 According to the instructions Vinyl Matt paint covers about 13 m²/litre.
   (a) Estimate the area covered with 2 gallons of paint.
   (b) Estimate how many litres of paint are needed to cover a wall which measures
        18 feet by 10 feet.

17 Francis can run at an average speed of 6 miles per hour.
Alistair can run at an average speed of 3 metres per second.
Francis says, "I can run faster than Alistair."
Is he correct? Show your working.

18 A blue whale weighs 140 tonnes to the nearest 10 tonnes.
What is the smallest possible weight of a blue whale?

19 Jean gives her height as 166 cm, to the nearest centimetre.
What are the limits between which her true height lies?

20 Bob weighs 76.9 kg, to the nearest 100 g.
What is Bob's minimum weight?

21 One of these formulae gives the volume of an egg of
height H cm and width w cm.

    Formula **A**: $\frac{1}{6}\pi H^2 w^2$ Formula **B**: $\frac{1}{6}\pi H w^2$ Formula **C**: $\frac{1}{6}\pi H w$

   (a) Which is the correct formula?
   (b) Explain how you can tell that this is the correct formula.

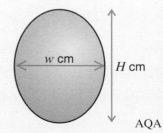

    AQA

22 In the following expressions, r, a and b represent lengths.
For each expression, state whether it represents a **length**, an **area**, a **volume**, or **none** of these.

   (a) πab (b) $\pi r^2 a + 2\pi r$ (c) $\dfrac{\pi ra^3}{b}$ AQA

Understanding and Using Measures

Do not use a calculator for this exercise.

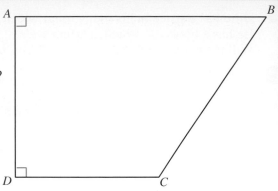

1 The diagram shows a scale drawing of a field.
Angle BAD = angle ADC = 90°.

(a) Which two lines are parallel to each other?
(b) Which line is perpendicular to DC?
(c) Measure angle ABC.
(d) Measure the length of the line CB.

2 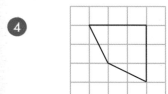 Copy the diagram and draw a reflection of the shape in the mirror line AB.

3 This shape has been drawn on 1 cm squared paper.

(a) What is the area of the shape?
(b) What is the perimeter of the shape?

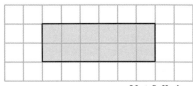

Not full size

4 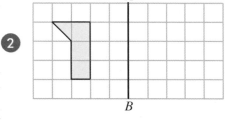 Copy the diagram and draw the line of symmetry.

5 (a) This shape is made using 1 cm cubes. What is the volume of the shape?

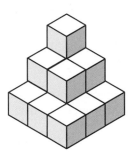

(b) 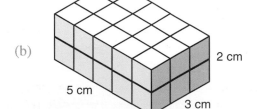 Calculate the volume of this cuboid.

5 cm
2 cm
3 cm

(c) Which of these diagrams is the net of a cube?

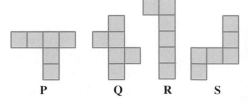

P   Q   R   S

6 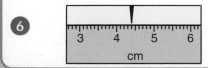 Write down the length shown by the scale.

3   4   5   6
cm

7 Which of these shapes are congruent to each other?

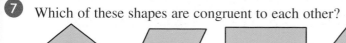

8 For each shape write down its order of rotational symmetry.

(a)   (b)   (c)

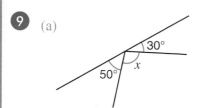 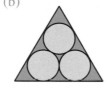 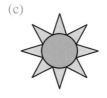

9 (a)

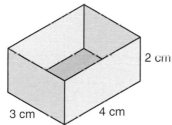

    (i) Find the size of the angle marked x.
         Give a reason for your answer.
    (ii) Which word describes angle x?
         acute   obtuse   reflex   right-angled

(b) In the diagram, lines *PQ* and RS are parallel.
They are crossed by two other straight lines.
Find the size of the angle marked y.

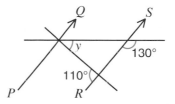

10

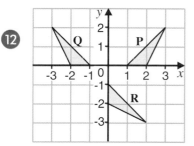

The diagram shows an open box.
(a) Draw an accurate net of the open box.
(b) Calculate the total outside surface area of the open box.

AQA

11 This 3-dimensional shape has been made using
linking cubes of side 1 cm.
On squared paper, draw diagrams to show:
(a) the plan of the shape,
(b) the elevation of the shape from **X**.

12

The diagram shows triangles **P**, **Q** and **R**.
(a) Describe the single transformation which takes **P** onto **Q**.
(b) Describe the single transformation which takes **P** onto **R**.

Copy triangle **P** onto squared paper.
(c) Draw an enlargement of triangle **P** with scale factor 2,
centre (0, 0).

13 The diagram shows a parallelogram and a rectangle.
Both shapes have the same area.
Calculate the length of the rectangle.

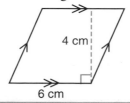 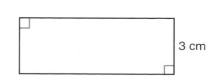

14 A building plot is a triangle with sides of length 60 m, 90 m and 100 m.
Construct an accurate scale drawing of the building plot.
Use a scale of 1 cm to 20 cm.

15 (a) Which of these shapes are similar to each other?

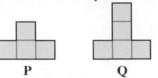
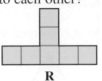
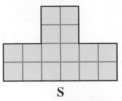

| **P** | **Q** | **R** | **S** |

(b) This shape is enlarged with scale factor 2.
  (i) What is the perimeter of the enlarged shape?
  (ii) What is the area of the enlarged shape?

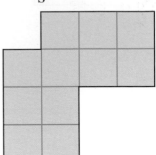

16 (a) 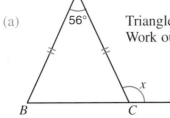 Triangle *ABC* is isosceles. *AB* = *AC*.
Work out the size of angle *x*.

(b) *BE* is parallel to *CD*.
  (i) Write down the size of angle *p*.
  (ii) Work out the size of angle *q*.

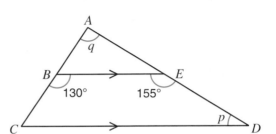

AQA

17 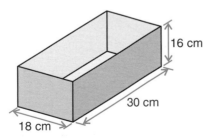 A shoe box is in the shape of a cuboid.
Calculate the volume of the box.

AQA

18 The diagram shows a rectangle *ABCD*.
The coordinates of *A*, *B* and *C* are given.
(a) Write down the equations of the lines of
symmetry of the rectangle *ABCD*.

(b) The rectangle *ABCD* has been translated **9 units to the left and 4 units down**.
Its new position is shown by the rectangle *EFGH*.

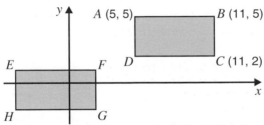

  (i) Write down the coordinates of *E*.
  (ii) Describe the translation that would move the rectangle back to its original position.

AQA

364

19 The diagram shows the shape of a car park.

(a) Calculate the perimeter of the car park.
(b) Calculate the area of the car park.

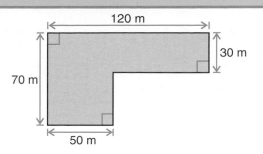

120 m
30 m
70 m
50 m

20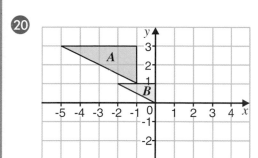

(a) Describe fully the single transformation which maps *A* onto *B*.
(b) *B* is mapped onto *C* by a translation with vector $\begin{pmatrix} 3 \\ -2 \end{pmatrix}$.
Draw a diagram to show the positions of *B* and *C*.
(c) *B* is mapped onto *D* by a rotation, 90° clockwise about (2, 1).
Show the position of *D* on your diagram.

21 *Y* is 50 m from *X* on a bearing of 080°.
Z is 70 m from *Y* on a bearing of 110°.
(a) Make a scale drawing to show the positions of *X*, *Y* and *Z*. Use a scale of 1 cm to 10 m.
(b) Find by measurement the distance and bearing of *X* from *Z*.

22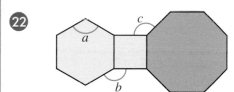

Each shape is a regular polygon.
Work out the size of each lettered angle.

23 Two straight roads are shown on the diagram.
A new gas pipe is to be laid from
Bere equidistant from the two roads.
The diagram is drawn to a scale of 1 cm to 1 km.
(a) Copy the diagram and construct the path of the gas pipe.

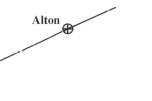

Alton
Bere
Cole

(b) The gas board needs a construction site depot.
The depot must be equidistant from Bere and Cole.
The depot must be less than 2 km from Alton.
Draw loci on your diagram to represent this information.
(c) The depot must be nearer the road through Cole than the road through Alton.
Mark on your diagram, with a cross, a possible position for the site depot.

24 Colin and Dexter are standing next to each other in the sunshine.
Colin is 150 cm tall and his shadow is 240 cm long.
Dexter is 165 cm tall. How long is his shadow?

25 The diagram shows the plan of a children's playground.
(a) What is the area of the playground?
(b) (i) Find the length of *BC*.
(ii) A fence is to be put around the perimeter of the playground.
Find the length of the fence.

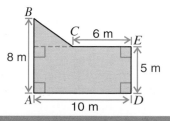

B
C 6 m
E
8 m
5 m
A
10 m
D

Shape, Space and Measures
Calculator Paper

You may use a calculator for this exercise.

1 Here is a map of Green Island.

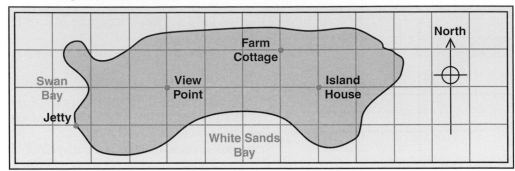

(a)   What is the name of the place which is West of View Point?
(b)   In which direction is Farm Cottage from Island House?

The length of side of each square on the map is 1 km.
(c)   How far is it from the Jetty to Island House?
(d)   Estimate the area of the island.

2 Look at this list of metric units.

centimetre	kilogram	metre	centilitre	gram	square metre
litre	tonne	square centimetre	millimetre	kilometre	

Choose the unit that would be best to use for measuring:
(a)   the distance from Poole to Manchester,
(b)   the weight of a double decker bus,
(c)   the height of a tree,
(d)   the area of carpet needed to cover a floor.

3 (a)   How many vertices has a cuboid?
(b)   How many faces has a square-based pyramid?
(c)   Here are the nets of some 3-D shapes.
      Write down the mathematical name of each 3-D shape.

(i)   (ii)   (iii)

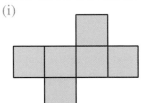   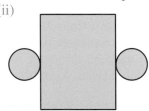   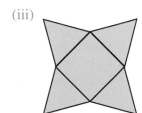

4 Here is a rectangle made from 36 square tiles. Its length is 18 tiles and its width is 2 tiles.

(a)   Write the length and width of two different rectangles which could be made using
      36 square tiles.
(b)   I have a single row of 72 square tiles.
      Give the length and width of two more rectangles which could be made using all
      these 72 tiles.

AQA

5 Find the size of the angles *a*, *b* and *c*.

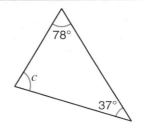

6

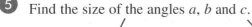

P Q R S

(a) Which of these diagrams has only one line of symmetry?
(b) Which of these diagrams has rotational symmetry but no lines of symmetry?   AQA

7 Draw an enlargement of shape *Q* using a scale factor of 3.

Q

8 (a) Anna measures her hand span to be 18 cm.
What is this in metres?

18 cm

(b) An exercise book is 0.3 cm thick.
What is this in millimetres?

0.3 cm   AQA

9 (a) Which of the following is **NOT** the net of a cuboid?

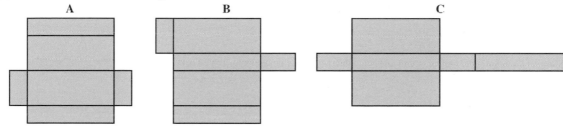

A   B   C

(b) A cuboid measures 2.5 cm by 4 cm by 1.5 cm.
Calculate the volume of the cuboid.

10 A triangle with one side extended is shown.
Explain why the angle marked *a* is 105°.

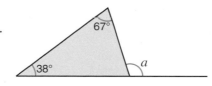

67°

38°   *a*

11

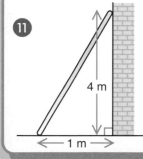

4 m

1 m

The ladder is put against the wall, as shown in this sketch.
The bottom of the ladder is 1 metre away from the wall.
It reaches 4 metres up the wall.
(a) Make a scale drawing to show the position of the ladder.
Use a scale of 4 centimetres to represent 1 metre.
(b) Use your scale drawing to work out the length of the ladder.
(c) Measure and write down the angle between the ladder and the ground.   AQA

12 By drawing at least six shapes, show how this shape can be used to make a tessellation.

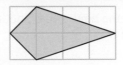

13 The map shows the positions of two towns, Akari and Borland.
 (a) What is the actual distance between Akari and Borland?
 (b) What is the bearing of Borland from Akari?

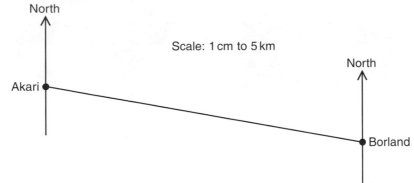

North

Scale: 1 cm to 5 km

Akari ●

North

● Borland

14

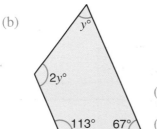

Work out the area of the shaded part of this rectangle.

15 (a) Work out the size of the angle marked x.
 Give a reason for your answer.

 (b)

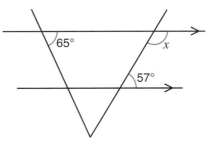

65° x

57°

(i) What special name is given to this shape?
 Give a reason for your answer.
(ii) Find the value of y.

16 Copy the diagram onto a grid. Allow values for x and y from -5 to 5.

 (a) Translate shape L by 2 units right and 3 units down.
 (b) Enlarge shape L with centre $(2, 1)$ and scale factor 2.

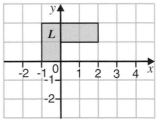

17 What is the area of the triangle?

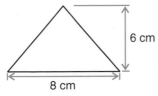

6 cm

8 cm

18 John uses a piece of string to measure the perimeter of shapes.
 It fits exactly round a rectangle 10 cm by 8 cm.
 He fits it exactly round a square. How long is one side of the square?

AQA

19 The diagram shows a rectangular doormat.
What is the area of the doormat
  (a) in cm²,
  (b) in m²?

1.5 m
WELCOME 80 cm

20 A triangle has sides of length 7 cm, 5 cm and 4 cm.
  (a) Make an accurate drawing of the triangle.
  (b) Work out the area of the triangle.

21 (a) Adrian is 6 feet 3 inches tall. Work out Adrian's height in centimetres.
  (b) Adrian weighs 78 kg. Work out Adrian's weight in pounds.

22 A table top is a circle of radius 50 cm.
  (a) Calculate the circumference of the table top.
  (b) Calculate the area of the table top.

23 What is the size of one exterior angle of a regular pentagon?

24 The front wheels of a tractor each have diameter 100 cm. The tractor is driven 100 metres.
How many complete turns do each of the front wheels make?

25 The diagram shows a rectangular field *ABCD*.
The side *AB* is 80 m long. The side *BC* is 50 m long.
Draw the diagram using a scale of 1 cm to 10 m.
Treasure is hidden in the field.
  (a) The treasure is equidistant from the sides *AB* and *AD*.
       Construct the locus of points for which this is true.
  (b) The treasure is 60 m from corner *C*.
       Construct the locus of points for which this is true.
  (c) Mark with an *X* the position of the treasure.

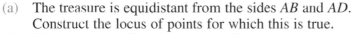
A   B

D   C

AQA

26 (a) Calculate the area of this trapezium.
  (b) Calculate the perimeter of this trapezium.

17 cm
6 cm
8 cm
AQA

27 Describe the single transformation which maps
  (a) *A* onto *B*,
  (b) *A* onto *C*.

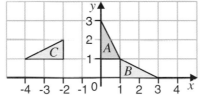

28 The diagram shows a circular paddling pool with a vertical side.
The radius of the pool is 35 cm.
The water in the pool is 8 cm deep.
Calculate the volume of water in the pool.

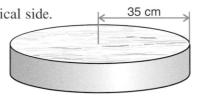
35 cm

AQA

29

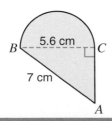
5.6 cm
B ------ C
7 cm
A

This shape is made up of a right-angled triangle and a semi-circle.
Calculate the total area of the shape.
Give your answer to a suitable degree of accuracy.

CHAPTER 34

Collection and Organisation of Data

To answer questions such as: Which is the most popular colour of car?
Is it going to rain tomorrow?
Which team won the World Cup in 2002?

we need to collect data.

Primary and secondary data

When data is collected by an individual or organisation to use for a particular purpose it is called **primary data**.
Primary data is obtained from experiments, investigations, surveys and by using questionnaires.

Data which is already available or has been collected by someone else for a different purpose is called **secondary data**.
Sources of secondary data include the Annual Abstract of Statistics, Social Trends and the Internet.

Data

Data is made up of a collection of **variables**. Each variable can be described, numbered or measured.

Data which can only be **described** in words is **qualitative**.
Such data is often organised into categories, such as make of car, colour of hair, etc.

Data which is given **numerical** values, such as shoe size or height, is **quantitative**.
Quantitative data is either **discrete** or **continuous**.
   Discrete data can only take certain values, usually whole numbers, but may include fractions (e.g. shoe sizes).
   Continuous data can take any value within a range and is measurable
   (e.g. height, weight, temperature, etc.).

EXAMPLES

The taste of an orange is a qualitative variable.
The number of pips in an orange is a discrete quantitative variable.
The surface area of an orange is a continuous quantitative variable.

Exercise 34.1

State whether the following data is qualitative or quantitative.
If the data is quantitative state whether it is discrete or continuous.

1. The colours of cars in a car park.
2. The weights of eggs in a carton.
3. The numbers of desks in classrooms.
4. The names of students in a class.
5. The sizes of spanners in a toolbox.
6. The depths that fish swim in the sea.
7. The numbers of goals scored by football teams on a Saturday.
8. The brands of toothpaste on sale in supermarkets.
9. The sizes of ladies dresses in a store.
10. The heights of trees in a wood.

Collection of data

Data can be collected in a variety of ways:

by observation, by interviewing people and by using questionnaires.

The method of collection will often depend on the type of data to be collected.

Data collection sheets

Data collection sheets are used to record data.

To answer the question, "Which is the most popular colour of car?" we could draw up a simple data collection sheet and record the colours of passing cars by observation.

EXAMPLE

A **data collection sheet** for colour of car is shown, with some cars recorded.

Colour of car	Tally	Frequency
Black	||	2
Blue	⫽⫽⫽ ⫽⫽⫽ |||	13
Green	||||	4
Red	⫽⫽⫽ ⫽⫽⫽ |	
Silver	⫽⫽⫽ ||	
White	⫽⫽⫽ ⫽⫽⫽ ||||	
	Total	

The colour of each car is recorded in the **tally** column by a single stroke.

To make counting easier, groups of 5 are recorded as ⫽⫽⫽.

How many red cars are recorded?
How many cars are recorded altogether?

The total number of times each colour appears is called its **frequency**.
A table for data with the totals included is called a **frequency distribution**.

For large amounts of discrete data, or for continuous data, we organise the data into **groups** or **classes**. When data is collected in groups it is called a **grouped frequency distribution** and the groups you put the data into are called **class intervals**.

EXAMPLE

The weights of 20 boys are recorded in the grouped frequency table shown below.

Weight w kg	Tally	Frequency
$50 \leqslant w < 55$	|	1
$55 \leqslant w < 60$	|||	3
$60 \leqslant w < 65$	⫽⫽⫽ ||||	9
$65 \leqslant w < 70$	⫽⫽⫽ |	6
$70 \leqslant w < 75$	|	1
	Total	20

Weights are grouped into class intervals of equal width.

$55 \leqslant w < 60$

means 55 kg, or more, but less than 60 kg.

John weighs 54.9 kg. *In which class interval is he recorded?*
David weighs 55.0 kg. *In which class interval is he recorded?*

What is the width of each class interval?

1 The tally chart shows the number of glass bottles put into a bottle bank one day.

Colour of glass	Tally
Clear	ЖЖ ЖЖ ЖЖ ||
Brown	ЖЖ ЖЖ ЖЖ ЖЖ |
Green	ЖЖ ЖЖ ЖЖ

  (a) How many green bottles were put into the bottle bank?
  (b) How many brown bottles were put into the bottle bank?
  (c) How many **more** brown bottles than clear bottles were put into the bottle bank?

2 John is doing a project about sport.
He asks people which sport they like best.
The data collection sheet shows his results.

Sport	Tally
Cricket	||
Football	ЖЖ ЖЖ ЖЖ ЖЖ ЖЖ |
Hockey	||||
Rugby	ЖЖ |||
Tennis	ЖЖ

  (a) How many people like rugby best?
  (b) How many people did John ask?
  (c) John says, "More than half of the people I asked liked football best."
       Is he correct?
       Give a reason for your answer.

3 Helen throws a dice 50 times.
The result of each throw is shown.

1	4	3	6	5	4	3	2	1	6
4	5	2	3	4	5	6	4	5	3
1	2	3	4	2	3	5	1	1	4
5	6	4	3	2	5	4	6	5	6
2	3	1	3	4	1	6	5	2	2

  (a) Copy and complete this table to record her results.

Number on dice	Tally
1	
2	
3	
4	
5	
6	

  (b) Which number occurred most frequently?

4 The colours of 40 cars in a car park are shown.

red	red	blue	green	white	grey
blue	red	red	grey	white	green
red	white	white	blue	red	white
blue	blue	green	black	white	blue
red	silver	silver	blue	red	red
silver	white	white	red	blue	green
red	blue	silver	white		

(a) Make a frequency table for the data.
(b) Which colour of car is most popular?
(c) How many **more** white cars than black cars are in the car park?

5 The days of the week on which some students were born are recorded.

Monday	Monday	Sunday	Wednesday	Thursday
Friday	Saturday	Tuesday	Monday	Friday
Thursday	Sunday	Monday	Friday	Tuesday
Thursday	Wednesday	Tuesday	Monday	Wednesday
Friday	Monday	Saturday	Friday	Thursday
Tuesday	Thursday	Monday	Sunday	Tuesday
Saturday	Wednesday	Friday	Thursday	Tuesday
Monday	Wednesday	Friday	Sunday	Thursday
Tuesday	Wednesday	Sunday		

(a) Make a frequency table for the data.
(b) How many students are included?
(c) On which day of the week did the most births occur?
(d) How many of these students were born on either a Saturday or a Sunday?

6 The ages, in years, of 40 people are shown below.

27	34	54	57	3	12	15	19
29	30	33	47	35	20	39	28
9	11	26	42	50	26	10	7
33	49	21	18	1	25	24	34
19	20	27	37	43	56	37	34

(a) Copy and complete the grouped frequency table for the data given.

Age (years)	Tally	Frequency
0 - 9		
10 - 19		
20 - 29		

(b) What is the width of each class interval?
(c) How many people are in the class interval 30 - 39?
(d) How many people are less than 20 years old?
(e) How many people are 40 years old, or older?

The heights, in centimetres, of 36 girls are recorded as follows.

148	161	175	156	155	160	178	159	170
163	147	150	173	169	170	174	166	163
162	158	155	165	168	154	156	163	167
172	170	165	160	164	172	157	173	161

(a) Copy and complete the grouped frequency table for the data.

Height h cm	Tally	Frequency
$145 \leqslant h < 150$		
$150 \leqslant h < 155$		

(b) What is the width of each class interval?
(c) How many girls are in the class interval $155 \leqslant h < 160$?
(d) How many girls are less than 160 cm?
(e) How many girls are 155 cm or taller?

8 Draw up a data collection sheet to record the month in which people were born.
Collect data from 50 people.
(a) Make a frequency table for the data.
(b) In which month did the most births occur?

Databases

If we need to collect data for more than one type of information, for example: the make, colour, number of doors and mileage of cars, we will need to collect data in a different way.

We could create a **data collection card** for each car.

Car	1
Make	Vauxhall
Colour	Grey
Number of doors	3
Mileage	18 604

Alternatively, we could use a data collection sheet and record all the information about each car on a separate line.

This is an example of a simple **database**.

Car	Make	Colour	Number of doors	Mileage
1	Vauxhall	Grey	3	18 604
2	Ford	Blue	2	33 216
3	Ford	White	5	27 435
4	Nissan	Red	4	32 006

When all the data has been collected, separate frequency or grouped frequency tables can be drawn up.

1 The database gives information about the babies born at a maternity hospital one day.

Baby's name	Time of birth	Weight (kg)	Length (cm)
Alistair	0348	3.2	44
Francis	0819	3.5	48
Louisa	1401	3.7	47

(a) Which baby was the longest?
(b) Which baby was the heaviest?
(c) Which baby was born first?

2 Part of a database on some students is shown.

Student	Gender	Day of birth	Month of birth
Alex	F	Tuesday	January
Brian	M	Thursday	June
Cody	F	Monday	October
David	M	Friday	May
Evelyn	F	Saturday	September
Fay	F	Monday	February
George	M	Tuesday	May
Harry	M	Wednesday	September
Irene	F	Monday	September
Jay	M	Thursday	April

(a) How many students were born on a Tuesday?
(b) How many female students were born in September?
(c) Which student was born on a Monday in February?

3 The following database gives information about ski resorts for one day in January.

Resort	Temp °C	Depth of snow Lower (cm)	Depth of snow Upper (cm)	Piste	Weather
Aspen	2	50	70	good	fair
Cervinia	−7	80	170	hard	windy
Cortina	−2	10	60	good	fine
Kitzbuhel	−2	40	90	good	cloudy
Klosters	0	45	136	good	cloudy
Meribel	1	55	150	fair	fine
Soldeu	−4	50	90	good	sunny
Val d'Isere	−9	100	130	good	sunny

(a) Which resort was coldest?
(b) Which resort was windy?
(c) At which resort was the piste hard?
(d) What was the depth of snow on the upper slopes at Klosters?
(e) At which resort was the depth of snow between 50 cm and 90 cm?

4 The following database gives information about a group of 16 year old students.

Student	Gender	Height (cm)	Shoe size	Pulse rate (beats/min)
Mary	F	162	6	72
Alan	M	170	8	64
Jim	M	186	10	72
Tony	M	180	10	68
Laura	F	172	8	70
Jane	F	168	7	82
Wendy	F	155	5	72
Mark	M	180	9	68
Peter	M	168	8	62
Beryl	F	166	7	72

(a) Which student has the smallest shoe size?
(b) What is the gender of the student with the highest pulse rate?
(c) Which students are the same height?
(d) How many students are taller than Jane?
(e) Which students have a pulse rate of 72?
(f) What is the difference between the highest pulse rate and the lowest pulse rate?

5 A database of cars is shown.

Car	Make	Colour	Number of doors	Mileage
1	Vauxhall	Grey	3	18 604
2	Ford	Blue	2	33 216
3	Ford	White	5	27 435
4	Nissan	Red	4	32 006
5	Vauxhall	Blue	4	31 598
6	Ford	Green	3	37 685
7	Vauxhall	Red	3	21 640
8	Nissan	White	2	28 763
9	Ford	White	3	30 498
10	Vauxhall	White	5	9 865
11	Nissan	Red	3	7 520
12	Vauxhall	Grey	5	16 482

(a) (i) Draw up separate frequency tables for make, colour and number of doors.
   (ii) Draw up a grouped frequency table for mileage.
       Use class intervals of 5000 miles,
       starting at $0 \leqslant m < 5000$, $5000 \leqslant m < 10\,000$,   ...

(b) (i) Which make of car is the most popular?
   (ii) How many Ford cars are white?
   (iii) How many cars have a mileage of 30 000 or more?
   (iv) How many cars have exactly 3 doors?

6   (a)   By using copies of the data collection card for cars or by using a copy of the data collection sheet, record information about the cars in your school car park.

    (b)   Draw up frequency tables for make, colour and number of doors and a grouped frequency table for mileage.

    (c)   (i)   Which make of car is the most popular?

          (ii)   Which colour of car is the most popular?

         (iii)   How many cars have a mileage of 30 000 or more?

         (iv)   How many cars have exactly 3 doors?

7   Use data collection cards to collect information about students in your class.
Include gender, height, shoe size and pulse rate.

    (a)   What is the smallest shoe size for students in your class?

    (b)   What is the gender of the student with the highest pulse rate?

    (c)   What is the difference between the highest pulse rate and the lowest pulse rate?

    (d)   What is the height of the tallest student?

    (e)   What differences are there in the data collected for male and female students?

8   (a)   Design a data collection card to collect information on the leisure time activities of students.

    (b)   Draw up frequency or grouped frequency tables for the data.

    (c)   Which leisure time activity is the most popular?

    (d)   What differences are there in the leisure time activities of male and female students?

Questionnaires

Questionnaires are frequently used to collect data.
In business they are used to get information about products or services and in politics they are frequently used to test opinion on a range of issues and personalities.

When constructing questions for a questionnaire you should:

  (1)   use simple language, so that everyone can understand the question;

  (2)   ask short questions which can be answered precisely, with a "yes" or "no" answer, a number, or a response from a choice of answers;

  (3)   provide tick boxes, so that questions can be answered easily;

  (4)   avoid open-ended questions, like: "What do you think of education?" which might produce long rambling answers which would be difficult to collate or process;

  (5)   avoid leading questions, like: "Don't you agree that there is too much bad language on television?" and ask instead:
"Do you think that there is too much bad language on television?"   Yes ☐   No ☐

  (6)   ask questions in a logical order.

Multiple-response questions

In many instances a choice of responses should be provided.

Instead of asking, "How old are you?" which does not indicate the degree of accuracy required and many people might consider personal, we could ask instead:

Which is your age group?

          under 18 ☐   18 to 40 ☐   41 to 65 ☐   over 65 ☐

Notice there are no gaps and only **one** response applies to each person.

Sometimes we invite **multiple responses** by asking questions, such as:

Which of these soaps do you watch?

    Coronation Street ☐   EastEnders ☐   Emmerdale ☐   Hollyoaks ☐

Tick as many as you wish.

1 John wants to find out what students think about the library service at his college.
Part of the questionnaire he has written is shown.

> Q1. What is your name? . . . .
>
> Q2. How many times a week do you go to the library?
>
> ☐ Often ☐ Sometimes ☐ Never

(a) Why should Q1 not be asked?
(b) What is wrong with the choices offered in Q2.

2 A questionnaire contained the following question:

**Don't you agree that reading a newspaper is better
for a child's eduction that watching TV?**

Give a reason why this question is not suitable for the questionnaire.

3 Susan wants to find out what people think about the Health Service.
Part of the questionnaire she has written is shown.

> Q4. What is your date of birth?
>
> Q5. Don't you agree that waiting lists for operations are too long?
>
> Q6. How many times did you visit your doctor last year?
>
> ☐ less than 5 ☐ 5 - 10 ☐ 10 or more

(a) Why should Q4 not be asked?
(b) Give a reason why Q5 is unsuitable.
(c) (i) Explain why Q6 is unsuitable in its present form.
    (ii) Rewrite the question so that it could be included in the questionnaire.

4 In preparing the questions for a questionnaire on radio listening habits the following questions
were rejected.
(a) When do you listen to the radio?
(b) What do you like about radio programmes?
(c) Don't you agree that the radio gives the best news reports?
Explain why each question is unsuitable and rewrite the question so that it could be included
in the questionnaire.

5 In preparing questions for a survey on the use of a library the following questions
were considered.
Explain why each question in its present form is unsuitable and rewrite the question.
(a) How old are you?
(b) How many times have you used the library?
(c) Which books do you read?
(d) How could the library be improved?

6 A mobile phone company wants to carry out a survey.
It wants to find out the distribution of the age and sex of customers and the frequency with
which they use the phone.
The company intends to use a questionnaire.
Write three questions and responses that will enable the company to carry out the survey.

7 A school is to conduct a homework survey. Suggest five questions which could be included.

8 A survey of reading habits is to be conducted. Suggest five questions which could be included.

9 A survey of eating habits is to be conducted. Suggest five questions which could be included.

Hypothesis

A **hypothesis** is a statement that may or may not be true.
To test a hypothesis we can construct a questionnaire, carry out a survey and analyse the results.

EXAMPLE

A questionnaire to test the hypothesis,
        "People think it is better to give than to receive,"
could include questions like these.

---

1. **Gender:**   male ☐   female ☐
2. **Age (years):** 11 - 16 ☐   17 - 21 ☐   22 - 59 ☐   60 & over ☐
3. **Do you think it is better to give than to receive?**
            Yes ☐   No ☐
4. **To which of the following have you given in the last year?**

    School ☐   Charities ☐   Church ☐
    Hospital ☐   Special appeals ☐   Homeless ☐

    Other (please list) _____

Suggest another question which could be included.

Sampling

When information is required about a small group of people it is possible to survey everyone.
When information is required about a large group of people it is not always possible to survey everyone and only a **sample** may be asked.
The sample chosen should be large enough to make the results meaningful and representative of the whole group or the results may be **biased**.
For example, to test the hypothesis,
            "Girls are more intelligent than boys,"
you would need to ask equal numbers of boys and girls from various age groups.

Exercise 34.5

1   George is investigating the cost of return journeys by train.
    He plans to ask ten passengers who are waiting at a station at midday the cost of their return journeys.
    Give two reasons why his sample may not be representative of all passengers.

2   Judy is investigating shopping habits.
    She plans to interview 50 women at her local supermarket on a Tuesday morning.
    Give three reasons why her sample may not be representative of all shoppers.

3   To investigate the hypothesis:
            "Children watch more television than adults."
    Harry asked 5 boys in his class and 5 teachers how much television they watched last night.
    Give three reasons why his sample is unsuitable.

4   Sam is investigating the hypothesis:
            "Men watch more football than women."
    Describe a suitable sample you could use to test this hypothesis.

Collection and Organisation of Data

5 Design a questionnaire to test the hypothesis:
"People think that everyone should take part in sport."
Describe the sample you could use to test this hypothesis.

6 Design a questionnaire to test the hypothesis:
"People think that animals should not be used to test drugs."
Describe the sample you could use to test this hypothesis.

7 Design a questionnaire to test the hypothesis:
"Children have too much homework."
Describe the sample you could use to test this hypothesis.

8 Explain how you could test the hypothesis:
"Boys are better at estimating than girls."

9 State one advantage and one disadvantage of a postal survey.

Two-way tables

We have already seen that the results of a survey can be recorded on data collection sheets and then collated using frequency or grouped frequency tables. We can also illustrate data using **two-way tables**.

A two-way table is used to illustrate the data for two different features (variables) in a survey.

EXAMPLE

The two-way table shows the results of a survey.

(a) How many boys wear glasses?
(b) How many children wear glasses?
(c) How many children were surveyed?
(d) Do the results prove or disprove the hypothesis:
"More boys wear glasses than girls"?

	Wear Glasses	
	Yes	No
Boys	4	14
Girls	3	9

(a) 4   (b) $4 + 3 = 7$ (c) $4 + 3 + 14 + 9 = 30$

(d) Disprove.   Boys: $\frac{4}{18} \times 100 = 22\%$ Girls: $\frac{3}{12} \times 100 = 25\%$

Exercise 34.6

1 A group of 25 students were each asked how many brothers and sisters they had. The table shows the results.

Number of sisters

		0	1	2	3
Number of brothers	0	5	1	2	0
	1	4	3	2	1
	2	2	3	1	0
	3	0	1	0	0

(a) How many students have no brothers or sisters?
(b) How many students have one sister?
(c) How many students have one brother and one sister?
(d) How many students have more brothers than sisters?
(e) How many students have the same number of brothers as sisters?

2   A tennis club has 37 members.

> 15 of the members are under 16 years old.
> 9 of the members are boys under 16 years old.
> 12 of the members are girls, 16 years old or over.

(a)   Copy and complete the two-way table.

	Under 16 years old	16 years old or over	Totals
Girls		12	
Boys	9		
Totals	15		37

(b)   How many members of the tennis club are boys, 16 years old or over?

3   Meera read a newspaper report which stated:

> **There is no connection between the number of bedrooms
> and the number of children in a house.**

Design a two-way table to record the number of bedrooms and the number of children in a sample of houses.

4   The two-way table shows information about the ages of people in a retirement home.

Age (years)

	60 - 64	65 - 69	70 - 74	75 - 79	80 and over
Men	0	2	5	8	1
Women	2	5	6	5	6

(a)   How many men are aged 75 - 79?
(b)   How many men are included?
(c)   How many people are aged 75 or more?
(d)   How many people are included?
(e)   What percentage of these people are aged 75 or more?

5   The two-way table shows information about a class of pupils.

	Can swim	Cannot swim
Boys	14	6
Girls	8	2

(a)   How many boys can swim?
(b)   How many boys are in the class?
(c)   What percentage of the boys can swim?
(d)   What percentage of the girls can swim?
(e)   Do the results prove or disprove the hypothesis:
> *"More boys can swim than girls"*?

Explain your answer.

6   The two-way table shows the results of a survey to test the hypothesis:
> *"More girls are left-handed than boys."*

	Left-handed	
	Yes	No
Boys	3	18
Girls	2	12

Do the results prove or disprove the hypothesis? Explain your answer.

7 The two-way table shows the results of a spelling test.

Number of spellings correct

	1 to 5	6 to 10	11 to 15	16 to 20
Male	1	3	6	5
Female	0	5	6	9

(a) How many females took the test?

(b) How many females got less than 11 spellings correct?

(c) John says, "Males are better at spelling because fewer males got less than 11 spellings correct."
Is he right?
Give a reason for your answer.

8 The two-way table shows the number of boys and girls in families taking part in a survey.

Number of girls	3	1		2		
	2	1	2	3		
	1	5	9		1	1
	0		3		2	
		0	1	2	3	4

Number of boys

(a) (i) How many families have two children?

     (ii) Does the data support the hypothesis:
       "More families have less than 2 children than more than 2 children"?
       Explain your answer.

(b) (i) How many girls are included in the survey?

     (ii) Does the data support the hypothesis:
       "More boys are born than girls"?
       Explain your answer.

9 The two-way table shows the age and gender of people taking part in a survey.

Age (years)

	Under 18	18 - 25	26 - 40	41 - 64	65 and over
Female	0	2	7	9	7
Male	0	4	17	19	10

Give two reasons why the data collected may not be representative of the whole population.

10 In a survey, 100 people were asked: "Would you like to be taller?"

> 58 of the people asked were men.
>
> 65 of the people asked said, "Yes."
>
> 24 of the women asked said, "No."

(a) Construct a two-way table to show the results of the survey.

(b) A newspaper headline stated:

> Over 80% of men would like to be taller.

Do the results of the survey support this headline?
Give a reason for your answer.

What you need to know

- **Primary data** is data collected by an individual or organisation to use for a particular purpose. Primary data is obtained from experiments, investigations, surveys and by using questionnaires.
- **Secondary data** is data which is already available or has been collected by someone else for a different purpose. Sources of secondary data include the Annual Abstract of Statistics, Social Trends and the Internet.
- **Qualitative** data – Data which can only be described in words.
- **Quantitative** data – Data that has a numerical value.
  Quantitative data is either **discrete** or **continuous**.
  Discrete data can only take certain values.
  Continuous data has no exact value and is measurable.
- **Data Collection Sheets** – Used to record data during a survey.
- **Tally** – A way of recording each item of data on a data collection sheet.
       A group of five is recorded as 卌.
- **Frequency Table** – A way of collating the information recorded on a data collection sheet.
- **Grouped Frequency Table** – Used for continuous data or for discrete data when a lot of data has to be recorded.
- **Database** – A collection of data.
- **Class Interval** – The width of the groups used in a grouped frequency distribution.
- **Questionnaire** – A set of questions used to collect data for a survey.
  Questionnaires should:
  (1) use simple language,   (4) avoid open-ended questions,
  (2) ask short questions which can be answered precisely,   (5) avoid leading questions,
  (3) provide tick boxes,   (6) ask questions in a logical order.
- **Hypothesis** – A hypothesis is a statement which may or may not be true.
- When information is required about a large group of people it is not always possible to survey everyone and only a **sample** may be asked.
  The sample chosen should be large enough to make the results meaningful and representative of the whole group (population) or the results may be **biased**.
- **Two-way Tables** – A way of illustrating two features of a survey.

Review Exercise 34

1 Here is some information about six students in a school.

No.	Name	Date of birth	Village	Stays for school lunch
1	Bright M.	12/02/95	Swinton	Yes
2	Patel D.	24/04/95	Bolton	Yes
3	Pearson M.	05/11/94	Bolton	No
4	Thompson A.	30/09/94	Bolton	Yes
5	Williams C.	15/07/95	Bolton	No
6	Yip D.	21/10/94	Goldthorpe	No

(a) What is the name of the oldest student shown in the list?
(b) Which students live in the village of Bolton **and** do not stay for school lunch?   AQA

2 Jane does a survey about vehicles passing her school.
She wants to know about the types of vehicles and their colours.
Design a suitable observation sheet to record this information.
Fill in your observation sheet as if you had carried out this survey.
You should invent suitable data for 25 vehicles.   AQA

3 The number of medals won by the top five countries in a Winter Olympics is shown.

	Gold	Silver	Bronze
Germany	9	9	6
Soviet Union	6	10	9
United States	4	4	0
Finland	4	3	6
Sweden	4	2	2

(a) How many Silver medals did Finland win?
(b) How many medals did Germany win altogether?
(c) Which two countries won the same number of medals?

4 The eye colour of a group of 12 students is shown.

blue	brown	green	blue	brown	brown
brown	blue	brown	green	brown	blue

(a) Copy and complete the frequency table.

Eye colour	Tally	Frequency
blue		
brown		
green		

(b) Which eye colour is the most frequent?

5 Laura writes a questionnaire to survey opinion on whether cars should be banned from the town centre.

(a) Which two of the following points are important when she is deciding which people to ask?
  A Ask people who look friendly.
  B Ask people at different times of the day.
  C Ask some men and some women.
  D Ask the first people she sees.

(b) Which two of the following points are important when she is writing the questionnaire?
  E Write polite questions.
  F Write as many questions as she can think of.
  G Write questions for car drivers only.
  H Write questions that do not require long answers.

AQA

6 Kathryn is conducting a survey on television viewing habits.
She thinks of two questions for the questionnaire.

Question 1. How old are you?	*Question 2.* When do you watch television?

(a) Explain why each of these questions is unsuitable.
(b) Rewrite each of these questions so that she could include them in her questionnaire. AQA

7 The table shows the age and gender of people taking part in a survey to test the hypothesis:
"Children have too much homework."

	Age (years)				
	Under 11	11 - 16	17 - 25	26 - 50	Over 50
Male	0	4	6	5	5
Female	0	0	0	0	0

Give three reasons why this is not a suitable sample.

8 This statement is made on a television programme about health.

"Three in every eight pupils do not take any exercise outside school."

(a) A school has 584 pupils.
According to the television programme, how many of these pupils do not take any exercise outside school?

(b) Claire says, "I go to the gym twice a week after school."
She decides to do a survey to investigate what exercise other pupils do outside school.
Write down **two** questions that she could ask.

(c) Matthew decides to do a survey in his school about the benefits of exercise.
He decides to ask the girls' netball team for their opinion.
Give **two** reasons why this is **not** a suitable sample to take.

(d) This is part of Matthew's questionnaire.

Question	*Don't you agree that adults who were sportsmen when they were younger suffer more from injuries as they get older?*
Response	*Tick one box*
	☐ *Yes* ☐ *Usually* ☐ *Sometimes* ☐ *Occasionally*

(i) Write down one criticism of Matthew's question.
(ii) Write down one criticism of Matthew's response section.

AQA

9 A travel agent says, "More women prefer holidays abroad than men."
The table shows the results of a survey to test this statement.

	Men	Women
Prefer holidays abroad	18	21
Prefer holidays in the UK	6	7

Do these results support the statement made by the travel agent?
Explain your answer.

10 The number of shops and the number of churches are recorded for each of twenty small villages.
The results are shown in the table.

Number of churches

		0	1	2	3
Number of shops	0	0	0	0	1
	1	0	1	0	5
	2	0	2	8	0
	3	2	1	0	0

(a) How many of the villages have 3 churches?
(b) How many of the villages have more shops than churches?
(c) Calculate the total number of churches in the twenty villages.

AQA

Pictograms and Bar Charts

Most people find numerical data easier to understand if it is presented in a pictorial or diagrammatical form. For this reason television reports, newspapers and advertisements frequently use graphs and diagrams to present data.

Pictograms

A **pictogram** uses symbols to represent information.
Each symbol can represent one or more items of data.

EXAMPLE

The table shows the number of drinks sold by a cafe in one day.

Coffee	. . . . .	120
Tea	. . . .	105
Milk Shake	.	20
Lemonade	. .	17

To draw a pictogram of this information we can use one cup symbol to represent 10 drinks.

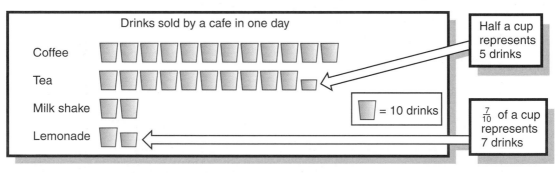

Drinks sold by a cafe in one day

Coffee
Tea
Milk shake
Lemonade

Half a cup represents 5 drinks

= 10 drinks

$\frac{7}{10}$ of a cup represents 7 drinks

Explain why some pictograms are difficult to read accurately.

Exercise 35.1

1 The pictogram shows the number of first class and second class stamps sold by a post office in one hour.

Number of stamps sold

= 10 stamps

First Class

Second Class

(a) How many first class stamps were sold?
(b) Estimate how many second class stamps were sold.

2 The pictogram shows the results of a survey of how students travel to college.
  (a) Fifteen students cycle to college.
       How many students walk?
  (b) How many students were included in the survey?

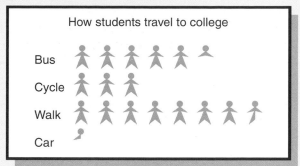

How students travel to college

3 One hundred boys were asked which sport they preferred.
The table shows the results.

Sport	Football	Cricket	Rugby	Hockey	Basketball
Number of boys	45	3	15	10	27

Draw a pictogram to represent this information.

Use ![person] = 5 boys.

4 The pictogram shows the colour of cars in a car park survey.

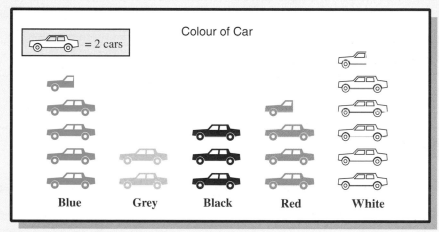

  (a) How many cars were blue?
  (b) How many cars were black?
  (c) Which colour is the most popular?
  (d) How many cars were included in the survey?

5 Fifty students were asked which european country they would visit next year.
The table shows the results.

Country	France	Germany	Spain	Italy
Number of students	23	15	7	5

Draw a pictogram to represent this information.

Use ![person] = 5 students.

Bar charts

Bar charts are a simple but effective way of displaying data.
Bars can be drawn either horizontally or vertically.

EXAMPLE

The table shows how a group of boys travelled to school one day.

Method of travel	Bus	Cycle	Car	Walk
Number of boys	2	7	1	5

(a)   Draw a bar chart to show this information.
(b)   Which method of travel is the mode?

(a)

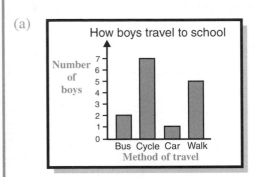

Notice that:
Bars are the same width.

There are gaps between the bars because data that can be counted is discrete.

The height of each bar gives the **frequency**.

The tallest bar represents the most frequent variable (category).

(b)   The most frequently occurring variable is called the **mode** or **modal category**.
Cycle is the modal category for these boys.

Bar-line graphs

Instead of drawing bars to show frequency we could draw vertical lines.
Such graphs are called **bar-line graphs**.
The lines can be drawn either horizontally or vertically.

EXAMPLE

The graph shows the number of goals scored by a football team in 10 matches.

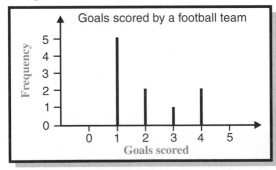

The frequency represents the number of matches played.

(a)   In how many matches was only one goal scored?
(b)   What is the difference between the largest number of goals scored in a match and the smallest number of goals scored?

(a)   5
(b)   The difference between the largest and smallest variable is called the **range**.
The range for the number of goals scored is $4 - 1 = 3$.

Exercise 35.2

1 The hair colour of all the students in a class is recorded.
The bar graph shows the results.

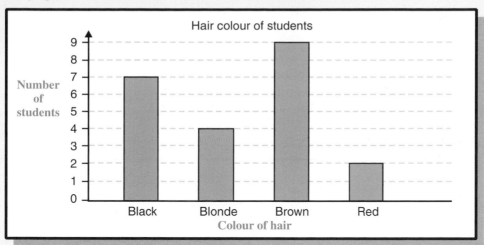

(a) How many students have black hair?
(b) Which hair colour is the mode?
(c) How many students are in the class?
(d) How many more students have brown hair than black hair?

2 The bar-line graph shows the sales of pairs of shoes by a shoe shop on one day.

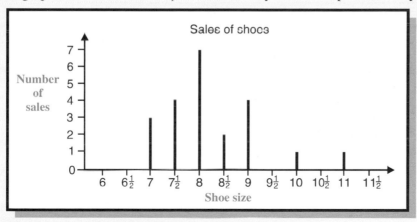

(a) Which shoe size is the mode?
(b) What is the range in the sizes of shoes sold?
(c) How many pairs of shoes were sold?

3 The result of throwing a dice 30 times is shown.

(a) Copy and complete the frequency table for these scores.

Score	Tally	Frequency
1		
2		
3		

1	3	5	1	5
2	5	2	1	2
6	6	3	3	6
2	3	2	6	4
4	3	3	4	6
3	5	1	4	5

(b) Draw a bar chart to show the data.
(c) Which score is the mode?

4 Some students were asked how many books they had carried to school that day. Their replies were:

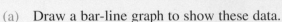

| 1 | 2 | 3 | 3 | 4 | 3 | 4 | 2 | 3 | 2 |
| 3 | 3 | 2 | 1 | 5 | 1 | 5 | 2 | 4 | 3 |

(a) Draw a bar-line graph to show these data.
(b) How many students carried less than 3 books?
(c) What is the range in the number of books carried by these students?

5 The bar chart shows the time Jim spent watching television each day last week.

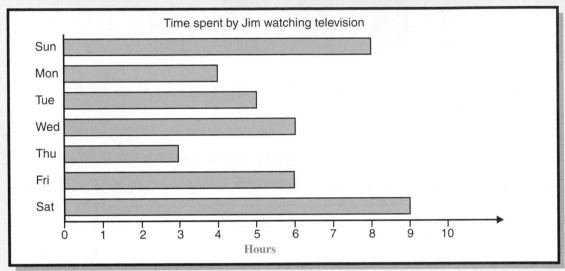

(a) On which day did Jim watch the most television?
(b) How many hours did Jim spend watching television on Tuesday?
(c) How many hours did Jim spend watching television last week?
(d) On which day did Jim spend a third of the day watching television?
(e) What fraction of the day did Jim spend watching television on Wednesday?
(f) What is the range of the number of hours per day Jim spent watching television?

6 The bar-line graph illustrates the number of goals scored per match by a hockey team.

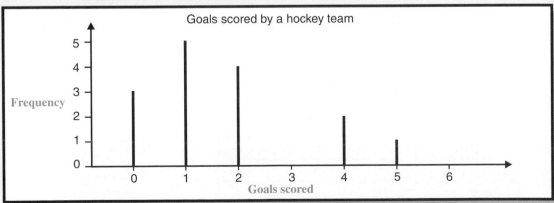

(a) How many matches have the team played?
(b) Which number of goals scored is the mode?
(c) What is the range of the number of goals scored?
(d) In what percentage of matches were no goals scored?

390

7 The table shows the amount of pocket money given each week to a number of girls.

Amount (£)	1	2	3	4	5	6	7	8	9	10
Number of girls	0	0	1	5	10	4	0	3	0	7

(a) Draw a bar-line graph of the data.
(b) What is the modal amount of pocket money?
(c) What is the range of the amount of pocket money given each week?
(d) What percentage of the girls got less than £5?

8 A group of senior citizens were asked how many children were in their families. The table shows the results.

Number of children	1	2	3	4	5	6	7
Number of families	3	6	11	8	4	1	2

(a) Draw a bar chart of this information.
(b) How many families had **more than** 4 children per family?
(c) What is the modal number of children per family?
(d) What is the range in the number of children per family?

9 The bar chart shows the day of birth for a group of children.

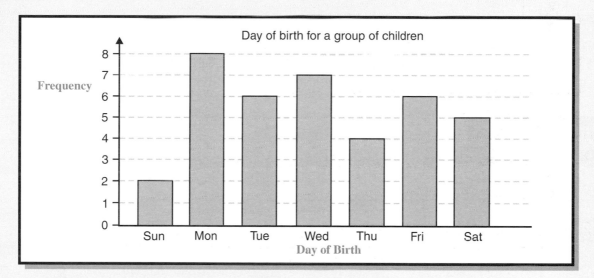

(a) How many children are in the group?
(b) How many **more** children were born on a Monday than on a Sunday?
(c) Which day of birth is the mode?

The table shows the day of birth for the girls.

Day of birth	Sun	Mon	Tue	Wed	Thu	Fri	Sat
Number of girls	1	5	2	3	3	1	4

(d) Draw up a table to show the day of birth for the boys.

10 Record the day of birth for all the students in your class.
Draw a bar chart of the data.
Compare your data with the data given in question 9.

Pictograms and Bar Charts

Bar charts can also be used to compare data.
The table shows how a class of children travelled to school one day.

Method of travel	Bus	Cycle	Car	Walk
Boys	2	7	1	5
Girls	3	1	5	6

To make it easier to compare the information given for boys and girls we can draw both bars on the same diagram, as shown.

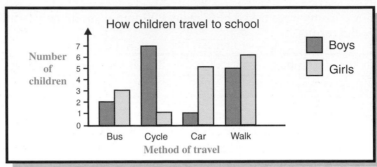

Seven boys cycle to school.

How many girls cycle to school?

There are 15 girls in the class and 6 walk to school.

The percentage of girls who walk to school is $\frac{6}{15} \times 100 = 40\%$.

What percentage of boys walk to school?

Compare and comment on the method of travel of these boys and girls.

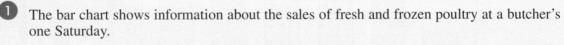

Exercise **35.3**

1 The bar chart shows information about the sales of fresh and frozen poultry at a butcher's one Saturday.

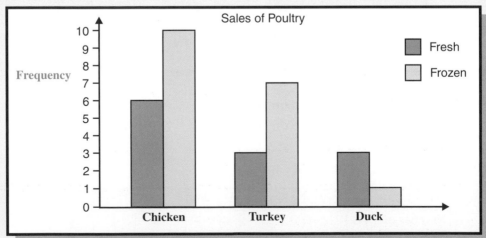

(a) How many frozen chickens were sold?
(b) How many fresh turkeys were sold?
(c) How many ducks were sold altogether?
(d) What fraction of the turkeys sold were frozen?
(e) What percentage of the ducks sold were fresh?

② The bar chart shows the reason given by students for being absent from school one day.

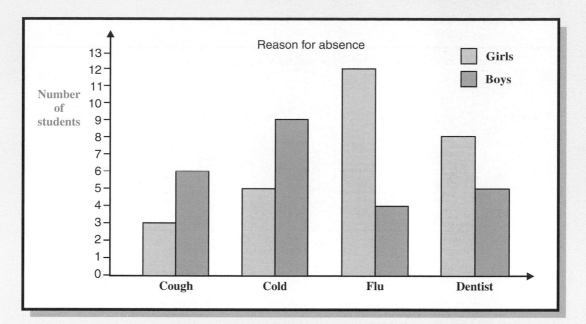

(a) How many girls were absent with flu?
(b) How many boys were absent with a cough?
(c) Which reason for absence was the mode?
(d) Compare and comment on the reasons for absence given by girls and boys.

③ The bar chart shows the marks scored by four different boys in both a numeracy test and an IQ test.

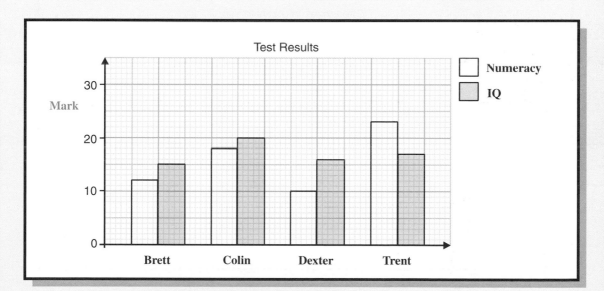

(a) Which boy scored the lowest IQ mark?
(b) Which boy scored the highest mark in numeracy?
(c) What is the range in the IQ marks?
(d) What is the range in the numeracy marks?
(e) The marks are added together to give each boy a total score.
    (i) Which boy had the highest score?
    (ii) Which boy had the lowest score?

Pictograms and Bar Charts

4 A group of children were asked how many hours they had spent watching television on a particular Sunday.
The bar chart shows the results.

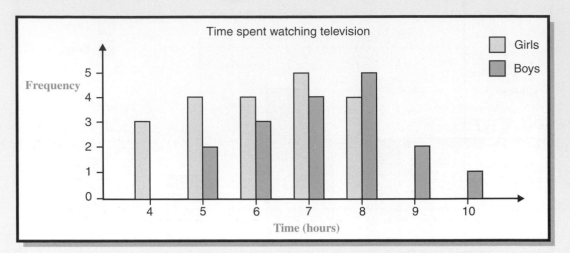

(a)   What was the modal time for the girls?
(b)   What was the range in time for the boys?
(c)   How many boys watched television for more than 8 hours?
(d)   (i)   How many girls were included in the survey?
        (ii)   What percentage of the girls watched television for 6 hours?
(e)   Compare and comment on the time spent watching television for these boys and girls.

5 The bar chart shows the results of a survey of the shoe sizes of pupils in a Year 9 class.

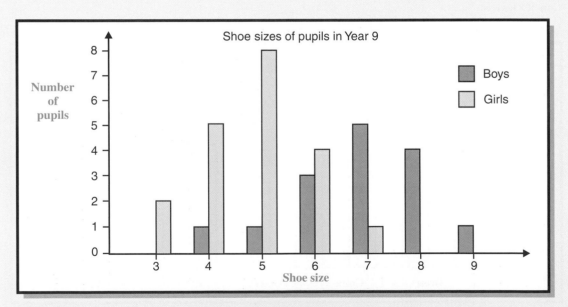

(a)   Which size of shoe is the mode for the girls?
(b)   Which size of shoe is the mode for the boys?
(c)   How many pupils have shoe size 6?
(d)   What percentage of boys have shoe size 8 or 9?
(e)   What percentage of girls have shoe size 3 or 4?
(f)   What is the range of shoe size for girls?
(g)   What is the range of shoe size for boys?
(h)   Compare and comment on the shoe sizes of boys and girls.

What you need to know

- **Pictogram**. Symbols are used to represent information.
  Each symbol can represent one or more items of data. E.g. 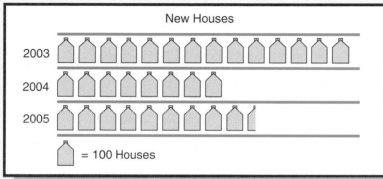 = 5 people
- **Bar chart**. Used for data which can be counted.
  Often used to compare quantities of data in a distribution.
  Bars can be drawn horizontally or vertically.
  Bars are the same width and there are gaps between bars.
  The length of each bar represents frequency.
  The longest bar represents the **mode**.
  The difference between the largest and smallest variable is called the **range**.
- **Bar-line graph**. Instead of drawing bars, horizontal or vertical lines are drawn to show frequency.

Review Exercise 35

1 The pictogram shows the number of new houses built in different years by a builder.

(a) How many houses were built in 2003?
(b) Estimate the number of houses built in 2005.

2 The usual midday temperatures, in °F, are shown for London and Athens.

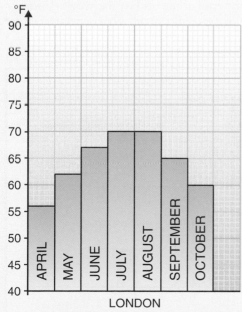
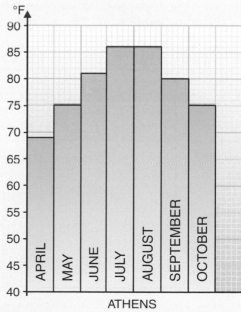

(a) What is the usual midday temperature in London in June?
(b) Find the difference in the temperatures in London and Athens in June.
(c) Which place has the greatest difference in temperature for the months shown?
     Justify your answer.

AQA

3 Sally did a survey of car colours.
The notebook shows all her results.

(a) Copy and complete her frequency table.

Colour	Tally	Frequency
White		
Blue		
Red		
Green		

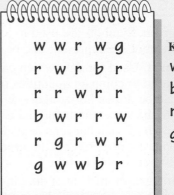

Key:
w white
b blue
r red
g green

w w r w g
r w r b r
r r w r r
b w r r w
r g r w r
g w w b r

(b) Show this information as a bar chart.
(c) Which colour is the mode?

AQA

4 The bar chart shows which days of the week shoppers went to a supermarket in 1995 and 2000.

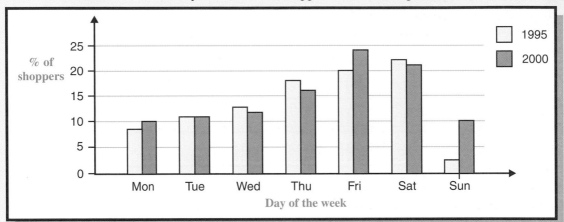

(a) Which day of the week was the most popular day for shopping in 1995?
(b) Did the shoppers choose different days to shop in 2000 compared with 1995?
Give a reason for your answer.
(c) "In 2000, about half the shoppers did their shopping at the end of the week (Friday, Saturday and Sunday)." Is this statement true or false? Show all your working.

AQA

5 The frequency diagram shows the distribution of shoe sizes for a class of 40 pupils.

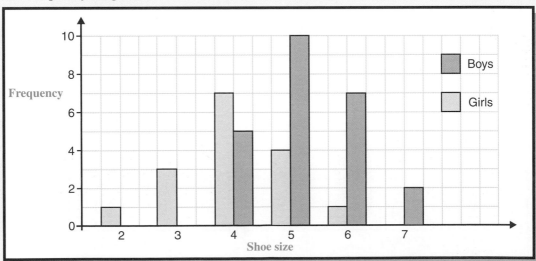

(a) What is the ratio of girls to boys in the class?
(b) By working out the range and mode for the boys and for the girls, compare and comment on the shoe sizes for boys and girls.

AQA

CHAPTER 36

Averages and Range

Activity

Some friends went on a school trip.
They each brought different amounts of spending money, as follows:

| Penny | £25 | Keith £35 | Nishpal £40 | Jayne £20 | Stephen £50 | Ben £60 |
| Charlotte £35 | | Suzie £50 | Dan £55 | Vicki £35 | Jack £55 | |

Ben brought the most and Jayne the least.
What was the difference in the amounts of spending money Ben and Jayne brought?
Which was the most common amount of money?
Who brought the middle amount of money?
How much was this?
If the friends shared out their money equally, how much would each person get?

Range

The difference between the highest and lowest amounts is called the **range**.
Range = highest amount − lowest amount

Types of average

The most common amount is called the **mode**.
When the amounts are arranged in order of size, the middle one is called the **median**.
If the money is shared out equally, the amount each person gets is called the **mean**.

EXAMPLE

The price, in pence, of a can of cola in eight different shops is shown.

$$35, \quad 39, \quad 39, \quad 32, \quad 37, \quad 35, \quad 35, \quad 40.$$

Find (a) the mode, (b) the median, (c) the mean price.

(a) The **mode** is the most common value.
The most common price is 35.
The mode is 35 pence.
We sometimes say the modal price is 35p.

(b) The **median** is found by arranging the values in order of size and taking the middle value.
Arrange the prices in order of size.
32, 35, 35, 35, 37, 39, 39, 40.

The middle price is $\frac{35 + 37}{2} = 36$

The median is 36 pence.

> Where there are an even number of values the median is the average of the middle two.

(c) The **mean** is found by finding the total of all the values and dividing the total by the number of values.
Add the prices.
35 + 39 + 39 + 32 + 37 + 35 + 35 + 40 = 292

The mean = $\frac{292}{8} = 36.5$

The mean is 36.5 pence.

Questions 1 to 6. Do not use a calculator. Show your working clearly.

1 Tony recorded the number of birthday cards sold each day.

$$3 \quad 4 \quad 8 \quad 1 \quad 4$$

(a) Work out the range of the number of cards sold each day.
(b) Write down the mode.
(c) Find the median number of cards sold each day.

2 Four students had the following number of books in their bags.

$$3 \quad 4 \quad 1 \quad 4$$

(a) What is the range of the number of books?
(b) What is the mean number of books?

3 Claire recorded the number of e-mail messages she received each day.

$$2 \quad 1 \quad 7 \quad 4 \quad 1$$

(a) Write down the mode.
(b) Find the median number of messages received each day.
(c) Calculate the mean number of messages received each day.

4 A postman delivers letters to a block of flats.
There are 8 flats in the block.
The number of letters he delivers to each flat is shown below.

$$1 \quad 3 \quad 4 \quad 4 \quad 2 \quad 2 \quad 6 \quad 2$$

(a) What is the range of the number of letters delivered?
(b) Write down the mode.
(c) Calculate the mean number of letters delivered to each flat.

5 The price, in pence, of a bar of chocolate in four different shops is shown.

$$20 \quad 17 \quad 22 \quad 25$$

Find the median of these prices.

6 Gail noted the number of stamps on 6 parcels delivered to her office.

$$3 \quad 2 \quad 4 \quad 5 \quad 7 \quad 3$$

(a) Write down the mode.
(b) Find the range of the number of stamps on a parcel.
(c) Find the median number of stamps on a parcel.
(d) Calculate the mean number of stamps on a parcel.

Questions 7 to 14. You may use a calculator.

7 Here is a list of the weights of some people, in kilograms.

$$68 \quad 74 \quad 63 \quad 81 \quad 76$$

(a) What is the range of the weights of the five people?
(b) Find the median weight.
(c) Calculate the mean weight.

8 Sanjay played a computer game 8 times and recorded his scores.

$$140 \quad 135 \quad 125 \quad 125 \quad 130 \quad 135 \quad 140 \quad 135$$

(a) Which score is the mode?
(b) Calculate the median of his scores.
(c) Find the mean score. Give your answer to the nearest whole number.

9 Seven people have an average of 9 computer games each.
How many computer games do they have altogether?

(10) Frank counted the number of books on different shelves in a library.
He recorded the following numbers.

$$38 \quad 40 \quad 26 \quad 49 \quad 37 \quad 43$$

(a) Find the median number of books on a shelf.
(b) What is the range of the number of books on a shelf?
(c) Calculate the mean number of books on a shelf.
    Give your answer correct to one decimal place.

(11) The mean of six numbers is 5.
Five of the numbers are 2, 3, 7, 8 and 6.
What is the other number?

(12) The mean of seven numbers is 6.
Six of the numbers are 2, 5, 7, 3, 7 and 10.
What is the other number?

(13) The mean length of 8 rods is 75 cm.
An extra rod is added.
The total length of the 9 rods is 729 cm.
What is the length of the extra rod?

(14) A group of 2 boys and 3 girls take a test.
The mean mark for the boys is 14.5.
The mean mark for the girls is 16.
Calculate the mean mark for the whole group.

Using the range and mean to compare data

In statistics we frequently need to compare two sets of data.
A simple comparison can be made by using the range to compare **spread** and the mean to compare **average**.

EXAMPLE

The weights of a sample of Cherry tomatoes have a range of 30 g and a mean of 45 g.
The weights of a sample of Moneymaker tomatoes have a range of 90 g and a mean of 105 g.
Compare and comment on the weights of Cherry and Moneymaker tomatoes.

Range: Cherry tomatoes 30 g, Moneymaker tomatoes 90 g.
The smaller range for Cherry tomatoes shows they are more consistent in weight.

Mean: Cherry tomatoes 45 g, Moneymaker tomatoes 105 g.
The smaller mean for Cherry tomatoes shows they have a lower average weight.

Comment: The average weight of Cherry tomatoes is lower but they are more consistent in weight.

Exercise 36.2

(1) The weights, in grams, of a sample of 10 economy potatoes are shown.

$$70 \quad 76 \quad 83 \quad 86 \quad 95 \quad 98 \quad 113 \quad 117 \quad 122 \quad 130$$

(a) (i) What is the range of these weights?
    (ii) Calculate the mean of these weights.

For a sample of 10 premium potatoes the range of their weights is 240 grams and the mean of their weights is 250 grams.

(b) Compare and comment on the weights of economy and premium potatoes.

2 The lateness of 12 buses is recorded.
The results, in minutes, are shown.

5   6   7   8   8   10   10   10   11   12   13   14

(a) (i) What is the range of lateness for these buses?

(ii) Calculate the mean lateness for these buses.

The lateness of 12 trains is also recorded.
The range in lateness for these trains is 14 minutes and the mean lateness is 5 minutes.

(b) Compare and comment on the lateness for these buses and trains.

3 The numbers of words typed per minute by a group of students are shown.

45   51   58   59   63   87   64   59   58   63   53

(a) (i) What is the range of their typing speeds?

(ii) Calculate the mean typing speed for these students.

For another group of students the mean of their typing speeds is 42 words per minute and the range is 67.

(b) Comment on the typing speeds of these two groups of students.

4 The numbers of goals scored in matches played by some first division football teams are shown.

1   5   0   2   2   3   0   4   2   0   1

(a) (i) What is the range of the number of goals scored?

(ii) Calculate the mean number of goals scored.

The numbers of goals scored in matches played by some third division football teams are shown.

3   2   4   1   1   3   2   1   7

(b) Compare the numbers of goals scored by these first and third division football teams.

5 The times, in minutes, taken by 8 boys to swim 50 metres are shown.

1.8   2.0   1.7   2.2   2.1   1.9   1.8   2.1

(a) (i) What is the range of these times?

(ii) Calculate the mean time.

The times, in minutes, taken by 8 girls to swim 50 m are shown.

2.1   1.9   1.8   2.3   1.6   2.0   2.6   1.9

(b) Comment on the times taken by these boys and girls to swim 50 m.

6 The table shows the percentage silver content of twenty ancient coins.

	Percentage silver content				
Roman coins	5.6	6.7	6.6	7.2	6.3
Chinese coins	6.8	6.7	6.2	5.4	7.3
Egyptian coins	5.1	7.0	5.8	6.9	7.6
Greek coins	5.6	7.2	6.6	6.8	5.7

(a) For each type of coin:

(i) calculate the range of the percentage silver content,

(ii) calculate the mean of the percentage silver content.

(b) Comment on your answers to (a).

7 Pupils in Year 7 are arranged in nine classes.
The class sizes are:

30   27   28   29   27   29   29   28   28

(a) Calculate the mean class size.

(b) The range of the class sizes for Year 11 is 12.
What does this tell you about the class sizes in Year 11 compared with those in Year 7?

Frequency distributions

After data is collected it is often presented using a **frequency distribution table**.

Johti measured the lengths of some twigs. He recorded the following results.

$$2 \quad 3 \quad 2 \quad 6 \quad 3 \quad 6 \quad 2 \quad 3 \quad 5 \quad 4 \quad 3$$
$$2 \quad 2 \quad 3 \quad 5 \quad 6 \quad 2 \quad 6 \quad 5 \quad 6 \quad 2$$

Johti then presented his results, as shown in this frequency distribution table.

Length (cm)	2	3	4	5	6
Number of twigs (frequency)	7	5	1	3	5

Using the frequency distribution table we can find the mode, median, mean and range of the lengths of the twigs that Johti measured.

To find the mode:

The mode is the length with the greatest frequency.
There were 7 twigs of length 2 cm.
This is more than any other length of twigs.
The mode of the lengths is 2 cm.

To find the median:

The median is the middle length when all the lengths are arranged in order of size.
We could list the 21 twigs in order of length and split them up like this:

10 shortest twigs	middle twig	10 longest twigs

This shows that the median length is the length of the 11th twig.
From the table we can see that:

the first 7 twigs are each 2 cm long,
and the next 5 twigs are each 3 cm long.

So, the 11th twig is 3 cm long.
The median length is 3 cm.

To find the mean:

$$\text{Mean} = \frac{\text{Total of all lengths}}{\text{Number of lengths}}$$

The best way to do this is to extend the frequency distribution table.

Length (cm) x	Number of twigs (frequency) f	Number of twigs × Length $f \times x$
2	7	14
3	5	15
4	1	4
5	3	15
6	5	30
Totals	$\Sigma f = 21$	$\Sigma fx = 78$

$$\text{Mean} = \frac{\text{Total of all lengths}}{\text{Number of lengths}} = \frac{\Sigma fx}{\Sigma f}$$

$$\text{Mean} = \frac{78}{21} = 3.714\ldots$$

Mean length = 3.7 cm, correct to 1 decimal place.

To find the range:

Range = longest length − shortest length
$$= 6 - 2 = 4 \text{ cm}$$

The range of the lengths is 4 cm.

Σ is the Greek letter 'sigma'.

Σf means the sum of frequencies.

Σfx means the sum of the values of fx.

$$\text{Mean} = \frac{\Sigma fx}{\Sigma f}$$

Do not use a calculator for questions 1 and 2.

1 A milkman delivers bottles of milk to 30 houses in a street.
The number of bottles of milk he delivers to each house is shown below.

1	4	2	1	2	1	2	1	1	2
2	2	1	3	1	2	1	2	2	1
3	2	3	1	3	1	4	3	1	4

(a) Copy and complete this frequency distribution table.

Number of bottles	1	2	3	4
Number of houses				

(b) What is the mode of the number of bottles delivered?
(c) What is the median number of bottles delivered?
(d) Calculate the mean number of bottles of milk delivered to each house.

2 Mark asked some students, "How many keys do you have on your key ring?"
The data collection sheet shows the responses he recorded.

Number of keys	Tally
2	||
3	|||
4	ⵌ |||
5	ⵌ
6	||

(a) Use Mark's data to make a frequency distribution table.
(b) How many students did Mark ask?
(c) Write down the mode of the number of keys on a key ring.
(d) Find the median number of keys.
(e) Calculate the mean number of keys on a key ring.

3 Pat recorded the weekly earnings of a group of students. Her results were as follows:

£25 £40 £30 £45 £25 £25 £30 £35 £45 £25
£35 £30 £35 £25 £35 £30 £35 £25 £45

(a) Show the data in a frequency distribution table.
(b) Find the range of the weekly earnings.
(c) What is the modal weekly earnings?
(d) How many people were in the group?
(e) What is the median weekly earnings?
(f) Calculate the total weekly earnings of the group.
(g) Calculate the mean weekly earnings. Give your answer to the nearest penny.

4 Find the mode, median and mean for the following data.

(a)
Number of letters delivered	1	2	3	4	5	6
Number of days taken to deliver	6	9	6	6	2	1

(b)
Number of books read last month	0	1	2	3	4	5
Number of students	1	4	10	4	1	1

(c)
Number of days absent in a year	0	1	2	3	4	5	6	7	8	9	10	11
Number of students	56	0	0	4	14	10	24	11	21	15	8	2

EXAMPLE

Find the range, mode, median and mean of the ages for the data shown in the bar chart.

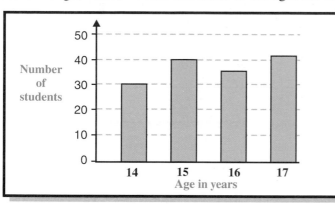

The range is the difference between the highest and lowest ages.

Range = 17 - 14
= 3 years

The most common age is shown by the tallest bar.

So, the modal age is 17 years.

Use a table to find the median and the mean.

Age (years) x	Frequency f	Frequency × Age $f \times x$
14	30	420
15	40	600
16	36	576
17	41	697
Totals	$\Sigma f = 147$	$\Sigma fx = 2293$

The middle student is given by: $\dfrac{147 + 1}{2} = 74$

The 74th student in the list has the median age.
The first 70 students are aged 14 or 15 years.
The 74th student has age 16 years.
Median age is 16 years.

$$\text{Mean} = \frac{\text{Total of all ages}}{\text{Number of students}} = \frac{\Sigma fx}{\Sigma f}$$

Mean = $\frac{2293}{147}$ = 15.598…

Mean age is 15.6 years, correct to 1 d.p.

Exercise 36.4   Try to do questions 1 and 4 without using a calculator.

1   Hilary observed customers using the express checkout at a supermarket.

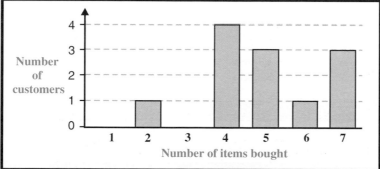

(a) Find the range of the number of items bought.
(b) What is the mode of the number of items bought?
(c) Work out the median number of items bought.
(d) Calculate the mean number of items bought.

2 During one week, the following numbers of various sizes of a particular style of shoe were sold.

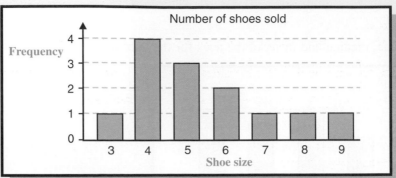

(a) Find the range of the shoe sizes sold.
(b) Which size is the median?
(c) Which size is the mode?
(d) Calculate the mean size.
    Comment on your answer.

3 (a) Find the range and mode of these prices.
(b) Calculate the median and mean price of a bottle of milk.

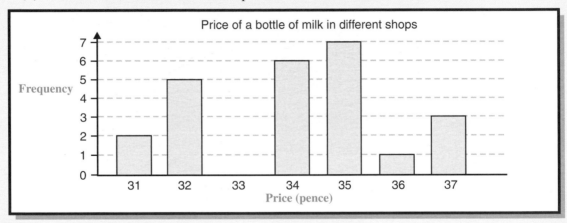

4 A group of students took part in a quiz on the Highway Code.
The bar chart shows their scores.

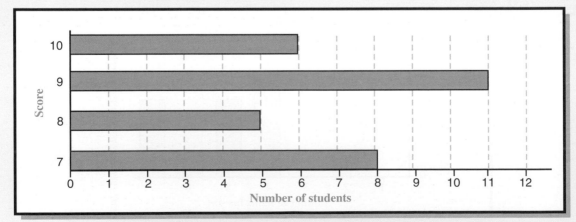

(a) Which score is the mode?
(b) What is the median score?
(c) How many students took part in the quiz?
(d) Calculate the mean score.

Grouped frequency distributions

When there is a lot of data, or the data is continuous, **grouped frequency distributions** are used.

Calculating the mean

For a grouped frequency distribution the true value of the mean cannot be found as the actual values of the data are not known.

To **estimate the mean**, we assume that all the values in each class are equal to the midpoint of the class.

$$\text{Estimated mean} = \frac{\Sigma\,(\text{frequency} \times \text{midpoint})}{\text{Total frequency}} = \frac{\Sigma fx}{\Sigma f}$$

Modal class

For a grouped frequency distribution with equal class width intervals, the **modal class** is the class (or group) with the highest frequency.

EXAMPLE

The table shows the masses of a group of children.
(a) Calculate an estimate of the mean mass.
(b) Find the modal class.

Mass (m kg)	Frequency
$40 \leqslant m < 50$	3
$50 \leqslant m < 60$	10
$60 \leqslant m < 70$	6
$70 \leqslant m < 80$	12

(a)

Mass (m kg)	Midpoint x	Frequency f	Frequency × Midpoint $f \times x$
$40 \leqslant m < 50$	45	3	135
$50 \leqslant m < 60$	55	10	550
$60 \leqslant m < 70$	65	6	390
$70 \leqslant m < 80$	75	12	900
	Totals	$\Sigma f = 31$	$\Sigma fx = 1975$

Midpoint of the class
$40 \leqslant m < 50$
is given by:
$$\frac{40 + 50}{2} = \frac{90}{2} = 45$$

Estimate of mean $= \dfrac{\Sigma fx}{\Sigma f} = \dfrac{1975}{31} = 63.709\ldots$

Estimate of mean mass $= 63.7$ kg, correct to 3 sig. figs.

(b) The modal class is $70\,\text{kg} \leqslant m < 80\,\text{kg}$

Exercise 36.5

1 Give the modal class and calculate an estimate of the mean for each of the following.

0 - means 0 or more but less than 10.

(a)

Salary (s) (£000's)	Number of employees
$10 \leqslant s < 20$	79
$20 \leqslant s < 30$	32
$30 \leqslant s < 40$	14
$40 \leqslant s < 50$	0
$50 \leqslant s < 60$	2

(b)

Time spent watching TV per week (hours)	Number of students
0 -	2
10 -	8
20 -	5
30 -	14
40 - 50	7

Averages and Range

2 The table shows the distribution of the weights of some turkeys.

Weight (w kg)	$2 \leqslant w < 4$	$4 \leqslant w < 6$	$6 \leqslant w < 8$	$8 \leqslant w < 10$
Frequency	7	9	5	3

Calculate an estimate of the mean weight of these turkeys.
Give your answer correct to one decimal place.

3 The table shows the distribution of the prices of houses for sale in a particular neighbourhood.

Price (p £000's)	$60 \leqslant p < 80$	$80 \leqslant p < 100$	$100 \leqslant p < 120$	$120 \leqslant p < 140$
Number of houses	3	7	4	1

Calculate an estimate of the mean price of these houses.
Give your answer to an appropriate degree of accuracy.

4 The table shows the distribution of the heights of trees in a wood.

Height in metres (to nearest 0.1 m)	3.0 - 3.4	3.5 - 3.9	4.0 - 4.4	4.5 - 4.9	5.0 - 5.4
Number of trees	12	10	23	18	13

Calculate an estimate of the mean height.

5 The table shows the distribution of marks in a test.

Mark	0 - 19	20 - 29	30 - 34	35 - 39	40 - 50
Number of students	12	23	25	14	3

Calculate an estimate of the mean mark.
Notice that the class intervals are not all equal.

Comparing distributions

The table shows the marks gained in a test.
Compare the marks obtained by the boys and the girls.

Mark (out of 10)	7	8	9	10
Number of boys	2	5	3	0
Number of girls	4	0	2	1

To compare the marks we can use the range and the mean.

Boys: Range = $9 - 7 = 2$
Girls: Range = $10 - 7 = 3$

The girls had the higher range of marks.

Boys: Mean = $\dfrac{2 \times 7 + 5 \times 8 + 3 \times 9 + 0 \times 10}{10} = \dfrac{81}{10} = 8.1$

Girls: Mean = $\dfrac{4 \times 7 + 0 \times 8 + 2 \times 9 + 1 \times 10}{7} = \dfrac{56}{7} = 8$

The boys had the higher mean mark.

Overall the boys did better as:
the girls' marks were more spread out with a lower average mark,
the boys' marks were closer together with a higher average mark.

To compare the overall standard, the median could be used instead of the mean.

1 Use the mean and the range to compare the number of goals scored per match by these teams.

Jays

Number of goals scored per hockey match	Number of matches
0	3
1	5
2	2
3	2
4	1
5	2

Wasps

Number of goals scored per hockey match	Number of matches
0	0
1	2
2	3
3	1
4	2
5	0

2 Use the mean and the range to compare the number of visits to the cinema by these women and men.

Number of visits to the cinema last month	0	1	2	3	4	5	6	More than 6
Number of women	8	9	7	3	2	1	1	0
Number of men	0	12	7	1	0	0	0	0

3 Use the mean and the range to compare the number of Valentine cards received by these boys and girls.

Number of Valentine cards	0	1	2	3	4	5	6
Number of boys	3	4	2	2	1	2	1
Number of girls	0	3	7	3	2	0	0

4 Deepak thought that the girls in his class wore smaller shoes than the boys on average, but that the boys' shoe sizes were less varied than the girls'.
He did a survey to test his ideas. The table shows his results. Was he correct?

Shoe size	$4\frac{1}{2}$	5	$5\frac{1}{2}$	6	$6\frac{1}{2}$	7	$7\frac{1}{2}$	8	$8\frac{1}{2}$	9	$9\frac{1}{2}$
Number of boys	1	0	5	4	4	2	1	0	1	0	0
Number of girls	0	2	0	2	3	0	2	0	3	1	1

5 (a) Find the modal class for the ages of customers in each of these two restaurants.
  (b) Which restaurant attracts more younger people?
  (c) Why is it only possible to find an approximate value for the age range of customers?

Age (years)	MacQuick	Pizza Pit
0 - 9	8	2
10 - 19	9	4
20 - 29	10	12
30 - 39	7	15
40 - 49	1	5
50 - 59	1	3
60 - 69	3	3
70 - 79	1	2
80 - 89	0	1

Averages and Range

6 The graphs show the monthly sales of bicycles before and after a marketing campaign.
Calculate the medians and the ranges.
Use your results to compare 'Before' with 'After'.

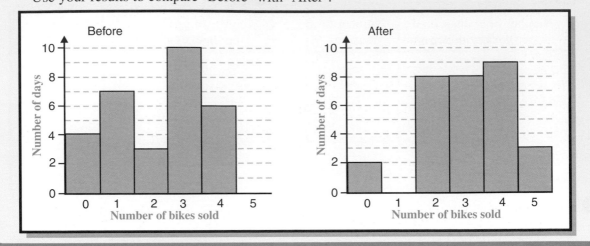

Which is the best average to use?

Many questions in mathematics have definite answers. This one does not.
Sometimes the mean is best, sometimes the median and sometimes the mode.
It all depends on the situation and what you want to use the average for.

EXAMPLE

A youth club leader gets a discount on cans of drinks if she buys all one size.
She took a vote on which size people wanted.
The results were as follows:

Size of can (ml)	100	200	330	500
Number of votes	9	12	19	1

Mode = 330 ml
Median = 200 ml
Mean = 245.6 ml, correct to one decimal place.

Which size should she buy?

The mean is no use at all because she can't buy cans of size 245.6 ml.
Even if the answer is rounded to the nearest whole number (246 ml), it's still no use.
The median is possible because there is an actual 200 ml can.
However, only 12 out of 41 people want this size.
In this case the **mode** is the best average to use, as it is the most popular size.

Exercise 36.7

In questions 1 to 3 find all the averages possible. State which is the most sensible and why.

1 On a bus: 23 people are wearing trainers,
    10 people are wearing boots,
    8 people are wearing lace-up shoes.

2 20 people complete a simple jigsaw. Their times, in seconds, are recorded.

       5,   6,   8,   8,   9,   10,   11,   11,   12,   12,
      12,   15,   15,   15,   15,   18,   19,   20,   22, 200.

3 Here are the marks obtained by a group of 11 students in a mock exam.
The exam was marked out of 100.

       5,   6,   81,   81,   82,   83,   84,   85,   86,   87,   88.

4 The times for two swimmers to complete each of ten 25 m lengths are shown below.

Swimmer A	30.1	30.1	30.1	30.6	30.7	31.1	31.1	31.5	31.7	31.8
Swimmer B	29.6	29.7	29.7	29.9	30.0	30.0	30.1	30.1	30.1	44.6

Which is the better swimmer?
Explain why.

5 The table shows the number of runs scored by two batsmen in several innings.

Batsman A	0	0	10	12	20	22	50	51	81	104		
Batsman B	0	24	25	27	28	30	33	34	44	45	46	96

Which is the better batsman?
Explain why.

6 A teacher sets a test.
He wants to choose a minimum mark for a distinction so that 50% of his students get this result.
Should he use the modal mark, the median mark or the mean mark?
Give a reason for your answer.

7 The cost of Bed and Breakfast at 10 different hotels is given.

£39.50   £55   £60   £50   £49   £42   £95   £59   £39.50   £45

(a) Wyn says, "The average cost of Bed and Breakfast is £39.50."
Which average is he using?
Give a reason why this is not a sensible average to use for this data.

(b) Which of the mode, median and mean best describes the average cost of Bed and Breakfast?
Give a reason for your answer.

What you need to know

- There are three types of **average**: the **mode**, the **median** and the **mean**.
  The **mode** is the most common value.
  The **median** is the middle value (or the mean of the two middle values) when the values are arranged in order of size.

  $$\textbf{Mean} = \frac{\text{Total of all values}}{\text{Number of values}}$$

- The **range** is a measure of **spread**.
  Range = highest value − lowest value

- To find the mean of a **frequency distribution** use:

  $$\text{Mean} = \frac{\text{Total of all values}}{\text{Number of values}} = \frac{\Sigma fx}{\Sigma f}$$

- To find the mean of a **grouped frequency distribution**, first find the value of the midpoint of each class.
  Then use: $\quad$ Estimated mean $= \dfrac{\text{Total of all values}}{\text{Number of values}} = \dfrac{\Sigma fx}{\Sigma f}$

- Choosing the best average to use:
  When the most **popular** value is wanted use the **mode**.
  When **half** of the values have to be above the average use the **median**.
  When a **typical** value is wanted use either the **mode** or the **median**.
  When all the **actual** values have to be taken into account use the **mean**.
  When the average should not be distorted by a few very small or very large values do **not** use the mean.

Averages and Range

1 Marcus asks 5 of his friends how many pets they have. The replies are:

1 2 0 3 1

Find the mean of these numbers.

2 The teacher in charge of an outdoor pursuits centre has 10 pairs of boots.
The sizes of the boots are as follows:

4 7 5 5 3 10 8 5 5 8

(a) Calculate the mean boot size.
(b) The teacher calculates the following:

median	mode	total	range	mean

Which of these will tell her how spread out the boot sizes are? *AQA*

3 The numbers of computers in eight classrooms are:

2 1 1 3 15 30 1 2

(a) What is the median number of computers?
(b) What is the modal number of computers?
(c) Why would the mean not be a sensible average for these data? *AQA*

4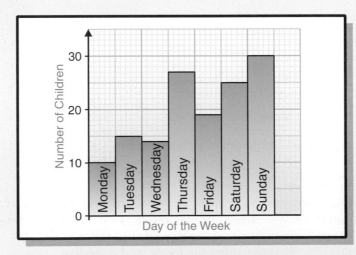

The weights, in kilograms, of a boat crew are:

96, 86, 94, 96, 91, 95, 90, 96, 43.

(a) Calculate
  (i) their median weight,
  (ii) the range of their weights,
  (iii) their mean weight.

(b) Which of the two averages, mean or median, best describes the data above?
  Give a reason for your answer. *AQA*

5 The diagram shows the number of children who played on the swings in a park during one week in June.

(a) How many children played on the swings on Wednesday?
(b) What was the total number of children who played on the swings in the week?
(c) Work out the mean number of children per day who played on the swings. *AQA*

6 Customers at a roller skating rink can hire skates.
The first seven customers on Monday morning take the following sizes.

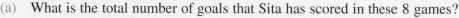

7   5   6   9   8   10   5

(a) For these seven customers find
    (i)   the median skate size,
    (ii) the modal skate size.

(b) The manager is going to buy some new skates.
He knows the median and modal skate size for 700 customers.
Which of the values would be the more useful to him?
Explain your answer.

AQA

7 Sita plays in the school netball team.
After 8 games her mean score is 6.5 goals.

(a) What is the total number of goals that Sita has scored in these 8 games?
(b) Selection for the County team is made after 9 games have been played.
Sita will be chosen for the County team if her mean score is
7 goals or more.
What is the smallest number of goals she must score in the
ninth game in order to be chosen?
(c) In the ninth game she actually scores 10 goals.
Does her mean score of goals increase or decrease?
Explain your answer.

AQA

8 Ros asks a group of students,
"How many children are there in your family?"
The graph shows the results of her survey.

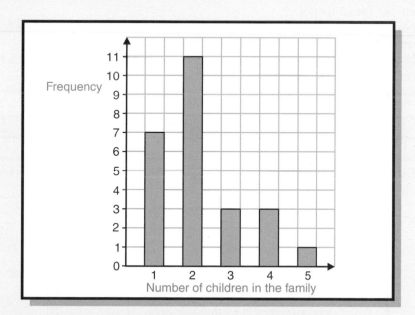

(a) How many students are in the group?
(b) What is the range in the number of children in the family for these students?
(c) What is the most common number of children in the family for these students?
(d) Calculate the mean number of children in the family for these students.
Give your answer to one decimal place.
(e) A national survey of 5000 families found the mean number of children per family
was 1.6 and the range in the number of children per family was 11.
Give a reason why the results obtained by Ros are different.

Averages and Range

9 (a) Pauline measures the lengths of some English cucumbers.
The lengths in centimetres are:

27, 28, 29, 30, 31, 31, 32, 33, 35, 37, 39.

   (i) What is the range of the lengths of these cucumbers?
   (ii) What is the mean length of these cucumbers?

(b) Pauline measures the lengths of some Spanish cucumbers.
The range of the lengths of these cucumbers is 6 cm and the mean is 30 cm.
Comment on the differences between these two varieties of cucumber. AQA

10 David is playing cricket.
The table shows the number of runs he has scored off each ball so far.

Number of runs	0	1	2	3	4	5	6
Number of balls	3	8	4	3	5	0	2

(a) (i) What is the median number of runs per ball?
   (ii) Calculate the mean number of runs per ball.

Off the next five balls, David scores the following runs: 4, 4, 5, 3 and 6.

(b) (i) Calculate the new median. (ii) Calculate the new mean.
(c) Give a reason why the mean is used, rather than the median, to give the average number of runs scored per ball. AQA

11 In an experiment 50 people were asked to estimate the length of a rod to the nearest centimetre. The results were recorded.

Length (cm)	20	21	22	23	24	25	26	27	28	29
Frequency	0	4	6	7	9	10	7	5	2	0

(a) Find the value of the median.
(b) Calculate the mean length.
(c) In a second experiment another 50 people were asked to estimate the length of the same rod.
The most common estimate was 23 cm. The range of the estimates was 13 cm.
Make **two** comparisons between the results of the two experiments. AQA

12 The weekly wages of employees are recorded.

Wage (£)	100 -	200 -	300 -	400 -	500 -	600 - 1000
Frequency	2	13	4	0	2	12

(a) Which is the modal group?
(b) In which group is the median value?
(c) Without calculating, state which of the mean, mode or median is the largest.
Explain your answer. AQA

13 On holiday Val records the length of time people stay in the pool.
The results are shown in the table.

Time (t mins)	Number of people
$0 < t \leq 10$	4
$10 < t \leq 20$	7
$20 < t \leq 30$	3
$30 < t \leq 40$	2
	16

Calculate an estimate of the mean time spent in the pool.
Give your answer to an appropriate degree of accuracy. AQA

Pie Charts and Stem and Leaf Diagrams

Pie charts

Bar charts are useful for comparing the various types of data (categories) with each other.
To compare each category with **all** the data collected we use a **pie chart**.

A pie chart is a circle which is divided up into sectors.
The whole circle represents the total frequency and each sector represents the frequency of one part (category) of the data.

Drawing pie charts

The table shows the ways in which some children like to eat eggs.

Method of cooking	Poached	Boiled	Scrambled	Fried
Number of children	5	8	6	11

To show this information in a pie chart we must find the angles of the sectors which represent each category. First calculate the angle which represents each child.

30 children are represented by 360°.
1 child is represented by 360° ÷ 30 = 12°.
Sector angle = Number of children in category × 12°

Method of cooking	Poached	Boiled	Scrambled	Fried	**Total**
Number of children	5	8	6	11	**30**
Sector angle	60°	96°	72°	132°	**360°**

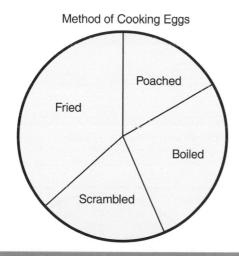

Method of Cooking Eggs

○○○○○○○○○○○○○○○
Using a table allows you to keep your work tidy and to make checks.

○○○○○○○○○○○○○○○
The whole circle represents the total frequency of 30.

Each sector represents the frequency of one category (method of cooking).

Exercise 37.1

 1 The table shows information about the trees in a wood.

Type of tree	Ash	Beech	Maple
Number of trees	20	25	15

Draw a pie chart for this data.

2 The colour of eyes of 90 people were recorded. The table shows the results.

Colour of eyes	Brown	Blue	Green	Other
Number of people	40	25	15	10

Draw a pie chart for this data.

3 The table shows information about the cars owned by a company.

Make of car	Ford	Saab	Vauxhall	BMW
Frequency	10	9	15	6

Draw a pie chart for this data.

4 The breakfast cereal preferred by some adults is shown.

Breakfast cereal	Corn flakes	Muesli	Porridge	Bran flakes
Number of adults	25	20	12	15

Show the information in a pie chart.

5 The table shows the sales of ice-cream cornets at a kiosk one day.

Ice-cream cornet	Vanilla	Strawberry	99
Frequency	94	37	49

Draw a pie chart for this data.

6 The table shows the results of a survey to find the most popular takeaway food.

Type of takeaway	Fish & chips	Chicken & chips	Chinese meal	Pizza
Number of people	165	204	78	93

Draw a pie chart for this data.

Interpreting pie charts

Pie charts are useful for showing and comparing proportions of data.
However, they do not show frequencies.
Such information can be found by interpreting the pie chart.

To interpret a pie chart we need to know:
- the sector angles (which can be measured from an accurately drawn pie chart), **and**
- the total frequency represented by the pie chart, **or**
  the frequency represented by one of the sectors.

EXAMPLE

The pie chart shows the makes of 120 cars.

(a) Which make of car is the mode?

(b) How many of the cars are Ford?

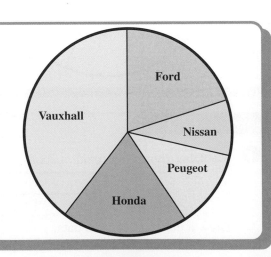

(a) The sector representing Vauxhall is the largest.
Therefore, Vauxhall is the mode.

(b) The angle of the sector representing Ford is 72°.
The number of Ford cars $= \frac{72}{360} \times 120 = 24$.

1 The pie chart shows the type of holiday chosen by 36 people.

   (a)   How many people chose a camping holiday?

   (b)   How many people chose a self-catering holiday?

   (c)   What type of holiday is the mode?

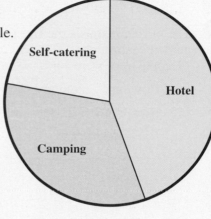

2 The pie chart shows the resorts chosen in Italy by 60 skiers.

   (a)   How many skiers chose Sauze d'Oulx?

   (b)   How many skiers chose Foppollo?

   (c)   Which resort is the mode?

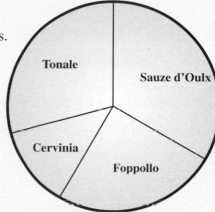

3 The pie chart shows the holiday destinations of 180 people.

   (a)   Which holiday destination is the mode?

   (b)   How many people went to Italy?

   (c)   How many people went to Germany?

   (d)   How many people went to Austria?

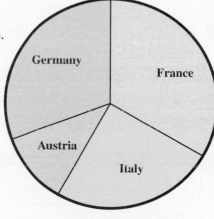

4 The pie chart shows the membership of an international committee.
The USA has 7 committee members.

   (a)   How many committee members has the EU?

   (b)   How many committee members has Canada?

   (c)   How many committee members are there altogether?

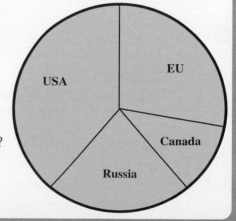

Pie Charts and Stem and Leaf Diagrams

5 The pie chart shows the five types of fruit tree sold by a garden centre.
21 apple trees were sold.

   (a)   How many plum trees were sold?

   (b)   How many fruit trees were sold altogether?

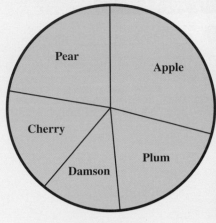

6 The pie chart shows the departure airports of some travellers.

   (a)   Which airport is the mode?

   (b)   360 travellers departed from Gatwick.
How many travellers departed from Heathrow?

   (c)   How many travellers are there altogether?

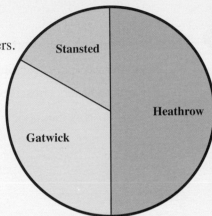

7 The pie chart illustrates the results of a survey of the colour of hair of 48 boys.

   (a)   Which colour of hair is the mode?

   (b)   How many boys have brown hair?

   (c)   What percentage of the boys have black hair?

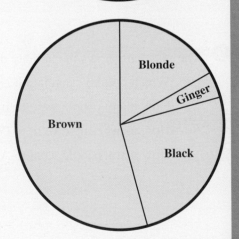

8 The pie chart shows the different types of tree in a forest.
There are 54 oak trees and these are represented by a sector with an angle of 27°.

   (a)   The pine trees are represented by an angle of 144°.
How many pine trees are there?

   (b)   There are 348 silver birch trees.
Calculate the angle of the sector representing silver birch trees.

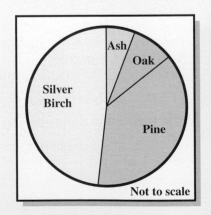

Not to scale

Stem and leaf diagrams

Data can also be represented using a **stem and leaf diagram**.

EXAMPLE

The times, in seconds, taken by 20 students to complete a puzzle are shown.

| 15 | 9 | 23 | 32 | 17 | 12 | 27 | 19 | 26 | 15 |
| 20 | 11 | 24 | 31 | 10 | 17 | 15 | 28 | 33 | 18 |

Construct a stem and leaf diagram to represent this information.

A stem and leaf diagram is made by splitting each number into two parts.
As the data uses 'tens' and 'units', the stem will represent the 'tens' and the leaf will represent the 'units'.
To draw the stem and leaf diagram begin by drawing a vertical line.
The digits to the left of the line make the **stem**.
The digits to the right of the line are the **leaves**.

9 is recorded as 0 9 →

The first number is 15.
Next to the stem of 1 record 5.

```
0 | 9
1 | 5  7  2  9  5  1  0  7  5  8
2 | 3  7  6  0  4  8
3 | 2  1  3
```

Once the data has been recorded, it is usual to redraw the diagram so that the leaves are in numerical order.

```
                              1 | 5   means 15 seconds
0 | 9
1 | 0  1  2  5  5  5  7  7  8  9
2 | 0  3  4  6  7  8
3 | 1  2  3
```

Stem and leaf diagrams are often drawn without column headings in which case a key is necessary.
e.g. 1 | 5 means 15 seconds

Exercise 37.3

1 The amount of petrol, in litres, bought by 20 motorists is shown.

| 16 | 23 | 27 | 10 | 35 | 42 | 26 | 25 | 24 | 17 |
| 23 | 41 | 33 | 35 | 25 | 19 | 16 | 31 | 12 | 29 |

Construct a stem and leaf diagram to represent this information.

2 The times, in seconds, taken to answer 24 telephone calls are shown.

| 3.2 | 5.6 | 2.4 | 3.5 | 4.3 | 3.6 | 2.8 | 5.8 | 3.3 | 2.6 | 3.2 | 2.8 |
| 5.6 | 3.5 | 4.2 | 1.5 | 2.7 | 2.5 | 3.7 | 3.1 | 2.9 | 4.2 | 2.4 | 3.0 |

Copy and complete the stem and leaf diagram to represent this information.

```
                    3 | 2   means 3.2 seconds
1 |
2 |
3 | 2
4 |
5 |
```

For this data
the stem represents 'units',
the leaf represents 'tenths'.

3 The number of press-ups completed by 18 students in one minute is shown.

$$\begin{array}{ccccccccc} 21 & 36 & 41 & 25 & 18 & 32 & 40 & 36 & 22 \\ 9 & 16 & 24 & 33 & 36 & 27 & 32 & 20 & 28 \end{array}$$

Draw a stem and leaf diagram to represent this information.

4 The heights, in centimetres, of the heels on 20 different pairs of shoes are shown.

$$\begin{array}{cccccccccc} 2.7 & 3.4 & 2.0 & 6.0 & 4.5 & 3.6 & 3.1 & 2.4 & 4.2 & 1.8 \\ 3.5 & 2.5 & 2.6 & 2.1 & 4.0 & 3.5 & 4.2 & 2.6 & 3.9 & 5.4 \end{array}$$

Construct a stem and leaf diagram to represent this information.

5 David did a survey to find the cost, in pence, of a loaf of bread.
The stem and leaf diagram shows the results of his survey.

```
                              2 | 7  means 27 pence
        2 | 7  9
        3 | 1  1  2  9  9  9
        4 | 2  5  9
        5 | 0
```

(a) How many loaves of bread are included in the survey?
(b) What is the range of the prices?
(c) Which price is the mode?

Back to back stem and leaf diagrams

Back to back stem and leaf diagrams can be used to compare two sets of data.

EXAMPLE

The results for examinations in Mathematics and English for a group of students are shown.
The marks are given as percentages.

Mathematics:	91	27	55	69	83	25	45	53	67	71
	30	52	45	59	86	73	65	47	54	38
English:	45	40	48	65	75	55	36	85	76	69
	64	58	47	64	67	72	83	74	62	51

(a) Construct a back to back stem and leaf diagram for this data.
(b) Compare and comment on the results in Mathematics and English.

(a)
```
        Mathematics    |        English        3 | 6  means 36%
              7  5 | 2 |
              8  0 | 3 | 6
         7  5  5 | 4 | 0  5  7  8
   9  5  4  3  2 | 5 | 1  5  8                  For Mathematics:
         9  7  5 | 6 | 2  4  4  5  7  9         5 | 2 means 25%
            3  1 | 7 | 2  4  5  6
            6  3 | 8 | 3  5
               1 | 9 |
```

(b) The range of marks in Mathematics is larger than in English.
    The modal group in English is 60 to 69, in Mathematics it is 50 to 59.

1 The stem and leaf diagram shows the distribution of marks for a test marked out of 50.

Boys					Girls							1 | 7 means 17 marks
				0	9							
		6	2	1	0	1	2	7				
	7 6 4	3	2	1	3	5	5	6	7	8		
9 5 3 2	0	3	2	5	9							
	5	1	4	1								
		0	5									

(a) What is the lowest mark for the girls?
(b) What is the highest mark for the boys?
(c) How many of these boys and girls scored more than 25 marks?
(d) Compare and comment on the marks for boys and girls.

2 The time taken to complete a computer game is recorded to the nearest tenth of a minute.
The times for a group of 20 adults and 20 children are shown.

Adults						Children			
7.9	8.2	7.3	9.2	6.4	6.4	5.4	4.9	6.6	7.1
6.5	6.1	8.2	7.8	7.0	5.1	6.5	6.3	7.4	6.5
9.4	8.0	7.3	5.4	7.7	8.2	7.7	5.9	6.8	7.6
10.1	5.9	6.7	7.3	6.0	5.3	6.2	8.0	4.7	7.9

(a) Construct a back to back stem and leaf diagram for this data.
(b) Compare and comment on the times for adults and children.

What you need to know

- **Pie chart**. Used for data which can be counted.
  Often used to compare proportions of data, usually with the total.
  The whole circle represents all the data.
  The size of each sector represents the frequency of data in that sector.
  The largest sector represents the **mode**.
- **Stem and leaf diagrams**. Used to represent data in its original form.
  Data is split into two parts. The part with the higher place value is the stem,
  e.g.   15   stem 1 leaf 5.
  The data is shown in numerical order on the diagram.
  e.g. 2 \| 3 5 9 represents 23, 25, 29. A key is given to show the value of the data.

  e.g. 3 \| 4 means 34 cm or   3 \| 4 means 3.4 cm, etc.
  Back to back stem and leaf diagrams can be used to compare two sets of data.

Review Exercise 37

1 In one week Ronnie rents out 90 items from his shop, as shown in the table below.

Item	Televisions	Videos	Computers	Other equipment
Frequency	35	30	17	8

Draw a pie chart to show all the week's rentals.

AQA

2 The number of letters received each day by a school is shown in the stem and leaf diagram.

					1	8 represents 18 letters
0	6	6	7	8		
1	2	2	2	4	7	8
2	0	3	7	7	9	

(a) On how many days did the school receive 6 letters?
(b) What was the highest number of letters received?
(c) When the number of letters for another day is included in the data, the range increases by 1.
How many letters did the school receive on that day?
Write down the **two** possible answers.

AQA

3 The pie chart shows the different types of trees in a wood.
(a) What fraction of the trees in the wood are ash?

There are 459 oak trees in the wood.
(b) How many trees are there in the wood altogether?
(c) How many elm trees are there?

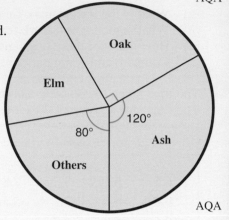

AQA

4 The weights in grams of 20 cherry tomatoes are shown.

| 5.4 | 4.6 | 6.7 | 3.9 | 4.2 | 5.0 | 6.3 | 5.4 | 4.8 | 3.5 |
| 4.6 | 5.6 | 5.8 | 6.0 | 2.8 | 4.4 | 4.7 | 5.6 | 5.1 | 4.8 |

Draw a stem and leaf diagram to represent this information.

5 A group of students took a test and their results were graded.

$\frac{1}{6}$ were awarded grade A, $\quad\quad$ $\frac{1}{4}$ were awarded grade B,

$\frac{3}{8}$ were awarded grade C, $\quad\quad$ and the rest were ungraded.

(a) Draw accurately a clearly labelled pie chart to represent this information.
(b) Twenty four students took the test.
How many students were awarded grade C?
The distribution of grades for boys in the group is shown in the bar chart.

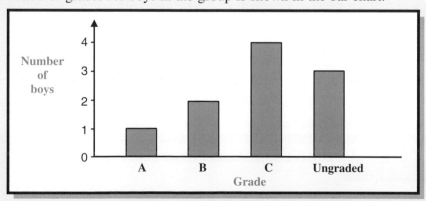

(c) Copy and complete the following table to show the distribution of grades for girls in the group.

Grade	A	B	C	Ungraded
Number of girls				

AQA

CHAPTER 38

Time Series and Frequency Diagrams

Time series

The money spent on shopping **each day**, the gas used **each quarter** and the rainfall **each month** are all examples of **time series**. A time series is a set of readings taken at time intervals.

A time series is often used to monitor progress and to show the **trend** (increases and decreases) so that future performance can be predicted. The type of graph used in this situation is called a **line graph**.

EXAMPLE

The table shows the temperature of a patient taken every half-hour.

Time	0930	1000	1030	1100	1130	1200
Temperature °C	36.9	37.1	37.6	37.2	36.5	37.0

Draw a line graph to illustrate the data.

To draw a line graph of this information, the given values are plotted and then joined to show the trend.

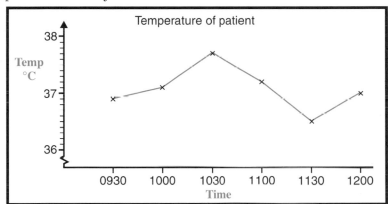

Only the plotted points show **known values**.

Lines are drawn to show the **trend**.

What is the highest temperature recorded?

*Explain why the graph can only be used to give an **estimate** of the patient's temperature at 1115.*

Exercise 38.1

1   The midday temperature at a seaside resort was recorded each day for one week.
    The line graph shows the results.

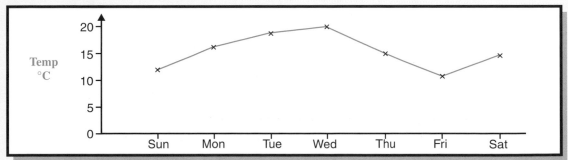

(a)   What was the midday temperature on Thursday?
(b)   Explain why you cannot use this line graph to estimate the temperature at midnight on Monday.

2 The number of cars sold by a car dealer is recorded each month.
The line graph shows the results for the first six months of a year.

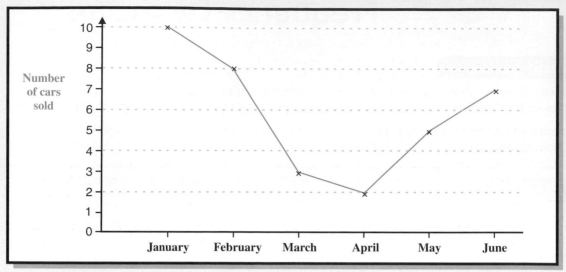

(a) How many cars were sold in February?

(b) In which months were more than 5 cars sold?

(c) Explain why you cannot estimate how many cars were sold halfway through April.

3 Each year, on his birthday, a teenager records his height.
The table shows the results.

Age (yrs)	13	14	15	16	17	18	19
Height (cm)	145	151	157	165	174	179	180

(a) Draw a line graph to represent this information.

(b) Use your graph to estimate:

    (i) the height of the teenager when he was $14\frac{1}{2}$ years of age,

    (ii) the age of the teenager when he reached 160 cm in height.

4 Joan is a member of Weight Watchers.
She records her weight at the beginning of each week.
The line graph shows a record of her weight for six weeks.

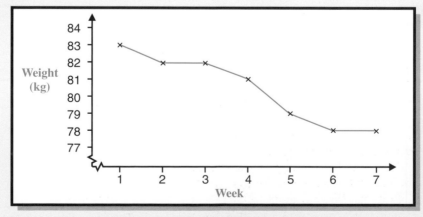

(a) What was Joan's weight at the beginning of Week 2?

(b) What was Joan's weight at the end of Week 2?

(c) How much weight did Joan lose in 6 weeks?

(d) In which week did Joan's weight first fall below 80 kg?

5 The table shows the amount of money in Jayne's savings account at the end of each month, for six months.

Month	January	February	March	April	May	June
Amount (£)	106	131	155	95	119	132

(a) Draw a line graph to represent this information.
(b) Use your graph to estimate the amount in her account in the middle of February.
(c) Explain what happened to the account between March and April.

Frequency diagrams

We use **bar charts** when data can be counted and there are only a few different items of data.
If there is a lot of data, or the data is continuous, we draw a **histogram** or **frequency polygon**.

Histograms

Histograms are used to present information contained in **grouped frequency distributions**. In this section we will only be drawing histograms for grouped frequency distributions that have equal class width intervals.

> Histograms with equal class width intervals look like bar charts with no gaps.

EXAMPLE

1 A supermarket opens at 0800.
The frequency diagram shows the distribution of the times employees arrive for work.
(a) How many employees arrive before 0730?
(b) How many employees arrive between 0730 and 0800?
(c) How many employees arrive after 0800?
(d) What is the modal class?

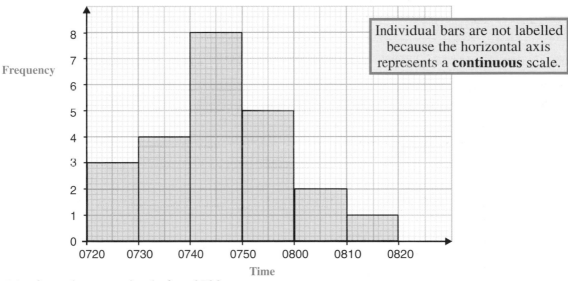

> Individual bars are not labelled because the horizontal axis represents a **continuous** scale.

(a) 3 employees arrive before 0730.
(b) Between 0730 and 0740, 4 employees arrive.
Between 0740 and 0750, 8 employees arrive.
Between 0750 and 0800, 5 employees arrive.
Employees arriving between 0730 and 0800 = 4 + 8 + 5 = 17.
(c) 3 employees arrive after 0800.
(d) The modal class is the time interval with the highest frequency.
The class interval 0740 to 0750 has the highest frequency.
The modal class is therefore, "0740 and less than 0750".

2 The frequency distribution of the heights of some boys is shown.

Height (h cm)	$130 \leq h < 140$	$140 \leq h < 150$	$150 \leq h < 160$	$160 \leq h < 170$	$170 \leq h < 180$
Frequency	1	7	12	9	3

Draw a histogram to illustrate the data.

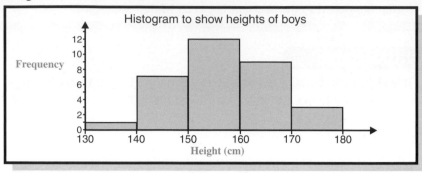

Exercise 38.2

1 The frequency diagram shows information about the weights of 100 people.

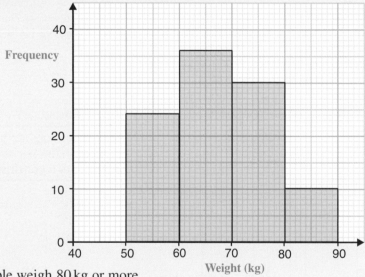

10 people weigh 80 kg or more.

(a) How many people weigh between 60 kg and 70 kg?
(b) How many people weigh less than 60 kg?
(c) How many people weigh 70 kg or more?
(d) Harry is included in the survey. He weighs 80 kg.
    In which class interval has his weight been recorded?

2 The distances, in metres, recorded in a long jump competition are shown.

5.46   5.80   5.97   5.43   6.72   5.93   6.26   6.64   5.13   6.05   6.36   6.88
6.11   5.50   6.38   5.71   6.55   6.10   5.84   5.49   6.20   5.67   6.34   6.00

(a) Copy and complete the following frequency distribution table.

Distance (m metres)	$5.00 \leq m < 5.50$	$5.50 \leq m < 6.00$	$6.00 \leq m < 6.50$	$6.50 \leq m < 7.00$
Frequency				

(b) Draw a histogram to illustrate the data.
(c) Which is the modal class?

3 Here are the mileages of cars in a roadside survey.

5442	2345	18561	16080	12500	10000	35001	34056	5156	37584
21243	36573	25057	18656	15209	29067	39893	6368	15987	24891
9999	3089	16724	25598	37151	436	4080	39949	27950	6543

(a) Copy and complete the frequency distribution table for these results.

Distance (m miles)	Tally	Frequency
$0 \leqslant m < 10\ 000$		
$10\ 000 \leqslant m < 20\ 000$		
$20\ 000 \leqslant m < 30\ 000$		
$30\ 000 \leqslant m < 40\ 000$		

(b) How many cars are included in the survey?
(c) Draw a frequency diagram to illustrate the data.
(d) Which is the modal class?

4 Aimee does a survey of the distances people travel to work.
The frequency diagram shows the results.

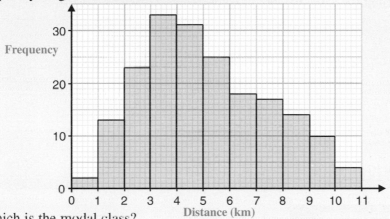

(a) Which is the modal class?
(b) How many people travel between 4 km and 5 km?
(c) How many people travel less than 2 km?
(d) How many people travel more than 10 km?
(e) How many people were included in the survey?

5 The frequency diagram illustrates the cost of holidays sold by a travel agent.

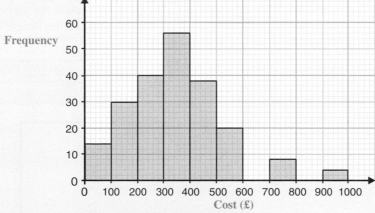

(a) How many holidays cost less than £100?
(b) How many holidays cost £400 or more?
(c) Which is the modal class?
(d) How many holidays were sold?

6 The table shows the distribution of the ages of people in a nursing home.

Age (years)	Number of people
60 and less than 70	3
70 and less than 80	13
80 and less than 90	7
90 and less than 100	6

Draw a histogram to show this information.

7 The table shows the grouped frequency distribution of the marks of 200 students.

Mark (%)	1 - 10	11 - 20	21 - 30	31 - 40	41 - 50	51 - 60	61 - 70	71 - 80	81 - 90	91 - 100
Number of Students	0	2	16	24	44	50	35	20	8	1

Draw a frequency diagram to show these results.

8 A frequency distribution of the heights of some girls is shown.

Height (h cm)	Frequency
$130 \leqslant h < 140$	3
$140 \leqslant h < 150$	5
$150 \leqslant h < 160$	12
$160 \leqslant h < 170$	4
$170 \leqslant h < 180$	1

Draw a histogram to illustrate the data.

Frequency polygons

Frequency polygons are often used instead of histograms when we need to compare two, or more, groups of data.

To draw a frequency polygon:
- plot the frequencies at the midpoint of each class interval,
- join successive points with straight lines.

To compare data, frequency polygons for different groups of data can be drawn on the same diagram. In the last section we drew a histogram to illustrate the frequency distribution of the heights of some boys. The same data can be illustrated using a **frequency polygon**.

Height (h cm)	$130 \leqslant h < 140$	$140 \leqslant h < 150$	$150 \leqslant h < 160$	$160 \leqslant h < 170$	$170 \leqslant h < 180$
Frequency	1	7	12	9	3

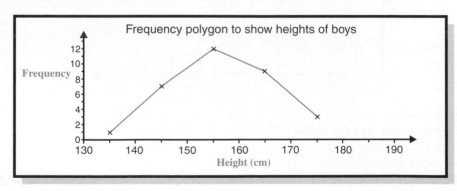

EXAMPLE

The frequency polygon shows the distribution of the distances travelled to work by the employees at a supermarket.

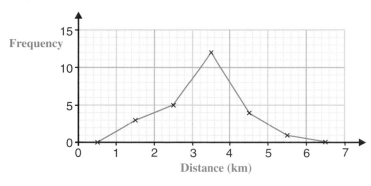

When drawing frequency polygons, frequencies are plotted at the midpoints of the class intervals.

(a)   How many employees travel between 2 km and 3 km to work?
(b)   How many employees travel more than 4 km to work?

(a)   The frequency for the midpoint of the class interval 2 km to 3 km is 5.
      So, 5 employees travel between 2 km and 3 km to work.

(b)   The frequency for the midpoint of the class interval 4 km to 5 km is 4.
      The frequency for the midpoint of the class interval 5 km to 6 km is 1.
      So, a total of 5 employees travel more than 4 km to work.

Can you work out how many people are employed at the supermarket?

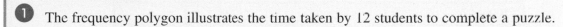

Exercise **38.3**

1   The frequency polygon illustrates the time taken by 12 students to complete a puzzle.

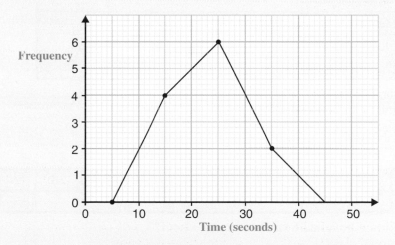

Copy and complete the table for the data.

Time (seconds)	Frequency
10 and less than 20	
20 and less than 30	
30 and less than 40	

Time Series and Frequency Diagrams

2 The frequency polygon shows the distribution of the ages of pupils who attend a village school.

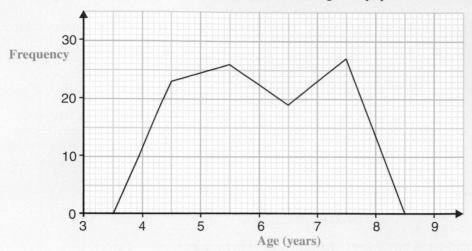

(a) How many pupils are between 5 and 6 years of age?
(b) How many pupils are under 4 years of age?
(c) How many pupils are over 7 years of age?
(d) How many pupils attend the school?

3 Here are the marks of pupils for a test in French.

44	23	36	60	50	45	35	56	41	37
31	57	43	29	67	45	34	54	29	25
46	52	27	36	39	45	41	54	49	37

(a) Copy and complete the frequency distribution table for these results.

Mark	Tally	Frequency
20 and less than 30		
30 and less than 40		
40 and less than 50		
50 and less than 60		
60 and less than 70		

(b) Which is the modal class?
(c) How many pupils scored less than 50?
(d) Draw a frequency polygon to illustrate the data.

4 The table shows the distances travelled to school by 100 children.

Distance (k km)	$0 \leqslant k < 2$	$2 \leqslant k < 4$	$4 \leqslant k < 6$	$6 \leqslant k < 8$	$8 \leqslant k < 10$
Frequency	27	35	22	10	6

(a) How many children travelled 6 km or more to school?
(b) Which is the modal class?
(c) Draw a frequency polygon to illustrate the data.

5 The numbers of fish caught by 50 anglers last month are shown.

Number of fish	1 to 10	11 to 20	21 to 30	31 to 40	41 to 50
Frequency	15	23	7	3	2

(a) Draw a frequency polygon for this information.
(b) How many anglers caught 20 fish or less?

6 The table shows the results of students in tests in English and Mathematics.

Marks	English	Mathematics
0 and less than 10	0	1
10 and less than 20	4	4
20 and less than 30	9	9
30 and less than 40	12	7
40 and less than 50	0	4

(a) Draw a frequency polygon for the English marks.
(b) On the same diagram draw a frequency polygon for the Mathematics marks.
(c) Compare and comment on the marks of the students in these two tests.

7 The table shows the results for competitors in the 2005 and 2006 Schools' Javelin Championship.
Only the best distance thrown by each competitor is shown.

Distance thrown (m metres)	Number of competitors 2005	Number of competitors 2006
$10 \leqslant m < 20$	0	1
$20 \leqslant m < 30$	3	4
$30 \leqslant m < 40$	14	19
$40 \leqslant m < 50$	21	13
$50 \leqslant m < 60$	7	11
$60 \leqslant m < 70$	0	2

(a) On the same diagram draw a frequency polygon for the 2005 results and then a frequency polygon for the 2006 results.
(b) Compare and comment on the results.

Misleading graphs

Television programmes, newspapers and advertisements frequently use graphs and diagrams to present information.
Many of the graphs and diagrams they use are well presented and give a fair interpretation of the facts, others are deliberately drawn to mislead.

Look at the graph below.
Why is it misleading?

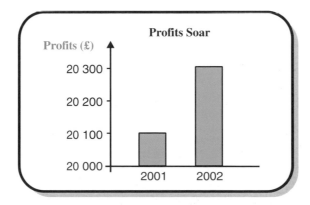

You should notice that the vertical scale does not begin at zero.
The actual increase in profits is only £200 but the graph makes it appear much more.

1 The following graph is misleading.
Explain why.

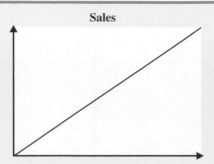

2 This graph is drawn to compare the money raised for charity by two schools.

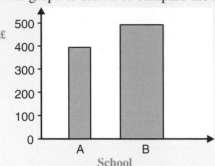

School A has raised £400.
School B has raised £500.
Why is the graph misleading?

3 This graph shows how the price of a litre of petrol has increased.
Why is the graph misleading?

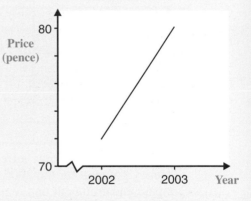

4
> **PASS WITH US**
> Our learners only need an average of 8 lessons before they can take the driving test.

Give a reason why this advertisement may be misleading.

5 This diagram is used to compare the average price of a house in two different years.
Why is the diagram misleading?

Big growth in house prices

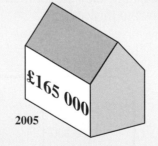

£150 000   2003

£165 000   2005

6 The graph shows the number of "Home" supporters and the number of "Away" supporters at a football match. Why is the graph misleading?

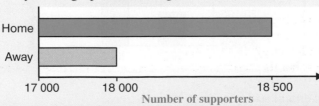

7

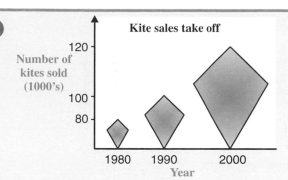

Kite sales take off

Number of kites sold (1000's)

120
100
80

1980   1990   2000
Year

Why is this diagram misleading?

8   The graph shows the votes cast for a political party in five elections.
Why is the graph misleading?

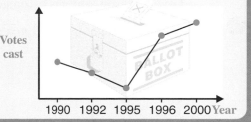

Votes cast

1990 1992 1995 1996 2000 Year

What you need to know

- A **time series** is a set of readings taken at time intervals.
- A **line graph** is used to show a time series. Only the plotted points represent actual values. Points are joined by lines to show the **trend**.
- **Histogram**. Used to illustrate **grouped frequency distributions.**
  The horizontal axis is a continuous scale.
- **Frequency polygon**. Used to illustrate grouped frequency distributions.
  Often used to compare two or more distributions on the same diagram.
  Frequencies are plotted at the midpoints of the class intervals and joined with straight lines.
  The horizontal axis is a continuous scale.
- **Misleading graphs**. Graphs may be misleading if:
  the scales are not labelled,   the scales are not uniform,   the frequency does not begin at zero.

Review Exercise 38

1   The graph shows the percentage of homes with a computer.

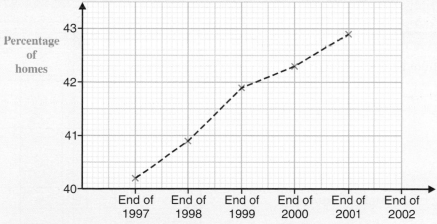

Percentage of homes

43
42
41
40

End of 1997   End of 1998   End of 1999   End of 2000   End of 2001   End of 2002

(a)   What percentage of homes had a computer by the end of 2001?
(b)   43.4% of homes had a computer by the end of 2002.
      Mark this point on a copy of the graph.
(c)   The graph shows that the percentage of homes with a computer increased every year.
      Which year had the greatest increase?
      Explain how the graph tells you this.

2 Jim bought his house in 1990.
The table shows the value of Jim's house on January 1st at 5-yearly intervals.

Year	1990	1995	2000	2005
Value of house (£)	95 000	68 000	84 000	136 000

(a) Draw a line graph to show this information.
(b) Estimate the value of Jim's house on July 1st 2003.
(c) Jim uses the graph to estimate the value of his house on January 1st 2010.
Give a reason why his estimate may not be very accurate.

3 The following graph appeared in a newspaper advert.

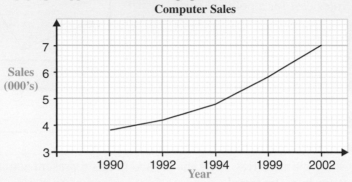

Computer Sales

Write down **two** ways in which the graph is misleading.

AQA

4 The height of some pupils is recorded.

Height h (cm)	Frequency
$120 \leqslant h < 125$	2
$125 \leqslant h < 130$	5
$130 \leqslant h < 135$	8
$135 \leqslant h < 140$	14
$140 \leqslant h < 145$	11
$145 \leqslant h < 150$	9
$150 \leqslant h < 155$	3
$155 \leqslant h < 160$	1

Ann records the data using class intervals of 10 cm.
(a) Copy and complete Ann's table.

Height h (cm)	Frequency
$120 \leqslant h < 130$	
$130 \leqslant h < 140$	
$140 \leqslant h < 150$	
$150 \leqslant h < 160$	

Ann draws a frequency diagram of her data.

Ann has made two mistakes in drawing her diagram.
(b) What are the two mistakes?

Another pupil is included.
The pupil has a height of 150 cm.
(c) Into which of Ann's class intervals should the pupil be placed?

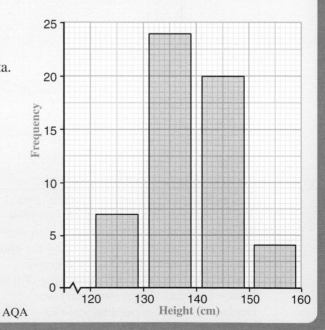

AQA

432

5 The frequency polygon shows the distribution of the marks scored in a science test.

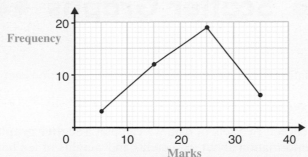

Copy and complete the frequency distribution table for these results.

Mark	Frequency
0 and less than 10	
10 and less than 20	
20 and less than 30	
30 and less than 40	

6 The grouped frequency table shows the results of a survey about the distance travelled to work by people each day.

Distance (d km)	$0 \leq d < 4$	$4 \leq d < 8$	$8 \leq d < 12$	$12 \leq d < 16$	$16 \leq d < 20$
Frequency	15	24	36	10	5

(a) How many people travelled less than 8 km to work?
(b) Which is the modal class?
(c) Draw a frequency diagram to illustrate the data.

7 A school holds a 'mini marathon'.
Every pupil who finishes the marathon gains points.
The table shows the times taken, the number of pupils and the points gained.

Time interval	Number of pupils	Points gained
1 hour or less	0	60
More than 1 hour but less than or equal to 2 hours	10	50
More than 2 hours but less than or equal to 3 hours	15	40
More than 3 hours but less than or equal to 4 hours	40	30
More than 4 hours but less than or equal to 5 hours	57	20
More than 5 hours but less than or equal to 6 hours	23	10

(a) What was the most common time interval?
(b) Ann and her brother Ben both take part.
Ann finishes only 5 minutes before Ben but gains twice as many points.
Explain how this could have happened.
(c) Draw a frequency diagram to show the number of pupils and the time taken.

AQA

Time Series and Frequency Diagrams

When we investigate statistical information we often find there are connections between sets of data, for example height and weight.
In general taller people weigh more than shorter people.

To see if there is a connection between two sets of data we can plot a **scatter graph**.
The scatter graph below shows information about the heights and weights of ten boys.

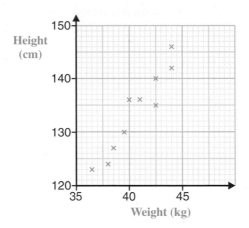

Each cross plotted on the graph represents the weight and height of one boy.

The diagram shows that taller boys generally weigh more than shorter boys.

Exercise **39.1**

1 The scatter graph shows the shoe sizes and heights of a group of girls.

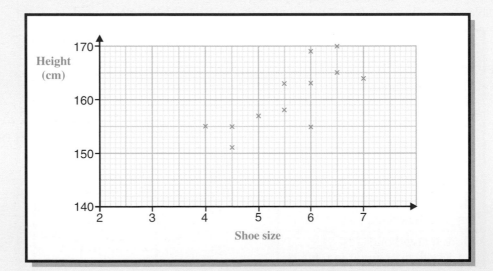

(a) How many girls wear size $6\frac{1}{2}$ shoes?

(b) How tall is the girl with the largest shoe size?

(c) Does the shortest girl wear the smallest shoes?

(d) What do you notice about the shoe sizes of taller girls compared to shorter girls?

2 The scatter graph shows the marks obtained by a group of students in a test in English and a test in French.

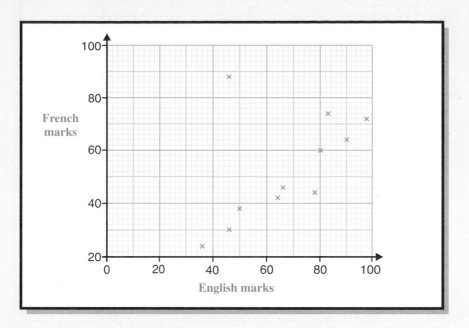

(a) Janice got the top mark in English.
What mark did she get in French?
(b) The results of one student look out of place.
    (i) What marks did the student get in English and in French?
    (ii) Give a possible reason why this student has different results from the rest of the group.

3 The scatter graph shows the pulse rates of a group of women after doing aerobics for one minute and their weight.

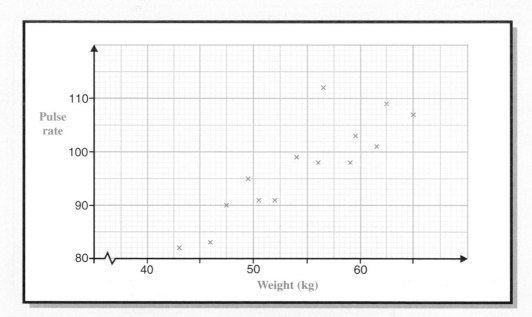

(a) How many of these women weigh less than 50 kg?
(b) What is the weight of the woman with the lowest pulse rate?
(c) What do you notice about the pulse rates of heavier women compared to lighter women?

4 The scatter graph shows the age and mileage of a number of cars.

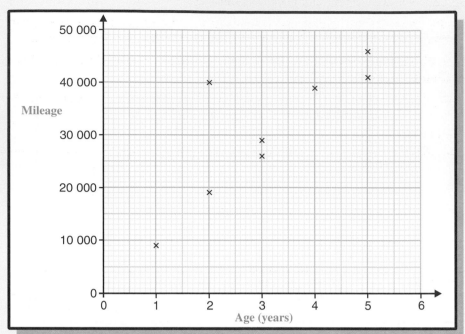

(a) One of these cars is 4 years old.
    What is the mileage of this car?

(b) Describe the relationship shown by the scatter graph.

(c) The age and mileage of one of these cars looks out of place.
    (i) What is the age and mileage of this car?
    (ii) Give a possible reason why the results for this car are different from the rest of the group.

5 The scatter graph shows the number of books read by some children and the reading ages of these children.

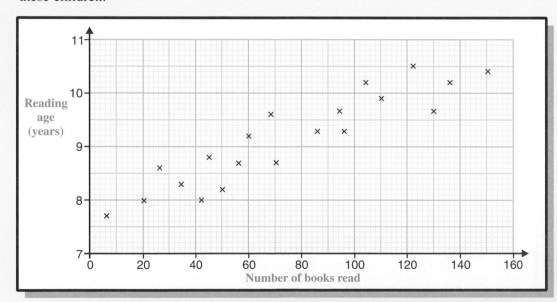

(a) How many children have read more than 100 books?

(b) One of these children has read 50 books.
    What is the reading age of this child?

(c) Describe the relationship shown by the scatter graph.

Correlation

The relationship between two sets of data is called **correlation**.

In general the scatter graph of the heights and weights shows that as height increases, weight increases. This type of relationship shows there is a **positive correlation** between height and weight.

But if as the value of one variable increases the value of the other variable decreases, then there is a **negative correlation** between the variables.

When no linear relationship exists between two variables there is **zero correlation**. This does not necessarily imply "no relationship", but merely "no linear relationship".

The following graphs show types of correlation.

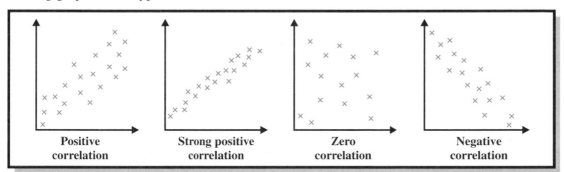

As points get closer to a straight line the stronger the correlation.
Perfect correlation is when all the points lie on a straight linc.

Exercise **39.2**

1
  (a) Which of these graphs shows the strongest positive correlation?
  (b) Which of these graphs shows perfect negative correlation?
  (c) Which of thesc graphs shows the weakest correlation?

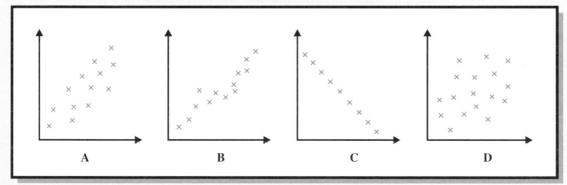

2
Describe the type of correlation you would expect between:
  (a) the age of a car and its secondhand selling price,
  (b) the heights of children and their ages,
  (c) the shoe sizes of children and the distance they travel to school,
  (d) the number of cars on the road and the number of road accidents,
  (e) the engine size of a car and the number of kilometres it can travel on one litre of fuel.

3
The table shows the distance travelled and time taken by motorists on different journeys.

Distance travelled (km)	30	45	48	80	90	100	125
Time taken (hours)	0.6	0.9	1.2	1.2	1.3	2.0	1.5

  (a) Draw a scatter graph for the data.
  (b) What do you notice about distance travelled and time taken?
  (c) Give one reason why the distance travelled and the time taken are not perfectly correlated.

Scatter Graphs · · · Scatter Graphs · · · Scatter Graphs · · ·

4 Tyres were collected from a number of different cars.
The table shows the distance travelled and depth of tread for each tyre.

Distance travelled (1000 km)	4	5	9	10	12	15	18	25	30
Depth of tread (mm)	9.2	8.4	7.6	8	6.5	7.4	7	6.2	5

(a) Draw a scatter graph for the data.
(b) What do you notice about the distance travelled and the depth of tread?
(c) Explain how you can tell that the relationship is quite strong.

5 Some students took a mathematics test and a test in Spanish.
The table shows their results.

Mark in mathematics test	10	15	18	20	26	30	34	36	46	51
Mark in Spanish test	15	50	30	12	40	19	28	48	21	42

(a) Plot a scatter graph of this information.
(b) Describe the correlation between the marks in the mathematics test and the marks in the
Spanish test.

Line of best fit

We have seen that **scatter graphs** can be used to illustrate two sets of data and, from the distribution
of points plotted, an indication of the relationship which exists between the data can be seen.

The scatter graph of heights and weights has been redrawn below and a **line of best fit** has been drawn,
by eye, to show the relationship between height and weight.

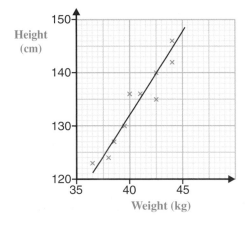

Lines of best fit:

• The slope of the line shows the trend of the points.

• A line is only drawn if the correlation (positive or negative) is strong.

• The line does not have to go through the origin of the graph.

Where there is a relationship between the two sets of data the line of best fit can be used to estimate
other values.

A boy is 132 cm tall.
Using the line of best fit an estimate of his weight is 40 kg.

In a similar way we can use the line to estimate the height of a boy when we know his weight.
A boy weighs 43 kg. Estimate his height.

1 The table shows the ages and weights of ten babies.

Age (weeks)	2	4	9	7	13	5	6	1	10	12
Weight (kg)	3.5	3.3	4.2	4.7	5	3.8	4	3	5	5.5

(a) Use this information to draw a scatter graph.
(b) What type of correlation is shown on the scatter graph?
(c) Draw a line of best fit.
(d) Mrs Wilson's baby is 11 weeks old.
     Use the graph to estimate the weight of her baby.

2 The table shows the temperature of water as it cools in a freezer.

Time (minutes)	5	10	15	20	25	30
Temperature (°C)	36	29	25	20	15	8

(a) Use this information to draw a scatter graph.
(b) What type of correlation is shown?
(c) Draw a line of best fit.
(d) Use the graph to estimate the time when the temperature of the water reaches 0°C.

3 The table shows the weights and fitness factors for a number of women.
The higher the fitness factor the fitter a person is.

Weight (kg)	45	48	50	54	56	60	64	72	99	112
Fitness Factor	41	48	40	40	35	40	34	30	17	15

(a) Use this information to draw a scatter graph.
(b) What type of correlation is shown on the scatter graph?
(c) Draw a line of best fit.
(d) Use the graph to estimate:
    (i) the fitness factor for a woman whose weight is 80 kg,
    (ii) the weight of a woman whose fitness factor is 22.

4 The following table gives the marks obtained by some candidates taking examinations in French and German.

Mark in French	53	35	39	53	50	59	36	43
Mark in German	64	32	44	70	56	68	40	48

(a) (i) Use this information to draw a scatter graph.
    (ii) Draw the line of best fit by eye.
(b) Use the graph to estimate:
    (i) the mark in German for a candidate who got 70 in French,
    (ii) the mark in French for a candidate who got 58 in German.
(c) Which of the two estimates in (b) is likely to be more reliable?
     Give a reason for your answer.

5 The table shows the times taken by some boys to run 200 metres and their inside-leg measurements.

Time (seconds)	31	33	34	38	38	38	42	43	45	47
Inside-leg (cm)	69	65	72	63	69	75	70	65	74	69

(a) Plot a scatter graph of these data.
(b) Explain why a line of best fit for these data would not be useful in estimating the time for a different boy to run 200 metres by taking his inside-leg measurement.

- A **scatter graph** can be used to show the relationship between two sets of data.
- The relationship between two sets of data is referred to as **correlation**.
- You should be able to recognise **positive** and **negative** correlation.

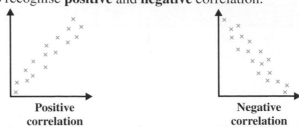

- When there is a relationship between two sets of data a **line of best fit** can be drawn on the scatter graph.
  The correlation is stronger as points get closer to a straight line.
- **Perfect correlation** is when all the points lie on a straight line.
- The line of best fit can be used to **estimate** the value from one set of the data when the corresponding value of the other set is known.

Review Exercise 39

1 Gordon wants to compare the cost of 10 paperback books with the number of pages in each book.

Cost (£)	4	5.50	3	9	8	2.50	6	10	7.50	5
Number of pages	120	240	75	100	500	80	350	550	400	220

The scatter graph shows the first five pairs of values.

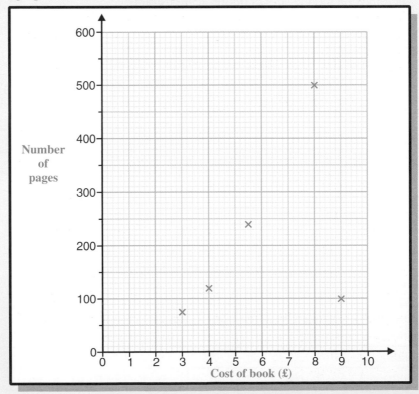

(a) Copy the diagram and plot the other five pairs of values.
(b) Which book does **not** follow the general pattern?
(c) Describe the relationship between the cost of a book and the number of pages.

AQA

2 The data from a survey of cars was used to plot several scatter graphs.

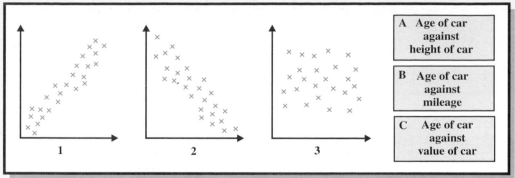

A	Age of car against height of car
B	Age of car against mileage
C	Age of car against value of car

Match each scatter graph to the correct description.

3 The table shows the ages and average heights of azalea plants.

Age (years)	1	2	2.5	3	5	5.5
Height (cm)	13	24	30	38	60	70

(a) Plot a scatter graph.
(b) Describe the relationship between the age and height of azalea plants.
(c) Use your scatter graph to estimate the average height of a four year old azalea plant. AQA

4 A number of women do aerobics for one minute.
Their ages and pulse rates are shown in the table.

Age (years)	16	17	22	25	38	42	43	50
Pulse rate (per minute)	82	78	83	90	99	97	108	107

(a) Use this information to draw a scatter graph.
Label the horizontal axis **Age (years)** from 0 to 50.
Label the vertical axis **Pulse rate (per minute)** from 60 to 120.
(b) What type of correlation is there between the ages and pulse rates of these women?
(c) Draw a line of best fit.
(d) Betty is 35 years old.
Use the line of best fit to estimate her pulse rate after doing aerobics for one minute. AQA

5 The table gives information about the age and value of a number of cars of the same type.

Age (years)	1	$4\frac{1}{2}$	6	3	5	2	4
Value (£)	8200	4900	3800	6200	4500	7600	5200

(a) Use the information to draw a scatter graph.
(b) What type of correlation is there between the age and value of these cars?
(c) Draw a line of best fit.
(d) Jo has a car of this type which is 7 years old and is in average condition.
Use the graph to estimate its value.

6 The table gives information about the petrol used for car journeys.

Petrol (litres)	3	5	6	8	4	11	10	9	2
Distance (km)	28	50	70	110	50	110	120	130	24

(a) Draw a scatter graph for this information.
(b) Draw a line of best fit.
(c) Use your graph to estimate:
  (i) the distance a car travels on 7 litres of petrol,
  (ii) the number of litres of petrol used by a car travelling a distance of 150 km.
(d) Which of the estimates in (c) is likely to be more reliable?
Give a reason for your answer.

AQA

CHAPTER 40

Probability

What is probability?

Probability, or **chance**, involves describing how likely something is to happen.

For example: How likely is it to rain tomorrow?

We often make forecasts, or judgements, about how likely things are to happen.
When trying to forecast tomorrow's weather the following **outcomes** are possible:

> sun, cloud, wind, rain, snow, …

We are interested in the particular **event**, rain tomorrow.

> In any situation, the possible things that can happen are called **outcomes**.
> An outcome of particular interest is called an **event**.

> The chance of an event happening can be described using these words:
> **Impossible   Unlikely   Evens   Likely   Certain**

Use one of the words in the box to describe the chance of rain tomorrow.

EXAMPLE

Describe the chance of each of the following events happening as:

> **Impossible   Unlikely   Evens   Likely   Certain**

(a) The next person who enters your classroom has green hair.
(b) The next person who enters the classroom is male.
(c) A number less than 5 is scored when a normal dice is rolled.

(a) Very few people have green hair.
So, the chance of the event happening is nearly **impossible**.
(b) Nearly impossible in an all girls school.
Nearly certain in an all boys school.
The probability is close to **evens** in a mixed school.
(c) The outcomes in the event 'less than 5' are 1, 2, 3 and 4.
The outcomes **not** in the event 'less than 5' are 5 and 6.
There are more outcomes in the event than not in the event.
So, the chance of the event happening is **likely**.

Exercise 40.1

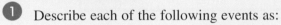

1 Describe each of the following events as:

> **Impossible   Evens   Certain**

(a) Christmas Day will be on 25th December next year.
(b) You will be 5 centimetres shorter on your next birthday.
(c) The next coin you drop will land 'tails' up.
(d) A fairy lives at the bottom of your garden.
(e) The next baby to be born will be a girl.

2 Describe each of the following events as:

Unlikely   Likely

(a)   A 6 is scored at least 500 times when a normal dice is rolled 600 times.
(b)   A 6 is scored at least 80 times when a normal dice is rolled 600 times.
(c)   It will rain on three days running in April.
(d)   It will rain on three days running in August.
(e)   A coin is tossed five times and lands heads up on each occasion.
(f)   A coin is tossed five times and lands heads up at least once.

3 Describe each of the following events as:

Impossible   Unlikely   Evens   Likely   Certain

(a)   Somewhere in the world it is raining today.
(b)   You roll a normal dice and get a 7.
(c)   You roll a normal dice and get an odd number.
(d)   A coin is tossed and it lands heads.
(e)   An apple will grow on a banana tree.
(f)   You win the next time you enter the lottery.

Probability and the probability scale

Estimates of probabilities can be shown on a **probability scale**. The scale goes from 0 to 1.
A **probability of 0** means that an event is **impossible**.
A **probability of 1** means that an event is **certain**.
Probabilities are written as a fraction, a decimal or a percentage.

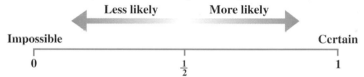

Describe the likelihood that an event will occur if it has a probability of $\frac{1}{2}$.

EXAMPLES

Estimate the probability of each of the following events happening.
Show your estimate on a probability scale.

1   It will snow in London next July.

This is possible but **very unlikely**.
So, the probability is very close to 0.

2   You will eat some vegetables today.

This is **very likely**.
So, the probability is close to 1.

3   A coin lands heads when it is tossed.

There is one outcome in the event (Heads)
and one outcome not in the event (Tails).
So, there is **an even chance** that the coin lands heads.
So, probability $= \frac{1}{2}$.

4   The next person to enter a shop will be left-handed.

There are a lot less left-handed people than
right-handed people.
The probability is between 0 and $\frac{1}{4}$.

1 Look at the events **A**, **B**, **C**, **D** and **E** listed below.

A   The next person you see will be less than 10 cm tall.

B   1, 2 or 3 is scored when an ordinary dice is rolled.

C   A day of the week ends with the letter Y.

D   It will snow on Christmas Day in London.

E   The school bus will be late tomorrow.

(a)   Which event has a probability of 0?

(b)   Which event has a probability of 1?

(c)   Which event has a probability of $\frac{1}{2}$?

2 The probability scale shows the probabilities of events **P**, **Q**, **R**, **S** and **T**.

Which of the five events

(a)   is certain to happen,

(b)   is impossible,

(c)   has an evens chance of happening,

(d)   is more likely to happen than to not happen, but is not certain to happen?

3 The probabilities of five events have been marked on a probability scale.

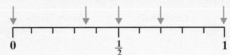

Copy the probability scale.

Event V   A coin lands 'heads' up.

Event W   A person is over 3 metres tall.

Event X   Picking a yellow sweet from a bag containing 7 yellow and 3 red sweets.

Event Y   Rolling an ordinary dice and getting a score less than 7.

Event Z   There is a 35% chance that it will rain tomorrow.

Label the arrows on your diagram to show which event they represent.

Calculating probabilities using equally likely outcomes

Probabilities can be **calculated** where all outcomes are **equally likely**.

> The probability of an event X happening is given by:
>
> Probability (X) = $\dfrac{\text{Number of outcomes in the event}}{\text{Total number of possible outcomes}}$

Remember

Probabilities have values which lie between 0 and 1.

You must write probabilities as a fraction, a decimal or a percentage.

Random and Fair

In many probability questions words such as '**random**' and '**fair**' are used.

These are ways of saying that all outcomes are equally likely.

For example:

A card is taken at **random** from a pack of cards.

This means that each card has an equal chance of being taken.

A **fair** dice is rolled.

This means that the outcomes 1, 2, 3, 4, 5 and 6 are equally likely.

EXAMPLES

1 A fair dice is rolled. What is the probability of getting:
  (a) a 6,
  (b) an odd number,
  (c) a 2 or a 3?

Total number of possible outcomes is 6.   (1, 2, 3, 4, 5 and 6)
The dice is fair, so each of these outcomes is equally likely.

  (a) 1 of the possible outcomes is a 6. $P(6) = \frac{1}{6}$

  (b) 3 of the possible outcomes are odd numbers. $P(\text{an odd number}) = \frac{3}{6} = \frac{1}{2}$

  (c) 2 of the possible outcomes are 2 or 3. $P(2 \text{ or } 3) = \frac{2}{6} = \frac{1}{3}$

2 This table shows how 100 counters are coloured red or blue and numbered 1 or 2.

	Red	Blue
1	23	19
2	32	26

The 100 counters are put in a bag and a counter is taken from the bag at random.
  (a) Calculate the probability that the counter is red.
  (b) Calculate the probability that the counter is blue and numbered 1.

Total number of possible outcomes = 100.

  (a) Red counters = 23 + 32 = 55 $P(\text{red}) = \frac{55}{100} = \frac{11}{20}$
      This could be written as 0.55 or 55%.

  (b) There are 19 counters that are blue and numbered 1. $P(\text{blue and } 1) = \frac{19}{100}$
      This could be written as 0.19 or 19%.

Exercise 40.3

1 A fair dice is rolled. What is the probability of getting:
  (a) a two,   (b) an even number,
  (c) a number less than five,   (d) a 3 or a 6?

2 A bag contains a red counter, a blue counter and a green counter.
  A counter is taken from the bag at random. What is the probability of taking:
  (a) a red counter,
  (b) a red or a green counter,
  (c) a counter that is not blue?

3 A bag contains 3 red sweets and 7 black sweets.
  A sweet is taken from the bag at random. What is the probability of taking:
  (a) a red sweet,   (b) a black sweet?

4 You toss a fair coin. What is the probability of getting:
  (a) a head,   (b) a tail?

5 The letters of the word T R I G O N O M E T R Y are written on separate cards.
  The cards are shuffled and dealt, face down, onto a table.
  A card is selected at random. What is the probability that the card shows:
  (a) the letter Y,   (b) the letter R?
  Write your answers in their simplest form.

6 This fair spinner is used in a game.

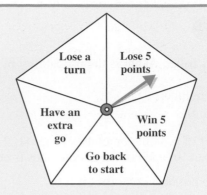

In the game a player spins the arrow.
What is the probability that the player:
(a) loses a turn,
(b) has an extra go,
(c) wins 5 points or loses 5 points?

7 The eleven letters of the word M I S S I S S I P P I are written on separate tiles.
The tiles are placed in a bag and mixed up. One tile is selected at random.
What is the probability that the tile selected shows:
(a) the letter M,   (b) the letter I,   (c) the letter P?

8 A card is taken at random from a full pack of 52 playing cards with no jokers.
What is the probability that the card:
(a) is red,   (b) is a heart,   (c) is the ace of hearts?

9 A bag contains 4 red counters, 3 white counters and 3 blue counters.
A counter is taken from the bag at random.
What is the probability that the counter is:
(a) red,   (b) white or blue,   (c) red, white or blue,   (d) green?

10 In a hat there are twelve numbered discs.

Nina takes a disc from the hat at random.
What is the probability that Nina takes a disc:
(a) with at least one 4 on it,
(b) that has not got a 4 on it,
(c) that has a 3 or a 4 on it?

43 44 45 46
47 48 49 50
51 52 53 54

11 This table shows how fifty counters are numbered either 1 or 2 and coloured red or blue.

One of the counters is chosen at random.
What is the probability that the counter is:
(a) a 1,   (b) blue,   (c) blue and a 1?

A blue counter is chosen at random.
(d) What is the probability that it is a 1?

A counter numbered 1 is chosen at random.
(e) What is the probability that it is blue?

	Red	Blue
1	12	8
2	8	22

12 Tim plays a friend at Noughts and Crosses.
He says: "I can win, draw or lose, so the probability that I will win must be $\frac{1}{3}$."
Explain why Tim is wrong.

13 The table shows the number of boys and girls in a class of 30 pupils who wear glasses.

A pupil from the class is picked at random.
(a) What is the probability that it is a boy?
(b) What is the probability that it is a girl who does not wear glasses?

A girl from the class is picked at random.
(c) What is the probability that she wears glasses?

A pupil who wears glasses is picked at random.
(d) What is the probability that it is a boy?

	Boy	Girl
Wears glasses	3	1
Does not wear glasses	11	15

The table shows the way that 120 pupils from Year 7 travel to Linfield School.

A pupil from Year 7 is chosen at random.
What is the probability that the pupil:
(a) walks to school,
(b) is a girl who travels by car,
(c) is a boy who does not travel by bus?

	Walk	Bus	Car	Bike
Boys	23	15	12	20
Girls	17	20	8	5

A girl from Year 7 is chosen at random.
What is the probability that:
(d) she walks to school,
(e) she does not travel by car?

A Year 7 pupil who travels by bike is chosen at random.
(f) What is the probability that the pupil is a boy?

Estimating probabilities using relative frequency

In question 12 in Exercise 40.3, probabilities **cannot** be calculated using equally likely outcomes.
In such situations probabilities can be estimated using the idea of **relative frequency**.

It is not always necessary to perform an experiment or make observations.
Sometimes the information required can be found in past records.

> The relative frequency of an event is given by:
>
> $$\text{Relative frequency} = \frac{\text{Number of times the event happens in an experiment (or in a survey)}}{\text{Total number of trials in the experiment (or observations in the survey)}}$$

EXAMPLES

① Jamie does the following experiment with a bag containing 2 red and 8 blue counters.

> **Take a counter from the bag at random.**
> **Record the colour then put the counter back in the bag. Repeat this for 100 trials.**

Jamie calculates the relative frequency of getting a red counter every 10 trials and shows his results on a graph. Draw a graph showing the results that Jamie might get.
This is the sort of graph that Jamie might get.

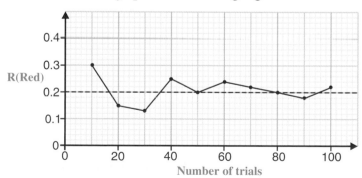

The dotted line shows the **calculated probability**.

$$P(\text{Red}) = \frac{2}{10} = 0.2$$

As the number of trials increases, relative frequency gives a better estimate of calculated probability.

Try Jamie's experiment yourself and see what sort of results you get.

② In an experiment a drawing pin is dropped for 100 trials.
The drawing pin lands "point up" 37 times.
What is the relative frequency of the drawing pin landing "point up"?

Relative frequency gives a better estimate of probability the larger the number of trials.

$$\text{Relative frequency} = \frac{37}{100} = 0.37$$

1 50 cars are observed passing the school gate. 14 red cars are observed.
What is the relative frequency of a red car passing the school gate?

2 In an experiment a gardener planted 40 daffodil bulbs of which 36 grew to produce flowers.
Use these results to find the relative frequency that a daffodil bulb will produce a flower.

3 The results from 40 spins of a numbered spinner are:

2	1	4	3	2	1	3	4	5	2
1	2	2	3	2	1	2	4	5	2
1	5	3	4	2	3	3	3	2	4
2	3	4	2	1	5	3	3	5	3

Use these results to estimate the probability of getting a 2 with the next spin.

4 A counter is taken from a bag at random.
Its colour is recorded and the counter is then put back in the bag.
This is repeated 300 times.
The number of red counters taken from the bag after every 100 trials is shown in the table.

Number of trials	Number of red counters
100	52
200	102
300	141

(a) Calculate the relative frequency after each 100 trials.
(b) Estimate the probability of taking a red counter from the bag.

5 Gemma keeps a record of her chess games with Helen.
Out of the first 10 games, Gemma wins 6. Out of the first 30 games Gemma wins 21.
Based on these results, estimate the probability that Gemma will win her next game of chess with Helen.

6 Rachel selects 40 holiday brochures at random.
The probability of a brochure being for a holiday in Italy is found to be 0.2.
How many brochures did Rachel select for holidays in Italy?

7 A counter was taken from a bag of counters and replaced.
The relative frequency of getting a red counter was found to be 0.3.
There are 60 counters in the bag.
Estimate the number of red counters.

8 500 tickets are sold for a prize draw.
Greg buys some tickets.
The probability that Greg wins first prize is $\frac{1}{20}$.
How many tickets did he buy?

9 A bypass is to be built to avoid a town.
There are three possible routes that the road can take.
A survey was carried out in the town.

Route	A	B	C
Relative frequency	0.4	0.5	0.1

30 people opted for Route C.
(a) How many people were surveyed altogether?
(b) How many people opted for Route A?
(c) How many people opted for Route B?

Mutually exclusive events

Events which **cannot happen at the same time** are called **mutually exclusive events**.
For example, the event 'Heads' cannot occur at the same time as the event 'Tails'.

> When A and B are events which cannot happen at the same time:
> $$P(A \text{ or } B) = P(A) + P(B)$$

The probability of an event not happening

> The events A and not A cannot happen at the same time.
> Because the events A and not A are certain to happen:
> $$P(\text{not } A) = 1 - P(A)$$

EXAMPLES

1   A bag contains 3 red (R) counters, 2 blue (B) counters and 5 green (G) counters.
A counter is taken from the bag at random.
What is the probability that the counter is:
(a) red,   (b) green,   (c) red or green?

Find the total number of counters in the bag.
$5 + 2 + 3 = 10$
Total number of possible outcomes = 10.

(a)   There are 3 red counters. $P(R) = \frac{3}{10}$

(b)   There are 5 green counters. $P(G) = \frac{5}{10} = \frac{1}{2}$

(c)   Events R and G cannot happen at the same time.
$$P(R \text{ or } G) = P(R) + P(G) = \frac{3}{10} + \frac{5}{10} = \frac{8}{10} = \frac{4}{5}$$

2   A bag contains 10 counters. 3 of the counters are red (R).
A counter is taken from the bag at random.
What is the probability that the counter is:
(a) red,   (b) not red?

Total number of possible outcomes = 10.

(a)   There are 3 red counters. $P(R) = \frac{3}{10}$

(b) $P(\text{not } R) = 1 - P(R) = 1 - \frac{3}{10} = \frac{7}{10}$

Exercise 40.5

1   A fish is taken at random from a tank.
The probability that the fish is black is $\frac{2}{5}$.
What is the probability that the fish is not black?

2   Tina has a bag of beads.
She takes a bead from the bag at random.
The probability that the bead is white is 0.6.
What is the probability that the bead is not white?

3   The probability of a switch working is 0.96.
What is the probability of a switch not working?

Probability . . . . Probability . . . . Probability . . . .

449

4 Six out of every 100 men are taller than 1.85 m.
A man is picked at random.
What is the probability that he is not taller than 1.85 m?

5 A bag contains red, white and blue balls.
A ball is taken from the bag at random.
The probability of taking a red ball is 0.4.
The probability of taking a white ball is 0.35.
What is the probability of taking a white ball or a blue ball?

6 Tom and Sam buy some tickets in a raffle.
The probability that Tom wins 1st prize is 0.03.
The probability that Sam wins 1st prize is 0.01.
(a) What is the probability that Tom or Sam win 1st prize?
(b) What is the probability that Tom does not win 1st prize?

7 A spinner can land on red, white or blue.
The probability of the spinner landing on red is 0.2.
The probability of the spinner landing on red or on blue is 0.7.
The spinner is spun once.
What is the probability that the spinner lands:
(a) on blue,
(b) on white?

8 A bag contains red, green, blue, yellow and white counters.
The table shows the probabilities of obtaining each colour when a counter is taken from the bag at random.

Red	Green	Blue	Yellow	White
30%	25%	20%	20%	10%

(a) (i)   How can you tell that there is a mistake in the table?
     (ii) The probability of getting a white counter is wrong.
           What should it be?

A counter is taken from the bag at random.
(b) (i)   What is the probability that it is either green or blue?
     (ii) What is the probability that it is red, green or blue?
     (iii) What is the probability that it is not yellow?

9 Some red, white and blue cubes are numbered 1 or 2.
The table shows the probabilities of obtaining each colour and number when a cube is taken at random.

	Red	White	Blue
1	0.1	0.3	0
2	0.3	0.1	0.2

A cube is taken at random.
(a) What is the probability of taking a red cube?
(b) What is the probability of taking a cube numbered 2?
(c) State whether or not the following pairs of events are mutually exclusive.
     Give a reason for each answer.
     (i)   Taking a cube numbered 1 and taking a blue cube.
     (ii) Taking a cube numbered 2 and taking a blue cube.
(d) (i)   What is the probability of taking a cube which is blue or numbered 1?
     (ii) What is the probability of taking a cube which is blue or numbered 2?
     (iii) What is the probability of taking a cube which is numbered 2 or red?

EXAMPLES

1 A fair coin is thrown twice.
Identify all of the possible outcomes and write down their probabilities.

Method 1
List the outcomes systematically.

1st throw	2nd throw
Head (H)	Head (H)
Head (H)	Tail (T)
Tail (T)	Head (H)
Tail (T)	Tail (T)

Method 2
Use a **possibility space diagram**.

2nd throw	T	H & T	T & T
	H	H & H	T & H
		H	T

1st throw

When a fair coin is tossed twice, there are four possible outcomes.
Because the coin is fair all the possible outcomes are **equally likely**.
Because all the outcomes are equally likely their probabilities can be worked out.

$P(\text{H and II}) = P(\text{H and T}) = P(\text{T and H}) = P(\text{T and T}) = \frac{1}{4}$

2 A fair dice is rolled twice.
Use a possibility space diagram to show all the possible outcomes.
What is the probability of getting a 'double six'?
What is the probability of getting any 'double'?
What is the probability that exactly one 'six' is obtained?

The dice is fair, so there are 36 equally likely outcomes.

2nd roll						
6	1 and 6	2 and 6	3 and 6	4 and 6	5 and 6	6 and 6
5	1 and 5	2 and 5	3 and 5	4 and 5	5 and 5	6 and 5
4	1 and 4	2 and 4	3 and 4	4 and 4	5 and 4	6 and 4
3	1 and 3	2 and 3	3 and 3	4 and 3	5 and 3	6 and 3
2	1 and 2	2 and 2	3 and 2	4 and 2	5 and 2	6 and 2
1	1 and 1	2 and 1	3 and 1	4 and 1	5 and 1	6 and 1
	1	**2**	**3**	**4**	**5**	**6**

1st roll

P(double 6)
There is one outcome in the event (6 and 6).

$P(\text{double 6}) = \frac{1}{36}$

P(any double)
The 6 outcomes in the event are shaded blue.

$P(\text{any double}) = \frac{6}{36} = \frac{1}{6}$

P(exactly one six)
The 10 outcomes in the event are shaded grey.

$P(\text{exactly one six}) = \frac{10}{36} = \frac{5}{18}$

Exercise 40.6

1 A red car (R), a blue car (B) and a green car (G) are parked on a narrow drive, one behind the other.
(a) List all the possible orders in which the three cars could be parked.

The cars are parked on the drive at random.
(b) What is the probability that the blue car is first on the drive?

2 Two fair dice are rolled and the numbers obtained are added.
  (a) Draw a possibility space diagram to show all of the possible outcomes.
  (b) Use your diagram to work out:
      (i)   the probability of obtaining a total of 10,
      (ii) the probability of obtaining a total greater than 10,
      (iii) the probability of obtaining a total less than 10.
  (c) Explain why the probabilities you worked out in (b) should add up to 1.

3 A fair coin is tossed and a fair dice is rolled.
Copy and complete the table to show all the possible outcomes.

Dice

		1	2	3	4	5	6
Coin	H		H2				
	T						

What is the probability of obtaining:
  (a) a head and a 5,
  (b) a tail and an even number,
  (c) a tail and a 6,
  (d) a tail and an odd number,
  (e) a head and a number more than 4,
  (f) an odd number?

4 Sanjay has to travel to school in two stages.
Stage 1: he can go by bus or train or he can get a lift.
Stage 2: he can go by bus or he can walk.
  (a) List all the different ways that Sanjay can travel to school.

Sanjay decides the way that he travels on each stage at random.
  (b) What is the probability that he goes by bus in both stages?

5 The diagram shows a fair spinner.
It is divided into four equal sections, numbered, as shown.
The spinner is spun twice and the numbers the arrow lands on each
time are added to obtain a score.
  (a) Copy and complete this table to show all the possible scores.

2nd spin

		1	2	3	4
	1	2	3		
1st spin	2	3			
	3				
	4				

  (b) Calculate the probability of getting a score of:
      (i) 2,   (ii) 3,   (iii) 6.

6 Students at a college must choose to study two subjects from the list:

Maths   English   Science   Art

  (a) Write down all the possible pairs of subjects that the students can choose.

David chooses both subjects at random.
  (b) What is the probability that one of the subjects he chooses is Maths?

James chooses Maths and one other subject at random.
  (c) What is the probability that he chooses Maths and Science?

7 Bag A contains 2 red balls and 1 white ball.
Bag B contains 2 white balls and 1 red ball.
A ball is drawn at random from each bag.

(a) Copy and complete the table to show all possible pairs of colours.

Bag A

		R	R	W
	W	RW		
Bag B	W			
	R			

(b) Explain why the probability of each outcome is $\frac{1}{9}$.

(c) Calculate the probability that the two balls are the same colour.

8 The diagram shows two sets of cards A and B.

One card is taken at random from set A.
One card is taken at random from set B.
(a) List all the possible outcomes.

The two numbers are added together.
(b) (i) What is the probability of getting a total of 5?
    (ii) What is the probability of getting a total that is not 5?

All the cards are put together and one of them is taken at random.
(c) What is the probability that it is labelled A or 2?

9 A spinner has an equal probability of landing on red, green, blue, yellow or white.
The spinner is spun twice.
(a) List all the possible outcomes.
(b) (i) What is the probability that, on both spins, the spinner lands on white?
    (ii) What is the probability that, on both spins, the spinner lands on white at least once?
    (iii) What is the probability that, on both spins, the spinner lands on the same colour?

10 The diagram shows two fair spinners.
Each spinner is divided into equal sections and numbered, as shown.
Each spinner is spun and the numbers that each arrow lands on are added together.

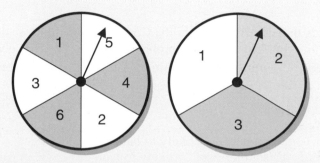

(a) Draw a possibility space diagram to show all the possible outcomes.
(b) Calculate the probability of getting a total of 2.
(c) Calculate the probability of getting a total of 6.

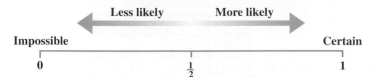

- You need to know the meaning of these terms: impossible, unlikely, evens, likely, certain.
- **Probability** describes how likely or unlikely it is that an event will occur.
  Probabilities can be shown on a probability scale.
  Probability **must** be written as a fraction, a decimal or a percentage.

Less likely   **More likely**

Impossible   Certain

0                    $\frac{1}{2}$ 1

- How to work out probabilities using **equally likely outcomes**.
  The probability of an event X happening is given by:

$$\text{Probability (X)} = \frac{\text{Number of outcomes in the event}}{\text{Total number of possible outcomes}}$$

- How to estimate probabilities using **relative frequency**.
  The relative frequency of an event is given by:

$$\text{Relative frequency} = \frac{\text{Number of times the event happens in an experiment (or in a survey)}}{\text{Total number of trials in the experiment (or observations in the survey)}}$$

- How to use probabilities to **estimate** the number of times an event occurs in an **experiment** or **observation**.

  Estimate = total number of trials (or observations) × probability of event

- **Mutually exclusive events** cannot occur at the same time.
  When A and B are mutually exclusive events:

$$\text{P(A or B)} = \text{P(A)} + \text{P(B)}$$

- A general rule for working out the probability of an event, A, **not happening** is:

$$\text{P(not A)} = 1 - \text{P(A)}$$

- How to find all the possible outcomes when two events are combined.
  By **listing** the outcomes systematically.
  By using a **possibility space diagram**.

Review Exercise 40

1

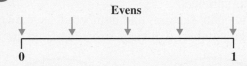

Evens

0   1

Below are five words that describe probability. Copy the probability scale.
Write each word in the right place on your probability scale. The first one is done for you.

Evens   Unlikely   Certain   Likely   Impossible

AQA

2 This spinner is used in a game.
The spinner is spun once.
  (a) Which number is the spinner most likely to land on?
  (b) Which numbers is the spinner equally likely to land on?
  (c) What is the probability that the spinner will land on 5?

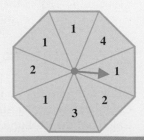

AQA

3 The probability that someone gets flu next winter is 0.3.
What is the probability that someone **does not** get flu next winter?

AQA

4 The probability scale below shows the probability of 5 events, labelled A, B, C, D and E.

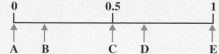

Match **4** of the letters to the probabilities of the events described below.
(a) The next child born in Rotherham Maternity Hospital will be a boy.
(b) Throwing a 4 with an ordinary 6 sided dice.
(c) Picking a toffee from a bag of sweets that contains only mints.
(d) Picking a black or blue ball from a bag that contains 5 red, 5 blue and 5 black balls.

AQA

5 Dennis has five coins in his pocket.
He has two 10p coins and three 2p coins.

He chooses a coin at random.
(a) What is the probability that Dennis chooses a 10p coin?

Dennis puts the coin back in his pocket.
He chooses another coin at random.
(b) Is it more likely to be a 10p coin or a 2p coin?
Explain your answer.

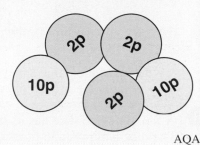

AQA

6 Mandy has a bag containing packets of crisps.
The bag contains: 5 packets of plain crisps;
                     3 packets of salt and vinegar crisps;
                     4 packets of bacon flavour crisps.

Mandy takes one packet of crisps out of the bag at random.
(a) What is the probability that it is a packet of salt and vinegar crisps?
(b) What is the probability that the packet will **not** be bacon flavour crisps?

AQA

7 In a raffle 100 tickets are sold. Only one prize can be won.
(a) Nicola buys one ticket. What is the probability that she wins the prize?
(b) Dee buys five tickets. What is the probability that she wins the prize?
(c) Keith buys some tickets. The probability that he wins the prize is $\frac{3}{20}$.
    (i) What is the probability that he does **not** win the prize?
    (ii) How many tickets did he buy?

AQA

8 A fair spinner has eight sides. The sides are numbered 1, 2, 2, 3, 3, 4, 5 and 6.
The spinner is spun once.
(a) What is the probability that the spinner lands on a 3?

Michael and Sheila play a game using the spinner.
The spinner is spun once.
Michael wins if the spinner lands on 1 or 2 or 3.
Sheila wins if it lands on 4 or 5 or 6.
(b) What is the probability that Sheila will win the game?
(c) Explain why this game is **not** fair.

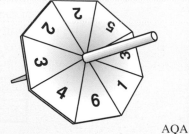

AQA

9 A card is chosen at random from Set 1 and another card is chosen at random from Set 2.

Set 1   Set 2

(a) List all the possible outcomes.
(b) What is the probability that a C is not one of the two cards chosen?

AQA

10 A game is played with two fair spinners.
In each turn of the game both spinners are spun and the numbers are added to get a score.

(a) Copy and complete the table to show each possible score.

	1	2	3	4
1				
2				
3				

(b) What is the probability of:
   (i) scoring 6, (ii) not scoring 6?

(c) To start the game a player needs to score either 2 or 5.
What is the probability that the game starts on the first throw?

11 A fair six-sided dice has its faces painted red (R), white (W) or blue (B).
The dice is thrown 36 times. Here are the results.

B B R W B W B R W B R W
R B B W W B W B B R B B
W W B B B R B B W R W B

(a) Copy and complete the table below for these results.

Outcome	Tally	Frequency
Red		
White		
Blue		
	Total	36

(b) Use these results to estimate the probability that the next throw will be red.
(c) Based on the results, how many faces do you think are painted each colour? AQA

12 (a) A fair coin is thrown.
What is the probability of getting a head?
(b) A bag contains 4 blue balls and 7 red balls.
A ball is chosen at random.
What is the probability that it is blue?
(c) Another bag contains only yellow, green and black counters.
   (i) Copy and complete the table to show the probability of getting a green counter.

Counter	Yellow	Green	Black
Probability	0.4		0.2

   (ii) What does this tell you about the numbers of counters in the bag? AQA

13 Brenda has a bag of fruit sweets.
There are 4 lemon, 5 orange, 8 strawberry and 3 blackcurrant sweets.
Brenda chooses one sweet at random. What is the probability that the sweet is
(a) a lemon sweet,
(b) not an orange sweet? AQA

Handling Data
Non-calculator Paper

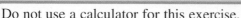

Do not use a calculator for this exercise.

1 Esther asks her friends which type of berry they like best.
Their replies are:

Strawberry	Blackberry	Raspberry	Raspberry	Strawberry
NONE	Strawberry	Raspberry	Strawberry	Strawberry
Strawberry	Blackberry	Strawberry	NONE	Raspberry

(a) Copy and complete the table below.

Berry	Tally	Frequency
Strawberry		
Blackberry		

(b) Draw a pictogram to show the results.
Use the symbol ● to represent 2 replies.

AQA

2 A group of students were asked which sport they did most.
The bar chart shows the results.

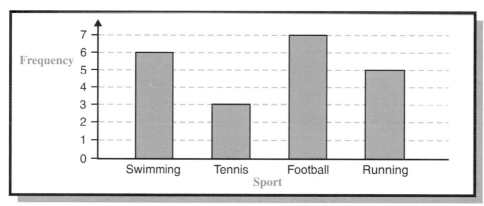

(a) How many students said, "swimming"?
(b) How many students were asked?
(c) Which sport is the mode?

3 Nazeem counted how many times his teacher said, "OK," during each of five lessons.
His results were: 5, 3, 2, 8, 2.
(a) Find the median.
(b) Calculate the mean.
(c) Work out the range.

AQA

4 Sam is a car salesman.
He records the number of cars of each make that he sells in a month.

Make	Ford	Nissan	Vauxhall	BMW
Frequency	16	10	6	4

Draw and label a pie chart to represent this information.

AQA

5 Danny has two fair spinners.
Spinner A has four equal sections, two are red, one is yellow and one is green.
Spinner B has six equal sections, three are red, one is yellow and two are green.

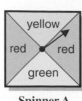

Spinner A   **Spinner B**

Danny spins each spinner once.
(a) Which colour is Spinner A most likely to land on?
(b) Which spinner is more likely to land on yellow, Spinner A or Spinner B?
    Give a reason for your answer.
(c) What is the probability that Spinner A lands on green?
(d) The probabilities of three events have been marked on the probability scale below.

R: Spinner B lands on red
G: Spinner B lands on green
Y: Spinner B lands on yellow

Copy the diagram and label each arrow with a letter to show which event it represents.

AQA

6 The graph shows the results of a survey of the times at which pupils arrived at school one day.

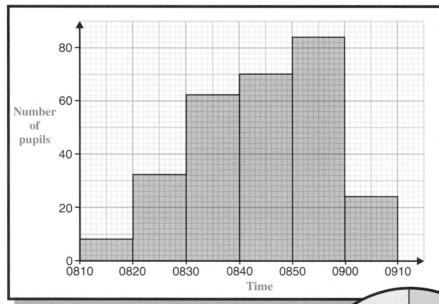

(a) (i) How many pupils arrived for school between 0830 and 0850?
    (ii) How many pupils attended school that day?

There were 24 pupils who were late for school.
The reasons given for being late are shown in
the pie chart.
(b) (i) Which reason for lateness is the mode?
    (ii) Use the pie chart to work out how many of
         these pupils overslept.

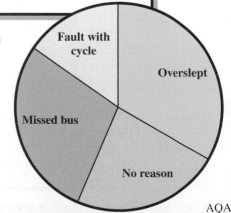

AQA

458

7 A fair six-sided dice and a fair coin are thrown together.
  (a) List all the possible outcomes.
  (b) What is the probability of getting a tail and a six?

8 Jo measures the lengths, in centimetres, of a sample of runner beans.
  The stem and leaf diagram shows the results.
  (a) What is the length of the longest runner bean?
  (b) How many runner beans are included in the sample?
  (c) What is the range of their lengths?
  (d) Which length is the median?

								1│3 means 13 cm
1	3	5	7	9				
2	0	3	4	6	7	7	7	8
3	0	1	1					

9 (a) Write down four whole numbers whose mode is 10 and whose range is 8.
  (b) Write down four whole numbers whose mode is 10 and whose mean is 9.
  (c) Write down four whole numbers whose mode is 10, whose mean is 9 and whose range is 8.

AQA

10 The table shows the height and trunk diameter of each of 8 trees.

Height (m)	1.5	2.5	4	4.5	5.5	6	6.5	7
Trunk diameter (cm)	10	10	15	20	20	25	25	30

  (a) Plot a scatter graph of these data.
  (b) Draw a line of best fit through the points on the scatter graph.
  (c) Describe the relationship shown in the scatter graph.
  (d) A tree is 3 metres tall. The trunk diameter is given as 31 centimetres.
  Explain why 13 centimetres is more likely.

AQA

11 Sue does a survey about the time that people spend watching TV and reading.
Two questions on her questionnaire are:

> **Question 1**   How much do you read?
> **Question 2**   There are more and more TV channels.
>   More and more TV programmes are being made.
>   We are watching more and more TV, so now we don't read enough.
>   Don't you agree?

  (a) Explain why each question is **not** a good one for this questionnaire.
  (b) Rewrite Question 1 so that it is more suitable for the questionnaire.

AQA

12 The table shows the number of Valentine cards received by 20 students.

Number of cards	0	1	2	3	4	5
Number of students	3	7	6	3	1	0

  (a) What is the range in the number of cards received?
  (b) What is the median number of cards per student?
  (c) Calculate the mean number of cards per student.

AQA

13 A box contains 20 plastic ducks.
3 of the ducks are green, 10 are blue and the rest are yellow.
A duck is taken from the box at random.
  (a) What is the probability that it is green?
  (b) What is the probability that it is yellow?

14 The diagram shows two sets of cards A and B.

One card is taken from set A and one card is taken from set B. List all the possible outcomes.

15 A group of students were each asked how many books they had read last month.
The frequency diagram shows the results.

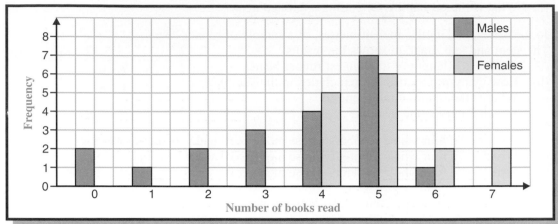

(a) How many students were included?
(b) What is the range in the number of books read by females?
(c) Calculate the mean number of books read by females.
(d) Compare and comment on the number of books read by males and the number of books read by females.

16 A spinner gives a score of 1, 2, 4 or 6 with the following probabilities.

Score	1	2	4	6
Probability	0.4	0.2	0.1	...

(a) Calculate the probability that the score is 6.
(b) Calculate the probability that the score is either 1 or 2.

AQA

17 The table shows information about a group of children.

	Can swim	Cannot swim
Boys	16	4
Girls	19	6

(a) One of these children is chosen at random.
What is the probability that the child can swim?
(b) A girl in the group is chosen at random.
What is the probability that she cannot swim?
(c) Tony says, "These results show a higher proportion of girls can swim."
Is he correct? Give reasons for your answer.

18 A spinner is labelled, as shown.
The results of the first 30 spins are given below.

1	2	3	3	5	1	3	2	2	4	5	3	2	1	2
5	2	4	1	5	1	5	2	2	4	2	5	4	2	3

(a) What is the relative frequency of getting the number 1?
(b) Is the spinner fair?
Give a reason for your answer

19 The maximum load for a lift is 1200 kg.
The table shows the distribution of the
weights of 22 people waiting for the lift.
Will the lift be overloaded if all of these people get in?
You **must** show working to support your answer.

Weight (w kg)	Frequency
$30 \leqslant w < 50$	8
$50 \leqslant w < 70$	10
$70 \leqslant w < 90$	4

AQA

Handling Data
Calculator Paper

You may use a calculator for this exercise.

1 The pictogram shows the eye colours of the pupils in Class 1.

Eye colour	Key: ⬮ = 2 pupils
Brown	⬮ ⬮ ⬮ ⬮ ⬮ ⬮ ⬮ ⬮
Blue	⬮ ⬮ ⬮ ⬮ ◖
Grey	⬮ ◖
Other	⬮ ⬮

(a) Which is the most common eye colour?
(b) How many children have blue eyes?
(c) How many more children have blue eyes than have grey eyes?   AQA

2 The number of newspapers delivered to each of 30 houses is shown.

1	1	0	2	1	2	2	0	1	1
3	1	1	2	1	1	0	0	2	1
0	0	1	2	1	2	1	0	3	2

(a) (i) Draw a bar chart to represent this information.
    (ii) Which number of newspapers is the mode?

A house in this street is picked at random.

(b) (i) Choose the word which best describes the probability that the house has
      3 newspapers delivered.

      evens   certain   likely   impossible   unlikely

    (ii) Copy and place a mark on the scale which shows the probability that the house has
      4 newspapers delivered.

      ```
      ⌐————————————————————————┐
      0                          1
      ```
                                                                                   AQA

3 Matthew conducts a survey of the favourite types of music of pupils in his class.
He records his results, as shown.

Rock	Heavy metal	Classical
Rap	Pop	Classical
Rock	Pop	Rock
Heavy metal	Pop	Pop
Pop	Rock	Soul
Pop	Rap	Pop
Rock	Heavy metal	Rock
Rap	Pop	Heavy metal

(a) State a better method of recording the data.
(b) Use your method to record these results.
(c) State the mode for the data.
(d) A pupil is picked at random.
    What is the probability that the pupil's favourite music is classical?   AQA

4 A youth club leader records the attendances at his youth club over a period of 12 weeks. The attendances are:

$$60, \quad 36, \quad 38, \quad 42, \quad 25, \quad 17, \quad 24, \quad 17, \quad 75, \quad 41, \quad 33, \quad 26.$$

(a) Calculate the median attendance.
(b) Calculate the mean attendance.
(c) (i) Which attendance is the mode?
  (ii) Why is the mode not a good measure of average for these attendances?
(d) Calculate the range of these attendances.   AQA

5 A fair coin is thrown and a fair dice is rolled.

> If the coin shows heads, the score is the number shown on the dice.
> If the coin shows tails, the score is double the number shown on the dice.

(a) Copy and complete the table to show each possible score.

Dice

		1	2	3	4	5	6
Coin	Heads	1					
	Tails				8		

(b) What is the probability of getting a score of 10?
(c) What is the probability of getting a score of less than 6?   AQA

6 The lengths, in centimetres, of a sample of leaves are shown.

4.7	5.0	6.4	5.6	6.1	4.9	5.8	6.5	7.2	6.7	6.5	5.3

(a) Calculate the mean length of these leaves.
(b) Draw a stem and leaf diagram to represent this information.

7 Julie counted the number of people in each car that passed her on her way to school.

Number of people	1	2	3	4
Frequency	24	8	3	1

Calculate the total number of people in the cars that passed Julie.   AQA

8 Here are three methods which can be used for estimating probabilities.

> **Method A**: A calculation based on equally likely outcomes.
> **Method B**: Carry out an experiment to collect data.
> **Method C**: Refer to past records.

Which method, **A**, **B** or **C**, is the **most** suitable for estimating each of the following?
(a) The probability that the 0830 train will arrive late tomorrow.
(b) The probability that when a drawing pin is thrown on the floor it will land point up.
(c) The probability that when a card is taken from a pack it will be a picture card.   AQA

9 A fire station was called out to 300 fires in the home last year.
The table shows information about where fires started.

Where fires started	Kitchen	Living room	Bedroom	Other
Number of fires	150	75	60	15

(a) Draw a clearly labelled pie chart to represent this information.
(b) What percentage of these fires started in the bedroom?   AQA

10 A class survey found the probability of having brown eyes is 0.6.
  (a) What is the probability of not having brown eyes?
  (b) There are 30 children in the class.
      Estimate the number with brown eyes.

11 The two-way table shows the age and sex of a sample of 20 pupils at a school.

Age (years)

	12	13	14	15
Number of boys	4	3	3	2
Number of girls	2	2	1	3

There are 1000 pupils in the school altogether.
  (a) Use the values in the table to estimate the number of boys in the school.
  (b) How could a better estimate be obtained?   AQA

12 The table shows some of the probabilities of getting different numbers of peas in a pod.

Number of peas	less than 5	5	6	7	more than 7
Probability	0.25	0.20	0.25	0.20	

  (a) Calculate the probability of getting 6 or less peas in a pod.
  (b) Calculate the probability of getting more than 7 peas in a pod.   AQA

13 The table shows the ages and weights of chickens.

Age (days)	10	20	40	50	70	80	100
Weight (g)	100	300	1000	1300	2000	2000	2400

  (a) Use this information to draw a scatter graph.
  (b) Describe the correlation between the age and weight of these chickens.
  (c) Draw a line of best fit.
  (d) Explain how you know the relationship is quite strong.

14 Greg watched 50 trains arrive at a station.
He recorded how many minutes late each train arrived.
The table gives a summary of his results.

Minutes late	0 to less than 10	10 to less than 20	20 to less than 30	30 to less than 40
Frequency	22	14	10	4

Draw a frequency polygon to represent these data.   AQA

15 A factory manager asked each of his 36 employees how many cups of tea and cups of coffee they drank at work yesterday.
The results are shown in the two-way table.

Number of cups of tea

		0	1	2	3
	0	0	3	7	2
Number of cups of coffee	1	4	2	3	1
	2	5	3	2	0
	3	0	1	2	1

  (a) How many employees drank 2 cups of tea?
  (b) How many employees drank the same number of cups of tea as cups of coffee?
  (c) How many employees drank more cups of tea than cups of coffee?
  (d) Calculate the total number of cups of tea drunk by these employees.   AQA

16. The pie chart shows the results of a survey about the takeaway foods students like best.

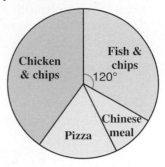

(a) Fish and chips is liked best by 80 students.
How many students like Pizza best?

(b) The table shows the results of a survey about the takeaway foods liked best by 100 pensioners.

Takeaway	Frequency
Fish and chips	40
Chicken and chips	43
Pizza	9
Chinese meal	8

Do these surveys support the hypothesis:
"More pensioners than students like fish and chips"?
Explain your answer.   AQA

17. Jamal, Des, William and Sue play a board game. There is only one winner.
The probability that Jamal wins is 0.3.
The probability that Des wins is 0.1.
(a) Calculate the probability that either William or Sue wins.
(b) William is twice as likely to win as Sue.
What is the probability that Sue wins?   AQA

18. The two-way table shows the results of a survey among adults at an out of town supermarket.

	Can drive	Cannot drive
Male	22	4
Female	15	4

A town has 27 953 adults.
(a) Give a reason why the sample may not be representative of the town.
(b) Use the results of the survey to estimate the number of adults in the town who can drive.
Give your answer to a suitable degree of accuracy.   AQA

19. A fair coin is thrown 20 times. It lands heads 12 times.
(a) What is the relative frequency of throwing a head?

The coin continues to be thrown.
The table shows the number of heads recorded for 20, 40, 60, 80 and 100 throws.

Number of throws	20	40	60	80	100
Number of heads	12	18	30	42	49

(b) Draw a graph to show the relative frequency of throwing a head for these data.
(c) Estimate the relative frequency of throwing a head for 1000 throws.   AQA

20. A survey was made of the time taken by 100 people to do their shopping at a local supermarket on a Monday.

Time taken (t minutes)	Number of people
$0 \leqslant t < 10$	43
$10 \leqslant t < 20$	35
$20 \leqslant t < 30$	17
$30 \leqslant t < 40$	5

Calculate an estimate of the mean time taken by these people to do their shopping.

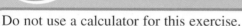
Do not use a calculator for this exercise.

1 (a) Write the number two thousand one hundred and thirty-nine in figures.
 (b) Write 573 to the nearest 10.
 (c) Work out 1542 + 468.

2 Place these numbers in order, starting with the smallest. 35, 305, 41, 4001, 587. AQA

3 Six numbers are shown. │ 23 38 45 61 73 96 │
 (a) Which numbers are even?
 (b) What is the difference between the largest number and the smallest number?
 (c) Which number must be added to 38 to get 61? AQA

4 (a) The number 3482 is multiplied by 100. What is the value of 8 in the answer?
 (b) The number 3482 is divided by 10. What is the value of 4 in the answer? AQA

5 (a) Shape *PQRST* is shown.
 (i) Which side is parallel to *PQ*?
 (ii) Which of these terms describes angle *x*?
   acute   reflex   obtuse   right angle
 (iii) Which word describes shape *PQRST*?
   pentagon   oblong   rhombus   hexagon
 (iv) What are the coordinates of point *P*?

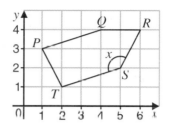

 (b)

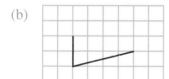

This diagram shows two sides of a parallelogram.
Copy the diagram and complete the parallelogram.

6 These patterns show the first three terms in a sequence.

 Pattern 1   **Pattern 2**   **Pattern 3**

 (a) Draw the next pattern in the sequence.
 (b) Copy and complete the table.

Pattern number	1	2	3	4
Number of squares	1	3	5	

 (c) How many squares will be in Pattern 5?
 (d) Explain why a pattern in this sequence cannot have 20 squares.

7

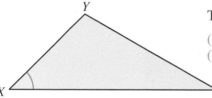

Triangle *XYZ* is shown.

 (a) Measure the length of *XZ*.
 (b) Measure the size of angle *X*.

8 Dave buys these items.
 (a) How much does he have to pay altogether?
 (b) He pays with a £5 note.
  How much change is he given?

75 pence   **49 pence**   **62 pence**

9 (a) Write $\frac{1}{2}$ as a percentage.

(b) Write $\frac{2}{5}$ as a decimal.

(c) Put these fractions in order, smallest first.

$$\frac{1}{2} \qquad \frac{2}{3} \qquad \frac{3}{4} \qquad \frac{2}{5}$$

(d) Work out $\frac{3}{4}$ of 36.

(e) What fraction of this rectangle is shaded?
Give your answer in its simplest form.

10 What number must be put in the box to make each of these statements true?

(a) $\boxed{} \times 3 = 15$ (b) $3 + \boxed{} = 11$ (c) $\boxed{} - 3 = 7$ (d) $3 - \boxed{} = -5$

11 (a) This shape has been drawn on 1 cm squared paper.
(i) What is the area of the shape?
(ii) What is the perimeter of the shape?

Not full size

(b) This cuboid is made from centimetre cubes.
Calculate the volume of this cuboid.

12 The set of numbers 3, 3, 4 has a product of 36 because $3 \times 3 \times 4 = 36$.
Write down two more sets of **three** numbers that have a product of 36.   AQA

13 Which of the numbers 2, 4, 6, 8, 9 and 24 are common factors of 12 and 24?

14 Work out. (a) $236 - 149$ (b) $23.6 \div 10$ (c) 16×27 (d) $0.2 \div 4$

15 Here are the six faces of a dice.

(a) Write down the order of rotational symmetry of the face with 3 spots.
(b) Write down **two** faces that have rotational symmetry of order 4.   AQA

16 (a) What is the next number in this sequence?
Explain how you found your answer.

$$\boxed{1, \quad 5, \quad 9, \quad 13, \quad \cdots}$$

(b) Is 100 in the sequence? Give a reason for your answer.

17 (a) Simplify $b \times b \times b$.
(b) Drinks cost 35 pence each. How much will d drinks cost?
(c) What is the value of $ab - 2c$ when $a = 4$, $b = 5$ and $c = 6$?

18 (a) What is the square of 7?
(b) What is the square root of 81?
(c) What is the reciprocal of 2?

19 An artist showed four paintings to 60 people and asked them which one they liked best.
The table shows how many people chose each painting.

Painting	A	B	C	D
Frequency	15	20	10	15

(a) Draw and label a pie chart to represent the information in the table.
(b) The artist says that the probability that a person likes picture **B** best is $\frac{1}{4}$.
Show that this is not true for the above data.   AQA

20 Work out. (a) $\frac{5}{6} - \frac{2}{3}$ (b) $\frac{4}{5} + \frac{3}{4}$ (c) $\frac{2}{5} \div 4$

21 Some of the rail services from Bournemouth to Waterloo are shown.

Bournemouth	0558	0616	—	0654	0715	0754
Brockenhurst	0619	0639	—	0722	0738	0823
Southampton	0634	0655	0714	0738	0754	0838
So'ton Parkway	0642	0703	0722	0746	0802	0848
Winchester	0656	0715	0733	0759	0812	0901
Woking	0738	—	0815	0845	—	0938
Waterloo	0804	0810	0844	0901	0908	1005

(a)   Tony catches the 0616 from Bournemouth to Waterloo.
      How long does the journey take?
(b)   Mrs Harrison catches the 0738 from Southampton to Winchester.
      What time does it arrive in Winchester?
(c)   Kath arrives at Brockenhurst station at 0730.
      What is the time of the next train to Woking?

22 (a)   Solve the equations.   (i) $x - 3 = 7$ (ii) $6x = 42$ (iii) $\frac{x}{2} = 6$
     (b)   Multiply out.   (i) $2(a - 3)$ (ii) $y(y + 1)$

23 If $m = -3$ and $n = 5$,   find the value of: (a) $2m - n$ (b) n^2 (c) $\frac{n(m + 9)}{3}$

24 A bag contains 20 beads.
    There are 7 red beads in the bag and the rest are blue or white.
    A bead is taken from the bag at random.
    (a)   What is the probability it is red?
    (b)   What is the probability it is not red?

25 Joyce wants to calculate $\frac{59 \cdot 6}{20.2 - 4.9}$.

    By writing each of the numbers in Joyce's calculation to the nearest whole number estimate
    the answer.

26 Brenda has seven cards.
    The cards are numbered, as shown.   **4 5 1 4 2 3 6**
    (a)   Brenda chooses one card at random.
          What is the probability of her getting an even number?
    Brenda gets an extra numbered card, so she has eight cards altogether.
    She chooses one at random and replaces it.
    She does this **fifty** times. Her results are shown in the table.

Number on card	1	2	3	4	5	6
Frequency	6	12	6	13	8	5

    (b)   Draw a bar chart to show Brenda's results.
    (c)   (i)   What do you think was the number on Brenda's extra card?
          (ii)   Explain why you decided it was this number.

AQA

27 The road distances between London and three other towns are shown on the diagram.

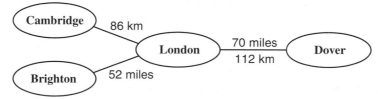

    (a)   Using the data from the London - Dover route draw a graph for converting miles
          into kilometres.
    (b)   How many kilometres is it from London to Brighton?
    (c)   How many miles is it from London to Cambridge?

AQA

28 In the diagram, *PQR* is an isosceles triangle.
The lines *PQ* and *RS* are parallel.
(a) Work out the size of angle *x*.
(b) (i) What is the size of angle *y*?
     (ii) Give a reason for your answer.

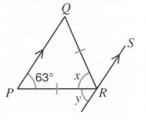

29 (a) What is the value of $2x^2$ when $x = 3$?
(b) Simplify $3t - t - 3$.
(c) Solve (i) $2(x - 3) = 8$, (ii) $3t + 1 = 7 - t$.

30

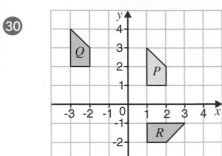

(a) Describe fully the single transformation which maps Shape *P* onto Shape *Q*.
(b) Describe fully the single transformation which maps Shape *P* onto Shape *R*.
(c) Copy Shape *P* onto squared paper. Enlarge Shape *P* by a scale factor of 3.

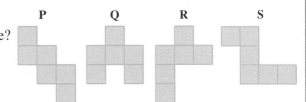

P   **Q**   **R**   **S**

31 (a) Which of the following is a net of a cube?

(b) A cube has edges of length 3 cm.
Calculate the surface area of the cube.

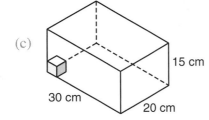

(c) Cubes of edge 3 cm are stored in a box, as shown.
The box is a cuboid with dimensions
15 cm by 20 cm by 30 cm.
What is the maximum number of cubes
that can be stored inside the box?   AQA

32 An approximate rule to convert feet into metres is:

| Multiply by 3 then divide by 10. |

(a) Use this approximate rule to convert 12 feet into metres.
(b) Write down an approximate rule that could be used to convert metres into feet.
(c) A family buys a carpet measuring 12 feet by 12 feet.
What is the area of the carpet in square metres?   AQA

33 The graph shows Zak's journey to work.
He starts on the motorway but is held up by an accident.
The police move traffic onto side roads where Zak continues his journey.
He arrives at work at 8.45 am.

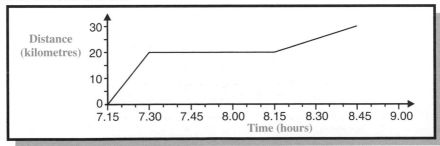

(a) What is his average speed before he is held up by the accident?
(b) How long was he held up by the accident?
(c) How far was his journey in kilometres?
(d) What was Zak's average speed for the whole journey?   AQA

34 Use approximations to estimate the value of: $\dfrac{316 \times 4.03}{0.198}$ AQA

35 A sequence begins 5, 8, 11, 14, …
Write an expression in terms of n for the nth term in the sequence.

36 The table gives information about the age and mileage of a number of cars.
The mileages are given to the nearest thousand miles.

Age (years)	1	3	5	3	5	4	7	$4\frac{1}{2}$
Mileage (nearest 1000 miles)	9000	24 000	46 000	27 000	41 000	39 000	62 000	40 000

(a) Use this information to draw a scatter graph.
(b) What type of correlation is there between the age and mileage of these cars?
(c) (i)   Draw a line of best fit.
      (ii)   Use your graph to estimate the age of the car with a mileage of 54 000.   AQA

37 A circle has a radius of 4.87 cm.
Use approximation to estimate the area of the circle in terms of π.

38 (a) Write 28 as the product of its prime factors.
(b) Find the least common multiple (LCM) of 28 and 42.   AQA

39 Two girls, Anne and Margaret, and two boys, Brian and Colin, play a computer game.
Each game has only one winner.

The probability that Anne wins is 0.3.
The probability that Brian wins is 0.15.
The probability that Colin wins is 0.45.

(a) What is the probability that Colin does not win?
(b) What is the probability that Margaret wins?
(c) What is the probability that one of the boys wins?   AQA

40 (a) Simplify.   (i) $c^2 \times c^3$ (ii) $d^6 \div d^2$
(b) Factorise.   (i) $2m - 4n$,   (ii) $t^2 - 2t$.

41 Phil has 80 birds; some are blue, and the rest are yellow.
Phil sells 30% of his birds.
The new ratio of blue birds to yellow birds is 4 : 3.
How many blue birds has he got left?   AQA

42 (a) List the values of n, where n is a whole number, such that $3 < 2n + 1 \leqslant 7$.
(b) Multiply out and simplify. $3(2x - 5) - 4(x - 2)$

43 The diagram shows the positions of shapes T, M and N.

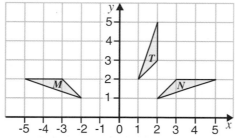

(a) Describe fully the single transformation which maps T onto M.
(b) Describe fully the single transformation which maps T onto N.
(c) M is mapped onto P by an enlargement, scale factor 2, centre $(-5, 1)$.
      Draw a diagram to show the positions of M and P.

44 (a) Draw the graph of $y = 6 - x^2$ for values of x from -3 to 3.
(b) Use your graph to solve the equation $6 - x^2 = 0$.

Calculator Paper ●●●●●●●●●●●●○

You may use a calculator for this exercise.

1 A number of people were asked where they had stayed on holiday.
The pictogram shows some of the results.

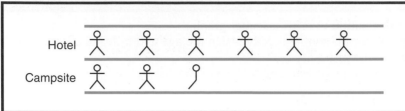

(a) 30 people said they had stayed at a hotel. What does 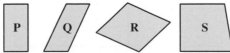 represent?
(b) Estimate how many people stayed at a campsite.
(c) 60 people were asked altogether.
The other people stayed at a bed and breakfast.
Draw a pictogram for those staying at a bed and breakfast.

2 Chloe orders the following items at a burger bar.

3 burgers at £1.49 each

4 large fries at £1.19 each

2 milkshakes at £1.10 each

3 colas at 69p each.

Copy and complete the bill.

BEST BURGERS		
3 burgers at £1.49	£4	47p
4 large fries at £1.19		
2 milkshakes at £1.10		
3 colas at 69p		
Total		

AQA

3 Place the following numbers in order of size, starting with the largest.

$4.7 \quad 4\frac{3}{5} \quad 4.58 \quad 4.629 \quad 4\frac{2}{3} \quad 4.8$

AQA

4 Les writes down six numbers.

| 2 | 3 | 6 | 7 | 9 | 24 |

(a) Which of these numbers are odd numbers?
(b) Which of these numbers are factors of 12?
(c) Which of these numbers is a multiple of 12?
(d) Which of these numbers is a square number?

5 (a) Which of these shapes is a rhombus?

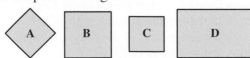

(b) Which two of these shapes are congruent?

6 Kim thinks of a number. She doubles it and adds 3. The answer is 17.
What is her number?

7

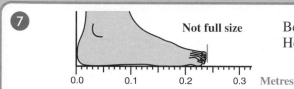

Not full size

Belinda measures her foot.
How long is it?

0.0   0.1   0.2   0.3   **Metres**

AQA

8 (a) Errol orders 3 pizzas to be delivered to his house.
How much will he pay altogether?

(b) Mr Patel pays £37 to have some pizzas delivered to his house.
How many pizzas did he order?

AQA

PETE'S PIZZA
£5.75 per pizza
We deliver to your
house for £2.50

9 The three patterns are made out of matchsticks.

Pattern 1   Pattern 2   Pattern 3

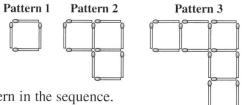

(a) Draw the next pattern in the sequence.

(b) Complete this table to show the number of matchsticks used for each pattern.

Pattern number	1	2	3	4	5
Number of matchsticks	4	10	16		

(c) How many matchsticks would be needed for the 20th pattern?
Show clearly how you worked out your answer.

AQA

10 Part of a sequence of numbers is: …   37   31   25   19   13   …

(a) Write down the number that comes before 37 in this sequence.
(b) Write down the number that comes after 13.
(c) Write down the rule for continuing this sequence.
(d) Write down the first negative number that appears when you continue this sequence.

AQA

11 Which of these shapes **cannot** be used to form a tessellation?
Explain why.

A   **B**   **C**

12 (a) Simplify $t + 2t + 3t$.
(b) Solve the equations   (i) $2w = 6$,   (ii) $x + 3 = 5$.
(c) What number must go in the box to make this statement true?   ☐ $\times\, 3 + 2 = 17$

13 (a) A fence is 2.3 metres high. What is the height of the fence in centimetres?
(b) A cycle race is 26 miles. How many kilometres is this?
(c) A rule for changing litres into gallons is:

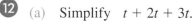

Divide by 4.5

A lorry's fuel tank holds 180 litres when it is full.
Use this rule to work out how many gallons the lorry's fuel tank holds.

14 The diagram shows the positions of points A, B and C.

(a) What are the coordinates of A?
(b) (i)   Copy the diagram and mark the position of D
so that $ABCD$ is a rectangle.
(ii)   What are the coordinates of D?

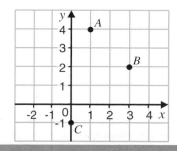

15 The number of pages in each chapter of a book are:

$$12, \quad 21, \quad 15, \quad 26, \quad 17, \quad 35, \quad 20, \quad 32.$$

Construct a stem and leaf diagram to show these numbers.

AQA

16 Given that $C = 3A + 2B$,
(a) work out the value of C when $A = 4$ and $B = 3$,
(b) work out the value of B when $C = 14$ and $A = 2$.

AQA

17 The diagram shows a rectangle which has been partly shaded.
(a) What fraction of the rectangle is shaded?
(b) Copy the rectangle.
Shade more of the rectangle so that the final diagram has rotational symmetry of order 2 but no lines of symmetry.

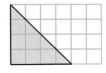

AQA

18 Starting with the numbers 4 and 5, Alan writes down the following number pattern:

$$4, \quad 5, \quad 9, \quad 14, \quad 23.$$

The number pattern is continued. What is the next number in the pattern?
Explain how you found your answer.

19 The diagram shows a rectangle.
Work out the size of angles x and y.

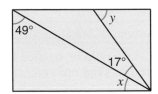

20 A caretaker is paid at a basic rate of £7.76 per hour for 36 hours a week.
Overtime is paid at one and a half times the basic rate.
One week the caretaker works 42 hours. How much is the caretaker paid that week?

21 A packet contains 15 biscuits.
(a) Harry eats 40% of the biscuits. How many biscuits does he eat?
(b) Lauren eats 4 of the biscuits. What percentage of the biscuits does she eat?
Give your answer to an appropriate degree of accuracy.

22 The sides of a rhombus are all of length 6 cm.
The shorter diagonal of the rhombus is 7 cm long.
Make an accurate drawing of the rhombus.

AQA

23 (a) In the diagram, $QR = RP$ and QP is parallel to RS.
Angle $PRS = 37°$.
What is the size of angle x?

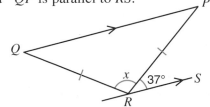

(b) The diagram shows the position of Shape P.

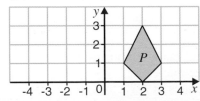

P is mapped onto Q by a rotation through 90° anticlockwise about (0, 0).
Draw a diagram to show the positions of shapes P and Q.

AQA

24 Jaspal has one white turban and one blue turban.
He also has one green shirt, one red shirt and one yellow shirt.
(a) List all the possible colour combinations if he wears one turban with one shirt.
(b) He chooses one shirt and one turban at random. What is the probability he chooses
    (i)   the blue turban,
    (ii) the white turban and the yellow shirt,
    (iii) the white turban but not the red shirt?

Dalia has 4 red saris, 5 blue saris and 2 green saris.
She chooses one sari at random.
(c) What is the probability that she chooses a blue sari?   AQA

25 (a) On graph paper draw and label the lines
$$y = 2x \quad \text{and} \quad y = 6 - x \quad \text{for values of } x \text{ from } -1 \text{ to } 4.$$
(b) Write down the coordinates of the point where the graphs cross.

26 The number x satisfies the equation $x^3 = 20$.
(a) Between which two consecutive whole numbers does x lie?
(b) Use a trial and improvement method to find this value of x correct to one decimal place.
    Show all your working clearly.   AQA

27 A garden is a rectangle measuring 26 m by 11.5 m. Grass covers 67% of the area of the garden.
Calculate the area of grass. Give your answer to a suitable degree of accuracy.

28 A shop sells flour in two sizes.
Which size gives better value for money?
You **must** show all your working.

    Size 1: weight 500 g, cost 39 pence.
    Size 2: weight 800 g, cost 59 pence.

29 Brian does a survey on what pupils eat for breakfast.
Here are three questions from his questionnaire.

> 1. What time do you get up?
>   Before 7.00 ☐   Between 7.00 and 8.00 ☐   After 8.00 ☐
> 2. Which of these do you have for breakfast?   Bacon ☐ Eggs ☐ Cereal ☐
> 3. How long does it take you to eat your breakfast? . . . .

Write down one **criticism** of each question.   AQA

30 A rucksack costs £36 plus VAT at $17\frac{1}{2}\%$. What is the total price of the rucksack?

31 Calculate $3.4^2 + \sqrt{3.4}$. Give your answer correct to 1 d.p.

32 A penny farthing bicycle is shown.
(a) The large wheel has a diameter of 160 cm.
    Find the circumference of the wheel.
(b) The small wheel has a circumference of 137 cm.
    The large wheel rotates once.
    How many complete rotations does the small wheel make?   AQA

33 Simon puts some red, green, blue and white beads in a bag.
He then calculates the probability of getting each colour when a bead is taken from the bag at random. The table shows his results.

Colour	Red	Green	Blue	White
Probability	0.1	0.2	0.35	0.25

Explain how you know that Simon has made a mistake.

34 A bowl of strawberries and cream weighs 210 grams.
The ratio, by weight, of strawberries to cream is 5 : 1.
What is the weight of the cream?

35 The diagram shows the shape of a playground in a park.
Calculate the area of the playground.

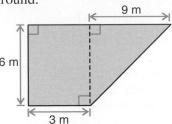

AQA

36 Carpet tiles measure 45 cm by 45 cm.
How many tiles are needed to cover a floor
which measures 4.05 m by 5.4 m?

37 (a) Calculate $\sqrt{8} \times 3.6^2$. Give your answer correct to 2 decimal places.
(b) Work out 8^3.
(c) Find the value of $\frac{1}{x} + y^3$ when $x = 5$ and $y = 0.5$.

38 (a) Solve (i) $5x + 7 = 2$, (ii) $3x - 9 = 7 - 2x$.
(b) Factorise $3a - 6b$.
(c) Simplify $5x - 2(x + 3)$.

39 A candle weighs x grams.
(a) Write an expression, in terms of x, for the weight of 20 candles.
(b) A box of 20 candles weighs 3800 g. The box weighs 200 g.
By forming an equation, find the value of x.

40 A police officer records the speeds of 60 cars on a dual carriageway.

Speed (mph)	40 to less than 50	50 to less than 60	60 to less than 70	70 to less than 80
Frequency	9	27	21	3

(a) Write down the modal class.
(b) Use the class midpoints to calculate an estimate of the mean speed of these cars. AQA

41 Mike took 400 books to sell at a Saturday market.
By 3 pm, he had sold 310 books at 80 pence each.
Mike then reduced the selling price of the remaining books to 50 pence each.
He was left with 24 unsold books which he gave away.
(a) Find the total amount Mike received from selling the books.
(b) Mike had spent £150 buying the books.
What was Mike's percentage profit? AQA

42 (a) Calculate $\dfrac{89.6 \times 10.3}{19.7 + 9.8}$
(b) By using approximations show that your answer to (a) is about right.
You must show all your working. AQA

43 Harry drives 182 miles. His average speed is 35 miles per hour.
How long does the journey take? Give your answer in hours and minutes. AQA

44 Write 36 as a product of its prime factors.

45 The table shows information about a group of adults.

	Can drive	Cannot drive
Male	32	8
Female	38	12

(a) One of these adults is chosen at random.
     What is the probability that the adult can drive?
(b) A man in the group is chosen at random.
     What is the probability that he can drive?
(c) A woman in the group is chosen at random.
     The probability that she can drive is 0.76.
     What is the probability that she cannot drive?
(d) Does the information given support the statement:

                   "More women can drive than men"?
     Explain your answer.

                                                             AQA

46 A sequence begins: 5,   3,   1, -1,   …
Write an expression, in terms of n, for the nth term of the sequence.

47 (a) This diagram shows a kite $ABCD$.
         $AB = 15$ cm, $BC = 36$ cm   and $AC = 39$ cm.
         Explain why angle $B = 90°$.

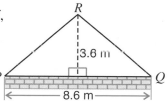

(b) The diagram shows two regular polygons.

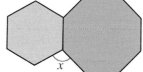

     Work out the size of angle x.   AQA

48 (a) Solve $8 - 3x = 2(x + 5)$.
     (b) Use trial and improvement to solve the equation $x^3 + 2x = 400$.
          Show all your trials. Give your answer correct to one decimal place.

49 PQR is the cross-section of a roof,
with $PR = RQ$. $PQ = 8.6$ m.
R is 3.6 m above the level of PQ.
Calculate the length RQ.

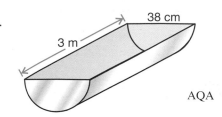

                                                             AQA

50 A water trough has a semi-circular cross-section, as shown.
The diameter of the end of the trough is 38 cm.
The trough is 3 m long.
Calculate the volume of water in the trough when it is full.
Give your answer in litres.

                                                             AQA

51 (a) Solve the inequality $3m - 5 > 7$.
     (b) Write down all the solutions of the inequality $-6 < 3n \leqslant 5$ where n is an integer.

52 a, b and c represent lengths. Which of these formulae represents a volume?
              $4(a + b + c)$, $2(ab + bc + ac)$, abc
Explain your answer.

Answers

Exercise 1.1   Page 1

1. **6, 12, 110**
2. (b) $54 = 50 + 4$
             $= 5 \times 10 + 4$
   (c) $456 = 400 + 50 + 6$
              $= 4 \times 100 + 5 \times 10 + 6$
   (d) $1872 = 1000 + 800 + 70 + 2$
              $= 1 \times 1000 + 8 \times 100 + 7 \times 10 + 2$
3. (a) 50, 3   (b) 300, 40, 1   (c) 600, 70, 3
   (d) 1000, 800, 90, 7   (e) 1000, 50, 2
4. (a) 30   (b) 3 000 000   (c) 600
   (d) 2000   (e) 4000   (f) 6
5. (a) 1 in 512   (b) 7 in 745   (c) 5 in 599
6. (a) **97, 23, 203**   (b) **302**   (c) **23**
7. (a) 39, 74, 168, 421
   (b) 544, 545, 554, 555
   (c) 3801, 3842, 3874, 4765, 5814
8. (a) 429, 425, 399, 103, 84
   (b) 349, 324, 239, 234
   (c) 9951, 9653, 9646, 9434
9. Even number. It ends with 0.
10. 732, 723, 372, 327, 273, 237
    Two are even numbers, 732 and 372.
11. 3458   3485   3548   3584   3845   3854
     4358   4385   4538   4583   4835   4853
     5348   5384   5438   5483   5834   5843
     8345   8354   8435   8453   8534   8543
12. (a) 754 Put digits in order, largest to smallest.
    (b) 457 Put digits in order, smallest to largest.
13. (a) 6512   (b) 1265

Exercise 1.2   Page 2

1. (a) Seventeen
   (b) Eighty-eight
   (c) One hundred and eighty-seven
   (d) Two thousand and forty-five
   (e) Five thousand six hundred and twelve
   (f) Seven thousand eight hundred and two
   (g) Eight thousand eight hundred and eighty-eight
   (h) Ninety-two thousand
   (i) One hundred and thirty-two thousand and forty-five
   (j) One million five hundred thousand
2. (a) (i) 1   (ii) 10   (iii) 100
       (iv) 1000   (v) 10 000
   (b) Previous number multiplied by ten, nought added.
   (c) 100 000, one hundred thousand, 1 000 000, one million

3. (a) 546   (b) 607   (c) 1010
   (d) 70 200   (e) 1 200 052
4. (a) 2 000 000   (b) 10 000 000
   (c) 500 000   (d) 1 500 000
5. The attendance at a football match was 48 000.
   The pitch measured 119 yards by 62 yards.
   After 25 minutes the centre forward (who cost £15 000 000) scored from 18 yards.
6. (a) 7030   (b) 463   (c) 11 000

Exercise 1.3   Page 3

1. (a) 28   (b) 33
2. (a) 29   (b) 43   (c) 43   (d) 53
3. (a) 12   (b) 15   (c) 27   (d) 56
   (e) 35   (f) 74   (g) 70   (h) 112
4. (a) 91   (b) 4   (c) 55   (d) 63
   (e) 38   (f) 17   (g) 76   (h) 23
5. 112 miles   8. £222
6. (a) 89p   (b) 81p   9. 256
7. £306   10. $92 + 73 + 5 = 170$
11. (a) 788   (b) 83   (c) 174
    (d) 952   (e) 2002   (f) 12 203
    (g) 201   (h) 1541
12. 40 710   14. 1378   16. £670
13. 1030 grams   15. 81 030   17. 520 km

Exercise 1.4   Page 6

1. (a) 33   (b) 27
2. (a) 32   (b) 27   (c) 18   (d) 26
3. (a) 5   (b) 92   (c) 43   (d) 68
   (e) 76   (f) 17   (g) 59   (h) 21
4. 31 runs
5. £87
6. (a) 4   (b) 16   (c) 17   (d) 50
   (e) 49   (f) 90   (g) 105   (h) 150
   (i) 301   (j) 4
7. £26
8. £815
9. (a) 354   (b) 428   (c) 1284   (d) 158
   (e) 2224   (f) 469   (g) 6268   (h) 3277
10. 48
11. 89
12. 384
13. (a) 17p   (b) 6p
14. (a) Tomato 39, Oxtail 18, Chicken 55
    (b) 112
15. Car C.   A: 8479   B: 11643   C: 13859

Exercise 1.5   Page 7

1. 40   3. 48   5. 144
2. 28   4. 72   6. £5.52
7. (a) 60p   (b) £1.14   (c) £1.60

8. (a) 84   (b) 85   (c) 252
   (d) 549   (e) 2112   (f) 15 895
   (g) 24 072   (h) 42 084

9. (a) 68   (b) 275   (c) 666   (d) 46

10. 29

Exercise **1.6**   Page 9

1. (a) 1320   (b) 12 300
   (c) 47 000   (d) 38 400

2. (a) 2310   (b) 514   (c) 100

3. (a) £1200   (b) £5900   (c) £71 000

4. (a) 240   (b) £4.90

5. 150 euros

6. (a) £120   (b) £700   (c) £8200

7. (a) 210 chairs   (b) 140 tables

8. (a) 300 seconds   (b) 540 seconds

9. 300

10. £750   **12.** 1600 kg

11. 560 grams   **13.** £1450

14. (a) 7140   (b) 18 960   (c) 21 480
   (d) 13 000   (e) 13 020   (f) 21 510

15. (a) E.g. To multiply by 200, multiply by 2 and then by 100.
   (b) E.g. To multiply by 2000, multiply by 2 and then by 1000.
   (c) (i) 13 400   (ii) 105 000
      (iii) 69 600   (iv) 1 435 000

16. 10 000   **17.** 60 000 staples

Exercise **1.7**   Page 10

1. 49p   **3.** £14   **5.** 15 packets

2. 6   **4.** 12p   **6.** 15

7. (a) 17   (b) 157
   (c) 136   (d) 75 remainder 5
   (e) 393 remainder 2   (f) 206
   (g) 1098   (h) 20 140

8. (a) 13   (b) 4 pence

Exercise **1.8**   Page 11

1. (a) 456   (b) 465   (c) 64   (d) 654

2. 6

3. (a) 100   (b) 702 000   (c) 10

4. 12   **6.** 16 classes   **8.** 7

5. 7   **7.** 25   **9.** £35

10. (a) 253   (b) 79   (c) 537
   (d) 126   (e) 45   (f) 613

11. (a) E.g. To divide by 200, divide by 100 and then by 2.
   (b) (i) 26   (ii) 412

12. 20 minutes

Exercise **1.9**   Page 12

1. (a) 204   (b) 345   (c) 1344
   (d) 2432   (e) 4862   (f) 38 772

2. 216   **5.** £966   **8.** £1536

3. 255   **6.** 1316   **9.** £5191

4. £9   **7.** £1392   **10.** £43 848

Exercise **1.10**   Page 13

1. (a) 43   (b) 32   (c) 27   (d) 32
   (e) 48   (f) 41   (g) 21   (h) 29

2. (a) 16 remainder 10   (b) 25 remainder 7
   (c) 17 remainder 35   (d) 13 remainder 13

3. 14 pints, 10p change

4. (a) 41   (b) 16

5. (a) 20   (b) 35, 10p change
   (c) 11, 5p change

6. 23

Exercise **1.11**   Page 14

1. 2

2. (a) 37   (b) 3   (c) 9   (d) 58
   (e) 6   (f) 30   (g) 5   (h) 19
   (i) 14   (j) 20   (k) 24   (l) 4
   (m) 0   (n) 5   (o) 6

3. $(6 + 3) \times 5$

4. (a) $(7 - 2) \times 3 = 15$
   (b) $(3 + 5) \div 2 = 4$
   (c) $(4 + 1) \times (7 - 2) = 25$

5. (a) $5 \times 6 + 7 = 37$
   (b) $5 + 6 \times 7 = 47$
   (c) $15 + 8 \times 9 = 87$
   (d) $15 \times 8 + 9 = 129$
   (e) $15 \times 8 - 9 = 111$
   (f) $15 \div 5 + 3 = 6$
   (g) $5 - 24 \div 6 = 1$
   (h) $19 \div 19 + 7 = 8$
   (i) $4 \times 4 + 7 \times 2 = 30$

6. Many answers, for example:
   $6 - 3 \times 2 + 1 = 1$ $6 - 3 - 2 + 1 = 2$
   $6 \div 3 + 2 - 1 = 3$ $6 \div 3 + 2 \times 1 = 4$
   $6 - 3 + 2 \times 1 = 5$ $6 - 3 + 2 + 1 = 6$
   $6 + 3 - 2 \times 1 = 7$ $6 \times 3 \div 2 - 1 = 8$
   $6 \times 3 \div 2 \times 1 = 9$ $6 + 3 + 2 - 1 = 10$

Exercise **1.12**   Page 15

1. 148 cm   **6.** 75 g

2. 79 kg   **7.** 35 cm

3. 30 g   **8.** 155 cm

4. (a) 89p   (b) 24p   **9.** 15 g

5. 25   **10.** (a) 36   (b) 43

Review Exercise **1**   Page 16

1. (a) **4, 12**   (b) **5, 7, 15, 19**

2. (a) Eight hundred and seventy thousand three hundred and two
   (b) 3 027 409

3. (a) 6 hundreds, 600
   (b) (i) 6 thousands, 6000   (ii) 6 tens, 60

4. 408, 480, 2295, 92 345, 120 000
5. (a) (i) 1500 (ii) 150 000
  (b) (i) 10 (ii) 10 000
6. (a) 67
  (b) (i) 83 (ii) 45 (iii) 71
  (c) (i) 500 (ii) 25
7. (a) 234 (b) 26
8. 240
9. (a) 9541 (b) 1459
10. E.g. 3, 4, 5: $3 + 4 + 5 = 12$, which is even.
11. (a) 97, 404 (b) 114, 306
  (c) 92, 209
12. (a) 2059 (b) 587
13. (a) 73 500 (b) 6420
  (c) 3020 (d) 462
14. (a) $390 \rightarrow \boxed{\div 10} \rightarrow 39$

  (b) $390 \rightarrow \boxed{\div 10} \rightarrow \boxed{+10} \rightarrow 49$

  (c) $38 \rightarrow \boxed{\times 10} \rightarrow \boxed{-10} \rightarrow \boxed{\div 10} \rightarrow 37$

15. (a) 6462 (b) 241
16. 298 miles
17. (a) 166 (b) 4
18. (a) 50 (b) 14 (c) 10 (d) 50
19. (a) True (b) False (c) True
20. 184 cm
21. 120
22. (a) 25
  (b) (i) $(1 + 3) \times (5 + 5) = 40$
      (ii) $1 + 3 \times (5 + 5) = 31$
23. No. $84 + 16 \times 5 = 84 + 80 = 164$
24. (a) 450 (b) 10 120
25. 14 664   **27.** £3427   **29.** 32
26. £14 522   **28.** £17   **30.** 50

CHAPTER 2

Exercise 2.1 — Page 20

1. (a) $4.7 = 4 + 0.7$
  (b) $5.55 = 5 + 0.5 + 0.05$
  (c) $7.62 = 7 + 0.6 + 0.02$
  (d) $37.928 = 30 + 7 + 0.9 + 0.02 + 0.008$
  (e) $7.541 = 7 + 0.5 + 0.04 + 0.001$
2. (a) 0.7 (b) 0.02 (c) 0.4
  (d) 0.009 (e) 80
3. (a) 1.68 (b) 1.09
4. (a) A 3.2, B 3.5, C 3.9
  (b) D 5.6, E 6.3
  (c) F 7.2, G 7.6
  (d) H 10.5, I 11
  (e) J 0.52, K 0.54, L 0.59
  (f) M 0.751, N 0.755, P 0.757
5.

6. (a) Tony (b) Mike
7. (a) 0.9 kg (b) 2.1 kg
8. (a) **0.07** (b) **0.6**
9. 93.07
10. 47.5074
11. (a) 3.001, 3.01, 3.1, 3.15, 3.2
  (b) 3.567, 3.576, 3.657, 3.675
  (c) 0.1, 0.15, 0.45, 0.5, 0.55
12. (a) 9.87, 9.78, 8.97, 8.79
  (b) 1.5, 0.15, 0.015, 0.00015
  (c) 2.701, 2.7, 2.67, 2.599

Exercise 2.2 — Page 22

1. (a) 18.8 (b) 6.4 (c) 18.3
  (d) 33.1 (e) 8.86 (f) 13.1
  (g) 12.38 (h) 17.49 (i) 12.449
  (j) 26.02 (k) 32.36 (l) 18.163
2. 7.15 metres
3. 36.1 litres
4. (a) **3.2, 4.1** (b) **1.6, 0.8** (c) **4.1, 0.8**
5. (a) 6.84 (b) 3.07 (c) 86.33
  (d) 15.781 (e) 16.033 (f) 24.88
6. 39.99 seconds
7. 4.35 kg
8. (a) Team A 148.93s
      Team B 149.53s
      Team C 149.08s
  (b) Team A, Team C, Team B

Exercise 2.3 — Page 23

1. (a) (i) 4.4 (ii) 6.23 (iii) 4.6
      (iv) 14.8 (v) 4.96 (vi) 20.8
      (vii) 11.08 (viii) 24.68
2. (a) 2.14 (b) 5.22 (c) 5.003
  (d) 1.24 (e) 8.28 (f) 6.273
  (g) 9.04 (h) 1.896
3. (a) (i) £3.30 (ii) £1.70
  (b) (i) £11.24 (ii) £3.76
  (c) (i) 83p (ii) £9.17
  (d) (i) £16.24 (ii) £33.76
4. 4.88 m
5. 1.55 m
6. 0.719 seconds

Exercise 2.4 — Page 24

1. 1.3   **5.** 11   **9.** 1.7
2. 10.9   **6.** 17.78   **10.** 0.32
3. 1.65   **7.** 1.2   **11.** 9.44
4. 0.9   **8.** 0.6   **12.** 3.2

Exercise 2.5 — Page 25

1. (a) 250.6 (b) 2506 (c) 25 060
  (d) 9.3 (e) 93 (f) 930
  (g) 0.623 (h) 6.23 (i) 62.3
  (j) 94.51 (k) 945.1 (l) 9451

2. (a) 3.77   (b) 0.377   (c) 0.0377
    (d) 0.027   (e) 0.0027   (f) 0.00027
    (g) 18.902   (h) 1.8902   (i) 0.18902
    (j) 0.9   (k) 0.09   (l) 0.009

3. (a) (i) 0.64   (ii) 6.4   (iii) 64
    (b) (i) 0.64   (ii) 0.064   (iii) 0.0064

4. (a) £2.50   (b) £25   (c) £250

5. (a) 5.04 km   (b) 50.4 km   (c) 504 km

6. (a) £7.95   (b) £0.12   (c) 86.9p

7. 12.3×1000 and $12.3 \div 0.001$
    $12.3 \div 100$ and 12.3×0.01
    12.3×0.1 and $12.3 \div 10$
    $12.3 \div 0.01$ and 12.3×100
    12.3×10 and $12.3 \div 0.1$
    12.3×0.001 and $12.3 \div 1000$

Exercise 2.6   Page 27

1. (a) 1.2   (b) 8.5   (c) 12.8   (d) 3.6
    (e) 13   (f) 17.6   (g) 10.8   (h) 30.1
    (i) 28.8   (j) 34.8

2. £7.60

3. (a) £8.05   (b) £1.95

4. (a) £14.95   (b) £5.05

5. (a) £12.25   (b) £34.93   (c) £22.69

6. (a) 0.42   (b) 0.06   (c) 8.75
    (d) 16.53   (e) 19.44   (f) 1.025
    (g) 3.888   (h) 9.38   (i) 3.78
    (j) 0.0432   (k) 0.028   (l) 0.0014

7. (a) (i) 3   (ii) 1.5   (iii) 0.24   (iv) 15
    (b) Each answer is less than the original number.

8. (a) 21p   (b) £1.84   (c) 78p

9. (a) £20.93   (b) £60.75
    (c) £209.82   (d) £122.24

10. (a) (i) £5.32   (ii) £7.68
       (iii) £3.60   (iv) £2.79
    (b) £1.38

Exercise 2.7   Page 28

1. (a) 0.3   (b) 1.5   (c) 1.7
    (d) 3.2   (e) 4.4

2. (a) 1.75   (b) 1.6   (c) 0.15
    (d) 2.6   (e) 1.75

3. (a) 5   (b) 12   (c) 350

4. (a) 4   (b) 15   (c) 15
    (d) 4   (e) 50

5. (a) 1.7   (b) 17   (c) 1.7

6. (a) 12.3   (b) 2.92   (c) 6.05   (d) 1430
    (e) 0.05   (f) 12.5   (g) 6.54   (h) 37.5

7. (a) 37   (b) 5.6   (c) 43.75
    (d) 46.9   (e) 1.062

8. (a) (i) 10   (ii) 6   (iii) 0.3
    (b) Each answer is greater than the original number.

9. 47p   **11.** 11p   **13.** 90.8p   **15.** 67

10. £1.35   **12.** 9p   **14.** 120   **16.** 45

Exercise 2.8   Page 30

1. (a) $\frac{1}{4}$ (b) $\frac{1}{2}$ (c) $\frac{3}{4}$ (d) $\frac{1}{10}$

2. (a) $\frac{7}{10}$ (b) $\frac{2}{5}$ (c) $\frac{1}{100}$ (d) $\frac{1}{5}$
    (e) $\frac{1}{20}$ (f) $\frac{3}{20}$ (g) $\frac{13}{25}$ (h) $\frac{7}{100}$
    (i) $\frac{1}{8}$ (j) $\frac{13}{20}$ (k) $\frac{3}{5}$ (l) $\frac{19}{20}$

3. (a) $1\frac{7}{10}$ (b) $2\frac{3}{10}$ (c) $1\frac{2}{5}$ (d) $3\frac{1}{4}$
    (e) $4\frac{4}{5}$ (f) $12\frac{1}{10}$ (g) $16\frac{3}{4}$ (h) $5\frac{1}{20}$

4. (a) $\frac{2}{3}$ (b) $\frac{1}{9}$ (c) $\frac{5}{9}$

Review Exercise 2   Page 31

1. **0.5, 0.55, 0.7, 0.8, 0.85**

2. $\frac{9}{20}$

3. (a) 12.41   (b) 4.33
    (c) $4.33 + 5.67 = 10$

4. (a) 14.4   (b) 4.2   (c) 1.3   (d) 20

5. (a) 0.907 and 0.26   (b) 0.907 and 1.2

6. 4.2 kg   **8.** 2.37 m

7. 40 s   **9.** 40.5 litres

10. (a) 2619   (b) 2.91

11. (a) £247.50   (b) £7.99

12. (a) 1.1275   (b) 27.5

13. (a) 46.62   (b) 10.152   (c) 105.9

14. (a) 7.2   (b) 1235   (c) 13 700

15. 11

16. 28p   **18.** 240.159574…

17. 4.63   **19.** 24.86052632…

CHAPTER 3

Exercise 3.1   Page 33

1. (a) 50   (b) 50   (c) 70

2. (a) 4870   (b) 4900   (c) 5000

3. (a) 360   (b) 400

4. (a) 7480   (b) 7500   (c) 7000

5.

Number	Nearest 10	Nearest 100	Nearest 1000
7613	7610	7600	8000
977	980	1000	1000
61 115	61 120	61 100	61 000
9714	9710	9700	10 000
623	620	600	1000
9949	9950	9900	10 000
5762	5760	5800	6000
7509	7510	7500	8000
7499	7500	7500	7000

6. (a) 3456 (b) 3500
7. Nearest 1000.
8. (a) 19 000 (nearest thousand)
   (b) 260 (nearest ten)
   (c) £50 (nearest pound)
   (d) 24 100 (nearest hundred)
   (e) 309 000 km^2 (nearest thousand km^2)
   (f) 190 km (nearest ten kilometres)
9. (a) 745, 746, 747, 748, 749
   (b) 750, 751, 752, 753, 754
   (c) Any number from 8450 to 8499
   (d) Any number from 8500 to 8549
10. (a) 35 (b) 44
11. (a) 1950 (b) 2049
12. 42 500
13. Smallest: 135, greatest: 144
14. 2749

Exercise 3.2 — Page 34

1. 4   **4.** 5   **7.** 29
2. 5   **5.** 9   **8.** 16   **10.** 7
3. 10   **6.** 24   **9.** 5   **11.** 5

Exercise 3.3 — Page 35

1. (a) 3.962 (b) 3.96 (c) 4.0
2. (a) 567.65 (b) 567.7 (c) 568
3. 4.86
4. 68.8 kg
5. Missing entries are:
   0.96, 0.97, 15.281, 0.06, 4.99, 5.00
6. (a) (i) 46.1 (ii) 59.7
      (iii) 569.4 (iv) 17.1 (v) 0.7
   (b) (i) 46.14 (ii) 59.70
      (iii) 569.43 (iv) 17.06 (v) 0.66
   (c) (i) 46.145 (ii) 59.697
      (iii) 569.434 (iv) 17.059 (v) 0.662
7. (a) 40.9 litres, nearest tenth of a litre
   (b) £2.37, nearest penny
   (c) 35.7 cm, nearest millimetre
   (d) £1.33, nearest penny
   (e) £14.26, nearest penny

Exercise 3.4 — Page 37

1. (a) 20 (b) 500 (c) 400 (d) 2000 (e) 20
2.

Number	sig. fig.	Answer
456 000	2	460 000
454 000	2	450 000
7 981 234	3	7 980 000
7 981 234	2	8 000 000
1290	2	1300
19 602	1	20 000

3. (a) 0.08 (b) 0.09 (c) 0.009 (d) 0.01

4.

Number	sig. fig.	Answer
0.000567	2	0.00057
0.093748	2	0.094
0.093748	3	0.0937
0.093748	4	0.09375
0.010245	2	0.010
0.02994	2	0.030

5. 490
6. (a) (i) 80 000 (ii) 80 (iii) 1000
      (iv) 0.007 (v) 0.002
   (b) (i) 83 000 (ii) 83 (iii) 1000
      (iv) 0.0073 (v) 0.0019
   (c) (i) 82 700 (ii) 82.7 (iii) 1000
      (iv) 0.00728 (v) 0.00190
7. 472 m^2 (3 s.f.)
8. (a) 157 cm^2 (3 s.f.)
   (b) 6100 m^2 (2 s.f.)
   (c) 1.23 m (nearest cm)
   (d) 15.9 m^2 (round up 1 d.p.)

Exercise 3.5 — Page 39

1. £50 + £90 + £60 + £100 = £300
2. £8000 + £1000 = £9000
3. (a) (i) 40 × 20 = 800
      (ii) 100 × 20 = 2000
      (iii) 800 × 50 = 40 000
      (iv) 900 × 60 = 54 000
   (b) (i) 80 ÷ 20 = 4
      (ii) 600 ÷ 30 = 20
      (iii) 900 ÷ 60 = 15
      (iv) 4000 ÷ 80 = 50
4. 40 × 50p = £20
5. 20 × 30 = 600
6. 600 ÷ 30 = 20
7. 15 000 ÷ 1000 = 15 km/l
8. (a) 30 is bigger than 29 and 50 is bigger
      than 48.
      So, 30 × 50 is bigger than 29 × 48.
   (b) 200 ÷ 10 = 20, 14, estimate is bigger

Exercise 3.6 — Page 40

1. (a) 5 (b) 2 (c) 2
   (d) 3 (e) 2 (f) 5
2. 40 × 50 = 2000
3. (a) 200 × 300 (b) 60 000 (c) 956
4. (a) 30 × 40 = 1200, 1312
   (b) 10 × 70 = 700, 792
   (c) 60 × 30 = 1800, 1972
   (d) 70 × 50 = 3500, 3240
   (e) 4 × 2 = 8, 7.56
   (f) 9 × 3 = 27, 27.59
   (g) 50 × 4 = 200, 202.02
   (h) 100 × 3 = 300, 299.86

5. (a) $600 \div 20 = 30, \quad 33$
   (b) $600 \div 20 = 30, \quad 29$
   (c) $300 \div 20 = 15, \quad 16$
   (d) $800 \div 40 = 20, \quad 24$
   (e) $10 \div 5 = 2, \quad 2.2$
   (f) $20 \div 4 = 5, \quad 4.8$
   (g) $30 \div 3 = 10, \quad 9.5$
   (h) $200 \div 5 = 40, \quad 39.9$

6. (a) $\dfrac{50 \times 200}{20} = 500, \quad 461.3\ldots$
   (b) $\dfrac{600}{10 \times 30} = 2, \quad 1.6695\ldots$
   (c) $\dfrac{20 \times 60}{40 : 4} = 120, \quad 108$
   (d) $\dfrac{60}{90 \div 9} + 50 = 56, \quad 54.48\ldots$

7. (a) $10 \times 4 \div 5 = 8, \quad 8.45625$
   (b) $(10 + 50) \div 6 = 10, \quad 9.774\ldots$
   (c) $400 \times 0.3 \div 6 = 20, \quad 18.709\ldots$
   (d) $(80 \times 5) \div (2 \times 10) = 20, \quad 20.456\ldots$

Exercise 3.7   Page 42

1. 167.5 cm
2. 49.5 kg
3. $8.5\,m \leqslant$ height of building $< 9.5\,m$
4. 99.5 m
5. Minimum weight: 835 g
   Maximum weight: 845 g
6. 287.5 g
7. 11.55 seconds
8. 9.35 kg
9. Least length: 2.65 m,   greatest length: 2.75 m
10. 94 ml
11. Minimum weight: 0.75 kg
    Maximum weight: 0.85 kg
12. $217.5\,l \leqslant$ capacity of tank $< 222.5\,l$

Review Exercise 3   Page 43

1. (a) 8480   (b) 8500   (c) 8000
2. (a) (i) 6 700 000   (ii) 100 000
   (b) 109 500
3. (a) 29 080 feet   (b) Nearest 10 000 feet
4. (a) 8   (b) £2.32
5. (a) 70   (b) 100   (c) 800
   (d) 700   (e) 80   (f) 1000
6. (a) $800 + 30 - 300 = 530$ (b) 522
7. (a) £170 000 and £13 000
   (b) (i) No.
       (ii) £170 000 + £13 000 = £183 000
8. (a) $20 \times 30 = 600$ (b) $2000 \div 40 = 50$
9. (a) $90 \times 2 = 180$ (b) $2000 \div 50 = 40$
10. $£30 \times 150 = £4500$
11. 18.7 is less than 19, but rounds to 19.
    0.96 is less than 1, but rounds to 1.
    $19 \times 1 = 19$,
    so, 18.7×0.96 must be less than 19.

12. 15
13. $\dfrac{400 \times 80}{200} = 160$
14. $40 \times 0.03 = 1.2$
15. Weight of apples should be given to nearest gram.
16. Smallest: 85,   largest: 94
17. (a) 25.57   (b) 25.6   (c) 30
18. $\dfrac{500 + 100}{30} = 20$ Answer is wrong.
19. (a) $30 \times 40 \times 10 = £12\,000$
    (b) Too large. All values rounded up.
20. 24.5 cm
21. (a) 17.5 m
    (b) $17.5\,m \leqslant$ length of bus $< 18.5\,m$
22. 1.6
23. 4.13
24. (a) 24.8605263…   (b) $\dfrac{5 \times 20}{6 - 2} = \dfrac{100}{4} = 25$
25. (a) 19 m²
    (b) Round up to the nearest square metre for enough carpet to cover whole floor.
26. 3.15 kg

CHAPTER 4

Exercise 4.1   Page 46

1. (a) Warmer   (b) Colder
   (c) Warmer   (d) Colder
2. (a) Less   (b) More
   (c) Less   (d) More
3. (a) **Colombo**   (b) **Moscow**
   (c) **−22°C, −17°C, −7°C, 0°C, 3°C, 15°C, 21°C.**
4. (a) −28°C, −13°C, −3°C, 19°C, 23°C.
   (b) −11°C, −9°C, −7°C, 0°C, 10°C, 12°C.
   (c) −29°C, −15°C, 2°C, 18°C, 27°C.
   (d) −20°C, −15°C, −5°C, 0°C, 10°C, 20°C.
5. (a) 78, −39, −16, −9, 11, 31, 51.
   (b) −5, −3, −2, −1, 0, 1, 2, 4, 5.
   (c) −103, −63, −19, −3, 5, 52, 99, 104.
   (d) −50, −30, −20, 0, 10, 30, 40.
   (e) −30, −15, −10, 0, 8, 17, 27.

Exercise 4.2   Page 47

1. (a) 1   (b) −2   (c) −2
   (d) −4   (e) −3
3. (a) −3   (b) −2   (c) −3
   (d) −2   (e) −3   (f) −4
   (g) −7   (h) −12   (i) −10
   (j) −7   (k) −21   (l) −1
4. (a) 9   (b) 1   (c) 12
   (d) −10   (e) 30   (f) 15
5. −£75 (£75 overdrawn)
6. 11°C   **8.** 17 cm
7. 5°C   **9.** 8 kg

10. (a) $-80\,\text{m}$ (b) $-200\,\text{m}$ (c) $60\,\text{m}$
    (d) $60\,\text{m}$ (e) $120\,\text{m}$ (f) $70\,\text{m}$
    (g) $300\,\text{m}$ (h) $50\,\text{m}$ (i) $130\,\text{m}$
    (j) $240\,\text{m}$ (k) $250\,\text{m}$

Exercise 4.3   Page 48

1. (a) 2   (b) 1   (c) -9 (d) 8
   (e) 4   (f) -5 (g) -7 (h) 7
   (i) 1   (j) -15 (k) -3 (l) -6
2. (a) 13   (b) 6   (c) 7   (d) 7
   (e) 5   (f) -12 (g) -1 (h) 13
   (i) -11 (j) 11   (k) 9   (l) 0
3. (a) 5   (b) 3   (c) 3
   (d) -7 (e) -6 (f) -1
4. (a) 6   (b) -8 (c) -28
   (d) 0   (e) -35 (f) 19
5. (a) $10°\text{C}$ (b) $10°\text{C}$ (c) $5°\text{C}$ (d) $37°\text{C}$
6. $-15°\text{C}$
7. $-22°\text{C}$
8. $2°\text{C}$

Exercise 4.4   Page 50

1. (a) 35   (b) -35 (c) 35
   (d) 10   (e) -10 (f) 10
   (g) 1   (h) -24 (i) -24
   (j) -45 (k) 64   (l) -42
   (m) 42   (n) -80 (o) -80
   (p) 32
2. (a) -20 (b) 60   (c) -30
   (d) 60   (e) -60 (f) -100
3. (a) -4 (b) 4   (c) 5
   (d) -5 (e) -5 (f) 5
   (g) 6   (h) -6 (i) 4
   (j) -8 (k) 6   (l) -5
4. (a) Ahmed 26,   Bridget 21,   Chris -21,
    Dileep -19,   Evan -3
   (b) Ahmed, Bridget, Evan, Dileep, Chris

Review Exercise 4   Page 51

1. $-21°\text{C}, -15°\text{C}, -11°\text{C}, -3°\text{C}, 7°\text{C}, 11°\text{C}$
2. (a) Poole   (b) Selby
3. (a) -18 (b) 4   (c) 4
4. (a) 5   (b) -2
5. (a) 13 degrees   (b) $-8°\text{C}$
6. 5 degrees
7. $33\,\text{m}$
8. (a) 24   (b) -24 (c) -24 (d) 24
   (e) 40   (f) -40 (g) -40 (h) 40
9. (a) -5 (b) 2   (c) -6 (d) 4
10. (a) E.g. (i) $\boxed{-5} \times \boxed{2} = \boxed{-10}$
         (ii) $\boxed{1} \div \boxed{-1} = \boxed{-1}$
   (b) -28
11. (a) $21°\text{C}$ (b) $-0.4°\text{F}$
12. 4

CHAPTER   5

Exercise 5.1   Page 53

1. $\mathbf{W}: \frac{1}{3}$ $\mathbf{X}: \frac{5}{6}$ $\mathbf{Y}: \frac{7}{15}$ $\mathbf{Z}: \frac{6}{25}$
3. (a) $\frac{4}{9}$ (b) $\frac{2}{8} = \frac{1}{4}$
   (c) $\frac{8}{16} = \frac{1}{2}$ (d) $\frac{3}{18} = \frac{1}{6}$
4. (a) $\frac{1}{3}$ (b) $\frac{1}{3}$ (c) $\frac{1}{3}$
   (d) $\frac{1}{3}$ (e) $\frac{1}{2}$
5. (a) **P**   (b) **P**   (c) **R**
   (d) **R**   (e) **Q**   (f) **S**
6. (a) (ii) $\frac{1}{6}$ (b) (ii) $\frac{1}{12}$

Exercise 5.2   Page 56

1. E.g. $\frac{40}{50}, \frac{20}{25}, \frac{4}{5}$ $\frac{4}{5}$ is the simplest form.
2. E.g. $\frac{15}{24}, \frac{30}{48}, \frac{45}{72}, \ldots, \frac{10}{16}, \frac{20}{32}, \ldots$
   Simplest form $\frac{5}{8}$
5. (a) 1   (b) 2   (c) 8
6. (a) E.g. $\frac{2}{6} = \frac{3}{9} = \frac{4}{12}$
   (b) E.g. $\frac{4}{18} = \frac{6}{27} = \frac{8}{36}$
   (c) E.g. $\frac{10}{16} = \frac{15}{24} = \frac{20}{32}$
   (d) E.g. $\frac{8}{10} = \frac{12}{15} = \frac{16}{20}$
   (e) E.g. $\frac{6}{20} = \frac{9}{30} = \frac{12}{40}$
   (f) E.g. $\frac{14}{24} = \frac{21}{36} = \frac{28}{48}$
7. (a) 2   (b) 3   (c) 3
8. (a) 6   (b) 8   (c) 12
9. $\frac{7}{16}, \frac{5}{8}, \frac{3}{4}$
10. $\frac{7}{10}, \frac{2}{3}, \frac{3}{5}, \frac{8}{15}$
11. $\frac{4}{5}$
12. (a) $\frac{3}{4}$ (b) $\frac{4}{5}$ (c) $\frac{2}{3}$ (d) $\frac{2}{9}$
    (e) $\frac{2}{3}$ (f) $\frac{2}{5}$ (g) $\frac{6}{25}$ (h) $\frac{4}{5}$
13. (a) $\frac{4}{20} = \frac{1}{5}$ (b) $\frac{3}{12} = \frac{1}{4}$ (c) $\frac{8}{12} = \frac{2}{3}$
    (d) $\frac{24}{60} = \frac{2}{5}$ (e) $\frac{60}{105} = \frac{4}{7}$
14. $\frac{4}{32} = \frac{1}{8}$
15. $\frac{30}{50} = \frac{3}{5}$
16. (a) $\frac{48}{60} = \frac{4}{5}$ (b) $\frac{12}{60} = \frac{1}{5}$
17. (a) $\frac{7}{10}$ (b) $\frac{1}{5}$
18. (a) $\frac{1}{6}$ (b) 24

Exercise **5.3** —

1. (a) $1\frac{3}{10}$ (b) $1\frac{1}{2}$ (c) $2\frac{1}{8}$ (d) $3\frac{3}{4}$ (e) $4\frac{3}{5}$
   (f) $4\frac{6}{7}$ (g) $3\frac{1}{2}$ (h) $3\frac{2}{3}$ (i) $1\frac{7}{9}$

2. (a) $\frac{27}{10}$ (b) $\frac{8}{5}$ (c) $\frac{35}{6}$ (d) $\frac{63}{20}$ (e) $\frac{41}{9}$
   (f) $\frac{53}{7}$ (g) $\frac{13}{4}$ (h) $\frac{14}{3}$ (i) $\frac{19}{8}$

3. (a) 3 (b) 4 (c) 3 (d) 8 (e) 8
   (f) 9 (g) 12 (h) 20 (i) 40 (j) 12

4. (a) 6 (b) 24

5. 56   8. £7.70

6. 28   9. £148.40

7. £5   10. (a) 9 (b) 10 (c) $\frac{5}{24}$

Exercise **5.4** —

1. (a) $\frac{3}{8}$ (b) $\frac{7}{12}$ (c) $\frac{7}{10}$ (d) $\frac{8}{15}$ (e) $\frac{9}{14}$

2. (a) $\frac{1}{8}$ (b) $\frac{1}{12}$ (c) $\frac{3}{10}$ (d) $\frac{2}{15}$ (e) $\frac{5}{14}$

3. (a) $1\frac{1}{4}$ (b) $1\frac{1}{2}$ (c) $1\frac{11}{20}$ (d) $1\frac{8}{21}$ (e) $1\frac{5}{24}$

4. (a) $\frac{1}{8}$ (b) $\frac{8}{15}$ (c) $\frac{5}{8}$ (d) $\frac{1}{15}$ (e) $\frac{1}{3}$

5. (a) $4\frac{1}{4}$ (b) $3\frac{5}{6}$ (c) $4\frac{3}{8}$ (d) $5\frac{17}{20}$ (e) $6\frac{13}{30}$

6. (a) $1\frac{1}{10}$ (b) $\frac{5}{12}$ (c) $1\frac{3}{8}$ (d) $3\frac{3}{10}$ (e) $2\frac{1}{4}$

7. (a) $\frac{1}{20}$ (b) $1\frac{3}{20}$ (c) $\frac{5}{24}$ (d) $5\frac{9}{20}$ (e) $1\frac{1}{16}$

8. (a) $\frac{5}{6}$ (b) $\frac{1}{6}$ 11. (a) $\frac{1}{4}$ (b) Billy

9. $\frac{11}{20}$ 12. $\frac{7}{60}$

10. (a) $\frac{5}{12}$ (b) $\frac{11}{12}$ 13. $\frac{3}{10}$

Exercise **5.5** —

1. (a) $3\frac{1}{2}$ (b) $2\frac{2}{3}$ (c) $1\frac{4}{5}$ (d) $6\frac{1}{4}$ (e) $6\frac{6}{7}$
   (f) 6 (g) 6 (h) 9 (i) $4\frac{1}{2}$ (j) $10\frac{1}{2}$

2. (a) $\frac{1}{6}$ (b) $\frac{1}{20}$ (c) $\frac{1}{10}$ (d) $\frac{5}{21}$ (e) $\frac{1}{6}$
   (f) $\frac{3}{8}$ (g) $\frac{1}{10}$ (h) $\frac{1}{3}$ (i) $\frac{1}{16}$ (j) $\frac{1}{20}$

3. (a) $\frac{1}{2}$ (b) $\frac{3}{10}$ (c) $\frac{1}{3}$ (d) $\frac{1}{3}$ (e) $\frac{3}{16}$
   (f) $\frac{2}{7}$ (g) $\frac{1}{2}$ (h) $\frac{1}{4}$ (i) $\frac{3}{10}$ (j) $\frac{7}{15}$

4. (a) 16 (b) $5\frac{1}{4}$ (c) $\frac{1}{12}$ (d) $\frac{1}{12}$ (e) $\frac{1}{6}$

5. (a) $1\frac{1}{8}$ (b) 2 (c) $3\frac{3}{4}$ (d) $3\frac{17}{20}$ (e) $4\frac{7}{8}$

6. 100 g   9. (a) $\frac{1}{8}$ (b) $\frac{1}{8}$

7. $1\frac{1}{2}$ kg   10. (a) $\frac{4}{15}$ (b) $\frac{2}{5}$

8. 3   11. (a) $\frac{4}{15}$ (b) 15

Exercise **5.6** —

1. (a) $\frac{1}{10}$ (b) $\frac{1}{6}$ (c) $\frac{1}{20}$ (d) $\frac{2}{5}$ (e) $\frac{3}{10}$
   (f) $\frac{3}{8}$ (g) $\frac{1}{3}$ (h) $\frac{1}{10}$ (i) $\frac{1}{8}$ (j) $\frac{3}{7}$

2. (a) 2 (b) $\frac{2}{5}$ (c) $\frac{1}{2}$ (d) $\frac{1}{2}$ (e) $\frac{2}{5}$
   (f) $1\frac{1}{2}$ (g) $\frac{2}{3}$ (h) $2\frac{5}{8}$ (i) $3\frac{1}{3}$ (j) 6

3. (a) $\frac{5}{6}$ (b) $\frac{9}{16}$ (c) $\frac{4}{5}$ (d) $1\frac{1}{3}$ (e) $\frac{2}{3}$
   (f) $1\frac{1}{2}$ (g) $\frac{2}{3}$ (h) $1\frac{1}{6}$ (i) $1\frac{1}{2}$ (j) $1\frac{4}{5}$

4. (a) $\frac{1}{15}$ (b) $\frac{4}{5}$ (c) $\frac{1}{5}$ (d) $2\frac{2}{3}$ (e) $\frac{4}{5}$

5. 20   7. $\frac{1}{9}$

6. $\frac{1}{10}$ 8. (a) 100 (b) 25

Exercise **5.7** —

1. $0.5, \frac{1}{2}$; $0.2, \frac{1}{5}$; $0.75, \frac{3}{4}$; $0.7, \frac{7}{10}$; $0.01, \frac{1}{100}$

2. (a) $\frac{3}{25}$ (b) $\frac{3}{5}$ (c) $\frac{8}{25}$ (d) $\frac{7}{40}$ (e) $\frac{9}{20}$
   (f) $\frac{13}{20}$ (g) $\frac{11}{50}$ (h) $\frac{101}{500}$ (i) $\frac{7}{25}$ (j) $\frac{111}{200}$
   (k) $\frac{5}{8}$ (l) $\frac{21}{25}$

3. (a) (i) 0.25 (ii) 0.5 (iii) 0.75
   (b) (i) 0.1 (ii) 0.3 (iii) 0.7
   (c) (i) 0.4 (ii) 0.6 (iii) 0.8

4. (a) (i) 0.15 (ii) 0.35 (iii) 0.95
   (b) (i) 0.16 (ii) 0.36 (iii) 0.92
   (c) (i) 0.07 (ii) 0.23 (iii) 0.53

5. (a) 0.125 (b) 0.625
   (c) 0.225 (d) 0.725

6. (a) $0.\dot{7}$ (b) $0.\dot{3}\dot{6}$
   (c) $0.1\dot{3}\dot{5}$ (d) $0.1\dot{6}$

7. (a) $0.\dot{6}$ (b) $0.\dot{4}$ (c) $0.8\dot{3}$
   (d) $0.7\dot{2}$ (e) $0.2\dot{6}$

8. (a) 0.33 (b) 0.17 (c) 0.43
   (d) 0.45 (e) 0.78

Review Exercise **5** —

1. (a) $\frac{3}{12} = \frac{1}{4}$ (b) Shade any 4 squares.

2. $\frac{3}{4} = \frac{9}{12}$ and $\frac{2}{3} = \frac{8}{12}$ so, $\frac{3}{4}$ is bigger.

3. $\frac{8}{20}$ 5. (a) 24 (b) $\frac{3}{4}$

4. (a) $\frac{1}{4}$ (b) 4   6. $4\frac{3}{8}$

7. (a) $3\frac{11}{12}$ (b) $\frac{11}{40}$ (c) $\frac{3}{7}$ (d) $\frac{5}{6}$

8. (a) £270 (b) £146

9. $\frac{1}{4}$ 10. $\frac{1}{10}$ 11. $\frac{7}{10}$

12. (a) 0.143
    (b) 1.014, 1.14, $1\frac{1}{7}$, 1.41, 11.14

CHAPTER 6

Exercise 6.1 **Page 67**

1. (a) 10, 20, 30, 40, 50
   (b) 3, 6, 9, 12, 15
   (c) 7, 14, 21, 28, 35
   (d) 6, 12, 18, 24, 30
   (e) 9, 18, 27, 36, 45
   (f) 20, 40, 60, 80, 100
2. (a) 20   (b) 42
   (c) third   (d) seventh
   (e) 10   (f) 9
3. (a) second   (b) second
   (c) fifth   (d) eighth
   (e) third
4. (a) 35   (b) 48
5. (a) (i) Answers are even numbers.
       (ii) Answers are even numbers.
       (iii) Answers are even numbers.
       (iv) Answers are odd numbers.
   (b) (i)   (ii)

×	2	3	6	7	9
2	E	E	E	E	E
3	E	O	E	O	O
6	E	E	E	E	E
7	E	O	E	O	O
9	E	O	E	O	O

×	O	E
O	O	E
E	E	E

   (c) Only **O** × **O** gives an odd number.

Exercise 6.2 **Page 69**

1. (a) 1, 2, 3, 4, 6, 12
   (b) 12 ÷ 8 = 1.5
       To be a factor, 8 would need to divide into 12 a whole number of times.
2. (a) 1 × 18, 2 × 9, 3 × 6
   (b) 1, 2, 3, 6, 9, 18
3. (a) 1 × 20, 2 × 10, 4 × 5
   (b) 1, 2, 4, 5, 10, 20
4. (a) 1, 2, 4, 8, 16
   (b) 1, 2, 4, 7, 14, 28
   (c) 1, 2, 3, 4, 6, 9, 12, 18, 36
   (d) 1, 3, 5, 9, 15, 45
   (e) 1, 2, 3, 4, 6, 8, 12, 16, 24, 48
   (f) 1, 2, 5, 10, 25, 50
   (g) 1, 2, 3, 4, 5, 6, 10, 12, 15, 20, 30, 60
   (h) 1, 2, 4, 5, 8, 10, 16, 20, 40, 80
5. (a) (i) 1, 2   (ii) 1, 3   (iii) 1, 5
       (iv) 1, 7   (v) 1, 11   (vi) 1, 13
   (b) 17, 19, 23, 29, 31, …
6. (a) (i) 1, 2, 4,   (ii) 1, 3, 9
       (iii) 1, 5, 25   (iv) 1, 7, 49
   (b) 121, 169, 289, 361, …

7. (a) (i) 1, 2, 3, 6   (ii) 1, 2, 5, 10
       (iii) 1, 2, 7, 14   (iv) 1, 2, 13, 26
       (v) 1, 5, 11, 55   (vi) 1, 2, 19, 38
   (b) 15, 21, 22, 33, 35, …
8. 6, 36
9. (a) 1, 5   (b) 1, 2, 4   (c) 1, 2
   (d) 1, 2, 3, 4, 6, 12
   (e) 1, 2, 3, 6
10. (a) 5   (b) 27   (c) 3, 5
11. (a) 4 (6, 12, 18, 36)
    (b) 8 (5, 10, 15, 20, 30, 40, 60, 120)
    (c) 6 (2, 4, 10, 20, 50, 100)
    (d) 8 (4, 8, 12, 16, 24, 32, 48, 96)
12. (i) 2, 3, 5, 7, 11, 13, 17, 19, 23, 29, 31, 37, 41, 43, 47
    Each number has 2 factors.
    Prime numbers.

Exercise 6.3 **Page 70**

1. (a) $5^2 = 5 \times 5$
   (b) $2^3 = 2 \times 2 \times 2$
   (c) $8^3 = 8 \times 8 \times 8$
2. (a) 4^3 (b) 8^2 (c) 10^6
3.

	Expression	Index form	Value
	$10 \times 10 \times 10 \times 10 \times 10$	10^5	100 000
(a)	$10 \times 10 \times 10 \times 10$	10^4	10 000
(b)	$10 \times 10 \times 10$	10^3	1000
(c)	10×10	10^2	100
(d)	10	10^1	10

4. (a) 27   (b) 36   (c) 64
   (d) 144   (e) 125   (f) 1 000 000
5. (a) $2^2 \times 3^2$ (b) $2 \times 3^3 \times 5$
   (c) $2 \times 3 \times 5^2$ (d) $2^3 \times 3 \times 5^2$
   (e) $3^3 \times 5^3$

Exercise 6.4 **Page 71**

1. (a) 2, 3   (b) 2, 5   (c) 2, 7
   (d) 3, 5   (e) 2, 3, 11   (f) 2, 3
2. (a) $2^2 \times 3$ (b) $2^2 \times 5$ (c) $2^2 \times 7$
   (d) $3^2 \times 5$ (e) $2 \times 3 \times 11$ (f) $2^2 \times 3^3$
3. 2^7
4. $2^3 \times 5^3$
5. (a) 25 000   (b) $2^3 \times 5^6$

Exercise 6.5 **Page 72**

1. (a) 24   (b) 160   (c) 20   (d) 90
   (e) 90   (f) 24   (g) 40   (h) 630
2. (a) 6   (b) 8   (c) 2   (d) 4
   (e) 11   (f) 4   (g) 3   (h) 15
3. (a) $2^3 \times 3$ (b) 2×3^3 (c) 6   (d) 216
4. (a) 3   (b) 2×3^4 (c) 54   (d) 324
5. 9.18 am

Exercise 6.6 — Page 74

1. (a) 49 (a) 125 (c) 0.5
2. (a) 1, 4, 9, 16, 25, 36, 49, 64, 81, 100, 121, 144, 169, 196, 225, 256, 289, 324, 361, 400
   (c) $21^2 = 400 + 41 = 441$
3. 1, 8, 27, 64, 125, 216, 343, 512, 729, 1000
4. **64**
5. No. $2^2 = 4$, $3^2 = 9$ and $5^2 = 25$
   $2^2 + 3^2 = 4 + 9 = 13$
6. (a) (i) 9 (ii) -8 (iii) 16 (iv) -125
   (b) The result of squaring a negative number is a positive number.
   The result of cubing a negative number is a negative number.
7. (a) (i) 169 (ii) 289 (iii) 6.25
   (iv) 0.64 (v) 94.09
   (b) (i) 216 (ii) 3375 (iii) 13.824
   (iv) 0.343 (v) 175.616
8. (a) (i) 0.5 (ii) 0.2 (iii) 0.1
   (iv) 2 (v) 10 (vi) 5
   (b) (i) 0.25 (ii) 0.05 (iii) 0.04
   (iv) 4 (v) 2.5 (vi) 6.25
9. E.g. $5 \times \frac{1}{5} = 1$
10. (a) 27 000 (b) 800 (c) 2.98
    (d) 10.648 (e) 70.56 (f) 31.25
    (g) 25.215 (h) 1.28

Exercise 6.7 — Page 75

1. (a) 6 (b) 2
2. (a) 5 (b) 10 (c) 8 (d) 7
3. (a) 5 (b) 5
4. Yes. $7^2 = 49$ and $8^2 = 64$
   55 is between 49 and 64.
   So, $\sqrt{55}$ lies between 7 and 8.
5. 9.8
6. (a) 2.3 (b) 17.3 (c) 4.0
7. (a) (i) 4.5 (ii) 10.4 (iii) 2.8
8. 1.9 m
9. 7.4 m

Exercise 6.8 — Page 77

1. (a) 2^7 (b) 4^9 (c) 6^3
   (d) 8^7 (e) 9^5 (f) 2^8
   (g) 5^{12} (h) 3^3
2. (a) 2^3 (b) 4^2 (c) 6
   (d) 8 (e) 3^6 (f) $2^1 = 2$
   (g) 5^2 (h) 11^3
3. (a) 8^7 (b) 2^9 (c) 7^5
   (d) $5^0 = 1$ (e) $4^1 = 4$ (f) 6^9
   (g) 10^4 (h) 3^{12}

4. (a) 3^6 (b) 10^2 (c) 4^5
   (d) 5^3 (e) 2^3 (f) $5^0 = 1$
   (g) 7^6 (h) 3^4
5. (a) 10^5 (b) 10^4
   (c) 10^6 (d) 10^5

Exercise 6.9 — Page 78

1. (a) 600 000 (b) 2000
   (c) 50 000 000 (d) 900 000 000
   (e) 3 700 000 000 (f) 28
   (g) 99 000 000 000 (h) 71 000
2. (a) 4500 (b) 78 000 000
   (c) 530 000 (d) 32 500
3. (a) (i) | 6 | 13 | (ii) 60 000 000 000 000
   (b) (i) | 9.6 | 12 | (ii) 9 600 000 000 000
   (c) (i) | 1.05 | 13 | (ii) 10 500 000 000 000
   (d) (i) | 1.3 | 14 | (ii) 130 000 000 000 000
   (e) (i) | 2.4 | 14 | (ii) 240 000 000 000 000
   (f) (i) | 2.5 | 12 | (ii) 2 500 000 000 000

Exercise 6.10 — Page 78

1. (a) 0.35
   (b) 0.0005
   (c) 0.000 072
   (d) 0.0061
   (e) 0.000 000 000 117
   (f) 0.000 000 813 5
   (g) 0.064 62
   (h) 0.000 000 004 001
2. (a) 0.0034 (b) 0.000 056 5
   (c) 0.000 72 (d) 0.913
3. (a) (i) | 6 | -10 | (ii) 0.000 000 000 6
   (b) (i) | 1.5 | -08 | (ii) 0.000 000 015
   (c) (i) | 3 | -11 | (ii) 0.000 000 000 03
   (d) (i) | 4.6 | -10 | (ii) 0.000 000 000 46
   (e) (i) | 4.24 | -09 | (ii) 0.000 000 004 24
   (f) (i) | 9.6 | -11 | (ii) 0.000 000 000 096

Review Exercise 6 — Page 80

1. (a) 1, 2, 3, 4, 6, 12 (b) 1, 2, 3, 6
2. (a) E.g. 7, 14, 21
   (b) (i) **10, 20**
   (ii) It is a multiple of 10 and ends in 0.
3. 1, 2, 4, 8
4. (a) 19 has only two factors, 1 and 19.
   15 has more than two factors, 1, 3, 5 and 15.
   (b) 23

5. (a) 9, 18 (b) 3, 6 (c) 3

6. (a) 36 (b) 64

7. (a) **9, 25, 100** (b) **20, 25, 100**
(c) **3, 29**

8. (a) 8: 1, 2, 4, 8
9: 1, 3, 9
11: 1, 11
17: 1, 17
121: 1, 11, 121
(b) (i) 11, 17 (ii) prime numbers
(c) (i) 9, 121 (ii) square numbers

9. (a) 8 (b) $13^3 = 13 \times 13 \times 13 = 2197$

10. (a) 8 (b) 4

11. (a) 3 (b) 10 000

12. 3^3 greater, $\sqrt{625} = 25$, $3^3 = 27$

13. E.g. 2, 3, 4, 5: $2 + 3 + 4 + 5 = 14$
14 is not a multiple of 4.

14. (a) 10^6 (b) 0.25

15. (a) 34 500 000 000
(b) 0.000 000 543

16. (a) 81 (b) 7 (c) 125

17. $2^3 \times 3$

18. 12

19. (a) 4 (b) $2^3 \times 3^2$
(c) 24 (d) 144

20. 90 seconds

21. $2^3 \times 3 \times 5$

22. (a) 39 000 000 000
(b) 0.000 067

23. 4096

24. (a) 4444488889
(b) 44444448888889

25. (a) 5^{11} (b) 5^4 (c) 5^4

26. 0.01024

27. (a) 2.4 (b) 14.65

28. 7.1

29. 5.8

30. (a) 5.5 (b) 5.6

31. (a) 36 (b) 0.028

32. 46.8

33. (a) 4 (b) 0.95

CHAPTER 7

1. (a) 35% (b) 54% (c) 24%
(d) 16% (e) 84% (f) 42%
(g) 5% (h) 46%

2. (a) (i) 40 (ii) 40%
(b) (i) 60, 60% (ii) 70, 70%
(iii) 45, 45% (iv) 24, 24%
(v) 46, 46% (vi) 68, 68%

3. (a) Missing entries are: $\frac{1}{5}, \frac{1}{4}, \frac{1}{2}, \frac{3}{4}, \frac{4}{5}$
(b) Missing entries are:
0.2, 0.25, 0.5, 0.75, 0.8

4. (a) $\frac{3}{20}$ (b) $\frac{1}{20}$ (c) $\frac{9}{50}$
(d) $\frac{13}{25}$ (e) $\frac{23}{100}$ (f) $\frac{1}{8}$

5. (a) 0.15 (b) 0.05 (c) 0.47
(d) 0.72 (e) 0.875 (f) 1.5

1. Entries are: 30%, 40%, 12%, 35%

2. Entries are: 70%, 45%, 5%, 120%

3. $33\frac{1}{3}\%$

4. (a) 34% (b) 48% (c) 15% (d) 80%
(e) 27% (f) 65% (g) $66\frac{2}{3}\%$ (h) $22\frac{2}{9}\%$

5. (a) 15% (b) 32% (c) 12.5%
(d) 7% (e) 112% (f) 1.5%

6. (a) $\frac{2}{5}, \frac{1}{2}, 0.55, 60\%$
(b) $0.42, 43\%, \frac{11}{25}, \frac{9}{20}$
(c) $28\%, 0.2805, \frac{57}{200}, \frac{23}{80}$

7. 80%

8. (a) 90% (b) 85% (c) 88% (d) 80%

9. B

10. Team A

1. (a) 60% (b) 70% (c) 21%

2. (a) 10% (b) 60% (c) 15%

3. 32%

4. 30%

5. (a) 40% (b) 60%

6. 12.5%

7. 37.5%

8. 3%

9. (a) $33\frac{1}{3}\%$ (b) $12\frac{1}{2}\%$ (c) $56\frac{1}{2}\%$

10. 9.5%

11. 25%

1. (a) 50 (b) 40 (c) 140 (d) 60
(e) 19.5 (f) 17 (g) 60 (h) 64

2. (a) £16 (b) £15 (c) £66
(d) £52.50 (e) 24 kg (f) 280 m
(g) £11.25 (h) 12p

3. (a) (i) 60 (ii) 105 (b) 45%

4. £20

5. 270

6. £42

7. £2.70

8. £10.50

9. (a) 660 (b) 198

10. 24 g

1. (a) £480 (b) £420 (c) £2800
   (d) £1080 (e) £3450 (f) £1260
   (g) £80 (h) £13 (i) £16.50
   (j) £57.50
2. (a) £420 (b) £600 (c) £2000
   (d) £150 (e) £10 200 (f) £4550
   (g) £510 (h) £5.50 (i) £33.60
   (j) £40.95
3. 40p per minute
4. £215
5. 759 g
6. £18.45
7. £244.64
8. £4706
9. £5622.50
10. £14 560
11. £10 530
12. £492

1. 30%
2. 40p per pint
3. 87.7p per litre
4. $83\frac{1}{3}\%$
5. (a) 36.3%
   (b) 23.7%
   (c) 37.8%
6. 17.3%
7. £109 520

1. 20%
2. 6%
3. 20%
4. $12\frac{1}{2}\%$
5. 28%
6. (a) $12\frac{1}{2}\%$ (b) 12%
   Rent went up by a greater percentage.
7. Becky   Sam's increase = 32%
            Becky's increase = 40%
8. 30.8%
9. Car A 13.8%
   Car B 18.2%
10. 5.9% decrease
11. 8% increase
12. 19%

1. (a) 0.5 (b) $\frac{3}{10}$ (c) 29%
2. (a) 70% (b) Shade any 3 rectangles.
3. 40%
4. $\frac{3}{5}$, 0.6, $\frac{6}{10}$
5. 25%
6. (a) 47% (b) 35%
7. 75p
8. 39%, $\frac{2}{5}$, 0.41, $\frac{21}{50}$
9. (a) 4 (b) 45%
10. **25% of £10 → £2.50**
    5% of £2 → 10p
    15% of £4 → 60p
    20% of £5 → £1
11. (a) 18 (b) 20%
12. 45%
13. 15%
14. No. Deposit = $\frac{2}{5}$ of £1800 = £720.
    Father gives Jane 30% of £1800 = £540.

15. £32.50
16. (a) 360 (b) 65%
17. £780
18. 90%
19. 180
20. $37\frac{1}{2}\%$
21. 57.3%
22. £5080
23. (a) 1.66 m (b) 13.7%
24. (a) 8.80 m (b) 15%
25. (a) (i) 60 (ii) 0.4 (b) 5%
26. 44.4%
27. (a) £2434.50 (b) 73.9%
28. 40%
29. 35.4%

CHAPTER 8

1. (a) 1030 (b) 2230 (c) 0145
   (d) 1345 (e) 2350
2. (a) 2.15 pm (b) 5.25 am (c) 11.20 pm
   (d) 10.05 am (e) 5.05 pm
3. (a) 7 am (b) 0700
4. (a) Start 9 am, Finish 3.30 pm
   (b) Start 0900, Finish 1530
   (c) 6 hours 30 minutes
5. (a) Start 11.54 am, Finish 1.35 pm
   (b) 1 hour 41 minutes
6. (a) 1230, 1255, 1320, 1325 (b) 25 minutes
7. (a) 2.28 pm (b) 48 minutes
8. (a) 1.15 pm (b) 2 hours 50 minutes
9. (a) 1325 (b) 4 hours 15 minutes
10. (a) 1.30 pm (b) 1330
11. (a) 1343 (b) 1.43 pm
12. (a) 1441 (b) 2.41 pm

1. (a) (i) 35 minutes (ii) 2.50 pm
   (b) (i) 24 minutes (ii) 3.08 pm
2. (a) 49 minutes (b) 1 hour 35 minutes
   (c) 1.42 pm (d) 0815
3. (a) (i) 5 minutes (ii) 1.03 pm
   (b) 1233
4. (a) 36 minutes (b) 1503 (c) 1713
5. (a) (i) 4.03 pm (ii) 25 minutes
   (b) 1540
6. (a) (i) 1520 (ii) 1610
   (b) 1650

1. £15.99
2. 29p
3. 90p
4. 86p
5. 96p
6. 85p
7. £1.60
8. 90p
9. £5.60
10. (a) £47
    (b) 10 days
11. (a) £24
    (b) 8 days
12. (a) £39
    (b) 120 miles
13. £26
14. £638
15. £165
16. £40.50

Exercise 8.4 — Page 99

1. (a) 4p   (b) 4p
2. Small: 5.2 g per penny.
   Large: 5.1 g per penny.
   Small tin is better buy.
3. Small: 8.7 g per penny.
   Large: 8.5 g per penny.
   Small pot is better buy.
4. 700 g
5. Large: 2.3 g per penny.
   Small: 2.2 g per penny.
   Large pot is better buy.
6. 1 kg
7. Medium
8. 1.5 litre bottle
9. Small: 0.882 ml per penny.
   Medium: 0.862 ml per penny.
   Large: 0.878 ml per penny.
   Small size is better value for money.
10. (a) Daisy's £448, Alfie's £438
    (b) Alfie's

Exercise 8.5 — Page 100

1. £6
2. (a) £3.50   (b) £73.50
3. (a) £59.50   (b) £399.50
4. (a) £15.75   (b) £105.75
5. (a) £8.97   (b) £188.50
6. £291.40
7. £170.37   10. £27 025
8. £216.20   11. £428.87
9. £77.55   12. £92.82

Exercise 8.6 — Page 101

1. (a) 310 euros   (b) 34 600 yen
   (c) 126 liri   (d) 2496 krone
   (e) 460 francs   (f) 284 dollars
2. (a) £39.13   (b) £28.90
   (c) £48.39   (d) £18.03
   (e) £24.65
3. (a) 38.75 euros   (b) £15
4. (a) 276 francs   (b) £10.65
5. (a) £387.10   (b) 46.5 euros
6. (a) £240.38   (b) 561.6 krone
7. (a) 710 dollars   (b) £18.13
8. £131.92
9. France: £5806.45, Japan: £5780.35
   Cheaper in Japan by £26.10

Review Exercise 8 — Page 102

1. 10.15 pm
2. (a) 5.05 pm   (b) 1 hour 44 minutes
3. £2.56 + £1.19 + £2.24 = £5.99
4. £40.72

5. 19p
6. Large packet.
   Standard: 9.08p per biscuit.
   Large: 8.53p per biscuit
7. (a) £20   (b) $26.50
8. 98p
9. (a) £44.16   (b) 156 miles
10. (a) £1045   (b) £95
11. £157.50
12. Large: 9.34 g per penny.
    Small: 9.66 g per penny.
    Small pot is better value.
13. £3.66
14. £586.32
15. Rome: 75p, London: 95p.
    Rome is cheaper, by 20p.

CHAPTER 9

Exercise 9.1 — Page 104

1. £130   7. £120
2. £7.50   8. £9.60   13. £10.20
3. 38 hours   9. £6.60   14. £24 000
4. £10.80   10. £694.40   15. £34 200
5. £23   11. £416.50   16. £1750
6. £26.35   12. £347.47   17. £15 000

Exercise 9.2 — Page 106

1. £7785   5. £3947.90
2. £8783   6. £107.16
3. (a) £1185 (b) £118.50 7. £38.71
4. (a) £1703 (b) £170.30 8. £11 911.20
   (c) £14.19   9. £704.60

Exercise 9.3 — Page 107

1. (a) £300   (b) £25
2. £187.28
3. £83.28
4. £101.71
5. 678
6. 7108
7. £971.65
8. £198
9. £67.28
10. (a) £231   (b) £210 000
    (c) £120   (d) £221
11. £537.40
12. £12 400
13. (a) £8.59   (b) £1892.80

Exercise 9.4 — Page 109

1. £10   3. £2
2. (a) £30 (b) £15   4. £225   5. £48

1. £16
2. £198
3. (a) £900   (b) £90
4. £94.18
5. $900 \times 0.2 + 200 \times 1 = £380$
   So, Noreen's calculation is wrong.
6. £28   9. £2156.40
7. (a) 8 hours   (b) £360   10. £1619.75
8. £401.94   11. £5.76

CHAPTER 10

1. (a) 2 GLUMS   (b) 5 : 2
2. (a) 3 SMILERS   (b) 3 : 1
3. (a) 3 GLUMS   (b) 5 : 3
4. 7 : 3
5. (a) 8 SMILERS and 2 GLUMS
   (b) 3 SMILERS and 9 GLUMS
6. (a) 75   (b) 12   (c) (i) 56   (ii) 140
7. (a) 16   (b) 9   (c) (i) 84   (ii) 112
8. (a) 14 SMILERS and 6 GLUMS
   (b)   9 SMILERS and 6 GLUMS
9. (a) (i) $\frac{1}{5}$ (ii) $\frac{4}{5}$ (iii) 1 : 4   (b) 3 : 1
10. (a) (i) 30% (ii) 70%   (b) 3 : 2
11. $\frac{2}{5}$
12. 75%

1. (a) E.g. 12 : 2, 18 : 3, 24 : 4
   (b) E.g. 14 : 4, 21 : 6, 28 : 8
   (c) E.g. 6 : 10, 9 : 15, 12 : 20
2. (a) 1 : 2   (b) 1 : 3   (c) 3 : 4
   (d) 2 : 5   (e) 3 : 4   (f) 2 : 5
   (g) 3 : 7   (h) 9 : 4   (i) 4 : 9
   (j) 7 : 3
3. (a) 12   (b) 28   (c) 100   (d) 20
4. 198 cm   7. 18 years old
5. 400 g   8. 2 : 3   10. 1 : 250
6. 64   9. 5 : 2   11. 1 : 1500
12. (a) 4 : 1   (b) 2 : 25   (c) 11 : 2
    (d) 5 : 2   (e) 4 : 1   (f) 40 : 17
    (g) 9 : 20   (h) 25 : 1   (i) 1 : 15
    (j) 2 : 1
13. 40 : 9   15. 2 : 3
14. 1 : 3   16. 1 : 2   17. 3 : 2

1. (a) 6, 3 (b) 15, 5 (c) 7, 28 (d) 90, 10
2. 8   4. Sunny £36, Chandni £12
3. 18   5. £28

6.

	4 : 1	3 : 2
(a)	32, 8	24, 16
(b)	16, 4	12, 8
(c)	64 kg, 16 kg	48 kg, 32 kg
(d)	160 g, 40 g	120 g, 80 g
(e)	£960, £240	£720, £480

7. (a) £14, £21   11. $\frac{1}{4}$
   (b) £32, £24   12. 80%
   (c) £3.50, £2   13. $\frac{5}{8}$
   (d) £1.80, £3   14. 60%
8. 45   15. 48%
9. £192   16. 170 000 km²
10. 2 033 000
17. (a) Jenny 50, Tim 30   (b) 10
18. (a) 5 : 2   (b) 6   (c) 55
    (d) 32 is not a multiple of 5 + 2 = 7
19. 8 cm, 12 cm, 18 cm
20. 200 kg
21. 40°, 60°, 80°
22. 45%
23. (a) $12\frac{1}{2}$%, $37\frac{1}{2}$%, 50%   (b) $\frac{1}{3}$ (c) $\frac{3}{5}$

1. (a) 16p   (b) £1.28   14. (a) 180 g
2. (a) £7   (b) £70   (b) 637.5 ml
3. (a) 30p   (b) £2.40   (c) 210 g
4. (a) £6.50   (b) £130   15. (a) 12 minutes
5. £2.85   (b) 16 miles
6. £201.60   16. (a) 12 m²
7. £61   (b) 12 litres
8. £2.85   17. (a) £260
9. £16.20   (b) 32
10. (a) 50 g   (b) 720 g   18. $17\frac{1}{2}$ minutes
11. £89.28   19. (a) 168.75 g
12. £32.41   (b) 36
13. (a) £3.08   (c) 540 ml
    (b) 12 minutes   20. £8.35

1. 1 : 3   13. (a) 24
2. 3 : 4   (b) 5 : 3
3. 9   14. £276
4. 12   15. 1 : 250
5. 8   16. £31.50
6. (a) 25 g   (b) 750 g   17. £1.45
7. 50   18. £11.50
8. (a) 375 g   (b) 360   19. 1 : 200 000
9. £4   20. (a) $\frac{3}{4}$
10. £262.50, £157.50
11. 12   (b) $12\frac{1}{2}$%
12. (a) 18   (b) 35%   (c) 200 cm³

CHAPTER 11

Exercise 11.1 — Page 121

1. 8 miles per hour
2. 7 km/h
3. 25 metres per minute
4. (a) 20 km/h
   (b) 50 km/h
   (c) 4 km/h
5. 10 km/h
6. 8 km
7. 30 miles
8. 3 km
9. (a) 150 km
   (b) 86 km
   (c) 40 km
10. $\frac{1}{2}$ hour
11. 20 seconds
12. $1\frac{1}{2}$ hours
13. (a) 3 hours
    (b) 2 hours
    (c) $3\frac{1}{2}$ hours
14. $2\frac{1}{4}$ hours
15. 6 km
16. $1\frac{1}{4}$ hours
17. (a) $2\frac{1}{2}$ hours
    (b) $1\frac{1}{4}$ hours
18. (a) 60 km/h
    (b) 1 hour
19. 11.20 am

Exercise 11.2 — Page 122

1. (a) 300 km
   (b) 5 hours
   (c) 60 km/h
2. 5 m/s
3. 10 km/h
4. 10.30 am
5. 11.09 am
6. (a) 13.8 km/h
   (b) 1.10 pm
7. (a) 8 hours
   (b) 111 km/h
8. (a) 4.81 m/s
   (b) 17.3 km/h
9. 150 m
10. 300 000 000 m/s

Exercise 11.3 — Page 123

1. 8 g/cm³
2. 9 g/cm³
3. 2.5 g/cm³
4. 28.6 g/cm³
5. 7200 g
6. 2000 cm³
7. 0.8 g/cm³
8. 9360 g
9. 118.3 people/km²
10. (a) 30 530 km²
    (b) 104.2 people/km²
    (c) 57 400 000

Review Exercise 11 — Page 124

1. 94 km/h
2. $1\frac{1}{2}$ hours
3. 240 miles
4. 14 km
5. 80 km/h
6. 3 miles
7. 60 miles per hour
8. 27 miles per hour
9. 40 miles per hour
10. 34 km/h
11. 2 hours 40 minutes
12. 1315
13. 8 km/h
14. 36 minutes
15. 5 hours 48 minutes
16. 25 km/h
17. 24 km/h
18. 125 m
19. 2.92 g/cm³
20. 4.35 cm³
21. 40 minutes

Number

Non-calculator Paper — Page 126

1. (a) Five thousand four hundred and twenty-three
   (b) (i) 31, 38, 139, 316, 1310
       (ii) 31, 139
2. (a) (i) 87   (ii) 56   (iii) 61
   (b) (i) 7000   (ii) 9.5
3. 6
4. (a) (i) 320 miles   (ii) 357 miles
   (b) 486 miles   (c) 5
5. (a) 9800p   (b) £98
6. −4, 0, 3.05, 3.5, 4
7. (a) £6.87   (b) £3.13
8. (a) 2.25 pm   (b) 32 minutes
9. (a) 26, 74   (b) 23, 73
   (c) $73 \times 74 = 5402$
10. (a) 0.2   (b) $\frac{7}{10}$ (c) $\frac{1}{2}$
11. (a) 801   (b) 4   (c) 5
12. (a) 12 000 000   (b) 50   (c) 0.5
13. (a) Shade any 3 squares.
    (b) Shade any 2 squares.   (c) $\frac{1}{3}$
14. (a) 5   (b) $\frac{1}{2}$ (c) 6
15. (a) 11 degrees Celsius   (b) −2°C
16. 56
17. £25.35
18. (a) $\frac{3}{10}$ (b) 9
    (c) (i) 34.2   (ii) 34.22
    (d) 48
19. (a) (i) 500   (ii) 93 000
    (b) 25 796
20. (a) $\frac{1}{10}$ (b) £5
21. 14 pence
22. (a) 17.5   (b) 2.95   (c) 1.6
23. (a) 273   (b) 32   (c) 200
24. (a) (i) 1600 metres   (ii) 1.6 km
    (b) 20 lengths
25. £48
26. (a) 100 000   (b) 7   (c) 8
    (d) 0.09   (e) 0.45
27. (a) 40   (b) 16   (c) $\frac{3}{10}$
28. (a) 25   (b) 75%
29. (a) 36   (b) (i) $1\frac{5}{12}$ (ii) $\frac{1}{10}$
    (c) Any decimal between 0.25 and 0.33…
    (d) $\frac{3}{10}$
30. 6
31. Large (500 g) = 2 × Standard (250 g)
    2 × £1.39 = £2.78   and   £2.78 > £2.69
    Large size is better value for money.

32. $100 \times 7 = 700$

33. (a) $\dfrac{70}{10-3}$ (b)   10

34. 40 miles per hour

35. (a)   220   (b)   60%

36. (a)   2500   (b)   2490

37. 4^3 is bigger. $4^3 = 64$, $7^2 = 49$

38. £5.62

39. 1310

40. (a)   35 pence   (b)   84 pence

41. Minimum: 395,   maximum 404

42. £90

43. (a)   49   (b)   9   (c) $\frac{2}{3}$ (d)   9   (e)   650 000

44. 90%

45. (a)   17.1911   (b)   171.911

46. (a)   12 mph   (b)   13 mph

47. (a)   4.2   (b)   0.08   (c) $1\frac{17}{20}$

48. 15

49. Least: 74.5 m,   greatest: 75.5 m

50. 7 g/cm³

51. (a) $2^3 \times 3^2$ (b)   24

52. 240 mph

53. Yes.
Year 1: £500 $\times$ 0.04 = £20 paid out
Year 2: £500 $\times$ 0.04 = £20 paid out
        Total interest = £20 + £20 = £40

54. Andi is correct.
Increase of 10%   = 110% of original speed
                   = 1.1 $\times$ original speed
Decrease of 10% = 90% $\times$ 1.1 = 0.99
                       = 99% of original speed

55. (a) $10^2 \div 0.4 = 250$
    (b)   (i) $p = 2$, $q = 3$
          (ii) $2 \times 3 \times 3 = 2 \times 3^2$ (iii)   72

56. **A**: False.   The prime numbers will be factors
                    of the product.
    B: True.   Even number + Odd number
                 = Odd number + Even number
                 = Odd number.

57. 3 hours 30 minutes

Number

 Section Review

Calculator Paper   Page 131

1. (a)   Five thousand six hundred and
          twenty-four
    (b) -15, -4,   3,   39,   120

2.

BEAUTIFUL BLOOMS		
Basic bouquet	£9	50p
6 Roses @ 40p each	£2	40p
4 Carnations @ 30p each	£1	20p
Delivery	£3	50p
Total	**£16**	**60p**

3. (a)   21 minutes   (b)   6.06 pm

4. (a)   (i)   940   (ii)   900   (b)   £2202
    (c)   (i) $\frac{3}{10}$ (ii)   30%   (d)   18

5. (a)   £13.11   (b)   £6.89
    (c)   £5, £1, 50p, 20p, 10p, 5p, 2p, 2p

6. (a) $-4°C$ (b)   11°C

7. £27.90

8. (a)   87   (b)   86.7

9. (a)   8 thousand   (b)   forty

10. (a)   12   (b)   8 pence

11. (a)   2.25
    (b)   2.035, $\sqrt{4.78}$,   2.21, $2\frac{1}{4}$, 1.53^2

12. 6 degrees Celsius

13. (a)   15   (b)   48

14. (a)   12   (b)   £1140

15. £30

16. (a)   (i)   28 000   (ii)   29 602
    (b)   £19.55

17. (a)   0.375   (b) $\frac{4}{5}$

18. £65.45

19. (a)   77 acres   (b)   57.75 acres

20. 8

21. 39p

22. £22.05

23. (a)   3, 6   (b)   3

24. £3.51

25. (a)   (i)
$$11 = 11$$
$$11 \times 11 = 121$$
$$11 \times 11 \times 11 = 1331$$
$$11 \times 11 \times 11 \times 11 = 14\ 641$$
         (ii)   Numbers added in pairs to give the
              middle digits of the next result.
    (b) $11 \times 11 \times 11 \times 11 \times 11 = 161\ 051$
        When sum of pair is greater than 9, tens
        carried to next pair, adding right to left.

26. (a)   £539   (b)   £404.25

27. 45 miles per hour

28. 300 euros is £97.30 less than £300

29. 35 pence

30. (a)   5.29   (b)   30   (c)   6.2

31. (a)   62.4 kg   (b)   10%

32. (a)   Yum.   Yum 8.4 g/p,   Core 8.3 g/p
    (b)   39p

33. £1660

34. £107.50

35. £65.80

36. (a)   75 g   (b)   40%

37. £1.92

38. If 1st number is odd, last number is odd.
Odd + Odd = Even
If 1st number is even, last number is even.
Even + Even = Even

39. £70.21

40. E.g. $2 + 3 = 5$, $3 + 5 = 8$

41. 14 km/h

42. (a) £87.80 (b) 119.5%

43. (a) 39.520… (b) $\frac{20^2}{7+3} = \frac{400}{10} = 40$

44. £16.80

45. 95.8%

46. (a) 1 953 125 (b) 0.143 (c) 0.0074

47. 7.8

48. Yes, she does have enough money.
400 000 dollars = £291 971
450 000 euros = £288 462

49. 2 hours 40 minutes

50. 3.21

51. 31.6%

52. Angela £20 Fran £35 Dan £45

53. 250 g

54. 4.7

55. 51.8 cm³

CHAPTER 12

Exercise 12.1 — Page 136

1. $n + 4$

2. $n - 3$

3. $3n$

4. $m + 6$

5. $m - 12$

6. $8m$

7. $p - 1$

8. $p + 5$

9. $25p$

10. $6k$

11. $5b$ pence

12. $\frac{c}{3}$ pence

13. $\frac{a}{5}$ pence

14. $\frac{36}{g}$

15. (a) $2t$ (b) $10t$

Exercise 12.2 — Page 138

1. (a) $2y$ (b) $3c$ (c) $5x$ (d) $7p$
(e) $2t$ (f) $3d$ (g) $3n$ (h) $5y$
(i) $10g$ (j) $8m$ (k) $13z$ (l) $2r$
(m) $5t$ (n) $4y$ (o) $3j$ (p) $4c$
(q) $7x$ (r) w (s) 0 (t) $-5y$
(u) $-5x$ (v) $-14a$ (w) $6b$ (x) $2m$

2. (a) $4x$ (b) $6a$ (c) $9y$ (d) $6u$

3. (a) Can be simplified, $2v$.
(b) Cannot be simplified, different terms.
(c) Can be simplified to $3v + 4$.
(d) Cannot be simplified, different terms.

4. (a) $8x + y$ (b) $w + 2v$
(c) $2a - 2b$ (d) $5x + 3y$
(e) $3 + 7u$ (f) $p + 4q$
(g) $3d - 7c$ (h) $2y + 1$
(i) $a + b$ (j) $4m + n$
(k) $9c - d$ (l) $x + y$
(m) $6p$ (n) $5 - 5k$
(o) $a + 3$

5. (a) $8a + 3b$ (b) $3p + 3q$
(c) $3m + 2n$ (d) $x - 2y$
(e) $2x + 3y$ (f) $d + 3$
(g) $3b - 2a$ (h) 7
(i) $a + 2b$ (j) $-2f$
(k) $v - 4w$ (l) $-2 - 5t$
(m) $4q - 4p$ (n) $1 - 7k$
(o) $c - d + 11$

6. (a) $4x + 2$ (b) $4a + 6b$
(c) $3x$ (d) $6y + 9$

7. (a) $2xy$ (b) $2pq$
(c) $3ab$ (d) $2x^2$
(e) $9y^2$ (f) $4a^2$
(g) $2d^2 - 3g^2$ (h) $5t^2 - t$
(i) $6m - m^2$ (j) $2p^2 - p$

Exercise 12.3 — Page 139

1. (a) $3a$ (b) $7b$ (c) $8c$ (d) $9d$
(e) $4e$ (f) $8f$ (g) $6p$ (h) $15q$
(i) r^2 (j) g^2 (k) $2g^2$ (l) $6g^2$
(m) $4t^2$ (n) $12t^2$ (o) $15u^2$ (p) $15m^2$
(q) $9d^2$ (r) $15x^2$ (s) $12y^2$ (t) $16k^2$

2. (a) $-3y$ (b) $-5y$ (c) $2y$ (d) $-6y$
(e) $-t^2$ (f) $-2t^2$ (g) $-10t^2$ (h) $10t^2$

3. (a) $5a$ (b) $4b$ (c) 12 (d) 20
(e) $2y$ (f) 8 (g) 2 (h) 18
(i) $3p$ (j) 3 (k) 9 (l) 6
(m) 4 (n) 5 (o) 4 (p) 9

4. (a) $-2y$ (b) $-3y$ (c) $-m$ (d) m
(e) $-3a$ (f) $-5d$ (g) $-3g$ (h) k

5. (a) ab (b) xy (c) y^2
(d) $2pq$ (e) $2a^2$ (f) $3xy$
(g) $6ab$ (h) $12gh$ (i) $6d^2$
(j) $3g^2$ (k) $5ab$ (l) $6gh$
(m) abc (n) m^3 (o) $2d^3$
(p) $3g^3$ (q) $6x^3$ (r) $10m^2n$
(s) $3abc$ (t) $18pqr$

6. (a) a^2 (b) $4x^2$ (c) $6g^2$ (d) $10y^2$

Exercise 12.4 — Page 140

1. (a) y^3 (b) t^5 (c) a^6 (d) g^{10}
(e) x^{11} (f) m^7 (g) k^5 (h) h^8

2. (a) y^2 (b) a (c) 1 (d) t^4
(e) g^{-1} (f) h^{-2} (g) x^3 (h) m^{-1}

3. (a) a^3b (b) m^4n^2 (c) $2y^3$ (d) $6d^5$
(e) a^5b^2 (f) $6b^2$ (g) 5 (h) $4a^4$

4. (a) t (b) g^{-1} (c) m^2 (d) y
(e) y^2 (f) m^{-1} (g) $2t^2$ (h) $3g^3$

1. (a) $2x + 10$ (b) $3a + 18$
    (c) $4y + 12$ (d) $4a + 2$
    (e) $6y + 4$ (f) $3a + 3b$

2. (a) $3x + 6$ (b) $2y + 10$
    (c) $4x + 2$ (d) $3p + 3q$

3. $2(q + 2)$ and $2q + 4$ $2(q - 1)$ and $2q - 2$
    $2(q + 1)$ and $4q + 2$ $2(2 - q)$ and $4 - 2q$

4. (a) $a^2 + a$ (b) $2d + d^2$
    (c) $2x^2 + x$

5. (a) $2x + 8$ (b) $3t - 6$
    (c) $20 - 4a$ (d) $15 - 6d$
    (e) $6b + 12c$ (f) $6m - 15n$
    (g) $x^2 + 3x$ (h) $t^2 - 3t$
    (i) $2g^2 + 3g$ (j) $2m - 3m^2$
    (k) $3t^2 + 5t$ (l) $m^2 - mn$

6. (a) $2x + 5$ (b) $3a + 11$
    (c) $6w - 17$ (d) $10 + 2p$
    (e) $3q$ (f) $7 - 3t$
    (g) $5z + 8$ (h) $8t + 15$
    (i) $2c - 6$ (j) $5a - 9$
    (k) $3y - 10$ (l) $2x + 6$
    (m) $8a + 23$ (n) $10x - 12$
    (o) $2p - 11$ (p) $5a + 2b$
    (q) $3x + y$ (r) $2p - 5q$
    (s) $5x - x^2$ (t) $a^2 - 2a$
    (u) $2y$

7. (a) $5x + 8$ (b) $5a + 13$
    (c) $9y + 23$ (d) $9a + 5$
    (e) $26t + 30$ (f) $5z + 13$
    (g) $12q + 16$ (h) $11x - 3$
    (i) $20e - 16$ (j) $12d + 6$
    (k) $3m^2 - 3m$ (l) $4a^2 - a$

8. (a) $-3x + 6$ (b) $-3x + 6$
    (c) $-2y + 10$ (d) $-6 + 2x$
    (e) $-15 + 3y$ (f) $-4 - 4a$
    (g) $3 - 2a$ (h) $2d + 6$
    (i) $2b - 6$ (j) $-6p - 9$
    (k) $m - 6$ (l) $5d + 1$
    (m) $-a^2 + 2a$ (n) $d - d^2$
    (o) $2x^2$ (p) $-6g^2 - 9g$
    (q) $7t^2 - 6t$ (r) $8m - 2m^2$

9. (a) $3a - 1$ (b) $y - 5$
    (c) $5m - 1$ (d) $3x - 4$
    (e) $1 - d$ (f) $t - 10$
    (g) $4m + 9$ (h) $-x - 18$
    (i) $23 - 6a$

1. $x^2 + 7x + 12$ **13.** $x^2 - 5x + 6$
2. $x^2 + 6x + 5$ **14.** $x^2 - 5x + 4$
3. $x^2 - 3x - 10$ **15.** $x^2 - 5x - 14$
4. $x^2 - x - 2$ **16.** $x^2 + 2x - 3$
5. $x^2 - 8x + 12$ **17.** $x^2 + 4x - 5$
6. $x^2 + 3x + 2$ **18.** $x^2 + 3x - 10$
7. $x^2 + 5x + 6$ **19.** $x^2 - 9$
8. $x^2 - x - 6$ **20.** $x^2 - 25$
9. $x^2 - 5x + 6$ **21.** $x^2 - 49$
10. $x^2 + 6x - 16$ **22.** $x^2 - 100$
11. $x^2 + 3x - 10$ **23.** $x^2 + 6x + 9$
12. $x^2 + 2x - 3$ **24.** $x^2 - 6x + 9$

1. (a) $2(x + y)$ (b) $3(a - 2b)$
    (c) $2(3m + 4n)$ (d) $x(x - 2)$
    (e) $a(b + 1)$ (f) $x(2 - y)$
    (g) $2(b - 2a)$ (h) $x(2x + 3)$
    (i) $g(1 - g)$

2. (a) $2(a + b)$ (b) $5(x - y)$
    (c) $3(d + 2e)$ (d) $2(2m - n)$
    (e) $3(2a + 3b)$ (f) $2(3a - 4b)$
    (g) $4(2t + 3)$ (h) $5(a - 2)$
    (i) $2(2d - 1)$ (j) $3(1 - 3g)$
    (k) $5(1 - 4m)$ (l) $4(k + 1)$

3. (a) $x(y - z)$ (b) $g(f + h)$
    (c) $b(a - 2)$ (d) $q(3 + p)$
    (e) $a(1 + b)$ (f) $g(h - 1)$
    (g) $a(a + 3)$ (h) $t(5 - t)$
    (i) $d(1 - d)$ (j) $m(m + 1)$
    (k) $r(5r - 3)$ (l) $x(3x + 2)$

1. $6t$ pence
2. $(x + 3)$ years old
3. (a) $p + 5$ (b) $2p$ (c) $2p - 3$
4. (a) $(n - 2)$ years old   (b) $3n$ years old
    (c) $(3n + 4)$ years old   (d) $(8n + 2)$ years
5. (a) £1.50   (b) $25n$ pence
6. (a) $3w$ (b) $w + 2$ (c) w^2
7. (a) $7p + q$ (b) $4r - 12$
8. $7d + 3$
9. (a) (i) $5x$ (ii) $3x$ (iii) $3y^2$
    (b) (i) $4x + 4y$ (ii) $5x - 10y$
10. (a) $3ab$ (b) $a^2 + 2a$
    (c) $2x - 6$ (d) $5x + 2$

11. (a) $5a - 3b$ (b) $3y$ pence
12. (a) (i) $2x$ units (ii) $(2x - 3)$ units
(b) $(5x - 3)$ units
13. (a) $y^2 - 4y$ (b) $7y + 4$
14. (a) $x^2 + 3x$ (b) $x^3 - 3x^2$
15. (a) p^3 (b) $24abc$ (c) 1
16. (a) $5x + 10y$ (b) $6x^2 - 3x + 4$
17. (a) (i) $3(2x - 5)$ (ii) $y(y + 7)$
(b) $y + 12$
18. (a) t^8 (b) p^4 (c) a^4
19. (a) $x^2 - x - 2$ (b) a^5b^2

CHAPTER 13

Exercise 13.1 — Page 146

1. (a) 3 (b) 4 (c) 9 (d) 16
2. (a) $x = 4$ (b) $a = 3$
(c) $y = 8$ (d) $t = 6$
(e) $h = 22$ (f) $d = 1$
(g) $z = 30$ (h) $p = 0$
(i) $c = 99$
3. (a) 5 (b) 5 (c) 18 (d) 21
4. (a) $a = 4$ (b) $e = 6$
(c) $p = 4$ (d) $y = \frac{1}{2}$
(e) $d = 10$ (f) $t = 9$
(g) $m = 28$ (h) $x = 100$
5. (a) 1 (b) 4 (c) 4
(d) 2 (e) 3 (f) 5

Exercise 13.2 — Page 147

1. 3 **3.** 5 **5.** 11 **7.** 4 **9.** 4
2. 14 **4.** 6 **6.** 5 **8.** 3 **10.** 7
11. (a) 2 (b) $2(x + 3) = 2x + 6$
12. (a) 9 (b) $3(x - 2) = 3x - 6$
13. (a) 6 (b) $2x + 3$

Exercise 13.3 — Page 149

1. (a) $y = 3$ (b) $x = 6$
(c) $a = 10$ (d) $e = 15$
(e) $d = 11$ (f) $c = 20$
(g) $x = 2$ (h) $y = 19$
(i) $m = 7$
2. (a) $q = 7$ (b) $m = 10$
(c) $n = 16$ (d) $p = 18$
(e) $x = 31$ (f) $y = 17$
(g) $a = 2$ (h) $k = 4$
(i) $h = 12$

3. (a) $x = 14$ (b) $t = 28$
(c) $f = 18$ (d) $y = 19$
(e) $b = 7$ (f) $x = 29$
(g) $m = 4$ (h) $k = 5$
(i) $y = 7$
4. (a) $c = 4$ (b) $a = 4$
(c) $f = 3$ (d) $p = 3$
(e) $h = 5$ (f) $u = 2$
(g) $d = 30$ (h) $e = 14$
(i) $f = 20$
5. (a) $p = 4$ (b) $t = 3$
(c) $h = 7$ (d) $b = 2$
(e) $d = 10$ (f) $x = 6$
(g) $c = 5$ (h) $n = 3$
(i) $x = 2$
6. (a) $a = 12$ (b) $x = 4$
(c) $a = 27$ (d) $x = 2$
(e) $b = 6$ (f) $x = 4$
(g) $k = 2$ (h) $b = 3$
(i) $c = 2$ (j) $c = 7$
(k) $y = 4$ (l) $x = 6$
7. (a) $a = 3$ (b) $x = 9$
(c) $a = 0$ (d) $p = 1$
(e) $y = 5$ (f) $p = 4$
(g) $x = 9$ (h) $k = 4$
(i) $m = 3$

Exercise 13.4 — Page 150

1. (a) $k = \frac{1}{2}$ (b) $a = -3$
(c) $d = -4$ (d) $n = -\frac{1}{2}$
(e) $t = -5$ (f) $n = 1$
(g) $m = 1\frac{1}{2}$ (h) $x = 2\frac{1}{3}$
(i) $y = -\frac{1}{2}$
2. (a) $x = -2$ (b) $y = -3$
(c) $t = -2$ (d) $a = -2$
(e) $d = -3$ (f) $g = -3$
(g) $t = \frac{1}{2}$ (h) $x = 7\frac{1}{2}$
(i) $d = 1\frac{2}{5}$ (j) $a = 1\frac{1}{2}$
(k) $g = \frac{1}{5}$ (l) $b = 4\frac{1}{2}$
3. (a) $x = -2$ (b) $n = -\frac{1}{2}$
(c) $x = -1$ (d) $y = -3$
(e) $x = -1$ (f) $x = -3$
(g) $x = -1\frac{1}{2}$ (h) $x = -4$
(i) $x = -1\frac{1}{2}$

Review Exercise 13 — Page 151

1. (a) 4 (b) 3 (c) 6 (d) 18
2. 8
3. (a) 49 (b) 29
4. (a) 7 (b) 4
5. (a) $3x - 5$ (b) -8 (c) 7
6. (a) $x = 7$ (b) $x = 4$
7. (a) $x = 20$ (b) $x = 4$ (c) $n = 2$
8. (a) $y = 2$ (b) $t = -3$
   (c) $g = \frac{1}{2}$ (d) $x = \frac{3}{5}$
9. (a) $a = 5$ (b) $b = -3$ (c) $c = 2$
10. $x = 12$
11. (a) $x = -4$ (b) $y = \frac{1}{2}$
12. (a) $x = -\frac{4}{5}$ (b) $y = -\frac{1}{2}$

CHAPTER 14

Exercise 14.1 — Page 152

1. $a = 3$
2. $x = 2$
3. $m = 3$
4. $y = \frac{1}{2}$
5. $y = -5$
6. $k = 15$
7. $x = 3$
8. $w = \frac{1}{2}$
9. $n = -\frac{4}{5}$
10. $m = -2$
11. $g = -1\frac{1}{2}$
12. $p = 4\frac{1}{2}$
13. $n = -5$
14. $y = 1\frac{1}{2}$
15. $d = -2\frac{1}{2}$

Exercise 14.2 — Page 153

1. (a) $x = 3$ (b) $a = 2$
   (c) $t = 2$ (d) $y = 0$
   (e) $e = 5$ (f) $x = 2$
2. (a) $p = 5$ (b) $c = 6$
   (c) $x = 3$ (d) $y = 9$
   (e) $g = 11$ (f) $q = 8$
3. (a) $a - 4$ (b) $b = 6$
   (c) $c = 1$ (d) $d = 9$
   (e) $e = 5$ (f) $f - 4$
4. (a) $w = 2$ (b) $s = 3$
   (c) $x = 2$ (d) $g = 3$
   (e) $q = 4$ (f) $t = 3$
   (g) $w = 4$ (h) $x = 3$
   (i) $y = 5$
5. (a) $p = -1$ (b) $d = -2$
   (c) $g = -2$ (d) $x = 8\frac{1}{2}$
   (e) $y = \frac{2}{5}$ (f) $t = \frac{1}{2}$
   (g) $t = 1\frac{3}{4}$ (h) $a = 2\frac{1}{2}$
   (i) $m = 2\frac{3}{5}$

Exercise 14.3 — Page 153

1. (a) $x = 5$ (b) $q = 2$
   (c) $t = 3$ (d) $e = 3$
   (e) $g = 4$ (f) $y = 1$
   (g) $x = 2$ (h) $k = 1$
   (i) $a = 4$ (j) $p = 6$
   (k) $m = 2$ (l) $d = 5$
   (m) $y = 5$ (n) $u = 3$
   (o) $q = 0$
2. (a) $d = 8$ (b) $q = 3$
   (c) $c = 2$ (d) $t = 3$
   (e) $w = 2$ (f) $e = 3$
   (g) $g = 5$ (h) $z = 4$
   (i) $m = 6$ (j) $a = 5$
   (k) $x = 4$ (l) $y = 3$
3. (a) $m = -4$ (b) $t = -2$
   (c) $p = -2$ (d) $x = 3\frac{1}{2}$
   (e) $a = \frac{1}{2}$ (f) $b = \frac{4}{5}$
   (g) $y = \frac{4}{5}$ (h) $d = \frac{3}{4}$
   (i) $f = -3\frac{1}{2}$
4. (a) $x = 4$ (b) $a = 1\frac{2}{3}$
   (c) $m = \frac{1}{2}$ (d) $a = -5$
   (e) $y = 5\frac{1}{2}$ (f) $n = -1\frac{1}{2}$
   (g) $d = -4\frac{1}{2}$ (h) $k = -11$
   (i) $t = -4$ (j) $q = 2$
   (k) $x = \frac{2}{3}$ (l) $a = 2\frac{1}{2}$

Exercise 14.4 — Page 154

1. (a) $12x$ (b) $x = 15$
2. (a) $6k$ kg (b) $2\frac{1}{2}$ kg
3. (a) $2x$ pence (b) 64 pence
4. (a) $(3y + 15)$ cm (b) $y = 8$
5. (a) $6a + 3$ (b) 7
6. (a) $(n - 7)$ years old
   (b) Dominic is 25 years old.
       Marcie is 18 years old.
7. (a) $(4y - 2)$ cm (b) 19 cm
8. (a) (i) £$(p + 4)$ (ii) £$(p - 3)$
       (iii) £$(3p + 1)$
   (b) Aimee £12, Grace £8, Lydia £5
9. (a) $(x + 10)$ pence (b) $(3x + 20)$ pence
   (c) 15 pence

Exercise 14.5 — Page 156

1. $x = 3.8$
2. (a) $w = 4.2$
   (b) $x = 2.3$
3. $x = 3.2$
4. $x = 9.4$
5. $x = 4.7$ cm

1. (a) $x = 4$ (b) $y = -1\frac{1}{2}$

2. (a) $x = 9\frac{1}{2}$ (b) $x = 2$

3. (a) $x = 7$ (b) $x = 2\frac{1}{2}$

4. (a) $x = 8$ (b) $x = 1\frac{3}{5}$
    (c) $x = 5$

5. (a) $x = 5$ (b) $x = -1$

6. (a) $x = 1\frac{1}{4}$ (b) $x = 1\frac{1}{2}$

7. (a) $x = 7$ (b) $x = -2$

8. $x = -1$

9. (a) $q = -2$ (b) $t = 1\frac{1}{2}$

10. (a) 13 years old   (b) 26 years old

11. (a) $(9x + 4)$ cm   (b) $x = 6$
     (c) 24 cm

12. (a) (i) $(x + 7)$ pence   (ii) $(3x + 7)$ pence
     (b) $x = 30$, cake costs 37p

13. $2\frac{1}{2}$ litres   **14.** $x = 6.7$ **15.** $x = 2.59$

CHAPTER 15

1. (a) 5   (b) 2   (c) 12   (d) 9

2. (a) 10   (b) -2 (c) 10   (d) 25

3. (a) 8   (b) 0   (c) 12   (d) 32

4. (a) 24   (b) -3 (c) 2
    (d) 18   (e) 18

5. (a) 15   (b) 0   (c) 2
    (d) 50   (e) 15

6. (a) 27   (b) -3 (c) $2\frac{1}{2}$
    (d) 90   (e) 54

1. (a) -1 (b) -4 (c) -12 (d) 3

2. (a) 0   (b) -8 (c) -10 (d) -16

3. (a) -8 (b) -8 (c) -12 (d) 2

4. (a) 12   (b) -9 (c) -2
    (d) -18 (e) 6

5. (a) -5 (b) -20 (c) -2
    (d) -50 (e) -45

6. (a) 3   (b) -21 (c) $-2\frac{1}{2}$
    (d) -90 (e) 54

1. (a) $5y$ pence   (b) $(y + 8)$ pence

2. $12e$

3. (a) $(a + 1)$ years old   (b) $(a - 4)$ years old
    (c) $(a + n)$ years old

4. $b - 3$

5. $(h + 12)$ cm

6. (a) $2d$ (b) $2d + 5$

7. (a) $P = y + 5$ (b) $P = y - 2$
    (c) $P = 2y$

8. Ben: $A = d - 2$ Charlotte: $A = 2d$
    Erica: $A = \dfrac{d}{2}$

9. (a) $P = 4g$ (b) $P = 4y + 4$
    (c) $P = 3x - 1$ (d) $P = 2a + 2b$

10. $C = 25d$

11. (a) £44   (b) £80   (c) $C = 12x + 8$

12. (a) 115   (b) 175   (c) 45
    (d)

n	$n + 1$	$n + 2$
	$n + 11$	
	$n + 21$	

    (e) $S_n = 5n + 35$

1. £48   **7.** (a) $M = -7$ (b) $n = 4\frac{1}{2}$

2. 17 points   **8.** (a) $H = 2.5$ (b) $g = 6$

3. £190   **9.** 33

4. 26 cm   **10.** 96 m

5. $T = 97$ **11.** 86°F

6. $X = 8$ **12.** 138 minutes

1. (a) $F = 35$ (b) $F = 75$
    (c) $F = -15$

2. (a) $V = 26$ (b) $V = 2$
    (c) $V = 16$

3. (a) $P = 3$ (b) $P = -9$
    (c) $P = 12$

4. (a) $C = 104$ (b) $C = 32$
    (c) $C = -24$

5. (a) $S = 40$ (b) $S = -2$
    (c) $S = 6$

6. (a) $T = 35$ (b) $T = -2$
    (c) $T = -18$

7. (a) $K = 11$ (b) $K = 13$

8. (a) $L = 2$ (b) $L = -11$

9. (a) $S = 9$ (b) $S = 9$
    (c) $S = 100$

10. (a) $R = 15$ (b) $R = 3$

11. $K = 36$

12. (a) $S = 18$ (b) $S = 18$
    (c) $S = 200$

13. (a) $S = 36$ (b) $S = 36$
    (c) $S = 400$

14. $T = 39$

15. (a) $A = 8$ (b) $A = 27$
    (c) $A = 64$

16. (a) $S = 16$ (b) $S = 54$
    (c) $S = 128$

17. $25°C$ **19.** $F = 4100$

18. 240 volts   **20.** $F = 144$

1. (a) £13   (b) $C = 2k + 3$
    (c) 2 km

2. (a) 140 minutes   (b) $C = 40k + 20$
    (c) 2 kg

3. (a) $F = 42$ (b) $F = 2C + 30$
    (c) $C = 14$

4. (a) 410   (b) $b = 3n + 50$
    (c) 140

5. (a) £51   (b) $T = 15d + 6$
    (c) 6 days

6. (a) £295   (b) $C = 45n + 70$
    (c) 9 days

1. (a) $m = a - 5$ (b) $m = a - x$
    (c) $m = a + 2$ (d) $m = a + b$

2. (a) $x = \dfrac{y}{4}$ (b) $x = \dfrac{y}{a}$
    (c) $x = 2y$ (d) $x = ay$

    (e) $x = \dfrac{5y}{3}$

3. (a) $p = \frac{1}{2}y - 3$ (b) $p = \dfrac{t - q}{5}$

    (c) $p = \dfrac{m + 2}{3}$ (d) $p = \dfrac{q + r}{4}$

4. $n = \dfrac{C - 35}{24}$

5. $R = \dfrac{V}{I}$

6. (a) $d = \dfrac{P}{4}$ (b) $d = 0.7$ cm

7. (a) $l = \dfrac{A}{b}$ (b) $l = 6$ cm

8. (a) (i) $D = ST$ (ii) $D = 96$ km
    (b) (i) $T = \dfrac{D}{S}$ (ii) $T = 2.5$ hours

9. (a) $b = \frac{1}{2}P - l$ (b) $b = 4.2$ cm

10. (a) $x = \dfrac{y - c}{m}$ (b) $x = 0.5$

11. (a) $t = \dfrac{v - u}{a}$ (b) $t = 2$

12. (a) $b = \dfrac{2A}{h}$ (b) $b = 6.4$

1. 16   **6.** $S = -14$

2. (a) 7 (b) -1 (c) 12   **7.** $P = 7.5$

3. £56   **8.** 108

4. $V = 17$ **9.** 6

5. 4   **10.** $T = 60$

11. (a) (i) £130   (ii) 9 days
    (b) $T = 25d$
    (c) Andy's: £153, Belinda's: £150
      Belinda's is cheaper by £3.

12. (a) 380 euros   (b) $e = 1.52p$

13. (a) 340   (b) $N = 2T + 20$
    (c) 140

14. (a) £276   (b) £$(3M + 60)$
    (c) (i) $3M + 60 = 186$ (ii) 42 miles

15. 7 rolls

16. $p = \dfrac{t + 50}{7}$

17. (a) $y = 3$ (b) $x = \dfrac{y - c}{m}$

CHAPTER   **16**

1. (a) 17, 21, 25   (b) 14, 16, 18
    (c) 16, 13, 10   (d) 28, 33, 38
    (e) 48, 96, 192   (f) $1\frac{1}{2}$, $1\frac{3}{4}$, 2
    (g) 2, 1, $\frac{1}{2}$ (h) 0.9, 1.0, 1.1
    (i) 2, 0, -2 (j) 5, 2.5, 1.25
    (k) 21, 28, 36   (l) 29, 47, 76

2. (a) 8, 14   (b) 10, 22
    (c) 8, 32   (d) 16, -2
    (e) 16, 36   (f) 8, 21
    (g) 2, 20, 26

3. (a) Add 7; 37, 44
    (b) Add 2; 13, 15
    (c) Add 4; 21, 25
    (d) Subtract 5; 11, 6
    (e) Divide by 2; 2, 1
    (f) Multiply by 3; 81, 243
    (g) Subtract 2; -10, -12
    (h) Subtract 3; -5, -8

4. (a) 28
    (b) Keep on adding 3 to the last term until
      you get to the 10th term, 28.

5. (a) David multiplies the last term by 2,
      $4 \times 2 = 8$
      Tony adds the next counting number,
      $4 + 3 = 7$
    (b) 512   (c) 46

6. No.
    To find the next number, add 6 to the last term.
    All numbers in the sequence will be odd.

1. (a) 1, 5, 9, 13, 17
   (b) 1, 2, 4, 8, 16
   (c) 40, 35, 30, 25, 20
   (d) 4, 5, 7, 11, 19
   (e) 47, 23, 11, 5, 2
   (f) 2, 6, 4, 5, 4.5

2. (a) 2, 4, 8   (b) 6, 12, 24
   (c) -2, -4, -8

3. (a) 4, 10, 22   (b) 8, 18, 38
   (c) -4, -6, -10

4. (a) (i) 21 (ii) 3   (b) 37

5. (a) (i) 36 (ii) 123   (b) 10

6. (a) 37, 60   (b) 1, 4

7. -6, -27

8. (a) 49, 97   (b) 1537

1. (a) 3 (b) 3 (c) 6 (d) 8
   (e) -2 (f) -4

2. (a) $3n$ (b) $8n$ (c) $12n$ (d) $2n$

3. (a) 3 (c) $3n + 1$ (d) 25

4. (a) 2 (c) $2n + 7$ (d) 47

5. (a) -4 (c) $24 - 4n$

6. (a) $3n - 2$ (b) $22 - 3n$ (c) $4n + 1$
   (d) $4n$ (e) $2n - 1$ (f) $4n + 3$
   (g) $8 - 2n$ (h) $3n + 2$ (i) $5n - 2$
   (j) $45 - 5n$ (k) $n - 1$ (l) $2n - 3$

7. 1st term is: $1 \times 1 + 3 = 4$
   2nd term is: $2 \times 2 + 3 = 7$
   3rd term is: $3 \times 3 + 3 = 12$

1.

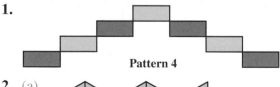

Pattern 4

2. (a)

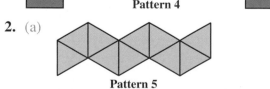

Pattern 5

   (b) Entries are: 2, 4, 6, 8, 10, 12
   (c) 27 is an odd number   (d) $2p$ (e) 24

3. (a)

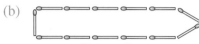

   (b) 4   **Pattern 4**
   (c) Entries are: 5, 9, 13, 17, 21
   (d) $4n + 1$ (e) 81

4. (a) (i) 15 (ii) 30 (iii) 300   (b) $3n$

5. (a) 11 (b) Pattern 7 (c) $m = 2p + 1$

6. (a) (i) 18 (ii) 82   (b) Pattern 7
   (c) Matches in length of pattern and number
   of verticals both increase by 2 every time.
   (d) $4n + 2$

7. (a) (i) 51   (ii) 100
   (b) (i) $x + 1$ (ii) $2x$

8. (a) $22 \, cm^2$ (b) $42 \, cm^2$
   (c) $4n + 2 \, cm^2$ (d) 9

9. (a) 27 (b) $T = 5n + 2$ (c) 52 (d) 15

1. (a) (i) 16 (ii) 32 (iii) -6
   (b) (ii) Double the last number.
   (iii) Subtract 4 from the last number.

2. (a) (i) 8, 4 (ii) Divide last term by 2.
   (b) 67, 202   (c) 5, 16, 8

3. (a) Missing entry is: 11
   (b)

Pattern 5

   (c) Pattern 11   (d) 203

4. (a) 38   (b) $\frac{1}{4}$

5. (a) (i) 31
   (ii) Add 6 to last number, $6n - 5$
   (iii) 55
   (b) (i) 29 (ii) 1

6. (a) 6, 7 (b) No. $\frac{8 + 4}{2} = 6$, $\frac{4 + 6}{2} = 5$

7. (a) 14 (b) 59 (c) $3n - 1$

8. (a) 33, 45
   (b) (i) 65
   (ii) Add twice the difference between
   the two previous terms.
   (c) $2n + 1$

9. (a) $3n$ (b) $3n - 2$

10. (a) 30 (b) 40 is not a multiple of 3.
   (c) $3p$

11. (a) 17 (b) 43 (c) $2n + 3$

12. (a) 28
   (b) Add on one more to last term each time,
   add on 2, 3, 4, 5, 6, 7, …
   (c) Triangular numbers.

CHAPTER   (17)

1. (a) $A(1, 3)$, $B(2, 1)$
2. $P(-2, -3)$, $Q(3, -2)$, $R(-3, 1)$
3. (b) $(2, 4)$
4. (b) $(-1, 1)$
5. (c) $D(-1, 2)$

Exercise **17.2** — Page 181

1. (1) $x = 1$ **2.**
   (2) $x = -3$
   (3) $y = 4$
   (4) $y = -1$
   (5) $y = x$

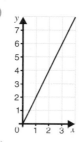

3. (a)

	x	0	1	2	3
(i)	y	2	3	4	5
(ii)	y	0	2	4	6
(iii)	y	0	-1	-2	-3
(iv)	y	2	1	0	-1

   (b) (i)   (ii)

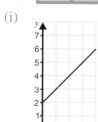

   (iii)   (iv)

4.

	x	0	1	2	3	4
(a)	y	-1	0	1	2	3
(b)	y	1	3	5	7	9

5. (b) $(0, -2)$

6. (a)

x	-1	1	3
y	5	1	-3

   (c) $(1.5, 0)$

7. (a) Missing entries are: 6, 4, 3
   (b) (i) $y = 2.5$ (ii) $y = 4.5$
8. (b) (i) $y = -5$ (ii) $x = 0.5$
9. (b) $x = 0.5$
10. (b) (i) $y = 5$ (ii) $x = 2.5$ (iii) $x = -1.5$

Exercise **17.3** — Page 183

1. (b) Same slope, parallel.
   y-intercept is different.
2. (a) gradient 3, y-intercept -1
3. $y = 3x$, $y = 3x + 2$
4. gradients: 4, 3, 2, -2, $\frac{1}{2}$, 2
   y-intercepts: 3, 5, -3, 4, 3, 0
5. (a) $y = 5x - 4$ (b) $y = -\frac{1}{2}x + 6$
6. (1) **C**   (2) **D**   (3) **B**   (4) **A**

7. (a) $y = x - 2$ (b) $y = 2x - 2$
   (c) $y = -2x - 2$
8. (c) 1   (d) $(0, -4)$ (e) $y = x - 4$
9. (c) $y = -2x - 1$
10. (a) Line slopes up from left to right.
   (b) Line slopes down from left to right.
   (c) Line is horizontal.
11. $y = 3x$
12. (a) £25   (b) £15 per hour
   (c) $y = 15x + 25$ (d) £145
13. (a) £3   (b) £2
   (c) $f = 2d + 3$ (d) £13
14. (a) 8 mm   (b) 0.06
   (c) $l = 0.06w + 8$ (d) 26 mm

Exercise **17.4** — Page 186

1. (a) $(0, 7)$ (c)
   (b) $(7, 0)$

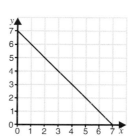

2. (a)

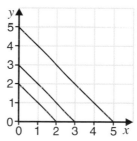

   (b) Parallel lines, same gradient.
3. (a) $(0, 6)$ (c)
   (b) $(2, 0)$

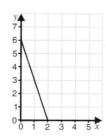

4. (a)   (b)

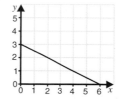
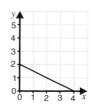

   (c)

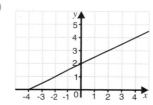

499

5. (a) (0, 5) (c)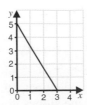
   (b) (3, 0)

6. (a) 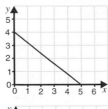 (b)

   (c)

Exercise 17.5   Page 187

1. (a)

x	1	2	3
$y = x + 2$	3	4	5
$y = 5 - x$	4	3	2

   (c) (1.5, 3.5) (d) $x = 1.5$
2. (b) (2.5, 8.5) (c) $x = 2.5$
3. (b) $x = 0.5$
4. $x = 1$
5. (b) $y = 9$ (c) $x = 3$

Review Exercise 17   Page 188

1. (a) $P(3, 2)$ (b) $Q(-1, 3)$
2. (a) $A(0, 2)$ (b) $D(-2, 0)$
3. (a) $M(1, -1)$ (b) $N(-3, -1)$
4. (a) (1) $y = 2$ (2) $x = -4$ (3) $y = x$
   (b) $(-4, 2)$
5. (b) $R(1, 1)$
6. A: R, B: Q, C: S, D: P
7. (a) 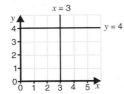 (b) (3, 4)

8. (a)

x	-3	0	2
y	0	3	5

   (b)
   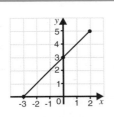

9. (b) $x = 1.75$
10. (a) Missing entries are: 3, 2, 0
   (b) 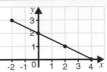 (c) $P(2.5, 1.5)$

11. (b) (2, 4)
12.

x	-2	0	2	4
y	3	2	1	0

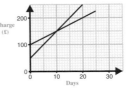

13. (a)

   (b) 10 days (c) Scaffold Ltd, £50

CHAPTER 18

Exercise 18.1   Page 190

1. (a) 4 inches (b) 25 centimetres
   (c) 40 centimetres
2. (a) 12.80 dollars (b) £6.25
3. (a) 5.6 km (b) 3.1 miles
4. (a) 8.8 pounds (b) 6.8 kg
5. (a) 10°C (b) 167°F

Exercise 18.2   Page 192

1. (a) 1005 (b) 24 miles (c) 3
2. (a) 1042 (b) 28 km
   (c) (i) 1115 (ii) 8 km
3. (a) 125 km
   (b) 1 hour
   (c) Faster going to Leeds - graph steeper.
4. (a) 50 km/h (b) 20 m/s
   (c) 9 miles/hour
5. (a) 0930 (b) 2 hours
   (c) 30 miles/hour
6. 18 km/h
7. (a) 10 km/h (b) 6.7 km/h
   (c) 8 km/h
8. (a)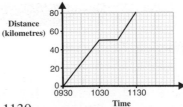

   (b) 1130

9. (a) 9 miles/hour
    (b) (i)

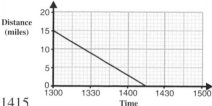

    (ii) 1415

10. (a) (b) (i)

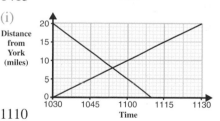

    (b) (ii) 1110

Exercise 18.3   **Page 195**

1. (a) 80   (b) 40
    (c) 1200, 1400
    (d) 1315 and 1400
    (e) 140
2. (a) 4 kg   (b) 6.5 cm
3. (a) Brian   (b) Afzal
    (c) 600 m   (d) Afzal
4. (a) £140   (b) 35 km   (c) £60
5. (a) 25 m   (b) 100 m
    (c) 2 minutes 20 seconds

Review Exercise 18   **Page 197**

1. (a) 17.5 cm   (b) 10 inches   (c) 90 cm
2. (b) £12.50   (c) 96 euros
3. (a) £70   (b) £40   (c) £30
4. (a) (i) 2   (ii) 45 minutes
    (b) (i) 12.5 km/h   (ii) 7 km/h
        (iii) 10 am and 12 noon

5. (a)

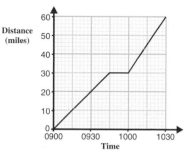

    (b) 40 miles/hour
6. (a) 0945   (b) 38 miles
    (c) 2   (d) 23 miles/hour

CHAPTER   19

Exercise 19.1   **Page 200**

1. (a) True   (b) True   (c) True
    (d) False   (e) True   (f) False
    (g) False   (h) True

2. (a) E.g. 5, 4, 3, …
    (b) E.g. $-2, -1, 0$ …
    (c) E.g. 5, 4, 3, …
    (d) E.g. 7, 6, 5, …
    (e) E.g. 11, 12, 13, …
    (f) -1
    (g) 5
    (h) One of: $-6, -5, -4, -3, -2$
3. (a) 2, 3, 4   (b) $-1, 0, 1, 2, 3$
    (c) $-4, -3, -2, -1$ (d) $-1, 0, 1, 2$
4. (a) $x \geqslant 2$ (b) $x < 1$
    (c) $-1 \leqslant x \leqslant 5$ (d) $-6 \leqslant x < -2$
    (e) $-2 < x < 1$ (f) $x < 5$ and $x \geqslant 8$
5. (a)
    (b)
    (c)
    (d)
    (e)
    (f)
    (g)
    (h)

Exercise 19.2   **Page 201**

1. (a) $n > 2$ (b) $x < -2$
    (c) $a < 4$ (d) $a < 4$
    (e) $d \leqslant 3$ (f) $t < -3$
    (g) $g > -2$ (h) $y \geqslant 0$
2. (a) $a < 4$ (b) $x \geqslant -2$
    (c) $y < -3$ (d) $c > 5$
    (e) $d < -3$ (f) $b \geqslant 1$
    (g) $b \leqslant 1$ (h) $c \leqslant 3$
    (i) $d > 4$ (j) $f < -2$
    (k) $g \leqslant \frac{1}{2}$ (l) $h < 2$
    (m) $x < -3$ (n) $j \geqslant 2\frac{1}{2}$
    (o) $k > -4$ (p) $m \leqslant 1\frac{2}{5}$

Exercise 19.3   **Page 202**

1. (a) $1 < x \leqslant 5$ (b) $-1 \leqslant x < 9$
    (c) $-7 < x \leqslant 4$
2. (a) $1 < x \leqslant 3$ (b) $-2 \leqslant x < 4$
    (c) $-2 \leqslant x \leqslant -1$ (d) $3 < x < 4\frac{1}{2}$
    (e) $-\frac{1}{3} \leqslant x \leqslant 4$ (f) $2 < x \leqslant 5$
3. (a) 6, 7, 8   (b) $-2, -1, 0, 1, 2, 3, 4$
    (c) 0, 1, 2   (d) 4, 5, 6, 7
    (e) 5   (f) 2, 3

Review Exercise **19** — Page 202

1. (a)

(b)

(c)

(d)

2. (a) $x < 3$ (b) $x \geqslant -5$
(c) $x \geqslant 4$ (d) $x \leqslant 3$

3. $x \geqslant 2$ **5.** $-3, -2, -1, 0, 1, 2, 3, 4$
4. $x < -1$ **6.** $-1, 0, 1$
7. (a) $x < 4$ (b) $-1 < x \leqslant 2$
(c) $1 \leqslant x < 3$ (d) $-3 < x < -1$
8. (a) $-1, 0, 1, 2, 3, 4$
(b) $-3, -2, -1, 0, 1$ (c) $-2, -1$
9. (a) $-4, -3, -2, -1, 0, 1$ (b) $x < \frac{1}{2}$
10. $-1 \leqslant x < 1$

CHAPTER **20**

Exercise **20.1** — Page 204

1. (a)

x	-3	-2	-1	0	1	2	3
y	8	3	0	-1	0	3	8

2. (a)

x	-3	-2	-1	0	1	2	3
y	7	2	-1	-2	-1	2	7

(c) $y = 0.25$ (d) $(-1.4, 0), (1.4, 0)$
3. (b) $y = -1.75$
4. Graphs have same shape, different positions.
5. (b) $y = 7.25$
(c) $x = -1.7$ and $x = 1.7$ (d) $(0, 1)$
6. (a)

x	-3	-2	-1	0	1	2	3
y	-3	2	5	6	5	2	-3

(c) $(-2.4, 0), (2.4, 0)$ (d) $(0, 6)$
7. (a)

x	0	1	2	3	4	5	6
y	0	2	6	12	20	30	42

(c) $x = 4.5$
9. (a)

x	-2	-1	0	1	2	3
y	8	4	2	2	4	8

10.

x	-5	-4	-3	-2	-1	0	1	2
y	8	2	-2	-4	-4	-2	2	8

Exercise **20.2** — Page 205

1. $x = 0$ or $x = 6$
2. $x = -3.6$ or $x = 0.6$
3. (a) Missing entries are: $1, -4, -7, -7, 1, 8$
(c) $x = \pm 2.8$

4. (b) $x = -1$ or $x = 0$ (c) $(-0.5, -0.25)$
5. (a) Entries are: 9, 4, 1, 0, 1, 4
(c) $x = 1$
6. (b) $x = -3$ or 1
7. (b) $x = \pm 3.2$ (c) $(0, 10)$
8. (b) $x = \pm 1.6$
9. (a) Entries are: $-3, 7, 13, 15, 13, 7, -3$
(c) $x = \pm 2.7$
10. (a) $x = \pm 3.2$ (b) $x = \pm 2.2$
(c) $x = 1$ or $x = 2$ (d) $x = \pm 2.4$
11. (a) Missing entries are: $4, -4, -2$
(c) $x = -0.4$ or $x = 2.4$
12. $x = -5$ or $x = 1$

Review Exercise **20** — Page 207

1. (a) $x = 0$ or $x = 4$ (b) $(2, -4)$
2. (a) Missing entries are: $4, -4, -4, -1$
(c) $x = \pm 2$
3. (b) $(0, 8)$ (c) $x = \pm 1.7$
4. (a) Missing entries are: $1, -1, 1, 5$
(c) $x = -0.6$ or 1.6
5. (b) Missing entries are: $1, -3, -2, 1, 6$
(c) $x = -0.7$ or $x = 2.7$
6. (b) $x = -4.2$ or $x = 1.2$
7. (a) Missing entries are: $5, -3$
(c) (i) Where the graph crosses the x axis.
(ii) $x = -0.2$
8. (b) $x = -0.5$ or $x = 1$

Algebra

Non-calculator Paper — Page 208

1. (a) (i) 39   (b)
(ii) -1

2. (a) 16   (b) 10
3. 120 minutes
4. (a)

Shape 4

(b) (i) Missing entries are: 21, 26
(ii) Add 5 to the last number.
(iii) 46. Multiply the shape number by 5, then add 1.
5. (a) $12t$ pence   (b) 1
6. (a) $6a$ (b) $a = 60°$
7. (b) $M(-1, 1)$
8. (a) $x = 5$ (b) $x = -2$ (c) $x = 3$
9. $(25x + 70y)$ pence

10. 7

11. (a) $x = 12$ (b) $x = 10$

12. (a) 18   (b) -10

13. (a) £22   (b) 9 days

14. (a) $12m$ (b) (i) $y = 8$ (ii) $a = 8$

15. (a) $3x$ pence   (b) $(x + 7)$ pence

16. (a) $6m + 2n$ (b) (i) $t = 15$ (ii) $p = 3$

17. (a) Missing entries are: 13, 15
    (b) 103. Double pattern number plus 3.

18. (a) -1 (b) 3

19. (a) 50   (b) 4   (c) 5

20. (a) (i) 2.5 miles   (ii) 6.4 kilometres
    (b) 5 miles = 8km, 20 × 8 km = 160 km

21. (a) $x = -1$ (b) $y = 5$

22. (a) (i) Missing entries are: 1, 5, 7, 9
    (c) $(-1, 1)$

23. (a) $5t$ pence   (b) $(t - 10)$ pence
    (c) $2t$ pence

24. (a) 24   (b) -1 (c) -5

25. (a) $4x + 11$ (b) $a = 54$

26. (a) $x = 2$ (b) $x = 4$

27. (a) (i) $3g$ (ii) g^3
    (b) (i) $g = 4.5$ (ii) $g = 4$ (iii) $g = 3.5$

28. (a) 18   (b) $2x - 1$ (c) $y = 4$

29. (a) 1000   (b) 5 km
    (c) 30 minutes   (d) 20 km/h

30. (a) Missing entries are: 6, 4, 3, 2, 1, 0

31. (a) (i) 25   (ii) -1.2
    (b) $3mn$
    (c) (i) $21x + 14y$ (ii) $a^2 - 3a$

32. (a) $x = 3$ (b) $y = 9$ (c) $z = 0.7$

33. (a) $2x$ pence   (b) $3(x + 5)$ pence
    (c) 16 pence

34. (c) $x = 3.5$

35. (a) $2(2a + b)$ (b) $t(3 - t)$

36. (a) (i) $x^2 - 5x$ (ii) $20p + 10$
    (b) (i) $3(2n + 3)$ (ii) $m(2m + 1)$
    (c) (i) $x = 100$ (ii) $x = -3\frac{1}{2}$

37. (a) $x < 4$ (b) $x \leqslant 1$

38. (a) c^4 (b) d^5 (c) e^7

39. (a) Missing entry is: -2 (c) $x = \pm 2.2$

40. (a) $y = 7$ (b) $x = \dfrac{y - c}{m}$

41. (a) Missing entries are: 3, 24
    (c) $x = 3.6$ cm

42. (a) $x = 3\frac{1}{3}$ (b) (i) 0, 1, 2 (ii) $x \geqslant \frac{1}{5}$

43. $x < 3$

1. (a) $(1, 2)$

2. (a) (i) 19   (ii) 39   (b) 5, -4

3. (a) (i) 31   (ii) Add 4 to the last number.
    (b) 3, -1 (c) 28

4. £42

5.

Input	Output
9	16
17	24

6. (a) (i) $x = 6$
          (ii) $x = 10$
    (b) $11g$
    (c) 0.5

7. 35. Take away the next counting number.

8. $P = 27.4$

9. (a) (i) 14, 11
        (ii) Subtract 3 from the last number
    (b) (i) 243, 729
        (ii) Multiply the last number by 3
        (iii) 6561

10. (a) 18   (b) $\frac{1}{2}$

11. (a) Missing entries are: 10, 13, 16
    (b) 3
    (c) 298. Three times pattern number minus 2.

12. (a) Even number.
    (b) Could be even or odd.
        When n is even, $3n$ is even, $3n + 1$ is odd.
        When n is odd, $3n$ is odd, $3n + 1$ is even.

13. (a) 70°F   (b) $F = 2C + 30$
    (c) $C = 35$

14. (a) (i) $4a$ (ii) $5a - b$ (iii) $3a^2$
    (b) $x - 6$ (c) $d - 29$

15. (a) 8 euros   (b) £75

16. (a) 8   (b) 7.5
    (c) Any two multiples of 4.
    (d) Starting numbers are multiples of 4.

17. (a) $18x$ pence   (b) $(3n + 2m)$ pence

18. **A** and **Q**, **B** and **P**, **C** and **S**, **D** and **R**

19. (a) (i) 40 miles.
        (ii) The cyclist was not moving.
        (iii) 10 miles per hour.
    (b) $2\frac{1}{2}$ hours.

20. (a) $7d - 4e$ (b) $12x + 28$ (c) $x = 0$

21. (a)

(b) $(3, 3)$

22. (a) $x = 2$ (b) $x = \frac{1}{2}$

23. (a) -1 (b) $2x - 7$

24. (a) $x = 6$ (b) $5x - 12$

25. (a) (i) 15 (ii) 1
     (b) $x = 3$

26. (a) $(5a + 3)\,$cm   (b) $5a + 3 = 23, \ a = 4$

27. (a) $2, 5$
     (b) 11th term
     (c) $(85 + 1)$ is not divisible by 3.

28. $G(0, 5), H(10, 0)$

29. (a) $3(2a + 1)$ (b) $t(1 - t)$

30. (a) 18 (b) $4n - 2$

31. (a) $x = -\frac{1}{2}$ (b) $m(m - 7)$ (c) $6, 7, 8$

32. $x = 2.6$

33. (a) Missing entries are: $-3, -6, -6$
     (c) $x = \pm 2.6$

34. (a) $3n$ (b) $3n + 1$

35. (a) $3x + 6$
     (b) (i) $x = 27$ (ii) $x = -\frac{4}{5}$
     (c) (i) m^5 (ii) n^3

36. $x = 4.1$

37. (a) $-1, 0, 1$ (b) $x \geqslant 5$

38. (a) (i) $4(2y + 1)$ (ii) $x(x^2 - 5)$
     (b) $x^5 y^2$
     (c) $x = \dfrac{y - 10}{5}$
     (d) (i) $t = 21$ (ii) $x = 8$

39. (a)

x	-1	0	1	2	3
y	5	2	1	2	5

     (c) $x = -0.4$ and 2.4

40. $x = 7.4$

CHAPTER 21

Exercise **21.1**   Page 216

1. $118°$.
Obtuse angles are greater than $90°$ but less than $180°$.

2. (a) $180°$ (b) $90°$ (c) $270°$
     (d) $90°$ (e) $120°$ (f) $6°$
     (g) $42°$ (h) $720°$ (i) $540°$

3. (a) $45°$ (b) $315°$

4. (a) $45°$ (b) $360°$

5. Acute: **A**, **G**
Obtuse: **B**, **C**, **F**, **H**
Reflex: **D**, **I**
Right: **E**

Exercise **21.2**   Page 217

1. (a) $39°$ (b) $118°$ (c) $42°$
3. (a) $217°$ (b) $234°$

Exercise **21.3**   Page 219

1. (a) $p = 65°$ (b) $p = 73°$ (c) $p = 36°$
2. (a) $q = 35°$ (b) $q = 144°$ (c) $q = 62°$
3. (a) $x = 110°$ (b) $x = 135°$ (c) $x = 30°$
4. (a) $y = 120°$ (b) $y = 150°$ (c) $y = 165°$
5. (a) $a = 35°$ (b) $b = 26°$ (c) $c = 105°$
6. (a) $a = 30°$ (b) $b = 150°$
     (c) $c = 42°$ (d) $d = 20°$
     (e) $e = 126°$ (f) $f = 203°$
     (g) $g = 133°, \ h = 47°$
     (h) $i = 112°$
     (i) $j = 127°, \ k = 53°, \ l = 37°$
     (j) $m = 96°$
     (k) $n = 47°, \ p = 43°$
7. (a) $x = 45°$ (b) $x = 30°$ (c) $x = 60°$
     (d) $x = 36°$ (e) $x = 40°$ (f) $x = 80°$
     (g) $x = 20°$ (h) $x = 30°$

Exercise **21.4**   Page 223

1. (a) $a = 65°$ (b) $b = 115°$
     (c) $c = 105°$ (d) $d = 100°$
2. (a) $a = 130°, \ b = 130°$
     (b) $c = 60°, \ d = 120°$
     (c) $e = 40°, \ f = 40°$
     (d) $g = 65°, \ h = 65°$
3. (a) $a = 63°$ (b) $b = 68°, \ c = 112°$
     (c) $d = 87°$ (d) $e = 124°$
     (e) $f = 65°$ (f) $g = 54°$
     (g) $h = 113°$ (h) $i = 124°$
4. (a) $a = 125°, \ b = 125°$
     (b) $c = 62°, \ d = 118°, \ e = 62°$
     (c) $f = 74°, \ g = 106°$
     (d) $h = 52°, \ i = 128°$
5. (a) $m = 84°, \ n = 116°$
     (b) $p = 56°, \ q = 116°$
     (c) $r = 61°$
     (d) $s = 270°$

Exercise **21.5**   Page 224

1. (a) $\angle BAC$ (b) $\angle RQS$ (c) $\angle XZY$
2. $a = \angle QPS, \ b = \angle PQS, \ c = \angle RQS,$
$d = \angle QRS, \ e = \angle QSR, \ f = \angle PSQ$
3. (a) $93°$ (b) $108°$ (c) $52°$
     (d) $110°$ (e) $65°$ (f) $295°$
     (g) $283°$ (h) $326°$
4. (a) (i) $43°$ (ii) supplementary angles
     (b) (i) $125°$ (ii) vertically opposite angles
     (c) (i) $63°$ (ii) corresponding angles
5. (a) $132°$ (b) $126°$ (c) $141°$
     (d) $85°$ (e) $65°$
     (f) $\angle QSP = 105°, \ \angle STU = 105°$
6. (a) $\angle AOB = 153°, \ \angle COD = 37°$
     (b) $\angle QTU = 48°, \ \angle QTS = 132°$
     (c) reflex $\angle TUV = 280°$

1. (a) (i) *d*   (ii) *a, c*   (iii) *b*   (iv) *e, f*
    (b) *a* = 106°, *b* = 90°, *c* = 111°,
        d = 53°, *e* = 307°, *f* = 254°

2. (a) *BC* and *DE*
    (b) *CD*
    (c) ∠*AED*, ∠*ABC*

3. (a) *a* = 57°, supplementary angles
    (b) *b* = 97°, angles at a point = 360°
    (c) *x* + 5*x* = 180° (supplementary angles)
              6*x* = 180°
               x = 30°

4. (a)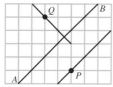

    (b) (i) *x* = 57°   (ii) obtuse

5. (a) *x* = 65°, alternate angles
    (b) *y* = 70°, allied angles

6. (a) *p* = 135°   (b) *q* = 45°
    (c) *r* = 55°

7. (a) ∠*AOC* = 78°   (b) ∠*AOD* = 75°

8. (a) *a* = 71°   (b) *b* = 63°
    (c) *c* = 117°

CHAPTER   22

1. (a) Yes   (b) Yes   (c) No
    (d) No   (e) Yes   (f) No

2. (a) Yes, obtuse-angled
    (b) No
    (c) Yes, acute-angled
    (d) Yes, right-angled
    (e) Yes, obtuse-angled
    (f) No

3. (a) *a* = 70°   (b) *b* = 37°
    (c) *c* = 114°   (d) *d* = 43°

1. *x* + 45° = 80° (ext. ∠ of a Δ)
    x = 80° − 45° = 35°

2. (a) *a* = 120°   (b) *b* = 110°
    (c) *c* = 50°   (d) *d* = 80°
    (e) *e* = 88°   (f) *f* = 55°

3. (a) *a* = 32°, *b* = 148°
    (b) *c* = 63°
    (c) *d* = 52°, *e* = 64°

4. Exterior angles of triangle are:
    180° − 27° = 153°, 180° − 85° = 95° and
    180° − 68° = 112°
    153°, 95° and 112° are all obtuse angles.

1. (a) Δ*PQT*   (b)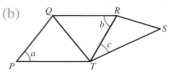
        Δ*QRT*
        Δ*RST*

2. (a) Isosceles   (b) Equilateral   (c) Δ*BCE*

3. (b) (i)   Acute-angled, scalene
        (ii) Right-angled, scalene
        (iii) Acute-angled, isosceles
        (iv) Obtuse-angled, isosceles

4. *AC* = *BC* ∠*BAC* = ∠*ABC*

5. (6, 2), (6, 8)

6. (a) *a* = 60°   (b) *a* = 85°
    (c) *a* = 75°   (d) *a* = 50°

7. *a* = 40°, *b* = 120°, *c* = 62°, *d* = 128°,
    e = 18°, *f* = 144°, *g* = 85°, *h* = 116°,
    i = 26.5°, *j* = 153.5°

8. (a) Equilateral   (b) 60°   (c) 78°

9. (a) Isosceles   (b) 74°   (c) 46°

10. (a) ∠*BCD* = 120°
    (b) ∠*PRQ* = 80°, ∠*QRS* = 160°
    (c) ∠*MNX* = 50°

5. (b) 9.3 cm   (c) 39°
6. (b) 5.9 cm   (c) 34°

1. 12 cm
2. (a) 13 cm   (b) 13.3 cm   (c) 19.9 cm
3. Δ*PQR* = 33 cm, Δ*QRS* = 32 cm,
    Δ*RST* = 35 cm. Δ*RST* has greatest perimeter
4. (a) *a* = 7 cm   (b) *b* = 3.8 cm   (c) *c* = 3 cm
5. (a) 4.5 cm²   (b) 10 cm²   (c) 6 cm²
6. (a) 9 cm²   (b) 6 cm²   (c) 3.6 cm²
7. (a) 7.2 cm²   (b) 4.16 cm²   (c) 11.52 cm²
8. (a) 3.8 cm²   (b) 8 cm²   (c) 3.24 cm²
9. (a) *h* = 6 cm   (b) *h* = 12 cm   (c) *h* = 4 cm
10. (a) *a* = 8 cm   (b) *b* = 4 cm   (c) *c* = 16 cm
11. 67.5 cm²
12. 60 cm

1. (b) (i) Isosceles   (ii) 8 cm²
    (c) Point *S* on line *y* = 5.
2. 132 cm²
3. (a) *x* = 116°   (b) *y* = 52°
4. (a) ∠*DCA* = 53°
    (b) ∠*BAC* = 50°
        ∠*ABC* = ∠*BCA* (Δ*ABC* is isosceles)
        ∠*BAC* = 180° − (2 × 65°)
6. (b) 16 cm²
7. *x* = 20°, ∠*DAC* = 40°
8. (a) 45 cm   (b) 75 cm²   (c) 10 cm

CHAPTER 23

3. (a) 5 (b) 2 (c) 0
4. 1, 1, 1, 2, 0
5. (a) 2 (b) 3 (c) 4 (d) 2 (e) 6 (f) 8
6. 1, 1, 3
7. (a)

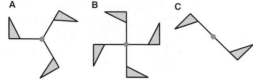

A   B   C

(b) **A**: 3, **B**: 4, **C**: 2
8. (a) MY (b) NZ (c) NXZ (d) JP
9. (a) 1 (b) (i)

10. (a) 1 (b) (i) (ii) 3

11. (a) (i) 0 (ii) 2
  (b) (i) 1 (ii) 1
  (c) (i) 2 (ii) 2
  (d) (i) 0 (ii) 4
  (e) (i) 1 (ii) 1

1. 9
2. (a) 4 (b) 4 (c) 2 (d) Infinite
3. (a) 2 (b) 2 (c) 4
4. (a) 4 (b) 1, 4
5. (a) 4 (b) 4

1. A, O; F, L; C, G; H, J; D, P
2. **D**, **E**
3. (a) $\triangle CED$ (b) $CBFE$
4. (a) $\triangle AXZ$ and $\triangle ZYC$, $\triangle BXZ$ and $\triangle ZYB$

1. **A**, **D** (SSS)
2. **A**, **D** (ASA)
3. (a) Yes, ASA (b) No
  (c) Yes, SAS (d) Yes, SSS
  (e) No (f) Yes, RHS
4. (a) No (b) No
  (c) Yes, RHS (d) No
  (e) Yes, ASA (f) Yes, ASA
5. **A** and **H**, ASA **B** and **F**, SSS
  D and **G**, RHS **E** and **I**, SAS

1.

2. (a) MATH (b) S
3. (a)

(b) (i) 0 (ii) 3

4. (a) **A** (b) **C**
5. (a) 5
6.

7. Triangles ABC and ADC are the same shape and size.
8. **A** and **G**, **D** and **F**
9. (a) $x = 70°$ (b) $y = 10\,cm$

1. (b) (i) Isosceles trapezium
    (ii) Parallelogram (iii) Rhombus
    (iv) Kite (v) Square
2. $M(1, 4)$
3. $S(3, 1)$
4. $C(5, 4)$
5. $Y(6, 4)$
6. $A(1, 3)$
7. $L(2, 3)$
8. $(3, 1), (3, 5)$
9. (a) $a = 90°$ (b) $a = 35°$
  (c) $a = 140°$ (d) $a = 114°$
10. (a) $a = 90°$, $b = 53°$, $c = 37°$
  (b) $d = 42°$, $e = 48°$
  (c) $f = 27°$, $g = 117°$
  (d) $h = 16°$, $i = 99°$, $j = 65°$
11. (a) $a = 36°$, $b = 144°$
  (b) $c = 124°$, $d = 56°$
  (c) $e = 130°$, $f = 23°$, $g = 23°$
  (d) $h = 42°$, $i = 84°$
12. (a) $a = 110°$, $b = 100°$ (b) $c = 95°$
  (c) $d = 46°$ (d) $e = 96°$
13. $a = 115°$, $b = 44°$
14. $\angle WXY = 72°$, $\angle XYZ = 108°$
15. (a) $a = 62°$ (b) $b = 54°$, $c = 36°$
  (c) $d = 62°$ (d) $e = 116°$, $f = 86°$
  (e) $g = 124°$ (f) $h = 75°$
  (g) $i = 38°$, $j = 42°$
  (h) $k = 55°$, $l = 45°$

Exercise 24.2

Page 254

1.

A	B	C	D	E	F	G	H	I
1	0	1	4	2	2	0	0	1
1	1	1	4	2	2	1	2	1

2. (a) 4   (b) 2   (c) 1
3. (a) 2   (b) 2   (c) 1
4. (b) 2   (c) 2

Exercise 24.3

Page 257

1. (a) (i) 12 cm   (ii) 8 cm²
    (b) (i) 14 cm   (ii) 12 cm²
    (c) (i) 16 cm   (ii) 15 cm²
2. (a) (i) 8 cm   (ii) 4 cm²
    (b) (i) 12 cm   (ii) 9 cm²
    (c) (i) 16 cm   (ii) 16 cm²
3. (a) **B** and **C**   (b) **B** and **D**
4. **A** 8 cm²   **B** 12 cm²   **C** 12 cm²
    D 9 cm²   **E** 8 cm²   **F** 12 cm²
5. (a) 8 cm   (b) 8 cm
    (c) 11.2 cm   (d) 5.6 cm
6. (a) 6 cm²   (b) 5.4 cm²
    (c) 11.5 cm²   (d) 7.92 cm²
7. (a) 49 cm²   (b) 5.76 cm²
    (c) 18.49 cm²   (d) 3.24 cm²
8. (a) 30 cm²   (b) 10 cm²   (c) 13.5 cm²
9. (a) 20 cm²   (b) 7.5 cm²   (c) 5.75 cm²
10. (a) $b = 4$ cm   (b) $b = 3$ cm   (c) $b = 2$ cm
11. (a) 3 cm   (b) 6 cm   (c) 8 cm
12. 14 m²   **14.** 4 cm
13. 84 cm²   **15.** 20 cm²   **16.** 4 cm

Review Exercise 24

Page 260

1. (a) **D**   (b) Kite   (c) 2   (d) **A**   (e) 4
2. (a) (i) 12 cm   (ii) **Q** and **S**
    (b) (i) 12 cm²   (ii) **P** and **S**
3. (b) (ii) $D(1, -4)$
4. $a = 82°$
5. (b) Trapezium
6. $Q(3, 6)$
7. (a) 1   (b) 56°
8. $a = 30°$, $b = 60°$, $c = 42°$, $d = 78°$
9. $\angle ADC = 180° - (87° + 44°) = 49°$
    $\angle ABC = \angle ADC$ (opp. ∠'s in a parallelogram)
    $\angle ABC = 49°$
10. 25 cm²
11. 7 cm
12. (a) 13.5 m²   (b) 8.1 m²
13. (a) 50 cm²   (b) 17.5 cm²
14. (a) If length is 10 cm then width is 5 cm.
      $A = lb$, so, $A = 5$ cm $\times$ 10 cm $= 50$ cm²
      So, length is 10 cm.
    (b) 30 cm

CHAPTER 25

Exercise 25.1

Page 264

1. (a) triangle   (b) quadrilateral
    (c) pentagon   (d) hexagon
2. (a) $x + 130° = 180°$ (supp. ∠'s)
        $x = 180° - 130° = 50°$
    (b) $y + 130° + 70° + 100° = 360°$
        (sum of ∠'s in a quad. = 360°)
        $y = 360° - 300° = 60°$
3. (a) $a = 63°$ (b) $b = 55°$, $c = 62°$
    (c) $d = 95°$, $e = 76°$
4. (a) $a = 48°$ (b) $b = 60°$
    (c) $c = 103°$ (d) $d = 100°$
5. (a) $a = 120°$ (b) $b = 62°$
    (c) $c = 76°$ (d) $d = 120°$, $e = 50°$
6. (a) 540°   (b) 720°   (c) 900°   (d) 1080°
7. (a) $a = 60°$ (b) $b = 120°$ (c) $c = 130°$
8. (a) $a = 199°$ (b) $b = 240°$, $c = 120°$
    (c) $d = 225°$, $e = 85°$

Exercise 25.2

Page 267

1. (a) (i) 120°   (ii) 90°   (iii) 60°   (iv) 45°
    (b) (i) 60°   (ii) 90°   (iii) 120°   (iv) 135°
2. 20
3. (a) 40   (b) 15   (c) 9   (d) 6
4. 8
5. (a) 5   (b) 20   (c) 40   (d) 4
6. (a) 72°   (b) 108°   (c) 540°
7. (a) $a = 90°$, $b = 60°$, $c = 210°$
    (b) $d = 90°$, $e = 120°$, $f = 150°$
    (c) $g = 90°$, $h = 135°$, $i = 135°$
    (d) $j = 105°$ (e) $k = 162°$
    (f) $l = 192°$ (g) $m = 132°$
    (h) $n = 96°$
8. (a) $a = 90°$, $b = 150°$
    (b) $c = 126°$, $d = 156°$, $e = 66°$
    (c) $f = 102°$
9. (a) $a = 60°$ (b) $b = 135°$, $c = 45°$
    (c) $d = 36°$, $e = 72°$

Exercise 25.3

Page 268

5. (b) (ii) 6   (c) 6
6. (a) (i) 5   (ii) 5
    (b) (i) 7   (ii) 7
    (c) (i) 10   (ii) 10

Review Exercise 25

Page 270

1. (a) Hexagon   (b)

    (c) 6

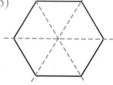

2. (a) Interior angle of equilateral triangle = 60°
At each vertex, 6 triangles will meet.
(c) $x = 135°$

3. (a) $x = 130°$ (b) $y = 100°$

4 (a) Line of symmetry is a perpendicular bisector.
(b) $y = 45°$ (c) $z = 125°$

5. (a) (i) Pentagon
(ii) Lengths of sides not all equal.
(b) (i)

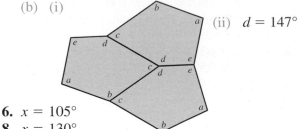

(ii) $d = 147°$

6. $x = 105°$

8. $x = 130°$

9. $\angle AED = 108°$ (int. $\angle$ of a pentagon)
$\angle CAE = 108° - 36° = 72°$
$\angle CAE + \angle AED = 180°$ (allied angles)
So, AC is parallel to ED.

10. (a) $x = 22.5°$, isosceles triangle
(b) Isosceles trapezium

11. (a) $q = 12°$ (b) 30 sides

12. (a) (i) $x = 36°$ (ii) $y = 18°$
(b) (i) parallel   (ii) $z = 36°$

CHAPTER 26

Exercise **26.1**   Page 272

1. (a) (i) Bike Hire Centre (ii) Bikers' Cafe
(iii) Bikers' Rest   (iv) High Peak
(b) (i) Highlands   (ii) North-East
(iii) Country Garden

2. (a) South   (b) East
(c) West   (d) South-West

3. (a) 45°   (b) 135°   (c) 90°
(d) 90°   (e) 180°

4. (a) South   (b) North

5. (a) North-West   (b) North-East

6. Missing entries are: South-East, South-East, South, North-West

Exercise **26.2**   Page 275

1. (a) 070°   (b) 180°   (c) 090°

2. (a) 065°   (b) 140°   (c) 249°
(d) 228°   (e) 300°   (f) 090°

3. (b) (i) 230°   (ii) 305°   (iii) 015°
(iv) 080°   (v) 125°   (vi) 355°

4.

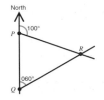

5. (a) 180°   (b) 070°   (c) 250°
(d) 043°   (e) 223°

6. (a) 315°   (b) 135°   (c) 285°
(d) 105°   (e) 230°   (f) 050°

7.

Exercise **26.3**   Page 277

1. 16.8 km

2. 3.2 cm

3. 1900 metres

4. (a) (i) 128°   (ii) 308°
(b) (i) 4.7 cm   (ii) 47 km

5. (a) 240 m   (b) 063°

6. (a) 48 km
(b)

7. 4.6 cm

8. (a) 2.5 cm   (b) 25 m

9. (a) 7 m   (b) 30 cm

10. (a) 13.5 km   (b) 072°   (c) 252°

11. (a) 065°   (b) 282°   (c) 1800 m

12. 70 cm

13. 19.8 km

14. (a) 8 km   (b) 114°   (c) 294°

15. (a) 6300 km   (b) 248°

16. (a) 50 km   (b) 347°   (c) 167°

Review Exercise 26   Page 281

1. (a) South   (b) North-East

2. (a) Orchard Lane   (b) West
(c) North-East   (d) North-West

3. 14.6 km

4. (a) $x = 133°$ (b) 047°
(c) (i) 295°   (ii) 227°

5. (a) 075°   (b) 120 km

6. (a) 230°   (b) 140°
(c) (i) 3.7 cm   (ii) 740 m

7. (b) 245°   (c) (i) 64 km   (ii) 197°

CHAPTER 27

Exercise **27.1**   Page 284

1. (a) 12 cm (b) 24 cm (c) 39 cm
2. (a) 15 cm (b) 30 cm (c) 38.4 cm
3. (a) 1p: 2 cm, 2p: 2.6 cm
   (b) 1p: 6 cm, 2p: 7.8 cm
4. (a) 37.7 cm (b) 22.0 cm (c) 47.1 cm
5. (a) 28.3 cm (b) 35.2 cm (c) 100.5 cm
6. 28 cm   10. 57.5 m
7. 75.4 cm   11. 5.03 m   14. 30 mm
8. 81.7 cm   12. 31.4 m   15. 60 cm
9. 40.8 cm   13. 3.8 cm   16. 67 m

Exercise **27.2**   Page 287

1. (a) 75 cm²   (b) 147 cm²   (c) 243 cm²
2. (a) 27 cm²   (b) 75 cm²   (c) 192 cm²
3. (a) 50 cm²   (b) 133 cm²   (c) 452 cm²
4. (a) 32.2 cm² (b) 45.4 cm² (c) 530.9 cm²
5. 22 167 cm²
6. 16 286 mm²   9. 0.79 m²
7. 491 cm²   10. 50.3 cm²   12. 7.14 m
8. 1.13 m²   11. 4.0 cm   13. 1.3 cm

Exercise **27.3**   Page 289

1. (a) 56.5 cm   8. (a) 27 m
   (b) 254.5 cm²   (b) 56.7 m²
2. (a) 26 cm   9. 31 cm
   (b) 55.4 cm²   10. (a) 207 cm
3. (a) 56.5 cm   (b) 4
   (b) 11.3 m   11. 24.5
4. 22.0 cm²   12. (a) 37.7 m
5. Circle: 50.3 cm²   (b) 17
   Semi-circle: 47.5 cm² 13. 18.8 cm
   The circle is bigger. 14. 18.2 cm²
6. (a) 225 π cm²   15. 38.6 cm
   (b) 30 π cm   16. 326 cm²
7. 21.5 cm²   17. 40.1 cm

Review Exercise **27**   Page 290

1. (a) 69.1 cm (b) (i) 11 cm (ii) 380 cm²
2. 707 cm²
3. π × 5² = 78.5 (less than 84)
   So, radius must be bigger than 5 m.
4. (a) 4.52 cm²   (b) 11.0 cm
5. (a) 13.8 m   (b) 15.2 m²
6. 2
7. 126 cm²
8. 1326
9. (a) 151 inches   (b) 1670 square inches
10. (a) 8.2 m   (b) 1.1 litres
11. (a) 16 π cm²   (b) 51.4 cm
12. (a) 50.9 cm   (b) 380 cm²

CHAPTER 28

Exercise **28.1**   Page 293

1. (a) M 17 cm²   (b) M
       A 12 cm²   (c) T
       T 9 cm²   (d) A and H
       H 12 cm²
       S 11 cm²
2. (a) 25 cm²   (b) 4 cm²   (c) 21 cm²
3. (a) 14 cm²   (b) 14 cm²   (c) 20 cm²
4. (a) 24 cm²   (b) 41 cm²   (c) 30 cm²
5. (a) 129 m²   (b) 105 cm² (c) 0.383 m²
6. 372 m²
7. (a) 58 cm²   (b) 13.7 cm²

Exercise **28.2**   Page 296

3. (a) 4   (b) 2 faces overlap
9. (a) Cube, 6, 8, 12   (b) Cuboid, 6, 8, 12
   (c) Pyramid, 5, 5, 8
   (d) Triangular prism, 5, 6, 9
10. (a) 8   (b) 5   (c) 12   (d) 120 cm

Exercise **28.3**   Page 298

1. (a) 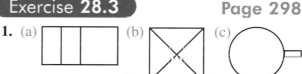 (b) (c)

2. (a)
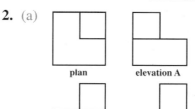
plan   elevation A
elevation B   elevation C

(b)
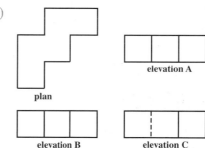
plan
elevation A
elevation B   elevation C

(c)
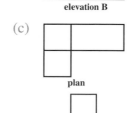
plan   elevation A
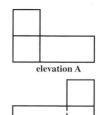
elevation B   elevation C

509

3.

 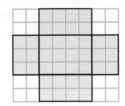

plan   elevation A

4. (a)

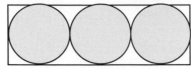

(b)

5. (a)   (b)

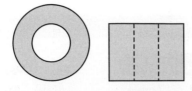

1. (b)   52 cm²   (c)   52 cm²
2. (a)   6 cm³   (b)   18 cm³   (c)   36 cm³
3. (a)   (i)   8   (ii)   27
       (iii)   64   (iv)   125
   (b)   (i)   24 cm²   (ii)   54 cm²
       (iii)   96 cm²   (iv)   150 cm²
4. (a)   27 cm³,   54 cm²   (b)   30 cm³,   62 cm²
   (c)   140 cm³,   166 cm²
5. (a)   4 cm³,   18 cm²   (b)   5 cm³,   20 cm²
   (c)   8 cm³,   28 cm²   (d)   7 cm³,   24 cm²
   (e)   14 cm³,   42 cm²   (f)   10 cm³,   32 cm²
6. (a)   150 cm³,   190 cm²
   (b)   51.8 cm³,   89.3 cm²
   (c)   19 440 cm³,   4644 cm²
   (d)   96.8 cm³,   132 cm²
   (e)   916 cm³,   612 cm²
7. 10 cm
8. 4.5 cm
9. 1155 cm²
10. 343 cm³

1. (a)   40 cm³   (b)   140 cm³   (c)   96 cm³
2. (a)   11 cm²,   15.4 cm³
   (b)   6 cm²,   15 cm³
   (c)   314 cm²,   6280 cm³
   (d)   3.14 cm²,   15.7 cm³
3. (a)   10 cm³   (b)   56 cm³   (c)   330 cm³
   (d)   113 cm³   (e)   393 cm³   (f)   120 cm³
   (g)   848 cm³   (h)   18 cm³
4. **P.**   **P** = 64π cm³,   **Q** = 63π cm³.
5. No.   π × 5 × 5 × 10 = 250π cm³
         π × 10 × 10 × 5 = 500π cm³

1. (a)   339 cm²   (b)   61.3 cm²   (c)   90.9 cm²
3. (a)   1260 cm²   (b)   6280 cm²
4. 274 cm²
5. (a)   14 100 cm²   (b)   18 800 cm²
6. 4 cm

1. **A** cube,   **B** cylinder,   **C** cone.
2. 5 faces, 9 edges, 6 vertices
3. **C**
4. (a)

 (b)   40 cm²

5. (a)   13   (b)   14 cm³
6.

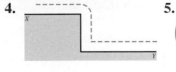

plan   elevation A

7. 20 cm²
8. (a)   26 cm³   (b)   3 cm
   (c)

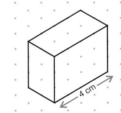

4 cm

   (d)   No.   Only two layers of 10 cubes will fit.
9. 787.5 m²   **14.** 1890 cm³
10. (a)   75 cm²   (b)   225 cm³   **15.** 216 cm³
11. (a)   54 cm³   (b)   12   **16.** 1040 cm³
12. 8510 cm²   **17.** 392 cm²
13. 350 cm³   **18.** 618

CHAPTER   **29**

4.   **5.**

6.

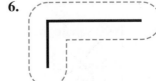

Page 312

5. Perpendicular bisectors pass through the centre of the circle.

7.

8.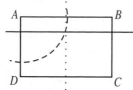

9.

12.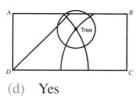

(d)   Yes

Review Exercise **29**

Page 317

1.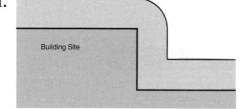

Building Site

2.

3. (a) (b)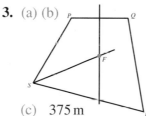

(c)   375 m

4. (b)   43 m

CHAPTER **30**

Exercise **30.1**

Page 319

1. (a) (b) (c)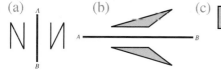

2. (a) 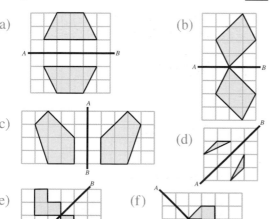 (b)

(c)

(d)

(e) (f)

3.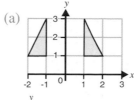

(a) (b)

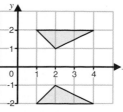

(c) 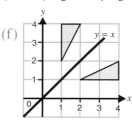 (d)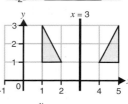

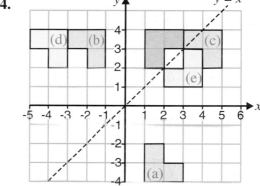

4.

5. (a)   (2, −1)   (b)   (−2, 1)   (c)   (0, 1)
   (d)   (2, −3)   (e)   (1, 2)

6. (a)   (3, −4)   (b)   (−3, 4)   (c)   (5, 4)
   (d)   (−5, 4)   (e)   (4, 3)

Exercise **30.2**

Page 321

1. (a) 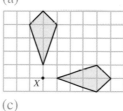 (b)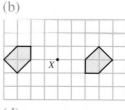

(c) 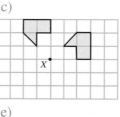 (d)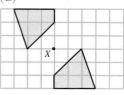

(e) 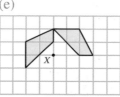 (f)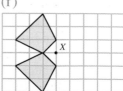

511

2. (a)

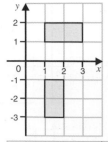

(b)

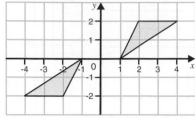

(c)

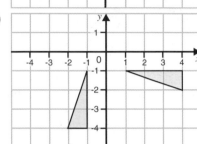

4. (a) $(4, -3)$ (b) $(-4, 3)$ (c) $(-3, -4)$
(d) $(6, 1)$ (e) $(0, 1)$ (f) $(3, -2)$

Exercise 30.3 Page 323

1.

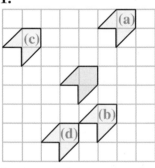

2.

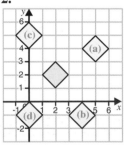

3. (a) $(5, 5)$ (b) $(1, 6)$ (c) $(4, 1)$ (d) $(1, 1)$
4. $T(7, 2)$

5. (a) $\binom{2}{1}$ (b) $\binom{1}{-2}$ (c) $\binom{-3}{1}$ (d) $\binom{-3}{-2}$

6. (b) $\binom{-3}{-2}$

Exercise 30.4 Page 325

1. (a) Translation (b) Reflection
(c) Rotation
2. (a) Rotation, $\frac{1}{2}$ turn about O
(b) Translation, 4 units right and
3 units down
3. (a) Reflection in x axis
(b) Rotation, 90° anticlockwise, about $(0, 0)$
(c) Translation, 3 units right and 2 units up
(d) Rotation, 180°, about $(0, 0)$
(e) Reflection in $x = 7$

4. (a) Reflection in y axis
(b) Rotation, 90° clockwise, about $(0, 0)$
(c) Translation, 5 units left and 5 units up
(d) Reflection in $y = x$
5. (a) Translation $\binom{-7}{-4}$
(b) Reflection in $y = -x$
(c) Rotation, 90° anticlockwise, about $(1, 0)$
6. (a) Reflection in $x = -1$
(b) Rotation, 90° clockwise, about $(1, -2)$
(c) Translation $\binom{5}{1}$

Exercise 30.5 Page 327

1. (c) Translation $\binom{8}{0}$
2. (c) Rotation, 90° anticlockwise, about $(5, 5)$
3. (d) Rotation, 90° anticlockwise, about $(0, 0)$
4. (c) Rotation, 180°, about $(1, -1)$
5. (c) Translation $\binom{-4}{3}$

Review Exercise 30 Page 328

1. (a) Translation (b) Rotation
2. (a) (b)

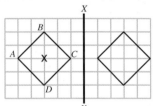

3.

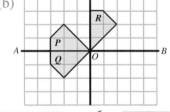

4. (a) (b)

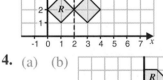

5.

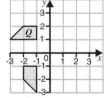

6.

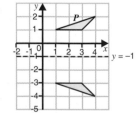

7. (a)

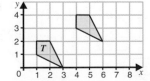

(b) Translation, 3 units left and 2 units down
8. (a) (i) **H** (ii) **G** (iii) **F**
(b) Reflection in the y axis

512

9. (a) Reflection in the *x* axis
(b) Rotation, 180°, about (0, 0)

10. (a) Translation $\begin{pmatrix} -6 \\ -3 \end{pmatrix}$
(b) Reflection in *y* = *x*
(c) Reflection in *x* = 3

11. (a) Reflection in *y* = −1
(b) Reflection, 90° clockwise, about (−1, −1)

12.

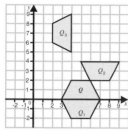

(d) Rotation, 90° anticlockwise, about (1, 3)

CHAPTER 31

Exercise 31.1 — Page 331

2. (a) 2   (b) 3   (c) $2\frac{1}{2}$ (d) $1\frac{1}{2}$

Exercise 31.2 — Page 332

3. (a) Enlarged shape: (4, 2), (6, 2), (2, 6)
(b) Enlarged shape: (3, 3), (9, 0), (9, 9)

5. (a) (6, 8)   (b) (9, 12)   (c) (6, 6)
(d) (3, 10)   (e) (5, 5)

Exercise 31.3 — Page 334

1. Scale factor 3, centre (0, 3)

2. (a) Scale factor 3, centre (0, 0)
(b) Scale factor 3, centre (5, 0)
(c) Scale factor 2.5, centre (0, 5)

3. Enlargement, scale factor 3, centre (0, 4)

4. Enlargement, scale factor 2, centre (1, 2)

Exercise 31.4 — Page 335

1. (a) Enlarged shape: (2, 2), (4, 1), (3, 4)
(b) Enlarged shape: (1, 1), (2, 0), (3, 2)

2. (a) Scale factor $\frac{1}{3}$, centre (5, 7)
(b) Scale factor $\frac{1}{3}$, centre (1, 7)
(c) Scale factor $\frac{2}{5}$, centre (5, 5)
(d) Scale factor $\frac{1}{2}$, centre (5, −2)
(e) Scale factor $\frac{1}{3}$, centre (0, 7)
(f) Scale factor $\frac{2}{3}$, centre (1, 1)

3. Enlargement, scale factor $\frac{1}{2}$, centre (0, 0)

Exercise 31.5 — Page 337

1. (a) Corresponding lengths not in same ratio.
(b) **P** and **R**

2. (a) Two circles   (d) Two squares

3. *a* = 4 cm, *b* = 24 cm

4. (a) Scale factor $= \frac{3}{2} = 1.5$ (b) *x* = 1.8 cm
(c) *a* = 120°

5. (a) ∠*ABC* = 77°, ∠*PRQ* = 35°
Both triangles have same angles.
(b) *AB* = 1.8 cm

6. 30 cm

7. (a) *x* = 1.5 cm, *y* = 2.4 cm, *a* = 70°
(b) *x* = 5 cm, *y* = 1.5 cm, *a* = 53°
(c) *x* = 30 cm, *y* = 17.5 cm, *z* = 10 cm

8. 15 cm

9. 2.8 cm

10. 18°

11. 5 cm

12. *x* = 16, *y* = 48

Review Exercise 31 — Page 339

2. Enlarged shape: (3, 3), (9, 3), (12, 6)

3. Enlargement, scale factor 2, centre (0, 0)

4. (a) Enlargement, scale factor 2, centre (0, 4)
(b) Enlargement, scale factor $\frac{1}{2}$, centre (0, 4)

6. 11.2 cm

7. (a) *x* = 4.5 cm   (b) *y* = 10 cm   (c) *z* = 46°

CHAPTER 32

Exercise 32.1 — Page 342

1. (a) 10 cm   (b) 25 cm   (c) 26 cm

2. (a) 7.8 cm   (b) 12.8 cm   (c) 10.3 cm

3. (a) $\sqrt{52}$ cm   (b) $\sqrt{20}$ cm

4. (a) 5   (b) 9.22   (c) 13
(d) 8.06   (e) 7.21

5. (a) *R* (9, 7)   (b) *X* (2, 3)
(c) *Y* (4, 2)   (d) 2.24

Exercise 32.2 — Page 343

1. (a) 8 cm   (b) 6 cm   (c) 2 cm

2. (a) 6.9 cm   (b) 10.9 cm   (c) 9.5 cm

3. 339 m

4. 36 cm²

5. 3.6 cm

Exercise 32.3 — Page 345

1. (a) 2.9 cm   **3.** 10 cm   **8.** 74.3 cm
(b) 5.7 cm   **4.** 8.5 cm   **9.** 13 cm
(c) 2.1 cm   **5.** 10.6 cm   **10.** 24 cm
(d) 2.0 cm   **6.** 15 cm   **11.** (a) 11.4 cm
2. 17 cm   **7.** 6.9 cm   (b) 43.5 cm²

Review Exercise 32 — Page 346

1. (a) 2.95 m   (b) *X*   **5.** 361 m

2. 28.3 km   **6.** (a) 154°

3. 8.60   (b) 2.81 km

4. 41² = 40² + 9²   **7.** 14.1 km
Δ*ABC* is right-angled at *C*.   **8.** 155 m

CHAPTER 33

Exercise **33.1** — Page 349

1. (a) 60 mm (b) 320 mm (c) 6320 mm
   (d) 86 mm (e) 8 mm (f) 0.8 mm
2. (a) 9 cm (b) 21 cm (c) 350 cm
   (d) 7.35 cm (e) 0.2 cm (f) 0.35 cm
3. (a) 2 m (b) 3.2 m (c) 45.5 m
   (d) 0.66 m (e) 0.08 m (f) 0.098 m
4. (a) 600 cm (b) 5600 cm (c) 760 cm
   (d) 2350 cm (e) 90 cm (f) 7 cm
5. (a) 4 km (b) 35 km
   (c) 6.5 km (d) 0.455 km
   (e) 0.075 km (f) 0.007 km
6. (a) 6000 m (b) 32 000 m
   (c) 650 000 m (d) 3310 m
   (e) 350 m (f) 85 m
7. (a) 20 000 cm² (b) 100 000 cm²
   (c) 5000 cm²
8. (a) 3 000 000 cm³ (b) 20 000 000 cm³
   (c) 400 000 cm³
9. (a) 2000 g (b) 45 000 g
   (c) 7500 g (d) 42 500 g
   (e) 600 g (f) 25 g
10. (a) 3 kg (b) 32 kg
    (c) 9.3 kg (d) 0.22 kg
    (e) 0.083 kg (f) 0.006 kg
11. (a) 320 000 ml = 320l
    (b) 0.32 t = 320 kg = 320 000 g
    (c) 3200 g = 3.2 kg = 0.0032 t
    (d) 320 mm = 32 cm = 0.32 m
    (e) 32 000 cm = 320 m = 0.32 km
    (f) 3.2 km = 3200 m = 320 000 cm
12. (a) 6000 kg (b) 8 kg
    (c) 0.8 kg (d) 650 kg
13. (a) 4000 m (b) 8 m
    (c) 0.086 m (d) 40 m
14. (a) 2000 ml (b) 500 ml
    (c) 850 ml (d) 30 ml
15. 2000 m and 2 km
16. 8 kg and 8000 g
17. 0.5 km
18. 0.3 t
19. (a) 3.123 m (b) 450 cm (c) 3240 km
    (d) 1 000 000 g (e) 0.4l
20. 0.5 m² is larger than 500 cm².
    0.5 m² = 5000 cm²
21. 800 000 cm³ is larger than 0.08 m³.
    0.08 m³ = 80 000 cm³
22. 1.98 l
23. 20
24. 50 g
25. 60
26. 50 ml

Exercise **33.2** — Page 352

4. 250 g
5. 30 cm
6. 200 ml
8. (a) kilometres (b) metres
   (c) centimetres (d) millimetres
   (e) kilograms (f) grams
   (g) litres (h) millilitres
9. 6.4 m, nearest 0.1 m
10. (a) 12 m, nearest metre,
    5.9 m, nearest 100 cm
    5 l, nearest litre
    500 m², nearest 50 m²
    (b) 1200 cm, nearest 100 cm
    5900 mm, nearest 100 mm
    5000 ml, nearest 1000 ml
    500 000 000 mm², nearest 50 000 000 mm²

Exercise **33.3** — Page 354

1. (a) 5 cm (b) 61 cm
2. (a) 78 inches (b) 8 inches
3. 5 feet 5 inches
4. (a) 8 km (b) 72 km
5. (a) 5 miles (b) 25 miles
6. 416 km
7. (a) 55 pounds (b) 2200 pounds
8. (a) 45 kg (b) 43 kg
9. 7 stones 8 pounds
10. (a) 3 litres (b) 11 litres
11. (a) 33 pounds (b) 35 pints
    (c) 200 inches (d) 150 mm
    (e) 20 inches (f) 180 cm
12. 22 pounds
13. 170 cm
14. 62.7 kg
15. (a) 600 m (b) 4.8 km (c) 5 feet
    (d) 2.75 pounds (e) 45.7 litres
16. 1500 cm²
17. 100 magazines
18. No. 10 kg is about 22 pounds.
19. No. 6 miles is about 9.6 km.
20. 12 stones or 76.4 kg
21. 136 g or 4.8 oz
22. 42.5 square feet
23. (a) 48 km/h (b) 80 km/h (c) 108 km/h
24. (a) 37.5 miles per hour
    (b) 90 miles per hour
25. 27 metres per second
26. (a) 12.8 km/litre (b) 30 miles per gallon
27. 3.35 kg, nearest 10 g
28. 6.8 m³, 1 d.p.

Exercise **33.4** — Page 355

1. (a) **X**: 8.5 cm **Y**: 3.2 cm
   (b) **X**: 85 mm **Y**: 32 mm

2. (a) **A**: 8.5 cm   **B**: 4.5 cm   **C**: 0.5 cm
   (b) **A**: 72 mm   **B**: 47 mm   **C**: 88 mm
   (c) **A**: 6.6 inches   **B**: 7.8 inches
      C: 8.7 inches

3. (a) (i) **A**: 1.7 kg   **B**: 2.6 kg   **C**: 0.45 kg
      (ii) **A**: 46.6 g   **B**: 45.8 g   **C**: 46.25 g
      (iii) **A**: 52.4 kg   **B**: 51.6 kg   **C**: 53.3 kg
      (iv) **A**: 300 ml   **B**: 650 ml   **C**: 450 ml
      (v) **A**: 45 ml   **B**: 29 ml   **C**: 12 ml
   (b) (i) 2.15 kg   (ii) 0.8 kg   (iii) 1.7 kg
      (iv) 350 ml   (v) 33 ml

4. **A**: 9°C,   **B**: − 8°C,   **C**: 18°C
5. (a) 9 gallons   (b) 40.5 litres
6. (a) **A**: 23 mph,   **B**: 48 mph,   **C**: 70 mph
   (b) **A**: 37 km/h,   **B**: 77 km/h,   **C**: 112 km/h

Exercise 33.5   Page 357

1. 263.5 m
2. $31.5 \text{ g} \leqslant$ weight of necklace $< 32.5 \text{ g}$
3. Minimum 1.55 m, maximum 1.65 m
4. 4.75 km
5. $645 \text{ g} \leqslant$ weight of block $< 655 \text{ g}$
6. 12.625 seconds

Exercise 33.6   Page 358

1. (a) area   (b) length   (c) length
   (d) length   (e) volume   (f) area
   (g) volume   (h) area
2. (a) perimeter   (b) area   (c) volume
   (d) none   (e) area   (f) perimeter
   (g) perimeter   (h) none   (i) volume
   (j) volume   (k) none   (l) area
3. (a) (i) $2\pi(x + y)$ (ii) $\pi(x^2 + y^2)$, πxy
4. $\frac{1}{2} pqs$, volume
   $2\left(p + q + r + \frac{3s}{2}\right)$, edge length
   $s(p + q + r) + pq$, surface area
5. (a) correct   (b) correct   (c) correct
   (d) wrong   (e) correct   (f) wrong

Review Exercise 33   Page 360

1. (a) 700 ml   (b) 1.5 litres
2. (a) 1.68 m   (b) 18 cm
3. 200
4. 8 full glasses
5. (a) **300 mm**   (b) **300 km**   (c) **300 m**
6. 4410 lb
7. (a) 3.9 cm   (b) 640 km
8. Taller: Ben by about 10 cm
   Heavier: Sam by about 1 kg
9. (a) 22.5 litres   (b) 5 gallons
10. (a) 227.2 ml
   (b) 8 fluid ounces = 227.2 ml
      $\frac{1}{4}$ litre = 250 ml
      So, 8 fluid ounces is less than $\frac{1}{4}$ litre.
11. 15 000 cm²

12. No. $2\frac{1}{2}$ feet = 30 inches $30 \times 2.5 = 75$ cm
13. (a) 0.04 m²   (b) 400 cm²
14. (a) 10 pounds   (b) 4.5 l
15. Susan is breaking the speed limit.
   35 miles/hour = 56 km/hour
16. (a) 120 m²   (b) 1.3 litres
17. No. Francis = 9.6 km/h, Alistair = 10.8 km/h
18. 135 tonnes
19. $165.5 \text{ cm} \leqslant$ Jean's height $< 166.5 \text{ cm}$
20. 76.85 kg
21. (a) **B**   (b) **B** has dimension 3.
               $\frac{1}{6} \pi H w^2 = \text{L} \times \text{L}^2 = \text{L}^3$
22. (a) area   (b) none   (c) volume

Shape, Space and Measures

Non-calculator Paper   Page 362

1. (a) *AB* and *DC*   (b) *AD*   (c) 56°   (d) 5.2 cm
2.

3. (a) 12 cm²   (b) 16 cm
4.

5. (a) 14 cm³   (b) 30 cm³   (c) **Q**
6. 4.4 cm
7. **A** and **D**
8. (a) 2   (b) 3   (c) 8
9. (a) (i) $x = 100°$ (ii) obtuse   (b) $y = 60$
10. (b) 40 cm²
11. (a)   (b)

        plan   elevation X

12. (a) Reflection in the *y* axis
   (b) Rotation, 90° clockwise, about (0, 0)
   (c)

13. 8 cm
15. (a) **P** and **S**   (b) (i) 32 cm   (ii) 44 cm²
16. (a) $x = 118°$
   (b) (i) $p = 25°$ (ii) $q = 105°$
17. 8640 cm³
18. (a) $x = 8$, $y = 3.5$
   (b) (i) $E(-4, 1)$
      (ii) 9 units to the right and 4 units up.

19. (a) 380 m (b) 5600 m²
20. (a) Enlargement, scale factor $\frac{1}{2}$, centre $(1, -1)$

(b) (c)
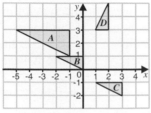

21. (b) 116 m, 276°
22. $a = 120°$, $b = 150°$, $c = 135°$
23.

24. 264 cm
25. (a) 56 m²
(b) (i) $BC = 5$ m
(ii) 34 m

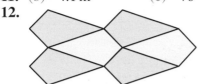
Shape, Space and Measures   Section Review

Calculator Paper   **Page 366**

1. (a) **Swan Bay** (b) **North-West**
(c) 6.5 km (d) 20 km²
2. (a) **kilometre** (b) **tonne**
(c) **metre** (d) **square metre**
3. (a) 8 (b) 5
(c) (i) cube (ii) cylinder
(iii) square-based pyramid
4. (a) 36 by 1, 12 by 3, 9 by 4, 6 by 6
(b) 36 by 2, 24 by 3, 18 by 4, 12 by 6, 9 by 8
5. $a = 153°$, $b = 52°$, $c = 65°$
6. (a) **P** (b) **R**
7.

8. (a) 0.18 m (b) 3 mm
9. (a) A (b) 15 cm³
10. $a = 38° + 67° = 105°$
11. (b) 4.1 m (c) 76°
12.

13. (a) 42.5 km (b) 100°
14. 12 cm²
15. (a) $x = 123°$, $x + 57° = 180°$
(b) (i) Trapezium (ii) $y = 60°$
16.

17. 24 cm²
18. 9 cm
19. (a) 12 000 cm²
(b) 1.2 m²
20. (b) 9.8 cm²
21. (a) 187.5 cm (b) 172 pounds
22. (a) 314 cm (b) 7854 cm²
23. 72°

24. 31
25.
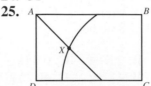

26. (a) 75 cm²
(b) 41.8 cm

27. (a) Reflection in $y = x$
(b) Rotation, 90° anticlockwise, about $(-1, 0)$
28. 30 788 cm³
29. 24 cm²

CHAPTER 34

Exercise **34.1**   **Page 370**

1. qualitative
2. quantitative, continuous
3. quantitative, discrete
4. qualitative
5. quantitative, discrete
6. quantitative, continuous
7. quantitative, discrete
8. qualitative
9. quantitative, discrete
10. quantitative, continuous

Exercise **34.2**   **Page 372**

1. (a) 15 (b) 21 (c) 4
2. (a) 8 (b) 45
(c) Yes. 26 like football, which is more than half of the 45 people asked.
3. (a)

Number on dice	Tally
1	◕
2	◕
3	◕
4	◕
5	◕
6	◕

(b) 4
4. (a)

Colour of car	Frequency
Black	1
Blue	9
Green	4
Grey	2
Red	11
Silver	4
White	9
Total	40

(b) red (c) 8 more

5. (a)

Day	Frequency
Monday	8
Tuesday	7
Wednesday	6
Thursday	7
Friday	7
Saturday	3
Sunday	5
Total	43

(b) 43   (c) Monday   (d) 8

6. (a)

Age (years)	Tally	Frequency
0 - 9	||||	4
10 - 19	++++ ||	7
20 - 29	++++ ++++ |	11
30 - 39	++++ ++++	10
40 - 49	||||	4
50 - 59	||||	4

(b) 10 years   (c) 10   (d) 11   (e) 8

7. (a)

Height h cm	Frequency
$145 \leqslant h < 150$	2
$150 \leqslant h < 155$	2
$155 \leqslant h < 160$	7
$160 \leqslant h < 165$	9
$165 \leqslant h < 170$	6
$170 \leqslant h < 175$	8
$175 \leqslant h < 180$	2
Total	36

(b) 5 cm   (c) 7   (d) 11   (e) 32

Exercise 34.3   Page 375

1. (a) Francis
(b) Louisa
(c) Alistair
2. (a) 2   (b) 2   (c) Fay
3. (a) Val d'Isere   (b) Cervinia
(c) Cervinia   (d) 136 cm
(e) Soldeu
4. (a) Wendy
(b) Female
(c) Tony and Mark 180 cm,
Peter and Jane 168 cm
(d) 5
(e) Mary, Jim, Wendy, Beryl
(f) 20 beats per minute

5. (a) (i)

Make	Frequency
Ford	4
Nissan	3
Vauxhall	5
Total	12

Colour	Frequency
Blue	2
Green	1
Grey	2
Red	3
White	4
Total	12

Number of doors	Frequency
2	2
3	5
4	2
5	3
Total	12

(ii)

Mileage (m)	Frequency
$0 \leqslant m < 5000$	0
$5000 \leqslant m < 10\,000$	2
$10\,000 \leqslant m < 15\,000$	0
$15\,000 \leqslant m < 20\,000$	2
$20\,000 \leqslant m < 25\,000$	1
$25\,000 \leqslant m < 30\,000$	2
$30\,000 \leqslant m < 35\,000$	4
$35\,000 \leqslant m < 40\,000$	1
Total	12

(b) (i) Vauxhall   (ii) 2   (iii) 5   (iv) 5

Exercise 34.4   Page 378

1. (a) Replies should be anonymous
(b) Not specific
2. Leading
3. (a) Too personal   (b) Leading
(c) (i) Groups overlap
(ii) ☐ less than 5 ☐ 5 to 9 ☐ 10 or more
4. (a) Too open   (b) Too open   (c) Leading
5. (a) Too personal   (b) Too open
(c) Too open   (d) Too open

Exercise 34.5   Page 379

1. Small sample. One data collection time.
2. Women only. One location.
One data collection time.

3. Small sample. Boys only. Adults all teachers.

4. Equal numbers of men and women from various age groups, chosen at different locations, at different times.

9. Advantage: confidential, wider circulation, etc.
Disadvantage: slow, non-response, etc.

Exercise 34.6   **Page 380**

1. (a) 5   (b) 8   (c) 3   (d) 10   (e) 9

2. (a)

	Under 16 years old	16 years old or over	Totals
Girls	6	12	18
Boys	9	10	19
Totals	15	22	37

(b) 10

3. For example:

Number of children

	0	1	2	3	4	5	6
1							
2							
3							
4							

Number of bedrooms

4. (a) 8   (b) 16   (c) 20   (d) 40   (e) 50%

5. (a) 14   (b) 20   (c) 70%   (d) 80%
(e) Disprove.   80% is greater than 70%.

6. Disprove.
Boys $\frac{3}{21} = \frac{1}{7}$,   Girls $\frac{2}{14} = \frac{1}{7}$,   same proportion.

7. (a) 20   (b) 5
(c) No.   Smaller proportion of females got less than 11 spellings correct.
Males: $\frac{4}{15} = 26.7\%$,   Females: $\frac{5}{20} = 25\%$

8. (a) (i) 10   (ii) No. Less 8, More 12
(b) (i) 37   (ii) No. Girls 37, Boys 37, same

9. E.g. Fewer females than males.
No-one under 18.

10. (a)

	Yes	No
Men	47	11
Women	18	24

(b) Yes.   Taller: $\frac{47}{58} \times 100 = 81\%$

Review Exercise 34   **Page 383**

1. (a) Thompson A.
(b) Pearson M., Williams C.

2. For example:
Colour

Make		Blue	Green	Red
	Ford			
	Nissan			
	Vauxhall			

---

3. (a) 3   (b) 24   (c) United States and Sweden

4. (a)

Eye colour	Tally	Frequency
blue	||||	4
brown	⊬⊦⊦ |	6
green	||	2

(b) brown

5. (a) **B, C**   (b) **E, H**

6. (a) *Question 1* - too personal
Question 2 - too open

7. E.g. Only males asked. No-one under 11.
Mainly adults surveyed.

8. (a) 219
(b) Do you take part in any exercise outside school?   YES/NO
How many times a week do you exercise?
0 ☐   1 - 5 ☐   6 - 10 ☐   more than 10 ☐
(c) Girls only. All take part in sport.
(d) (i) Leading
(ii) Response classes are not precise.

9. Do not support.
Women 75%,   Men 75%,   same proportion.

10. (a) 6   (b) 5   (c) 38

CHAPTER **35**

Exercise 35.1   **Page 386**

1. (a) 70   (b) 103
2. (a) 39   (b) 84
4. (a) 9   (b) 6   (c) White   (d) 37

Exercise 35.2   **Page 389**

1. (a) 7   (b) Brown   (c) 22   (d) 2
2. (a) 8   (b) 4   (c) 22
3. (a)

Score	Frequency
1	4
2	5
3	7
4	4
5	5
6	5
Total	30

(c) 3

4. (b) 8   (c) 4
5. (a) Saturday   (b) 5 hours   (c) 41 hours
(d) Sunday   (e) $\frac{1}{4}$ (f) 6 hours
6. (a) 15   (b) 1   (c) 5   (d) 20%
7. (b) £5   (c) £7   (d) 20%
8. (b) 7   (c) 3   (d) 6

9. (a) 38 (b) 6 (c) Monday

(d)

Day of birth	S	M	T	W	T	F	S
Number of boys	1	3	4	4	1	5	1

Exercise **35.3** Page 392

1. (a) 10 (b) 3 (c) 4 (d) $\frac{7}{10}$ (e) 75%
2. (a) 12 (b) 6 (c) Flu
3. (a) Brett (b) Trent
(c) 5 marks (d) 13 marks
(e) (i) Trent (ii) Dexter
4. (a) 7 hours (b) 5 hours
(c) 3 (d) (i) 20 (ii) 20%
(e) E.g. Boys have higher mode and larger range.
5. (a) 5 (b) 7 (c) 7 (d) $33\frac{1}{3}$%
(e) 35% (f) 4 (g) 5
(h) Boys have higher mode and larger range.

Review Exercise **35** Page 395

1. (a) 1400 (b) 950
2. (a) 67°F (b) 14°F
(c) Athens. London: 70 − 56 = 14°F
Athens: 86 − 69 = 17°F
3. (a)

Colour	Tally	Frequency
White	HHT HHT	10
Blue	|||	3
Red	HHT HHT ||||	14
Green	|||	3
Total		30

(c) red
4. (a) Saturday
(b) Yes. Friday most popular and more people shop on Sunday.
(c) True. Approximately 55% of people shop at the end of the week (Friday 24%, Saturday 21%, Sunday 10%).
5. (a) 16 : 24 = 2 : 3
(b) Boys: range = 3, mode = size 5.
Girls: range = 4, mode = size 4.
Girls: higher range. Boys: higher mode.

CHAPTER 36

Exercise **36.1** Page 398

1. (a) 7 (b) 4 (c) 4
2. (a) 3 (b) 3
3. (a) 1 (b) 2 (c) 3
4. (a) 5 (b) 2 (c) 3
5. 21p

6. (a) 3 (b) 5 (c) 3.5 (d) 4
7. (a) 18 kg (b) 74 kg (c) 72.4 kg
8. (a) 135 (b) 135 (c) 133
9. 63 **12.** 8
10. (a) 39 (b) 23 (c) 38.8 **13.** 129 cm
11. 4 **14.** 15.4

Exercise **36.2** Page 399

1. (a) (i) 60 g (ii) 99 g
(b) Premium more widely spread and heavier.
2. (a) (i) 9 minutes (ii) 9.5 minutes
(b) Buses are more variable, but less late on average.
3. (a) (i) 42 words per minute
(ii) 60 words per minute
(b) Second group equal on average, but a little more varied.
4. (a) (i) 5 (ii) 1.8
(b) Third division - more goals on average and more spread.
5. (a) (i) 0.5 minutes (ii) 1.95 minutes
(b) Girls a little slower on average and more varied.
6. (a) (i) Roman 1.6%, Chinese 1.9%, Egyptian 2.5%, Greek 1.6%
(ii) Roman 6.48%, Chinese 6.48%, Egyptian 6.48%, Greek 6.38%
(b) Mean silver content is very similar for all coins. Egyptian coins have largest variation in silver content.
7. (a) 28.3
(b) Much bigger variation in the sizes of classes in Year 11.

Exercise **36.3** Page 402

1. (a) Entries are: 12, 10, 5, 3
(b) 1 (c) 2 (d) 1.97
2. (a)

Number of keys	2	3	4	5	6
Frequency	2	3	8	5	2

(b) 20 (c) 4 (d) 4 (e) 4.1
3. (a)

Wage (£)	25	30	35	40	45
Frequency	6	4	5	1	3

(b) £20 (c) £25 (d) 19
(e) £30 (f) £620 (g) £32.63
4. (a) 2, 2.5, 2.7 (b) 2, 2, 2.1
(c) 0, 5, 4.5

Exercise **36.4** Page 403

1. (a) 5 (b) 4 (c) 5 (d) 5
2. (a) 6 (b) 5 (c) 4
(d) 5.38. No shoe of this size.
3. (a) Range 6p, mode 35p
(b) Median 34p, mean 34.1p
4. (a) 9 (b) 9 (c) 30 (d) 8.5

1. (a) £10 000 ⩽ s < £20 000, £20 350
   (b) 30 - 40 hours, 29.4 hours
2. 5.3 kg   4. 4.3 m
3. £94 000   5. 27.7

1. Jays: mean 1.9, range 5
   Wasps: mean 2.4, range 3
   Wasps scored more on average and had
   less spread.
2. Women: mean 1.6, range 6
   Men: mean 1.5, range 2
   Women made more visits to the cinema,
   though the number of visits is more spread.
3. Boys: mean 2.27, range 6
   Girls: mean 2.27, range 3
   Larger variation in the number of cards
   received by boys.
   Average for boys and girls is the same.
4. Average: Boys 6.2, Girls 7.2
   Range: Boys 4, Girls $4\frac{1}{2}$
   No. Girls' average greater than boys'.
   Correct about variation.
5. (a) MacQuick 20 - 29, Pizza Pit 30 - 39
   (b) MacQuick - mean 26 years
       (Pizza Pit 36.5 years)
   (c) Exact ages not known.
6. Before: median 3, range 4
   After: median 3, range 5
   Would have been better to calculate the means.
   Before 2.2. After 3.0

1. Mode trainers. Cannot calculate others.
2. Mode 15s, median 12s, mean 22.15s
   Median most sensible, not affected by 200 as
   is mean, mode not much use.
3. Mode 81, median 83, mean 69.8
   Median most sensible, not affected by 5 and 6
   as is mean, mode not much use.
4. Swimmer A.
   Mean is lower (A 30.88s, B 31.38s)
   Range less (A 1.7s, B 15s)
   Median is higher (A 30.9s, B 30.0s)
5. Batsman B.
   Higher median (B 31.5, A 21)
   Higher mean (B 36, A 35)
6. He should use the median mark. The median
   mark is the middle mark, so, half of the
   students will get the median mark or higher.
7. (a) Mode.
       Represents the lowest cost for these data.
   (b) Median.
       Mean affected by **one** much higher cost.
       Mode is equal to the lowest cost.

1. 1.4
2. (a) 6   (b) Range
3. (a) 2   (b) 1
   (c) Mean affected by 15 and 30.
4. (a) (i) 94 kg   (ii) 53 kg   (iii) 87.4 kg
   (b) Median, as mean is affected by the low
       value of 43 kg.
5. (a) 14   (b) 140   (c) 20
6. (a) (i) 7   (ii) 5
   (b) Mode - in greatest demand
7. (a) 52   (b) 11
   (c) Increase. 62 ÷ 9 = 6.9
8. (a) 25   (b) 4   (c) 2   (d) 2.2
   (e) Ros had a much smaller sample and there
       were no very large families in her survey.
9. (a) (i) 12 cm   (ii) 32 cm
   (b) English cucumbers are longer on average
       and slightly more varied in length.
10. (a) (i) 2   (ii) 2.28
    (b) (i) 2.5   (ii) 2.63
    (c) The calculation for the mean involves
        the number of runs scored off every ball.
11. (a) 24 cm   (b) 24.3 cm
    (c) 1st group had higher mode (1st 25,
        2nd 23), but lower range (1st 7, 2nd 13).
12. (a) £200 - £300
    (b) £300 - £400
    (c) Mean, influenced by 12 people earning
        £600 - £1000.
13. 16.9 minutes

CHAPTER 37

1.

Tree	Ash	Beech	Maple
Angle	120°	150°	90°

2.

Colour	Brown	Blue	Green	Other
Angle	160°	100°	60°	40°

3.

Car	Ford	Saab	Vauxhall	BMW
Angle	90°	81°	135°	54°

4.

Cereal	Corn flakes	Muesli	Porridge	Bran flakes
Angle	125°	100°	60°	75°

5.

Ice cream	Vanilla	Strawberry	99
Angle	188°	74°	98°

6.

Takeaway	Fish & Chips	Chicken & Chips	Chinese Meal	Pizza
Angle	110°	136°	52°	62°

1. (a) 12   (b) 8   (c) Hotel
2. (a) 20   (b) 15   (c) Sauze d'Oulx
3. (a) France   (b) 45   (c) 55   (d) 20
4. (a) 5   (b) 2   (c) 18
5. (a) 14   (b) 72
6. (a) Heathrow   (b) 540   (c) 1080
7. (a) Brown   (b) 26   (c) 25%
8. (a) 288   (b) 174°

1.

	1 | 0 means 10 litres
1	0 2 6 6 7 9
2	3 3 4 5 5 6 7 9
3	1 3 5 5
4	1 2

2.

	3 | 2 means 3.2 seconds
1	5
2	4 4 5 6 7 8 8 9
3	0 1 2 2 3 5 5 6 7
4	2 2 3
5	6 6 8

3.

	1 | 6 means 16 press-ups
0	9
1	6 8
2	0 1 2 4 5 7 8
3	2 2 3 6 6 6
4	0 1

4.

	2 | 7 means 2.7 cm
1	8
2	0 1 4 5 6 6 7
3	1 4 5 5 6 9
4	0 2 2 5
5	4
6	0

5. (a) 12   (b) 23 pence   (c) 39 pence

1. (a) 9   (b) 50   (c) 17
   (d) Highest mark scored by a boy.
   Lowest mark scored by a girl.
   Boys have a greater range of marks.

2. (a) Adults   Children   4 \| 7 means 4.7 mins

Adults		Children
	4	7 9
9 4	5	1 3 4 9
7 5 4 1 0	6	2 3 4 5 5 6 8
9 8 7 3 3 3 0	7	1 4 6 7 9
2 2 0	8	0 2
4 2	9	
1	10	

   (b) Adults have larger range.
   Fastest time recorded by child,
   slowest time recorded by adult.

1.

Item	TVs	VCRs	Computers	Other
Angle	140°	120°	68°	32°

2. (a) 2 days   (b) 29 letters   (c) 5 or 30
3. (a) $\frac{1}{3}$ (b) 1836   (c) 357
4.

	5 | 4 means 5.4 grams
2	8
3	5 9
4	2 4 6 6 7 8 8
5	0 1 4 4 6 6 8
6	0 3 7

5. (a)

Grade	A	B	C	Ungraded
Angle	60°	90°	135°	75°

   (b) 9
   (c)

Grade	A	B	C	Ungraded
Number of girls	3	4	5	2

CHAPTER 38

1. (a) 15°C
   (b) Temperature variations during each day
   are not known. Line only indicates trend
   in midday temperatures.
2. (a) 8   (b) Jan, Feb, June
   (c) No information given about when cars
   are sold during the month.
3. (b) (i) 154 cm   (ii) 15 years 4 months
4. (a) 82 kg   (b) 82 kg
   (c) 5 kg   (d) Week 4
5. (b) £118 - £119
   (c) Money was withdrawn

1. (a) 36 (b) 24 (c) 40 (d) 80 - 90 kg
2. (a) Entries are: 4, 7, 9, 4
   (c) 6.00 and less than 6.50
3. (a)

Distance (m miles)	Frequency
$0 \leqslant m < 10\,000$	9
$10\,000 \leqslant m < 20\,000$	8
$20\,000 \leqslant m < 30\,000$	6
$30\,000 \leqslant m < 40\,000$	7

   (b) 30   (d) $0 \leqslant m < 10\,000$
4. (a) 3 but less than 4   (b) 31
   (c) 15   (d) 4   (e) 190
5. (a) 14   (b) 70
   (c) £300 and less than £400
   (d) 210

Exercise 38.3 — Page 427

1.

Time (seconds)	Frequency
10 and less than 20	4
20 and less than 30	6
30 and less than 40	2

2. (a) 26   (b) 0   (c) 27   (d) 95

3. (a)

Mark	Frequency
20 and less than 30	5
30 and less than 40	8
40 and less than 50	9
50 and less than 60	6
60 and less than 70	2

    (b) 40 and less than 50   (c) 22

4. (a) 16   (b) $2 \leq k < 4$

5. (b) 38

6. (c) English results have smaller range.
English modal class is higher.

7. (b) 2006 results have a larger range.
2005 results have higher modal class.

Exercise 38.4 — Page 430

1. Axes not labelled.
2. Bars not equal width.
3. Vertical axis does not begin at zero.
4. Pass rate not given.
Advert implies you "pass" after 8 lessons.
5. 10% increase in price, disproportionate increase in diagram size.
6. Horizontal axis does not begin at zero and is not a uniform scale.
7. Vertical scale not uniform. Size of diagrams disproportionate to increase in sales.
8. Horizontal scale not uniform.
Vertical scale not calibrated.

Review Exercise 38 — Page 431

1. (a) 42.9%   (c) 1999 - steepest gradient
2. (b) £110 000
    (c) Prices of houses rise and fall.
Future prices unpredictable.
3. Vertical axis does not begin at zero.
Horizontal axis is not uniform.
4. (a)

Height h (cm)	Frequency
$120 \leq h < 130$	7
$130 \leq h < 140$	22
$140 \leq h < 150$	20
$150 \leq h < 160$	4

    (b) There should be no gaps.
$130 \leq h < 140$ should be 22 not 24.
    (c) $150 \leq h < 160$

5.

Mark	Frequency
0 and less than 10	3
10 and less than 20	12
20 and less than 30	19
30 and less than 40	6

6. (a) 39
    (b) $8 \leq d < 12$

7. (a) More than 4 hours but less than or equal to 5 hours.
    (b) Ann gained 20 points ($4 \leq t < 5$),
Ben gained 10 points ($5 \leq t < 6$).

CHAPTER 39

Exercise 39.1 — Page 434

1. (a) 2   (b) 164 cm   (c) No
    (d) Taller girls usually have larger shoe sizes than shorter girls.
2. (a) 72
    (b) (i) English 46, French 88
       (ii) French could be her first language.
3. (a) 4
    (b) 43 kg
    (c) Tend to be higher
4. (a) 39 000 miles
    (b) Older cars tend to have higher mileages.
    (c) (i) 2 years, 40 000 miles   (ii) Hire car
5. (a) 6
    (b) 8.2 years
    (c) Children who read more tend to have a higher reading age.

Exercise 39.2 — Page 437

1. (a) **B**   (b) **C**   (c) **D**
2. (a) Negative   (b) Positive
    (c) No correlation   (d) Positive
    (e) Negative
3. (b) Positive correlation
    (c) Different conditions, types of road, etc.
4. (b) Negative correlation
    (c) Points are close to a straight line.
5. (b) Scatter of points suggests that there is no correlation.

Exercise 39.3 — Page 439

1. (b) Positive correlation   (d) 4.8 to 4.9 kg
2. (b) Negative correlation   (d) 38 minutes
3. (b) Negative correlation
    (d) (i) 27 to 28   (ii) 91 to 92 kg
4. (b) (i) 88 - 89   (ii) 49 - 50
    (c) (ii), as estimated value is within the range of known values.
5. (b) Scatter of points suggests correlation is close to zero.

Review Exercise 39   Page 440

1. (b) £9, 100 pages   (c) Positive correlation
2. 1 B,   2 C,   3 A
3. (b) Positive correlation   (c) 50 cm
4. (b) Positive correlation.
   (d) 90 to 100 beats per minute.
5. (b) Negative correlation.
   (d) £2600 - £3000
6. (c) (i) 80 - 85 km   (ii) 11 - 13 litres
   (d) (i), as estimated value is between known values.

CHAPTER 40

Exercise 40.1   Page 442

1. (a) Certain   (b) Impossible
   (c) Evens   (d) Impossible
   (e) Evens
2. (a) Unlikely   (b) Likely
   (c) Likely   (d) Unlikely
   (e) Unlikely   (f) Likely
3. (a) Certain   (b) Impossible
   (c) Evens   (d) Evens
   (e) Impossible   (f) Unlikely

Exercise 40.2   Page 444

1. (a) **A** (b) **C** (c) **B**
2. (a) **T** (b) **P** (c) **R** (d) **S**
3. W   Z   V   X   Y

   0         $\frac{1}{2}$ 1

Exercise 40.3   Page 445

1. (a) $\frac{1}{6}$ (b) $\frac{1}{2}$ (c) $\frac{2}{3}$ (d) $\frac{1}{3}$
2. (a) $\frac{1}{3}$ (b) $\frac{2}{3}$ (c) $\frac{2}{3}$
3. (a) $\frac{3}{10}$ (b) $\frac{7}{10}$
4. (a) $\frac{1}{2}$ (b) $\frac{1}{2}$
5. (a) $\frac{1}{12}$ (b) $\frac{1}{6}$
6. (a) $\frac{1}{5}$ (b) $\frac{1}{5}$ (c) $\frac{2}{5}$
7. (a) $\frac{1}{11}$ (b) $\frac{4}{11}$ (c) $\frac{2}{11}$
8. (a) $\frac{1}{2}$ (b) $\frac{1}{4}$ (c) $\frac{1}{52}$
9. (a) $\frac{2}{5}$ (b) $\frac{3}{5}$ (c) 1 (d) 0
10. (a) $\frac{2}{3}$ (b) $\frac{1}{3}$ (c) $\frac{3}{4}$
11. (a) $\frac{2}{5}$ (b) $\frac{3}{5}$ (c) $\frac{4}{25}$ (d) $\frac{4}{15}$
    (e) $\frac{2}{5}$
12. The events are not equally likely.

13. (a) $\frac{7}{15}$ (b) $\frac{1}{2}$ (c) $\frac{1}{16}$ (d) $\frac{3}{4}$
14. (a) $\frac{1}{3}$ (b) $\frac{1}{15}$ (c) $\frac{11}{24}$ (d) $\frac{17}{50}$
    (e) $\frac{21}{25}$ (f) $\frac{4}{5}$

Exercise 40.4   Page 448

1. $\frac{7}{25}$
2. $\frac{9}{10}$
3. $\frac{3}{10}$
4. (a) $\frac{52}{100} = 0.52$ $\frac{102}{200} = 0.51$ $\frac{141}{300} = 0.47$
   (b) 0.47
5. $\frac{21}{30} = \frac{7}{10}$
6. 8
7. 18
8. 25
9. (a) 300 (b) 120 (c) 150

Exercise 40.5   Page 449

1. $\frac{3}{5}$
2. 0.4
3. 0.04
4. $\frac{47}{50}$
5. 0.6
6. (a) 0.04 (b) 0.97
7. (a) 0.5 (b) 0.3
8. (a) (i) The probabilities add to 105%
       (ii) 5%
   (b) (i) 45% (ii) 75% (iii) 80%
9. (a) 0.4 (b) 0.6
   (c) (i) Mutually exclusive - there are no blue cubes numbered 1
       (ii) Not mutually exclusive - probability = 0.2
   (d) (i) 0.6 (ii) 0.6 (iii) 0.7

Exercise 40.6   Page 451

1. (a) RBG, RGB, GBR, GRB, BGR, BRG
   (b) $\frac{1}{3}$
2. (a)

	2nd dice					
	1	**2**	**3**	**4**	**5**	**6**
1	2	3	4	5	6	7
2	3	4	5	6	7	8
3	4	5	6	7	8	9
4	5	6	7	8	9	10
5	6	7	8	9	10	11
6	7	8	9	10	11	12

(1st dice on left)

   (b) (i) $\frac{1}{12}$ (ii) $\frac{1}{12}$ (iii) $\frac{5}{6}$
   (c) They cover all possible scores.

3.

Dice

Coin		1	2	3	4	5	6
	H	H1	H2	H3	H4	H5	H6
	T	T1	T2	T3	T4	T5	T6

(a) $\frac{1}{12}$ (b) $\frac{1}{4}$ (c) $\frac{1}{12}$ (d) $\frac{1}{4}$ (e) $\frac{1}{6}$ (f) $\frac{1}{2}$

4. (a) Stage 1: Bus, Bus, Train, Train, Lift, Lift
Stage 2: Bus, Walk, Bus, Walk, Bus, Walk
(b) $\frac{1}{6}$

5. (a)

2nd spin

1st spin		1	2	3	4
	1	2	3	4	5
	2	3	4	5	6
	3	4	5	6	7
	4	5	6	7	8

(b) (i) $\frac{1}{16}$ (ii) $\frac{1}{8}$ (iii) $\frac{3}{16}$

6. (a) Maths, English   Maths, Science
Maths, Art   English, Science
English, Art   Science, Art
(b) $\frac{3}{6} = \frac{1}{2}$ (c) $\frac{1}{3}$

7. (a)

Bag A

Bag B		R	R	W
	W	RW	RW	WW
	W	RW	RW	WW
	R	RR	RR	WR

(c) $\frac{4}{9}$

8. (a)

A	1	1	2	2	3	3
B	2	3	2	3	2	3

(b) (i) $\frac{2}{6} = \frac{1}{3}$ (ii) $\frac{4}{6} = \frac{2}{3}$ (c) $\frac{4}{5}$

9. (a)

2nd spin	W	RW	GW	BW	YW	WW
	Y	RY	GY	BY	YY	WY
	B	RB	GB	BB	YB	WB
	G	RG	GG	BG	YG	WG
	R	RR	GR	BR	YR	WR
		R	G	B	Y	W

1st spin

(b) (i) $\frac{1}{25}$ (ii) $\frac{9}{25}$ (iii) $\frac{5}{25} = \frac{1}{5}$

10. (a)

	1	2	3	4	5	6
1	2	3	4	5	6	7
2	3	4	5	6	7	8
3	4	5	6	7	8	9

(b) $\frac{1}{18}$ (c) $\frac{1}{6}$

524

1.

0 ——————— 1

2. (a) 1 (b) 3, 4 (c) 0

3. 0.7

4. (a) **C** (b) **B** (c) **A** (d) **D**

5. (a) $\frac{2}{5}$
(b) 2p coin, because there are more.

6. (a) $\frac{1}{4}$ (b) $\frac{2}{3}$

7. (a) $\frac{1}{100}$ (b) $\frac{1}{20}$ (c) (i) $\frac{17}{20}$ (ii) 15

8. (a) $\frac{1}{4}$ (b) $\frac{3}{8}$
(c) They do not have equal chance of winning.

9. (a) **Set 1**: *A*, *A*, *A*, *B*, *B*, *B*, *C*, *C*, *C*
Set 2: *A*, *B*, *C*, *A*, *B*, *C*, *A*, *B*, *C*
(b) $\frac{4}{9}$

10. (a)

	1	2	3	4
1	2	3	4	5
2	3	4	5	6
3	4	5	6	7

(b) (i) $\frac{2}{12} = \frac{1}{6}$ (ii) $\frac{10}{12} = \frac{5}{6}$
(c) $\frac{4}{12} = \frac{1}{3}$

11. (a)

Outcome	Tally	Frequency
Red	⫲⫲⫲ ⫲⫲	7
White	⫲⫲⫲ ⫲⫲⫲ ⫲	11
Blue	⫲⫲⫲ ⫲⫲⫲ ⫲⫲⫲ ⫲⫲⫲	18
	Total	36

(b) $\frac{7}{36}$
(c) Red 1, White 2, Blue 3.

12. (a) $\frac{1}{2}$ (b) $\frac{4}{11}$
(c) (i)

Counter	Yellow	Green	Black
Probability	0.4	0.4	0.2

(ii) Same number of yellow and green counters.
Twice as many yellow and green counters as there are black counters.
Total number of counters is a multiple of 5.

13. (a) $\frac{1}{5}$ (b) $\frac{3}{4}$

Non-calculator Paper — Page 457

1. (a)

Berry	Tally	Frequency
Strawberry	|||| ||	7
Blackberry	||	2
Raspberry	||||	4
None	||	2

(b)

Berry	Key: ● = 2 replies
Strawberry	● ● ● ◖
Blackberry	●
Raspberry	● ●
None	●

2. (a) 6　(b) 21　(c) Football
3. (a) 3　(b) 4　(c) 6
4.

Make	Ford	Nissan	Vauxhall	BMW
Angle	160°	100°	60°	40°

5. (a) Red
(b) Spinner A. Probabilities A: $\frac{1}{4}$　B: $\frac{1}{6}$
(c) $\frac{1}{4}$
(d)

Y　G　R

0 — 1

6. (a) (i) 132　(ii) 280
(b) (i) Overslept　(ii) 8
7. (a) 1H, 2H, 3H, 4H, 5H, 6H, 1T, 2T, 3T, 4T, 5T, 6T.
(b) $\frac{1}{12}$
8. (a) 31 cm　(b) 15　(c) 18 cm　(d) 26 cm
9. (a) Highest − lowest = 8
10, 10, and 2 other numbers
(b) Sum of four numbers = 36
10, 10 and 2 other numbers
(c) E.g. 4, 10, 10, 12
Sum of four numbers = 36 **and**
highest − lowest = 8
10. (c) Positive correlation.
Taller trees have wider trunks.
(d) 13 cm is closer to the line of best fit for a tree which is 3 m tall.
11. (a) **Q1**: Too open, vague.
Q2: Too long **and** leading.
(b) How many hours per week do you read?
☐ less than 1　☐ 1 to 3　☐ more than 3
12. (a) 4　(b) 1.5　(c) 1.6
13. (a) $\frac{3}{20}$　(b) $\frac{7}{20}$
14. Set A: 1, 1, 2, 2, 3, 3
Set B: X, Y, X, Y, X, Y

15. (a) 35　(b) 3　(c) 5.1
(d) Males have greater range, 7 compared with 3. Females have greater average, 5.1 compared with 3.6.
16. (a) 0.3　(b) 0.6
17. (a) $\frac{7}{9}$　(b) $\frac{6}{25}$
(c) Can swim: girls 0.76, boys 0.8.
Not true. 0.8 > 0.76
18. (a) $\frac{1}{6}$
(b) Probably not, as 2 occurs twice as many times as any other number.
19. Estimated total weight
= $(8 \times 40) + (10 \times 60) + (4 \times 80) = 1240$ kg
Likely to be overloaded.

Handling Data

Calculator Paper — Page 461

1. (a) Brown　(b) 9　(c) 6
2. (a) (ii) 1
(b) (i) Unlikely
(ii)

0 — 1

3. (a) Use a data collection sheet.
(b)

Music	Tally	Frequency
Classical	||	2
Heavy metal	||||	4
Pop	|||| |||	8
Rap	|||	3
Rock	|||| |	6
Soul	|	1
	Total	24

(c) Pop　(d) $\frac{2}{24} = \frac{1}{12}$
4. (a) 34.5　(b) 36.17
(c) (i) 17　(ii) Lowest attendance
(d) 58
5. (a)

		Dice					
		1	2	3	4	5	6
Coin	Heads	1	2	3	4	5	6
	Tails	2	4	6	8	10	12

(b) $\frac{1}{12}$　(c) $\frac{7}{12}$
6. (a) 5.9 cm
(b)

4 | 7 means 4.7 cm

4	7 9
5	0 3 6 8
6	1 4 5 5 7
7	2

7. 53 people

8. (a) **C**   (b) **B**   (c) **A**

9. (a)

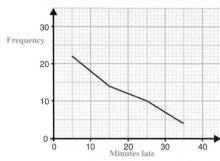

Where fires started	Angle
Kitchen	180°
Living room	90°
Bedroom	72°
Other	18°

(b) 20%

10. (a) 0.4   (b) 18

11. (a) 600   (b) Larger sample.

12. (a) 0.7   (b) 0.1

13. (b) Positive correlation
(d) Plotted points close to line of best fit.

14.

15. (a) 14   (b) 5   (c) 16   (d) 49

16. (a) 40
(b) Yes.   Pensioners 40%,   students $33\frac{1}{3}$%

17. (a) 0.6   (b) 0.2

18. (a) E.g. Most people travel to out-of-town shopping supermarkets by car and so a high proportion of drivers will be included.
(b) 23 000

19. (a) $\frac{12}{20} = \frac{3}{5}$ (c) 500

20. (a) 13.4 minutes

Exam Practice

Exam Practice

Non-calculator Paper Page 465

1. (a) 2139   (b) 570   (c) 2010

2. 35, 41, 305, 587, 4001

3. (a) **38, 96**   (b) **73**   (c) **23**

4. (a) 8000   (b) 40

5. (a) (i) *TS*   (b)
(ii) **obtuse**
(iii) **pentagon**
(iv) *P* (1, 3)

6. (a)

Pattern 4

(b)

Pattern number	1	2	3	4
Number of squares	1	3	5	7

(c) 9
(d) All patterns have an odd number of squares.

---

7. (a) *XZ* = 5.5 cm   (b) angle *X* = 45°

8. (a) £1.86   (b) £3.14

9. (a) 50%   (b) 0.4   (c) $\frac{2}{5}, \frac{1}{2}, \frac{2}{3}, \frac{3}{4}$
(d) 27   (e) $\frac{2}{3}$

10. (a) 5   (b) 8   (c) 10   (d) 8

11. (a) (i) 14 cm²   (ii) 22 cm   (b) 24 cm³

12. E.g. 2, 3, 6 and 2, 2, 9

13. 2, 4, 6

14. (a) 87   (b) 2.36   (c) 432   (d) 0.05

15. (a) 2   (b) Faces 1, 4 and 5 (any two)

16. (a) 17.   Add 4 to the last number.
(b) No.   All terms in the sequence are odd numbers.

17. (a) b^3 (b) 35*d* pence   (c) 8

18. (a) 49   (b) 9   (c) 0.5

19. (a)

Painting	A	B	C	D
Angle	90°	120°	60°	90°

(b) $\frac{20}{60} = \frac{1}{3}$

20. (a) $\frac{1}{6}$ (b) $1\frac{11}{20}$ (c) $\frac{1}{10}$

21. (a) 1 hour 54 minutes   (b) 0759   (c) 0823

22. (a) (i) *x* = 10   (ii) *x* = 7   (iii) *x* = 12
(b) (i) $2a - 6$ (ii) $y^2 + y$

23. (a) −11   (b) 25   (c) 10

24. (a) $\frac{7}{20}$ (b) $\frac{13}{20}$

25. $\frac{60}{20 - 5} = 4$

26. (a) $\frac{4}{7}$
(c) (i) 2   (ii) 2's and 4's nearly equal.

27. (b) 83 km   (c) 54 miles

28. (a) *x* = 54°
(b) (i) *y* = 63°   (ii) alternate angles

29. (a) 18   (b) $2t - 3$
(c) (i) *x* = 7   (ii) $t = 1\frac{1}{2}$

30. (a) Translation, 4 units left and 1 unit up.
(b) Rotation, through 90° clockwise, about (0, 0).

31. (a) **P**   (b) 54 cm²   (c) 300

32. (a) 3.6 m
(b) Multiply by 10 then divide by 3.
(c) 12.96 m²

33. (a) 80 km/h   (b) 45 minutes
(c) 30 km   (d) 20 km/h

34. $\frac{300 \times 4}{0.2} = 6000$

35. $3n + 2$

36. (b) Positive correlation
(c) (ii) 6 to 6.3 years

37. $5 \times 5 \times \pi = 25\pi$ cm²

38. (a) $2^2 \times 7$ (b) 84

39. (a) 0.55   (b) 0.1   (c) 0.6

40. (a) (i) c^5 (ii) d^4
(b) (i) $2(m - 2n)$ (ii) $t(t - 2)$

41. 32

42. (a) 2, 3   (b) $2x - 7$

43. (a) Rotation, 90° anticlockwise, centre (0, 0).
     (b) Reflection in $y = x$.
     (c)

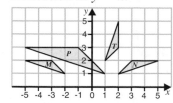

44. (b) $x = \pm 2.4$

Exam Practice

Calculator Paper   Page 470

1. (a) 5 people   (b) 13 people

Bed and Breakfast

     (c) 🧍 🧍 🧍   🧍 = 17 people

2. Missing entries are: £4.76, £2.20, £2.07
Total £13.50

3. 4.8, 4.7, $4\frac{2}{3}$, 4.629, $4\frac{3}{5}$, 4.58

4. (a) 3, 7, 9   (b) 2, 3, 6   (c) 24   (d) 9

5. (a) **R**   (b) **A** and **C**

6. 7

7. 0.24 m

8. (a) £19.75   (b) 6 pizzas

9. (a)

Pattern 4

     (b) Missing entries are: 22, 28
     (c) 118. Multiply the pattern number by 6, then subtract 2.

10. (a) 43   (b) 7
     (c) Subtract 6 from the last number.
     (d) -5

11. Shape **C**. The shape cannot be used to cover a surface without leaving gaps.

12. (a) $6t$ (b) (i) $w = 3$ (ii) $x = 2$ (c) 5

13. (a) 230 cm   (b) 41.6 km   (c) 40 gallons

14. (a) $A\,(1, 4)$ (b) (ii) $D\,(-2, 1)$

15.

	1	2 means 12 pages
1	2 5 7	
2	0 1 6	
3	2 5	

16. (a) $C = 18$ (b) $B = 4$

17. (a) $\frac{8}{24} = \frac{1}{3}$ (b)

18. 37. Add the last two numbers.

19. $x = 41°$, $y = 58°$

20. £349.20

21. (a) 6   (b) 27%

23. (a) $x = 106°$ (b)

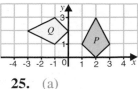

24. (a)

Turban	Shirt
white	green
white	red
white	yellow
blue	green
blue	red
blue	yellow

     (b) (i) $\frac{1}{2}$ (ii) $\frac{1}{6}$
         (iii) $\frac{1}{3}$
     (c) (ii) $\frac{5}{11}$

25. (a)

     (b) (2, 4)

26. (a) Between 2 and 3   (b) $x = 2.7$

27. 200 m²

28. **Size 1**: 12.8 grams/penny
Size 2: 13.6 grams/penny
Size 2 gives more grams per penny.

29. 1. No answer for 7 am and 8 am.
2. No response box for other items.
3. Too open. No time boxes given.

30. £42.30

31. 13.4

32. (a) 503 cm   (b) 3 complete turns

33. Sum of probabilities should equal 1.
$0.1 + 0.2 + 0.35 + 0.25 = 0.9$

34. 35 g

35. 45 m²

36. 108

37. (a) 36.66   (b) 512   (c) 0.325

38. (a) (i) $x = -1$ (ii) $x = 3.2$
     (b) $3(a - 2b)$ (c) $3x - 6$

39. (a) $20x$ grams   (b) 180 grams

40. (a) 50 mph to less than 60 mph
     (b) 58 mph

41. (a) £281   (b) $87\frac{1}{3}\%$

42. (a) 31.284   (b) $\frac{90 \times 10}{20 + 10} = 30$

43. 5 hours 12 minutes

44. $2^2 \times 3^2$

45. (a) $\frac{7}{9}$ (b) $\frac{4}{5}$ (c) 0.24
     (d) No. 0.8 of men can drive.
         0.76 of women can drive. $0.8 > 0.76$

46. $7 - 2n$

47. (a) If $\angle B$ is a right angle, then by Pythagoras on $\triangle ABC$. $AC^2 = AB^2 + BC^2$
               $(39^2 = 15^2 + 36^2)$
     (b) $x = 105°$

48. (a) $x = -\frac{2}{5}$ (b) $x = 7.3$

49. $RQ = 5.6$ m

50. 170 litres

51. (a) $m > 4$ (b) $-1, 0, 1$

52. abc. Only formula with dimension 3.

Index ●●●●●●●●●●●●●●●●●●●●●●●●●●●●●●●●